Internet of Things

Challenges, Advances, and Applications

CHAPMAN & HALL/CRC COMPUTER and INFORMATION SCIENCE SERIES

Series Editor: Sartaj Sahni

PUBLISHED TITLES

ADVERSARIAL REASONING: COMPUTATIONAL APPROACHES TO READING THE OPPONENT'S MIND
Alexander Kott and William M. McEneaney

COMPUTER-AIDED GRAPHING AND SIMULATION TOOLS FOR AUTOCAD USERS
P. A. Simionescu

COMPUTER SIMULATION: A FOUNDATIONAL APPROACH USING PYTHON
Yahya E. Osais

DELAUNAY MESH GENERATION
Siu-Wing Cheng, Tamal Krishna Dey, and Jonathan Richard Shewchuk

DISTRIBUTED SENSOR NETWORKS, SECOND EDITION
S. Sitharama Iyengar and Richard R. Brooks

DISTRIBUTED SYSTEMS: AN ALGORITHMIC APPROACH, SECOND EDITION
Sukumar Ghosh

ENERGY-AWARE MEMORY MANAGEMENT FOR EMBEDDED MULTIMEDIA SYSTEMS: A COMPUTER-AIDED DESIGN APPROACH
Florin Balasa and Dhiraj K. Pradhan

ENERGY EFFICIENT HARDWARE-SOFTWARE CO-SYNTHESIS USING RECONFIGURABLE HARDWARE
Jingzhao Ou and Viktor K. Prasanna

EVOLUTIONARY MULTI-OBJECTIVE SYSTEM DESIGN: THEORY AND APPLICATIONS
Nadia Nedjah, Luiza De Macedo Mourelle, and Heitor Silverio Lopes

FROM ACTION SYSTEMS TO DISTRIBUTED SYSTEMS: THE REFINEMENT APPROACH
Luigia Petre and Emil Sekerinski

FROM INTERNET OF THINGS TO SMART CITIES: ENABLING TECHNOLOGIES
Hongjian Sun, Chao Wang, and Bashar I. Ahmad

FUNDAMENTALS OF NATURAL COMPUTING: BASIC CONCEPTS, ALGORITHMS, AND APPLICATIONS
Leandro Nunes de Castro

HANDBOOK OF ALGORITHMS FOR WIRELESS NETWORKING AND MOBILE COMPUTING
Azzedine Boukerche

PUBLISHED TITLES CONTINUED

HANDBOOK OF APPROXIMATION ALGORITHMS AND METAHEURISTICS
Teofilo F. Gonzalez

HANDBOOK OF BIOINSPIRED ALGORITHMS AND APPLICATIONS
Stephan Olariu and Albert Y. Zomaya

HANDBOOK OF COMPUTATIONAL MOLECULAR BIOLOGY
Srinivas Aluru

HANDBOOK OF DATA STRUCTURES AND APPLICATIONS
Dinesh P. Mehta and Sartaj Sahni

HANDBOOK OF DYNAMIC SYSTEM MODELING
Paul A. Fishwick

HANDBOOK OF ENERGY-AWARE AND GREEN COMPUTING
Ishfaq Ahmad and Sanjay Ranka

HANDBOOK OF GRAPH THEORY, COMBINATORIAL OPTIMIZATION, AND ALGORITHMS
Krishnaiyan "KT" Thulasiraman, Subramanian Arumugam, Andreas Brandstädt, and Takao Nishizeki

HANDBOOK OF PARALLEL COMPUTING: MODELS, ALGORITHMS AND APPLICATIONS
Sanguthevar Rajasekaran and John Reif

HANDBOOK OF REAL-TIME AND EMBEDDED SYSTEMS
Insup Lee, Joseph Y-T. Leung, and Sang H. Son

HANDBOOK OF SCHEDULING: ALGORITHMS, MODELS, AND PERFORMANCE ANALYSIS
Joseph Y.-T. Leung

HIGH PERFORMANCE COMPUTING IN REMOTE SENSING
Antonio J. Plaza and Chein-I Chang

HUMAN ACTIVITY RECOGNITION: USING WEARABLE SENSORS AND SMARTPHONES
Miguel A. Labrador and Oscar D. Lara Yejas

IMPROVING THE PERFORMANCE OF WIRELESS LANs: A PRACTICAL GUIDE
Nurul Sarkar

INTEGRATION OF SERVICES INTO WORKFLOW APPLICATIONS
Paweł Czarnul

INTERNET OF THINGS: CHALLENGES, ADVANCES, AND APPLICATIONS
Qusay F. Hassan, Atta ur Rehman Khan, and Sajjad A. Madani

INTRODUCTION TO NETWORK SECURITY
Douglas Jacobson

PUBLISHED TITLES CONTINUED

LOCATION-BASED INFORMATION SYSTEMS: DEVELOPING REAL-TIME TRACKING APPLICATIONS
Miguel A. Labrador, Alfredo J. Pérez, and Pedro M. Wightman

METHODS IN ALGORITHMIC ANALYSIS
Vladimir A. Dobrushkin

MULTICORE COMPUTING: ALGORITHMS, ARCHITECTURES, AND APPLICATIONS
Sanguthevar Rajasekaran, Lance Fiondella, Mohamed Ahmed, and Reda A. Ammar

NETWORKS OF THE FUTURE: ARCHITECTURES, TECHNOLOGIES, AND IMPLEMENTATIONS
Mahmoud Elkhodr, Qusay F. Hassan, and Seyed Shahrestani

PERFORMANCE ANALYSIS OF QUEUING AND COMPUTER NETWORKS
G. R. Dattatreya

THE PRACTICAL HANDBOOK OF INTERNET COMPUTING
Munindar P. Singh

SCALABLE AND SECURE INTERNET SERVICES AND ARCHITECTURE
Cheng-Zhong Xu

SOFTWARE APPLICATION DEVELOPMENT: A VISUAL C++®, MFC, AND STL TUTORIAL
Bud Fox, Zhang Wenzu, and Tan May Ling

SPECULATIVE EXECUTION IN HIGH PERFORMANCE COMPUTER ARCHITECTURES
David Kaeli and Pen-Chung Yew

TRUSTWORTHY CYBER-PHYSICAL SYSTEMS ENGINEERING
Alexander Romanovsky and Fuyuki Ishikawa

VEHICULAR NETWORKS: FROM THEORY TO PRACTICE
Stephan Olariu and Michele C. Weigle

X-MACHINES FOR AGENT-BASED MODELING: FLAME PERSPECTIVES
Mariam Kiran

Internet of Things

Challenges, Advances, and Applications

Edited by
Qusay F. Hassan
Atta ur Rehman Khan
Sajjad A. Madani

CRC Press
Taylor & Francis Group
Boca Raton London New York

CRC Press is an imprint of the
Taylor & Francis Group, an **informa** business
A CHAPMAN & HALL BOOK

CRC Press
Taylor & Francis Group
6000 Broken Sound Parkway NW, Suite 300
Boca Raton, FL 33487-2742

First issued in paperback 2020

ISBN-13: 978-0-367-57236-5 (pbk)
ISBN-13: 978-1-4987-7851-0 (hbk)

Library of Congress Cataloging-in-Publication Data

Library of Congress Cataloging-in-Publication Data
Names: Hassan, Qusay F., 1982- editor. | Khan, Atta ur Rehman, editor. |
Madani, Sajjad A., 1976- editor.
Title: Internet of things : challenges, advances, and applications / edited
by Qusay F. Hassan, Atta ur Rehman Khan, Sajjad A. Madani.
Other titles: Internet of Things (Hassan)
Description: Boca Raton : Taylor & Francis, CRC Press, [2017] | Series:
Chapman & Hall/CRC computer & information science series | Includes
bibliographical references.
Identifiers: LCCN 2017055385 | ISBN 9781498778510 (hardback : alk. paper)
Subjects: LCSH: Internet of things.
Classification: LCC TK5105.8857 .I565 2017 | DDC 004.67/8--dc23
LC record available at https://lccn.loc.gov/2017055385

Visit the Taylor & Francis Web site at
http://www.taylorandfrancis.com

and the CRC Press Web site at
http://www.crcpress.com

Contents

Preface ix
Acknowledgments xiii
Reviewers xv
About the Editors xvii

PART I Concepts and Adoption Challenges

Chapter 1 Introduction to the Internet of Things 3
Karolina Baras and Lina M. L. P. Brito

Chapter 2 Organizational Implementation and Management Challenges in the Internet of Things 33
Marta Vos

PART II Technological Advances and Implementation Considerations

Chapter 3 Cooperative Networking Techniques in the IoT Age 51
Luigi Alfredo Grieco, Giuseppe Piro, Gennaro Boggia, and Domenico Striccoli

Chapter 4 Exploring Methods of Authentication for the Internet of Things 71
Fatemeh Tehranipoor, Nima Karimian, Paul A. Wortman, Asad Haque, Jim Fahrny, and John A. Chandy

Chapter 5 Energy-Efficient Routing Protocols for Ambient Energy Harvesting in the Internet of Things 91
Syed Asad Hussain, Muhammad Mohsin Mehdi, and Imran Raza

Chapter 6 IoT Hardware Development Platforms: Past, Present, and Future 107
Musa Gwani Samaila, João Bernardo Ferreira Sequeiros, Acácio Filipe Pereira Pinto Correia, Mário Marques Freire, and Pedro Ricardo Morais Inácio

Chapter 7 IoT System Development Methods 141
Görkem Giray, Bedir Tekinerdogan, and Eray Tüzün

Chapter 8 Design Considerations for Wireless Power Delivery Using RFID 161
Akaa Agbaeze Eteng, Sharul Kamal Abdul Rahim, and Chee Yen Leow

PART III Issues and Novel Solutions

Chapter 9 Overcoming Interoperability Barriers in IoT by Utilizing a Use Case–Based Protocol Selection Framework 181

Supriya Mitra and Shalaka Shinde

Chapter 10 Enabling Cloud-Centric IoT with Publish/Subscribe Systems 195

Daniel Happ, Niels Karowski, Thomas Menzel, Vlado Handziski, and Adam Wolisz

Chapter 11 The Emergence of Edge-Centric Distributed IoT Analytics Platforms 213

Muhammad Habib ur Rehman, Prem Prakash Jayaraman, and Charith Perera

PART IV IoT in Critical Application Domains

Chapter 12 The Internet of Things in Electric Distribution Networks: Control Architecture, Communication Infrastructure, and Smart Functionalities 231

Qiang Yang, Ali Ehsan, Le Jiang, Hailin Zhao, and Ming Cheng

Chapter 13 Satellite-Based Internet of Things Infrastructure for Management of Large-Scale Electric Distribution Networks 253

Qiang Yang and Dejian Meng

Chapter 14 IoT-Enabled Smart Gas and Water Grids from Communication Protocols to Data Analysis 273

Susanna Spinsante, Stefano Squartini, Paola Russo, Adelmo De Santis, Marco Severini, Marco Fagiani, Valentina Di Mattia, and Roberto Minerva

Chapter 15 The Internet of Things and e-Health: Remote Patients Monitoring 303

Assim Sagahyroon, Raafat Aburukba, and Fadi Aloul

Chapter 16 Security Considerations for IoT Support of e-Health Applications 321

Daniel Minoli, Kazem Sohraby, Benedict Occhiogrosso, and Jake Kouns

Chapter 17 IoT Considerations, Requirements, and Architectures for Insurance Applications 347

Daniel Minoli, Benedict Occhiogrosso, Kazem Sohraby, James Gleason, and Jake Kouns

Chapter 18 The Internet of Things and the Automotive Industry: A Shift from a Vehicle-Centric to Data-Centric Paradigm 363

Zahra Saleh and Steve Cayzer

Glossary 389

Index 397

About the Contributors 403

Preface

The Internet of Things (IoT) is a new technology boom that is affecting our lives in a positive way. Be it driverless cars, smart refrigerators, patient monitoring systems, smart grids, or industrial automation, IoT is changing everything. The notion of the Internet of Everything is already a commonplace and studies reveal that there will be more than 30 billion objects connected to the Internet by 2020. Looking at the phenomenal growth of the amount of Internet connected objects over the past decade, these figures seem realistic rather than exaggeration. The promising present and future of IoT motivated us to compile thoughts and current work of the leading researchers of the world and present them in the form of a book.

This edited book is an effort to cover IoT, related technologies, and common issues in adoption of IoT on a large scale. It surveys recent technological advances and novel solutions for the common issues in IoT environment. Moreover, it provides detailed discussion about the utilization of IoT and the underlying technologies in critical application areas, such as smart grids, healthcare, insurance, and automotive industry.

The chapters of this book are authored by several international researchers and industry experts. This book is composed of 18 self-contained chapters that can be read based on interest without having to read the entire book. These chapters were carefully selected after a rigorous review of more than 70 chapters which we received for possible inclusion in this book.

This book is an excellent reference for researchers and post-graduate students working in the area of IoT. It also targets IT professionals interested in gaining deeper knowledge of the IoT, its challenges, and application areas. This book is mainly for readers who have a good knowledge of IT and moderate knowledge of IoT. However, the chapters are organized in a way that provides a base to the readers by starting with basic concepts, followed by main challenges of IoT, and then the advanced topics. We also tried to include sufficient details and provide the necessary background information in each chapter to help the readers to easily understand the content. We hope the readers will enjoy this book.

ORGANIZATION OF THE BOOK

This book is divided into four parts, each of which is devoted to a distinctive area.

Part I: Concepts and Adoption Challenges

This part is composed of two chapters which cover IoT concepts and main adoption challenges.

Chapter 1 gives a general introduction to the IoT. The chapter starts with a brief overview of the history of IoT and its common definitions, and then it presents the main architectural and reference models proposed over the past few years to enable efficient building of IoT systems. This chapter also gives a good overview of the IoT key enabling technologies, main application domains as well as the challenges that would hinder its adoption.

Chapter 2 presents the challenges towards the global adoption of IoT. The chapter focuses on seven challenges, namely interoperability, standards, privacy, security, trust, data management, and legislation and governance.

Part II: Technological Advances and Implementation Considerations

This part has six chapters which address various IoT related technological advances.

Chapter 3 provides an overview of diverse cooperative networking techniques in various IoT environments. It presents cooperative approaches in cellular systems with focus on 4G and 5G networks, D2D communications, WLANs and WSNs, VANETs, and other wireless networks with energy harvesting capabilities.

Chapter 4 reviews current authentication techniques and discusses the possibility of utilizing them in IoT systems. Moreover, it evaluates the applicability of those techniques and highlights their pros and cons.

Chapter 5 discusses energy efficient routing protocols and scheduling techniques for IoT devices. It provides description of IPv6 over 6LoWPAN and presents performance comparison of the routing protocols. The chapter also briefly discusses ambient energy harvesting approaches available for IoT systems, and the related IoT routing challenges.

Chapter 6 presents a comprehensive survey of a variety of IoT hardware development platforms, showing how their technical specifications and capabilities have improved over the years. The chapter starts with a brief history of the IoT hardware development platforms that were available in the past, followed by discussion of the current platforms available in the market, and finally, it attempts to forecast the features of future platforms.

Chapter 7 briefly introduces the six IoT system development methods found in literature, and attempts to evaluate them against a number of criteria to present their elements, characteristics, and coverage. This chapter can be used as a guide by stakeholders involved in the development of IoT systems, including project managers and method engineers.

Chapter 8 discusses the design considerations for utilizing passive RFID for wireless power delivery. RFID is an integral part of IoT based systems and one of its possible uses is to deliver energy to IoT devices. The chapter presents comprehensive introduction of RFID principles, design considerations for wireless power delivery through passive RFID through load modulated RFID links, and considerations for radio frequency harvesting RFID power delivery.

Part III: Issues and Novel Solutions

This part provides three chapters that propose novel solutions for common implementation issues of IoT.

Chapter 9 explores inherent limitations and interoperability issues of IoT. It assesses protocols based on the OSI model, and proposes protocol selection criteria for different applications of IoT. The chapter presents eight example use cases to help users select the right communication protocols/technologies for their IoT systems.

Chapter 10 proposes a novel architecture that utilizes a publish/subscribe model to extend traditional cloud platforms that are widely used in IoT implementations. This chapter highlights the limitations of the traditional cloud-centric IoT and discusses how the extended model overcomes such limitations.

Chapter 11 highlights the role of data analytics in IoT systems and proposes a novel multi-tier architecture that enables distributed data analytics on IoT devices. This chapter also provides a survey of three device-centric platforms that inspired the creation of the proposed architecture.

Part IV: IoT in Critical Application Domains

This is the final part of the book and is composed of seven chapters that present the integration of IoT in key application domains. These chapters present the underlying technologies, implementation requirements, considerations, and impacts of IoT on highlighted applications.

Chapter 12 presents a system that utilizes IoT in electric power distribution networks. The chapter discusses communication standards, protocols, and requirements of such networks. Moreover, it provides two case studies of radial and meshed distribution networks in smart grids.

Chapter 13 discusses low orbit satellites in the management of large-scale electric power distribution networks. The chapter shows how this technology can enable effective communication in smart grids that are composed of large, geographically dispersed and heterogenous electric power distribution networks.

Chapter 14 discusses how the utilization of IoT in smart water and gas grids allows for efficient resource monitoring and consumption forecasting. The chapter covers different technological issues in this area including network architectures, communications protocols, and radio coverage estimation tools. In addition, the chapter highlights the importance of machine learning in such applications.

Chapter 15 emphasizes the role of remote patient monitoring in the future of healthcare, and describes how the different IoT technologies facilitate the collection, transfer, processing, visualization and storage of critical information about patients.

Chapter 16 reviews various IoT security challenges in e-Health applications. It discusses the need for IoT security architectures and presents some of the available mechanisms that can be used at the different IoT architecture layers for securing e-Health applications.

Chapter 17 shows how IoT technologies are applicable to industries that are outside the realm of infrastructure-based environments. Specifically, the chapter discusses possible IoT applications in the insurance industry. These applications enable personal and environment monitoring, with the goal of optimizing risk management. The chapter reviews some of the technical challenges that face the efficient integration of IoT in insurance applications, and covers the architectures, standards, and security mechanisms that can be leveraged by such applications.

Finally, Chapter 18 discusses the implications of the emergence and adoption of IoT in the automotive industry. The chapter is based on several interviews with representatives from leading companies, namely Google, IBM, BMW, Deloitte, and P3. The findings show the digital transformation and service innovation enabled by IoT and how the main focus is shifting from traditional car manufacturing to acquiring, using and selling data generated by the numerous sensors and systems embedded in modern vehicles.

Dr. Qusay F. Hassan
Dr. Atta ur Rehman Khan
Dr. Sajjad A. Madani

Acknowledgments

We would like express our gratitude to everyone who participated in this project and made this book a reality. In particular, we would like to acknowledge the hard work of the authors and their patience during the revisions of their chapters.

We would also like to acknowledge the outstanding comments of the reviewers, which enabled us to select these chapters out of the many we received and improve their quality. Some of the authors also served as referees; their double task is highly appreciated.

We also thank our family members for realizing the importance of this project and their consistent support throughout the project.

Lastly, we are very grateful to the editorial team at CRC Press for their support through the stages of this project. Special thanks to our editor, Randi Cohen, for her great support and encouragement. We enjoyed working with Randi, who was involved in all the phases, from the time this project was just an idea through the writing and editing of the chapters, and then during the production process. We would also like to thank Lara Silva McDonnell, from Deanta Global Publishing, for managing the production process of this book.

Dr. Qusay F. Hassan
Dr. Atta ur Rehman Khan
Dr. Sajjad A. Madani

Reviewers

Dr. Tahir Maqsood, *COMSATS Institute of Information Technology, Pakistan*
Dr. Osman Khalid, *COMSATS Institute of Information Technology, Pakistan*
Dr. Saad Mustafa, *COMSATS Institute of Information Technology, Pakistan*
Dr. Sushant Singh, *Cygnus Professionals Inc., New Jersey, USA*
Prof. Haijun Zhang, *University of Science and Technology, Beijing, China*
Dr. Marco Tiloca, *Research Institutes of Sweden*
Dr. Qiang Yang, *College of Electrical Engineering, Zhejiang University, Hangzhou, China*

About the Editors

Qusay F. Hassan, PhD, is an independent researcher and a technology evangelist with 15 years of professional experience in information and communications technology. He is currently a systems analyst at the U.S. Agency for International Development in Cairo, Egypt, where he deals with large-scale and complex systems. Dr. Hassan earned his BS, MS, and PhD from Mansoura University, Egypt, in computer science and information systems in 2003, 2008, and 2015, respectively. His research interests are varied and include the Internet of Things, Service-Oriented Architecture, high-performance computing, cloud computing, and grid computing. Dr. Hassan has authored and coauthored several journal and conference papers, as well as book chapters. He also published two books recently: *Networks of the Future: Architectures, Technologies, and Implementations* (CRC Press, 2017) and *Applications in Next-Generation High Performance Computing* (IGI Global, 2016). He is also currently editing a handbook on IoT to be published in 2018. Dr. Hassan is an IEEE senior member, and a member of the editorial board of several associations.

Dr. Atta ur Rehman Khan is the Head of AU Cybersecurity Center and Associate Professor at Department of Computer Science, Air University, Pakistan. He has extensive experience in teaching, research, and industry at key positions. He has completed renowned projects, published research articles in reputed journals/conferences, and edited/co-authored multiple books. Currently, he is an Associate Editor of IEEE Access, Springer Journal of Cluster Computing, Associate Technical Editor of IEEE Communications Magazine, Editor of Elsevier Journal of Network and Computer Applications, Oxford Computer Journal, IEEE SDN Newsletter, KSII Transactions on Internet and Information Systems, SpringerOpen Human-centric Computing and Information Sciences, SpringerPlus, and Ad hoc & Sensor Wireless Networks journal, member of IEEE R&D committee for Smart Grids, IEEE Smart Grids Education Committee, IEEE Smart Grids publications committee, IEEE Smart Grids Operations Committee, Amazon Web Services (AWS) Educate program, NVIDIA GPU Educators Program, Steering Committee Member/ Track Chair/ Technical Program Committee (TPC) member of over 60 international conferences, and Higher Education Commission (Pakistan) approved PhD supervisor. He also serves as a domain expert for multiple international research funding bodies, namely European Commission and CONICYT National Commission for Scientific and Technological Research, Santiago, Chile. Moreover, he is a Senior Member of IEEE and Professional Member of ACM. He has received multiple awards, fellowships, and research grants. His areas of research interest include mobile computing, cloud computing, ad hoc networks, IoT, and security. For more updated information, visit his website at http://attaurrehman.com.

Sajjad A. Madani, PhD, is an associate professor in the Department of Computer Science, at COMSATS Institute of Information Technology. Additionally, he is the director of the virtual campus of the same institute. Dr. Madani earned his PhD from Vienna University of Austria in 2008. He is a senior Institute of Electrical and Electronics Engineers member, a member of the Institute of Electrical and Electronics Engineers Industrial Electronics Society, and a member of the Pakistan Engineering Council. He is also a member of the academic council of the COMSATS Institute of Information Technology, a member of the board of the Faculty of Information Science and Technology, and a member of the board of studies of the Computer Science Department. Dr. Madani is actively involved in the research areas of cloud computing, low-power wireless sensor networks, and the application of industrial informatics to electrical energy networks. He has more than 70 refereed research articles to his name. Dr. Madani has also been actively involved in the International Conference on Frontiers of Information Technology (FIT) for the last several years as the program cochair. He has won several awards and scholarships during his career and studentship. He is the recipient of a prestigious 12th National Teradata IT Excellence Award. He was awarded a gold medal for earning the top position in the department at the University of Engineering and Technology, KPK, Pakistan.

Part I

Concepts and Adoption Challenges

1 Introduction to the Internet of Things

Karolina Baras and Lina M. P. L. Brito

CONTENTS

1.1 Introduction 3
1.2 Definition of IoT 5
1.3 Proposed Architectures and Reference Models 8
1.3.1 IoT-A 10
1.3.2 IoT RA 11
1.3.3 IEEE P2413 12
1.3.4 Industrial Reference Architectures 12
1.3.5 Other Reference Models and Architectures for IoT 14
1.3.5.1 Cisco Reference Model 14
1.3.5.2 Reference IoT Layered Architecture 15
1.4 Enabling Technologies 16
1.4.1 Identification and Discovery 16
1.4.2 Communication Patterns and Protocols 17
1.4.3 Devices and Test Beds 18
1.5 Application Areas: An Overview 21
1.5.1 Smart Cities 21
1.5.2 Healthcare 21
1.5.3 Smart Homes and Smart Buildings 23
1.5.4 Mobility and Transportation 23
1.5.5 Energy 24
1.5.6 Smart Manufacturing 24
1.5.7 Smart Agriculture 24
1.5.8 Environment/Smart Planet 24
1.6 Challenges 24
1.6.1 Interoperability 25
1.6.2 Openness 25
1.6.3 Security, Privacy, and Trust 26
1.6.4 Scalability 26
1.6.5 Failure Handling 27
1.7 Conclusion 27
References 28

1.1 INTRODUCTION

Back in 1989, there were around 100,000 hosts connected to the Internet (Zakon, 2016), and the World Wide Web (WWW) came to life a year later at CERN with the first and only site at the time.* Ten years after Tom Berners-Lee unleashed the WWW, a whole new world of possibilities started

* http://info.cern.ch/.

to emerge when Kevin Ashton, from the Massachusetts Institute of Technology's (MIT) Auto-ID Labs, coined the term *Internet of Things* (Ashton, 2009). In the same year, Neil Gershenfeld published his work on things that think, where he envisioned the evolution of the WWW as "things start to use the Net so that people don't need to" (Gershenfeld, 1999). Simultaneously, in Xerox PARC Laboratories in Palo Alto, California, the so-called third era of modern computing was dawning, with Mark Weiser introducing the concept of "ubiquitous computing" in his paper published in *Scientific American* (Weiser, 1991). Tabs, pads, and boards were proposed as the essential building blocks for the computing of the future. Wireless networking and seamless access to shared resources would make user experience with technology as enjoyable as "a walk in the woods."

In 1999, the number of hosts exceeded 2 million and the number of sites jumped to 4 million (Zakon, 2016). The Institute of Electrical and Electronics Engineers (IEEE) standard 802.11b (Wi-Fi) had just been published, with transmission rates of 11 Mbits/s. GSM was growing fast, but the phones were not at all smart yet. They (only) allowed for making phone calls and sending short messages. GPS signals for civil usage were still degraded with selective availability, and the receivers were heavy, huge, and expensive. The area of wireless sensor networks (WSNs) also emerged in the 1990s with the concept of smart dust, a big number of tiny devices scattered around an area capable of sensing, recording, and communicating sensed data wirelessly.

In the dawn of the eagerly expected twenty-first century, the technological growth accelerated at an unprecedented pace. Although the reports published 10 and 20 years after Weiser's vision showed that not everything turned out to be just as he had imagined, significant changes were introduced in the way we use technology and live with it. Our habits changed, our interaction with technology changed, and the way we grow, play, study, work, and communicate also changed.

In 2005, the International Telecommunications Union (ITU) published its first report on the Internet of Things (IoT), noting that

> *"Machine-to-machine communications and person-to-computer communications will be extended to things, from everyday household objects to sensors monitoring the movement of the Golden Gate Bridge or detecting earth tremors. Everything from tyres to toothbrushes will fall within communications range, heralding the dawn of a new era, one in which today's internet (of data and people) gives way to tomorrow's Internet of Things."* (ITU-T, 2005)

Three years later, in 2008, the number of devices connected to the Internet outnumbered the world's population for the first time. The introduction of Internet protocol version 6 (IPv6)* resolved the problem of the exhaustion of IP addresses, which was imminent near the end of the twentieth century. The first international conference on IoT† took place in March 2008 to gather industry and academia experts to share their knowledge, experience, and ideas on this emerging concept. In the following years, the number of IoT-related events and conferences grew enormously.

Open-source electronics such as Arduino,‡ which reached the market between 2005 and 2008, gave birth to millions of new ideas and projects for home and office automation, education, and leisure. Other examples of single-board computers (SBCs) followed: Raspberry Pi,§ BeagleBone Black,¶ Intel Edison,** and so on. Today, one can buy a dozen tiny but fairly powerful computers for less than $50 each, connect them to the Internet and to a plethora of sensors and actuators, collect and analyze gigabytes of data, and make interesting home or office automation projects with

---

* https://tools.ietf.org/html/rfc2460.
† http://www.iot-conference.org/iot2008/.
‡ https://www.arduino.cc/.
§ https://www.raspberrypi.org/.
¶ https://beagleboard.org/black.
** https://software.intel.com/en-us/articles/what-is-the-intel-edison-module.

real-time visualizations of information generated from the data on the go. Alternatively, one can use remote networks of intelligent devices deployed somewhere else, for example, OneLab.*

In 2009, the Commission of the European Communities published a report on the IoT action plan for Europe showing that the IoT had reached a very high level of importance among European politicians, commercial and industry partners, and researchers (Commission of the European Communities, 2009). Several global standard initiatives were created in recent years to discuss and define IoT-related issues and establish global agreement on standard technologies to be deployed in IoT projects. For example, oneM2M† was created in 2012 as a global standard initiative that covers machine-to-machine and IoT technologies, which go from requirements, architecture, and application programming interface (API) specifications, to security solutions and interoperability issues.

In 2015, the European Commission created the Alliance for the Internet of Things (AIOTI)‡ to foster interaction and collaboration between IoT stakeholders. The convergence of cloud computing, the miniaturization and lower cost of sensors and microcontrollers, and the omnipresence of digital connectivity all contributed to making the IoT a reality for years to come.

In fact, some sources consider that the four pillars of digital transformation are cloud, mobility, big data, and social networking, and that IoT is based on these (IDC, 2015; i-SCOOP, 2015).

Gartner forecasts that by 2020 there will be more than 20 billion "things" connected to the Internet (Gartner, Inc., 2013). This number excludes PCs, smartphones, and tablets.

Now that the IoT is finally becoming a reality, there is a need for a global understanding on its definition, a reference architecture (RA), requirements, and standards. In the following sections, an overview of the current IoT landscape will be given and some of the proposals that are on the table for discussion in several groups, alliances, and consortia focused on IoT will be highlighted. There is at least an agreement on some of the requirements that need to be addressed, but still there is space for improvement and even more collaboration among the stakeholders. For example, unique device identification, system modularity, security, privacy, and low cost are some of issues that need further discussion and action.

This chapter covers the fundamentals of IoT, its application domains, and the main challenges that still need to be surpassed. The rest of the chapter is organized as follows: Section 1.2 reviews the main definitions and concepts involved. Some of the proposed architectures and reference models (RMs) are described in Section 1.3. Section 1.4 includes an overview of IoT-enabling technologies and the efforts of several working groups and consortia to create standards for the IoT. Section 1.5 gives an overview of the main IoT application domains, and Section 1.6 highlights main IoT implementation challenges. The last section provides the main conclusions of the chapter and outlines current trends.

1.2 DEFINITION OF IoT

There have been several international organizations and research centers involved in the creation of common standards for the IoT. One of the first steps in this process has been to find a common definition. The first definitions of the IoT were tightly coupled to the radio-frequency identification (RFID)–related context of the Auto-ID Labs at MIT, where the term first emerged. As the concept became universal, the definition started to evolve to more general terms. For Kevin Ashton, the meaning of the IoT and the consequences of its implementation in our environments are the following (Ashton, 2009):

> *"If we had computers that knew everything there was to know about things—using data they gathered without any help from us—we would be able to track and count everything, and greatly reduce waste,*

---

* https://onelab.eu/.

† http://www.onem2m.org/.

‡ http://www.aioti.org/.

loss and cost. We would know when things needed replacing, repairing or recalling, and whether they were fresh or past their best. We need to empower computers with their own means of gathering information, so they can see, hear and smell the world for themselves."

Another definition of the IoT is the following (Atzori et al., 2010):

"The basic idea of this concept is the pervasive presence around us of a variety of things or objects – such as Radio-Frequency IDentification (RFID) tags, sensors, actuators, mobile phones, etc. – which, through unique addressing schemes, are able to interact with each other and cooperate with their neighbors to reach common goals."

The Study Group 20 (SG 20) was created in 2015 as a result of the 10-year experience period that followed the publication of the first ITU report on IoT in 2005 and the findings of the International Telecommunications Union Telecommunication Standardization Sector (ITU-T) Focus Group on Smart Sustainable Cities, which ceased to exist in 2015. In the SG 20 recommendation document Y.2060 (ITU-T, 2012), the following definition is given:

"Internet of things (IoT): A global infrastructure for the information society, enabling advanced services by interconnecting (physical and virtual) things based on existing and evolving interoperable information and communication technologies.

NOTE 1 – Through the exploitation of identification, data capture, processing and communication capabilities, the IoT makes full use of things to offer services to all kinds of applications, whilst ensuring that security and privacy requirements are fulfilled.

NOTE 2 – From a broader perspective, the IoT can be perceived as a vision with technological and societal implications."

The ITU-T document goes on to explain that IoT adds a new dimension ("any thing") to the already existing "any time" and "any place" communication provided by the digital connectivity expansion. In this context, "things" are defined as being physical or virtual identified objects capable of communicating. Physical objects are all kinds of everyday objects that are present in our environments and that can contain sensors, actuators, and communication capability. Examples of physical objects are electronic appliances, industrial machinery, and digitally enhanced everyday objects. Virtual objects exist in the information world and can be stored, accessed, and processed. Software is an example of a virtual object.

The European Union (EU) created the IoT European Research Cluster (IERC) as a platform for FP7 (7th Framework Programme for Research and Technological Development) projects on IoT. Currently, the IERC is Working Group 1 (WG 1) of the AIOTI, which was created to establish collaboration and communication between different entities involved in IoT development, standardization, and implementation.

The definition published on the IERC website* states that the IoT is

"A dynamic global network infrastructure with self-configuring capabilities based on standard and interoperable communication protocols where physical and virtual 'things' have identities, physical attributes, and virtual personalities and use intelligent interfaces, and are seamlessly integrated into the information network."

International Organization for Standardization/International Electrotechnical Commission (ISO/IEC) Joint Technical Committee 1 (JTC1) was created in 1987 and is responsible for standard development in the information technology (IT) area, having so far published more than 3000 standards.† WG 10 (former SWG 5) is one of JTC1 working groups responsible for IoT-related issues.

* http://www.internet-of-things-research.eu/about_iot.htm.

† https://www.iso.org/committee/45020.html.

In one of the SWG 5 reports published in 2015, the adopted definition for IoT is given in the following terms (ISO/IEC JTC1, 2015):

"An infrastructure of interconnected objects, people, systems and information resources together with intelligent services to allow them to process information of the physical and the virtual world and react."

In the Request for Comments (RFC) 7452,* which talks about the architectures for networks of smart objects, IoT is defined as follows:

"The term 'Internet of Things' (IoT) denotes a trend where a large number of embedded devices employ communication services offered by Internet protocols. Many of these devices, often called 'smart objects,' are not directly operated by humans but exist as components in buildings or vehicles, or are spread out in the environment."

The IEEE IoT initiative published a document (IEEE, 2015) with an overview of the IoT applications and a proposal of a definition in order to start a discussion and to give its community members an opportunity to contribute to the definition of the IoT.† The document presents two definitions, one for the small-scale scenarios:

"An IoT is a network that connects uniquely identifiable 'Things' to the Internet. The 'Things' have sensing/actuation and potential programmability capabilities. Through the exploitation of unique identification and sensing, information about the 'Thing' can be collected and the state of the 'Thing' can be changed from anywhere, anytime, by anything."

The other is for the large-scale scenarios:

"Internet of Things envisions a self-configuring, adaptive, complex network that interconnects 'things' to the Internet through the use of standard communication protocols. The interconnected things have physical or virtual representation in the digital world, sensing/actuation capability, a programmability feature and are uniquely identifiable. The representation contains information including the thing's identity, status, location or any other business, social or privately relevant information. The things offer services, with or without human intervention, through the exploitation of unique identification, data capture and communication, and actuation capability. The service is exploited through the use of intelligent interfaces and is made available anywhere, anytime, and for anything taking security into consideration."

A fairly complete collection of IoT definitions can be found in Minoli (2013), in which the definitions are organized into two categories: those that define IoT as a concept and those that define IoT as an infrastructure. The following definition that tries to encompass both the concept and the infrastructure can be found in the above-cited book:

"A broadly-deployed aggregate computing/communication application and/or application-consumption system, that is deployed over a local (L-IoT), metropolitan (M-IoT), regional (R-IoT), national (N-IoT), or global (G-IoT) geography, consisting of (i) dispersed instrumented objects ('things') with embedded one- or two-way communications and some (or, at times, no) computing capabilities, (ii) where objects are reachable over a variety of wireless or wired local area and/or wide area networks, and (iii) whose inbound data and/or outbound commands are pipelined to or issued by a(n application) system with a (high) degree of (human or computer-based) intelligence."

---

* https://tools.ietf.org/html/rfc7452.

† http://iot.ieee.org/definition.html.

Although the wording may be slightly different, it seems that there are several touching points among the definitions. For example, the IoT is made of (physical and virtual) objects that are uniquely identifiable, that are able to capture their context (sensors), and that are able to transmit and/or receive data over the Internet and, in the case of actuators, are able to change their own state or the state of their surroundings—all of which should ideally be done without or with very little direct human intervention.

1.3 PROPOSED ARCHITECTURES AND REFERENCE MODELS

The end goal of IoT systems is to achieve a synergy between different systems, meaning that they should interoperate and communicate automatically to provide innovative services to the users. Therefore, standardization is needed to ensure that IoT platforms will allow distinct systems to reliably interoperate.

While it is expected that the IoT will positively revolutionize all the different sectors of the economy in society, it will also produce a very large amount of data. This not only brings new challenges regarding the management, processing, and transmission of data, but above all, it also brings new concerns regarding data security. So, on top of standardization for interoperability, security standards are also needed to protect the individuals, businesses, and governments that will use the IoT systems (Dahmen-Lhuissier, 2016; ixia, 2016).

Recently, several attempts were made—and are still being made—to develop RAs, by either standards organizations or industries and universities. However, standardization is difficult to achieve in the real world. RAs are vital for standardization, as they define guidelines that can be used when planning the implementation of an IoT system (Weyrich and Ebert, 2016).

Therefore, with the intention of making interoperability between different IoT systems possible, several attempts have been made in recent years to create reference layered models for IoT (Bassi et al., 2013; Bauer et al., 2013; Gubbi et al., 2013). Several standards development organizations (SDOs) are also engaged in this process, as will be described below.

oneM2M specifications focus on the creation of a framework to support applications and services, such as smart grid, connected car, home automation, public safety, and health. During a workshop organized by the European Telecommunications Standards Institute (ETSI), which took place in November 2016, the European Commission highlighted the need for an open common RA for IoT, enabling the integration of different services, for the specific case of smart cities application. In fact, this is of critical importance not only to smart cities but also to all areas of application of IoT technologies. In addition to oneM2M standardization activities, ETSI has also created a working group on sustainable digital multiservice cities, specifically for the case of smart cities projects (Antipolis, 2016).

The IEEE has produced more than 80 standards* that relate to several areas of IoT systems and has around 60 ongoing projects to develop new standards also related to the IoT. Among all the standards, projects, and events promoted by IEEE, two important initiatives need to be emphasized: the IEEE Standards Association (IEEE-SA) engaged participants in key regions of the world to create the IoT Ecosystem Study, which encompasses three main areas—market, technology, and standards—but also examines the role of academia and research, and the importance of user acceptance; and the IEEE P2413 Working Group is focusing on the creation of a standard architecture for IoT, the "IoT architecture." The resulting draft standard basically defines an architectural framework for the IoT: it describes different IoT domains, gives definitions of IoT domain abstractions, and identifies commonalities between different IoT domains.†

* http://standards.ieee.org/innovate/iot/stds.html.

† http://standards.ieee.org/develop/project/2413.html.

The GSM Association (GSMA)* has gathered nearly 800 mobile operators and 300 companies worldwide to address four areas of the mobile industry: "Personal Data (Enabling trust through digital identity), Connected Living (Bringing the Internet of Things to life), Network 2020 (The future of mobile communications), Digital Commerce (Streamlining interactions and transactions)" (GSMA, 2016). In August 2015, the GSMA established a new project named Mobile IoT Initiative,† supported by a group of 26 of the world's leading mobile operators, equipment manufacturers, and module and infrastructure companies, to address the use of low-power wide area (LPWA) solutions in the licensed spectrum.

The GSMA Connected Living Programme (LP) is working with mobile operators to fasten the delivery of IoT solutions that exploit connectivity in innovative ways. In February 2016, the Connected LP also published new guidelines designed to promote the secure development and deployment of services in the IoT market. The result is the document entitled "GSMA IoT Security Guidelines," developed in conjunction with the mobile industry, which offers IoT service providers practical recommendations on handling common security and data privacy threats associated with IoT services (GSMA, 2016).

For 2017, the GSMA Connected LP focuses on four new goals: (1) Mobile IoT, which mainly addresses increasing the market awareness and support for licensed spectrum LPWA solutions; (2) completing the technical specification of the Consumer Remote SIM Provisioning; (3) positioning operators as key partners within the IoT big data market through the delivery of data sets and APIs; and (4) supporting operators in the provision of services that enable smart cities (GSMA, 2016). It is important to be aware of these new goals to get an idea of the directions in which efforts are being made and of what is happening at the moment in the area.

ITU also created the Internet of Things Global Standards Initiative (IoT-GSI),‡ which worked in detailing the requirements for developing the standards that are necessary to enable the deployment of IoT on a global scale, taking into account the work done in other SDOs. In July 2015, this group decided to create the SG 20, which focuses on "IoT and its applications including smart cities and communities." Therefore, all activities conducted by the IoT-GSI were transferred to the SG 20,§ which has produced around 300 related documents so far.

Both Industrial Internet Reference Architecture (IIRA)¶ and Reference Architecture Model for Industrie 4.0 (RAMI 4.0) were developed, focusing on taking advantage of IoT technology to increase the efficiency of the industrial processes, either improving manufacturing itself or making the supply chain from the suppliers to the customers more effective.

Sensor Network Reference Architecture (SNRA),** in turn, provides a general overview of the characteristics of a sensor network and the organization of the entities that comprise such a network. It also describes the general requirements that are identified for sensor networks, which relate to IoT systems since sensor networks are used by IoT systems as a tool for collecting data. A working group created by ISO/IEC,†† involving industry and commerce, academic and research bodies, and government, is working on the development of an RA for IoT (IoT RA—IoT Reference Architecture) that aims to describe the characteristics and aspects of IoT systems, define the IoT domains, describe the RM of IoT systems, and describe the interoperability of IoT entities (ISO/IEC JTC1, 2016). IoT RA intends to become the common reference for all the RAs already proposed by other organizations, including Internet of Things—Architecture (IoT-A) and the standards developed by ITU (Yoo, 2015).

Since there is no universally accepted definition of IoT, different groups have developed different approaches according to the domain in which they are active (ISO/IEC, 2014).

* http://www.gsma.com/.
† http://www.gsma.com/connectedliving/mobile-iot-initiative/.
‡ https://www.itu.int/en/ITU-T/gsi/iot/Pages/default.aspx.
§ https://www.itu.int/en/ITU-T/studygroups/2017-2020/20/Pages/default.aspx.
¶ http://www.iiconsortium.org/IIRA.htm.
** https://www.iso.org/obp/ui/#iso:std:iso-iec:29182:-1:ed-1:v1:en.
†† https://www.iso.org/standard/65695.html.

Additionally, GS1,* a nonprofit organization working in the area of barcoding standardization, claims that the evolution of standards for IoT has followed a path where no standards existed to a situation where too many standards are available, leading to difficult choices to be made when designing IoT applications (GS1, 2016, 1). These difficulties may be accentuated by manufacturers who try to protect their products and solutions and are not necessarily interested in adopting open standards or ensuring interoperable solutions.

In 2013, a consortium involving industry and university partners, like Alcatel-Lucent, IBM, NEC Siemens, Sapienza University of Rome, and University of Surrey, created an architectural reference model (ARM) for IoT, named IoT-A (Bassi et al., 2013). For the partners, achieving interoperability between solutions across various platforms could only be ensured through interoperability both at the communication level and at the service level. At the end of the project, funded by the EU, the benefits of the developed architecture were demonstrated through the implementation of real-life use cases.

In the following sections, some of the more relevant RAs are briefly described. As they are still being developed, there might be new updates in this field.

1.3.1 IoT-A

Currently, the IoT-A project is no longer active. However, IoT-A is described here since it is being used as a basis for developing other architectures, such as the IoT RA or Reference IoT Layered Architecture (RILA), which are also discussed in the following sections.

The IoT-A ARM was created in order to achieve interoperability between different IoT systems (Bassi et al., 2013). The IoT ARM is defined to be abstract so that it can be used as a reference for generating concrete system architectures. It consists of an RM and an RA.

The RM, presented in Figure 1.1, provides a common understanding of the IoT domain by modeling its concepts and their relationships. Similar to the Open Systems Interconnection (OSI) model, the IoT RM by itself does not specify the technical particularities of an IoT system.

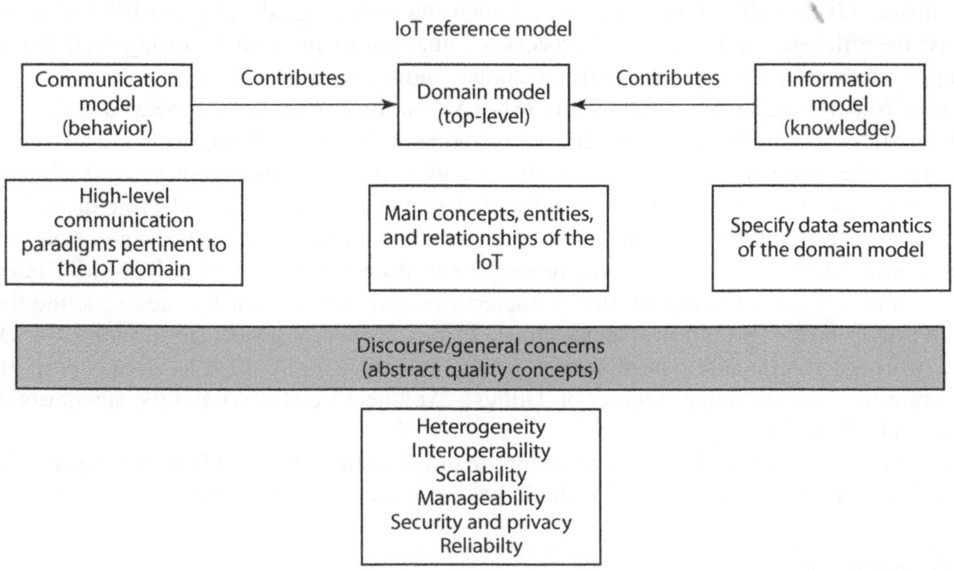

FIGURE 1.1 RM proposed by IoT-A. (Adapted from Bassi, A., et al., *Enabling Things to Talk: Designing IoT Solutions with the IoT Architectural Reference Model*, Springer, Berlin, 2013, 163–211.)

* http://www.gs1.org.

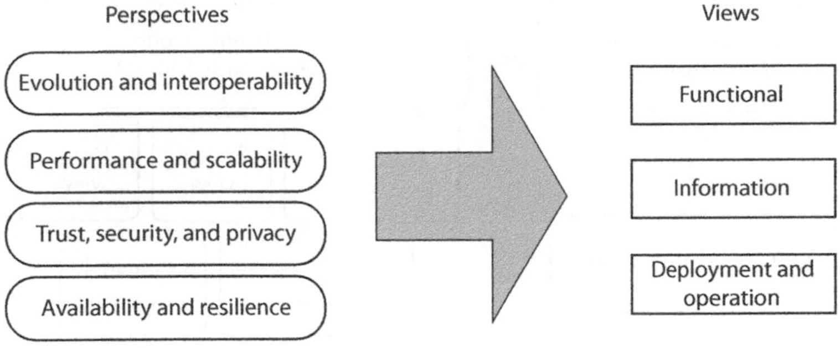

FIGURE 1.2 Perspectives and views of IoT-A. (Adapted from Bassi, A., et al., *Enabling Things to Talk: Designing IoT Solutions with the IoT Architectural Reference Model*, Springer, Berlin, 2013, 163–211.)

The *domain model* considers a top-level description of the concepts and entities (physical entities, devices, resources, and services) that represent particular aspects of the IoT domain, and defines their relations. Therefore, the domain model can also be used as a taxonomy of the IoT.

The *information model* specifies the data semantics of the domain model; that is, it refers to the knowledge and behavior of the entities considered in the domain model, since they are responsible for either keeping track of certain information or performing specific tasks (it describes which type of information the entities are responsible for).

The *communication model*, in turn, addresses the main communication paradigms necessary for connecting entities, ensuring interoperability between heterogeneous networks. The proposed communication model is structured in a seven-layer stack and describes how communication has to be managed, by each layer, in order to achieve the interoperability features required in the IoT. It also describes the actors (communicating elements) and the channel model for communication in IoT.

The RA of IoT-A mainly consists of "views" and "perspectives," which vary depending on the requirements of each specific application. Figure 1.2 illustrates that the perspectives "evolution and interoperability," "performance and scalability," "trust, security, and privacy," and "availability and resilience" are applied to all the views: the "functional" view, the "information" view, and the "deployment and operation" view, respectively.

While applying perspectives to views, not every view is impacted by the perspectives in the same manner or grade. For example, the perspectives have a high impact when applied to the operation view.

1.3.2 IoT RA

IoT RA, created by the ISO/IEC* (CD 30141), envisions the construction of an IoT system based on a generic IoT conceptual model (CM) that includes the most important characteristics and domains of IoT. Then, it uses the CM as a basis to create a high-level system-based RM. This reference model is, in turn, structured in five architectural views (functional view, system view, user view, information view, and communication view) from different perspectives, which compose the RA itself. Figure 1.3 shows the relation between these three components (CM, RM, and RA).

In essence, the IoT RA provides the basics to create a concrete system architecture. The IoT RA is considered an application-specific architecture or a "target system architecture" since the RA can adapt to the requirements of a specific system, like agricultural system, smart home/building, smart city, and so forth.

* https://www.iso.org/standard/65695.html.

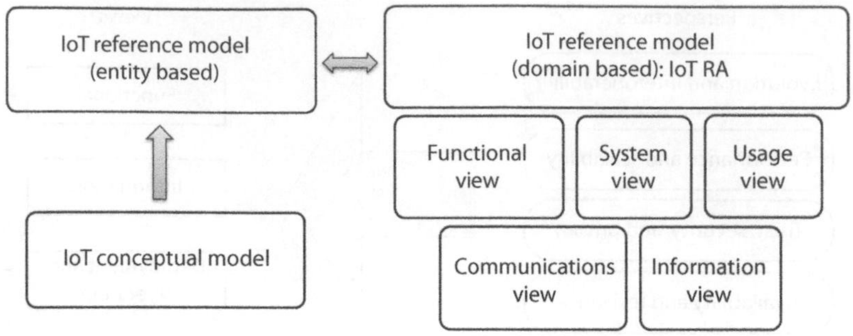

FIGURE 1.3 Relation between CM, RM, and RA. (Adapted from ISO/IEC, Information technology—Data structure—Unique identification for the Internet of things, ISO/IEC 29161:2016, August 2016, https://www.iso.org/obp/ui/#iso:std:iso-iec:29161:ed-1:v1:en.)

1.3.3 IEEE P2413

IEEE P2413 is based on ISO/IEC/IEEE 42010:2011: "Systems and Software Engineering Architecture Description."* The goal is not to create a new standard but to address common aspects of different application domains of the IoT. The IEEE working group is collaborating with ISO, ITU-T, and the Industrial Internet Consortium (IIC), among others, with the common goal of achieving better standards for the IoT in all its areas of application. The focus is on achieving interoperability, together with other quality attributes, such as protection, privacy, security, and safety. Some of these challenges are further discussed in Section 1.6.

1.3.4 Industrial Reference Architectures

The IIRA† is a standard-based open architecture for Industrial Internet Systems (IISs), proposed by the IIC Technology Working Group, whose members are companies like AT&T, Cisco, IBM, General Electric, and Intel. The Industrial Internet is considered an IoT system, enabling intelligent industrial operations and focusing on key characteristics for this type of systems: safety, security, and resilience. IISs cover energy, healthcare, manufacturing, the public sector, transportation, and related industrial systems.

The Industrial Internet Architecture Framework (IIAF) is based on ISO/IEC/IEEE 42010:2011, and as such, it uses the same constructs and common terms, such as viewpoints, concerns, and stakeholders, as well as views and models.‡ The IIRA is the result of applying IIAF to the Industrial IoT systems. Table 1.1 shows an overview of the IIRA. Each viewpoint influences the viewpoints below it. In turn, lower viewpoints validate and sometimes cause revisions in the higher viewpoints. There are some crosscutting concerns, such as security and safety, which are discussed in other reports from the IIC.

Initially designed for German industry, RAMI 4.0, in turn, is a result of cooperation between: Plattform Industrie 4.0 (Industrie 4.0 [I4.0] is considered a specialization within IoT); some German associations, like BITKOM, VDMA, and ZVEI; and several German companies (Adolphs et al., 2015). Their main goal was to achieve a common understanding of what is necessary to evolve from current industries and make I4.0 become a reality. To do that, it was necessary to develop an architecture model to be used as a reference in this migration.

* http://grouper.ieee.org/groups/2413/Intro-to-IEEE-P2413.pdf.
† http://www.iiconsortium.org/IIRA.htm.
‡ http://www.iiconsortium.org/IIC_PUB_G1_V1.80_2017-01-31.pdf.

TABLE 1.1
IIRA Overview: Viewpoints and Their Concerns and Stakeholders

Viewpoints	Concerns	Crosscutting Concerns		Stakeholders
Business	Identification of stakeholders, business vision, values, and objectives of an Industrial IoT system	Safety	Security	Decision makers, product managers, system engineers
Usage	Expected system usage			System engineers, product managers, other users
Functional	Functional components, interfaces and interactions between them			System architects, developers, integrators
Implementation	Technologies needed, communication protocols, life cycle			System architects, developers, integrators, system operators

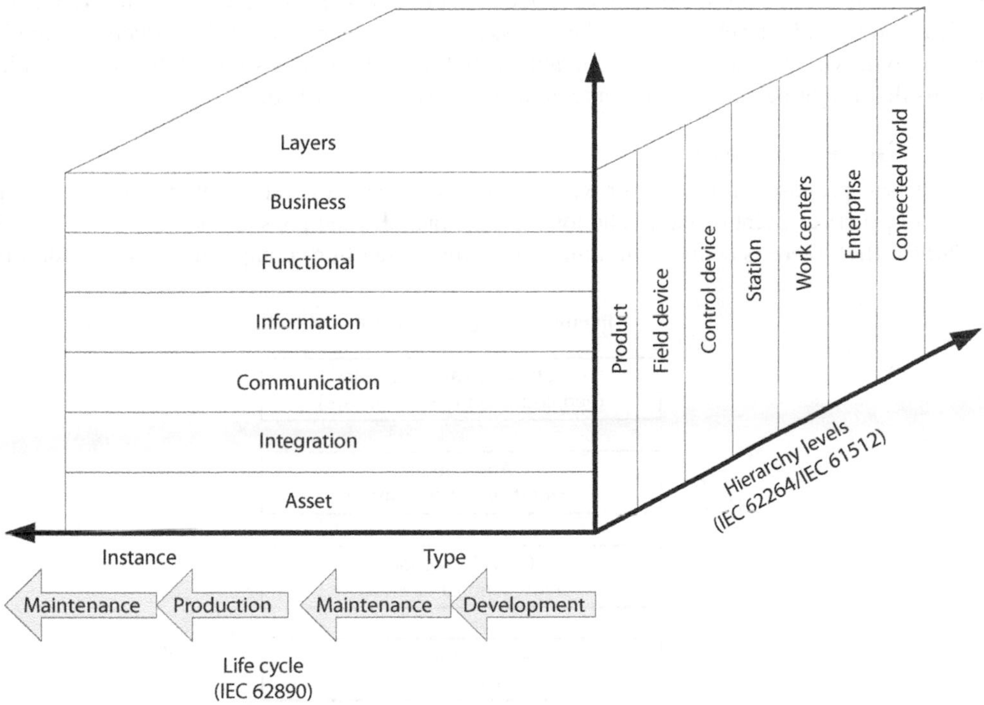

FIGURE 1.4 RAMI 4.0. (Adapted from Adolphs et al., Reference Architecture Model for Industrie 4.0 (RAMI 4.0.), 2015, https://www.zvei.org/fileadmin/user_upload/Presse_und_Medien/Publikationen/2016/januar/GMA_Status_Report__Reference_Archtitecture_Model_Industrie_4.0__RAMI_4.0_/GMA-Status-Report-RAMI-40-July-2015.pdf.)

RAMI 4.0 focuses on the optimization of central industrial processes, namely, research and development, production, logistics, and service. It describes the structures and functions of the I4.0 components, based on existing and relevant standards.

Figure 1.4 shows the RAMI 4.0 architecture model, which is a three-dimensional model, where the horizontal axis represents the life cycle of systems or products, distinguishing between "type" (life cycle of a product from the idea to the product, going through design, development, and testing) and "instance" (it represents the manufacturing of a type; its life cycle goes from manufacturing, selling, and delivering to the client, to being installed in a particular system). The vertical axis (six layers), in turn, corresponds to the IT perspective of an I4.0 component, meaning that it breaks complex projects into smaller parts, like business processes, functional descriptions, communications behavior, and hardware/assets. Finally, the third axis represents a functional hierarchy, which does not refer to equipment classes or hierarchical levels of the automation pyramid, but to grouping functionalities and responsibilities within the factories. It contains core aspects of I4.0, such as field device, control device, station, work centers, and enterprise, but expands the hierarchy levels of the IEC 62264 standard by adding "product" and "connected world."

1.3.5 Other Reference Models and Architectures for IoT

So far, there have been several contributions to create RMs for IoT, most of them based on IoT-A. In fact, up to now, several architectures have been proposed, but as they are designed for a specific IoT application, they cannot be used as a reference, since they do not adapt to other applications' requirements (Yin et al., 2015; Pang, 2013; Vlacheas et al., 2013; Domingo, 2012; Yun and Yuxin, 2010). An alternative architectural stack for the "web of things" consists of "levels of functionalities," with each level composed of a set of application protocols and tools (Guinard and Trifa, 2016). The idea is to provide developers with the necessary tools to implement IoT products and applications, and to maximize reuse and interoperability. In this section, only some of the most significant proposals that might be considered as a reference are briefly described.

1.3.5.1 Cisco Reference Model

In 2014, Cisco proposed a seven-layer RM (Cisco, 2014), which is represented in Figure 1.5, giving a more practical point of view. The lowest level includes the physical devices and controllers (the *things*); then there is connectivity and, above that, edge (fog) computing, where some initial

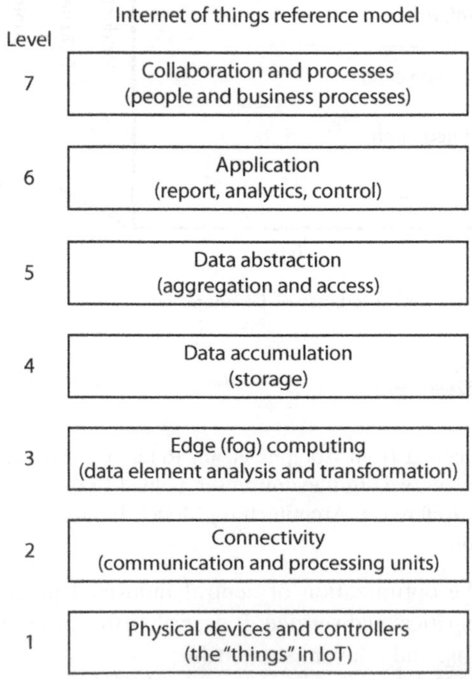

FIGURE 1.5 IoT RM proposed by Cisco. (Adapted from Cisco, The Internet of things reference model, 2014, http://cdn.iotwf.com/resources/71/IoT_Reference_Model_White_Paper_June_4_2014.pdf.)

aggregation, elimination of data duplication, and analysis can be carried out. The lower three levels, in turn, are considered operational technology (OT). The top four levels relate to the IT. The lowest level in the IT part of the stack is storage, and this is followed, going toward the top, by data abstraction, applications, and collaboration and (business) processes.

1.3.5.2 Reference IoT Layered Architecture

Every IoT RA must include some essential components, such as interoperability and integration components, context-aware computing techniques, and security guidelines for the whole architecture (Karzel et al., 2016). The resulting proposed architecture is RILA. RILA is a more concrete architecture, intended to be easier to comprehend for customers and industry than the high-level IoT-A. It not only provides guidelines of how to put IoT-A in practice but also demonstrates that this architecture can really be implemented using actual use cases. RILA acts between things, devices, and the user.

RILA consists of six layers, as depicted in Figure 1.6. Besides these layers, there are two cross section layers, "security" and "management," that affect all other layers.

The *device integration layer* includes all the different types of devices, receives their measurements, and communicates actions. This layer can be seen as a translator that speaks many languages (Karzel et al., 2016). The output of the sensors and tags, as well as the input of the actuators, depends on the protocol they implement.

The *device management layer* is responsible for receiving device registrations and sensor measurements from the device integration layer, and for communicating status changes for actuators to the device integration layer. Then, the device integration layer checks if the status change (i.e., the action) conforms with the respective actuator and translates the status change to the actuator. The device management layer controls the devices that are connected to the system; every change to a device's registration, as well as new measurement data, should be communicated from the device integration layer to the device management layer, so the information can be updated and stored.

Normally, the *data management layer* is a central database (but it can also be a data warehouse or even a complete data farm, in the case of larger IoT systems) that stores all data of a thing. Thus,

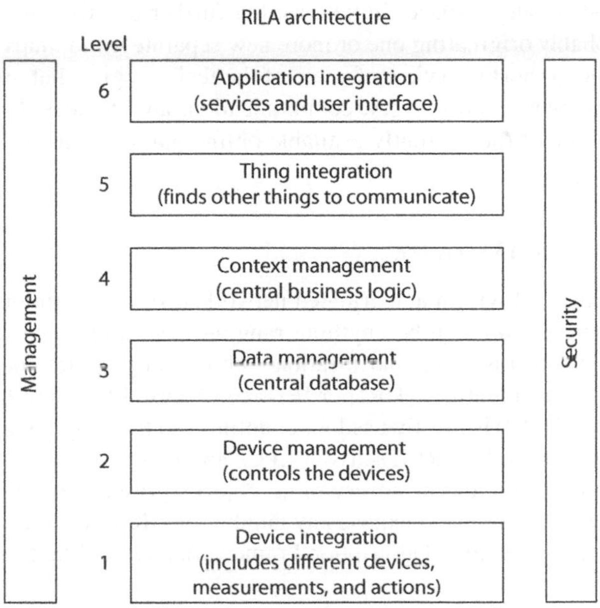

FIGURE 1.6 Reference IoT layered architecture. (Adapted from Karzel, D., et al., A reference architecture for the Internet of things, January 29, 2016, https://www.infoq.com/articles/internet-of-things-reference-architecture.)

the implementation of the data management layer strongly depends on the use case (Karzel et al., 2016).

The *context management layer* defines the central business logic and is responsible for tasks like defining the goals of the thing, consuming and producing the context situations of the things, evaluating the context situation toward the goals, triggering actions that will help to fulfill the goal according to the evaluated rules, and finally, publishing context situations for other things.

The *thing integration layer* is responsible for finding other things to communicate, verifies if communication with the new thing is possible, and is responsible for a registration mechanism.

The *application integration layer* connects the user to the thing, being considered the service layer, or even a simple user interface. The concrete implementation of the layer depends on the use case.

1.4 ENABLING TECHNOLOGIES

In the previous section, a few international organizations that are engaged in defining and developing common standards for the IoT were already mentioned. Many of these are integrated in alliances and consortia that include members from industry, SDOs, manufacturers, network and service providers, academia, and research laboratories.* oneM2M, for example, consists of eight important SDOs from all around the world: ARIB (Japan), ATIS (United States), CCSA (China), ETSI (Europe), TIA (United States), TSDSI (India), TTA (Korea), and TTC (Japan), which have come together with six industry partners and consortia (Broadband Forum, Continua Alliance, GlobalPlatform, HGI, Next Generation M2M Consortium, and OMA) and more than 200 member organizations, including companies (like Alcatel, Nokia, Huawei, Deutsche Telekom, and Qualcomm) and universities.

ISO WG 10 created a very interesting and comprehensive mind map with six IoT-related areas: requirements, technologies, application areas, stakeholders, standards, and other considerations. The map is available on their website as an appendix of their IoT report published in 2015 (ISO, 2017, 1). For technologies, a separate mind map was created due to such a great number of existing and developing technologies that all strive to be key enabling technologies for the IoT. The map is not completely exhaustive due to space limitations, but further details can be added easily to the existing branches, probably originating one or more new separate mind maps for each branch.

This section presents a short overview of some of the technologies that are being used for connected objects identification and discovery, communication, and devices. In the final part of this section, we will see some of the currently available online platforms for IoT project prototyping, development, and testing.

1.4.1 Identification and Discovery

For the things to be identified within a distributed networked system, a unique identifier is needed for each thing. And here a *thing* can be anything ranging from a physical to a virtual object, an event, or a person. Data such as time and location—either geographic (coordinates) or within a network (Uniform Resource Identifier [URI] or IP)—can be used for identification purposes. The Electronic Product Code (EPC) is mostly used for counting and tracking goods in the supply chains without human intervention. While RFID-driven EPCs are attached to the physical objects, URIs and IPs allow the identification and discovery of an object's presence on the web. HyperCat,† for example, is a solution that allows us to expose any number of URIs, together with additional information attached to them in Resource Description Framework (RDF)–like format.

* https://www.postscapes.com/internet-of-things-alliances-roundup/.

† http://www.hypercat.io/.

IP addresses are also used as identifiers for networked objects, together with name labels, which can be used for human readability and resource discovery using a naming service, such as mDNS.* mDNS is an IETF project that aims to adapt the well-established Domain Name System (DNS) programming interfaces and formats to small networks where there are no name servers available.

People, in turn, can be identified through the devices they carry along or through more sophisticated techniques, such as biometric data obtained through face or iris recognition, fingerprints, voice, and so on. The ISO IoT technology mind map includes more identification-related technologies and possible solutions.

In 2016, ISO/IEC published the 29161 standard (ISO/IEC, 2016), which aims to ensure full compatibility among different identification forms.

1.4.2 Communication Patterns and Protocols

RFC 7452† describes four basic communication patterns for IoT environments: device-to-device communication pattern, device-to-cloud communication pattern, device-to-gateway communication pattern, and back-end data sharing pattern.

The device-to-device pattern is applied when two devices communicate directly, normally using a wireless network. There are several protocol stacks available to carry out this type of communication. Depending on the usage scenario, the protocol stack may include, for instance, Bluetooth or IEEE 802.15.4, IPv6, User Datagram Protocol (UDP), and Constrained Application Protocol (CoAP).

The device-to-cloud communication pattern is used when data captured by the device from the environment is uploaded to an application service provider. Communication is based on IP, but when the device manufacturer and the application service provider are the same, the integration of other devices may be difficult. For this not to happen, the protocols used to communicate with the server need to be made available.

The device-to-gateway communication pattern may be used when the system contains non-IP devices, when support for legacy devices is needed, or when additional security functionality must be implemented. Gateways can also be mobile, providing only temporary connections to the Internet. Smartphones are an example of those.

The back-end data sharing pattern is used when there is a need to analyze combined data from several sources. RESTful APIs can be used, although they are not standardized. This pattern may allow users to move their data from one IoT service to another (Rose et al., 2015).

Beyond some well-known communication standards, like Bluetooth, Wi-Fi, or GSM, IoT systems use many more. It is worth pointing out that new communication standards are being developed specifically for some of the IoT scenarios. For example, when the system is composed of devices with constrained resources across wide area networks, low-power wide area network (LPWAN) solutions must be applied.

On the one hand, the last two decades or so registered immense growth in the field of mobile telecommunications, with devices that have more and more resources in terms of processors, memory, sensors, and connectivity. As such, 3GPP and the mobile operators have been making a huge effort toward new standards for mobile communications with even larger bandwidths, higher bit rates, and support for multimedia live streaming.

On the other hand, solutions for the scenarios that will not so much include resource-rich devices, but mostly resource-constrained devices, stayed behind and only appeared recently. In 2016, 3GPP concluded a new standard for the IoT called NB-IoT (3GPP, 2016), which is supported by leading manufacturers and by the world's 20 largest mobile operators (Vodafone Group, 2016). NB-IoT is a technology that allows us to bidirectionally and securely connect multiple sensors and devices with

* http://multicastdns.org/.

† https://tools.ietf.org/html/rfc7452.

low bandwidth requirements. It is a technology that requires low energy consumption (more than 10 years of autonomy) and allows strong penetration of the radio signal in indoor environments.

In the meantime, many other low-power solutions for wide area networks, such as LoRa, NWave, and Sigfox, have been developed and adopted by many IoT deployments. It is predictable that there will be a convergence among these technologies in the near future, and some of them will eventually prevail, while others will disappear or remain in use in legacy deployments. Table 1.2 summarizes some of the communication protocols and standards currently in use.

1.4.3 Devices and Test Beds

A wide range of specialized and multipurpose sensors are available on the market, as well as many SBCs with embedded sensors or with support for several sensors and actuators. Wearable devices are also acquiring more and more enthusiasts, especially in the area of sports and physical activity tracking, sometimes in addition to the already well-equipped smartphones with several embedded sensors.

Sensors enable devices to capture data from their environment. They can be categorized in different ways. Here is a nonexhaustive list of possible categories and some examples:

- *Location*: GPS, GLONASS, Galileo, Wi-Fi, Bluetooth, ultra-wideband (UWB)
- *Biometric*: Fingerprint, iris, face
- *Acoustic*: Microphone
- *Environmental*: Temperature, humidity, pressure
- *Motion*: Accelerometer, gyroscope

Actuators, in turn, allow devices to act on their environment and may be of different types, like hydraulic, pneumatic, electric, mechanical, or piezoelectric.

The SBCs are becoming more and more popular among both hobbyists and researchers. They are low cost (less than $50), provide a reasonable amount of processing power and RAM, sometimes have embedded sensors or support the connection of external sensors, and support wireless connectivity (Wi-Fi, Bluetooth Low Energy [BLE], and ZigBee). Mostly, these are Linux machines. The most popular examples are Arduino, Raspberry Pi, BeagleBone Black, Intel Edison, and Pine 64.* Their characteristics are summarized in Table 1.3.

As an alternative to deploying one's own hardware on site, there are laboratories around the world where the physical devices are located and can be tested remotely, especially for large-scale scenarios testing. Examples are the FIT IoT-lab,† located in France; the III-IoTLab,‡ in Taiwan; and iMinds,§ in Belgium. All of them are parts of OneLab,¶ which offers several test beds giving developers a means to quickly implement and test their projects in controlled environments, being able to evaluate them before the actual deployment takes place.

OneLab is not restricted to the IoT area; it also offers test beds for other areas, like cloud computing or software-defined networks. Regarding test beds for IoT, there is one test bed specifically targeted for *smart cities* applications and another for *connected commerce*, which addresses logistics monitoring (e.g., monitoring the quality of food and its transportation during the whole supply chain).

* https://www.pine64.org/.
† https://www.iot-lab.info/.
‡ https://iot.snsi.iii.org.tw/.
§ http://ilabt.iminds.be/.
¶ https://onelab.eu/.

TABLE 1.2
Overview of Communication Technologies and Standards for IoT

Name	Frequency	Range	Examples	Standards
BLE	2.4 GHz	1–100 m >100 m	Headsets, wearables, sports and fitness, healthcare, proximity, automotive	IEEE 802.15.1 Bluetooth SIG
EnOcean	315, 868, 902 MHz	300 m outdoor 30 m indoors	Monitoring and control systems, building automation, transportation, logistics	ISO/IEC 14543-3-10
GSM	Europe: 900 MHz and 1.8 GHz		Mobile phones, asset tracking, smart meter, M2M	3GPP
LTE	United States: 1.9 GHz and 850 MHz			
LoRa	Sub-1 GHz ISM band	2–5 km urban 15 km suburban 45 rural	Smart city, long range, M2M	LoRaWAN
NB-IoT	700–900 MHz	10–15 km rural deep indoor penetration	Smart meters, event detectors, smart city, smart home, industrial monitoring	3GPP LTE Release 13
NFC	13.56 MHz	Under 0.2 m	Smart wallets, smart cards, action tags, access control	ISO/IEC 18092 ISO/IEC 14443-2, -3, -4 JIS X6319-4
NWave	Sub-1 GHz ISM band	Up to 10 km	Agriculture, smart city, smart meter, logistics, environmental	Weightless
RFID	120–150 kHz (LF), 13.56 MHz (HF), 2450–5800 MHz (microwave), 3.1–10 GHz (microwave), 433 MHz (UHF), 865–868 MHz (Europe), 902–928 MHz (North America) (UHF)	10 cm–200 m	Road tolls, building access, inventory, goods tracking, building automation, smart energy, smart city logistics	ISO 18000
DASH7		0–5 km		
Sigfox	900 MHz	3–10 km urban 30–50 km rural	Smart meters, remote monitoring, security	
Weightless	470–790 MHz	Up to 10 km	Smart meters, traffic sensors, industrial monitoring	Weightless
Wi-Fi	2.4 GHz, 3.6 GHz, 4.9/5 GHz	Up to 100 m	Routers, tablets, smartphones, laptops	IEEE 802.11
Z-Wave	ISM band 865–926 MHz	100 m	Monitoring and control for home and light commercial environments	Z-Wave Recommendation ITU G.9959
ZigBee	2.4 GHz; 784 MHz in China, 868 MHz in Europe, and 915 MHz in the United States and Australia	10–20 m	Home and building automation, WSN, industrial control	IEEE 802.15.4

Source: Data from Postscapes, IoT technology guidebook, IoT Technology | 2017 Overview Guide on Protocols, Software, Hardware and Network Trends, 2017, https://www.postscapes.com//internet-of-things-technologies/; Opensensors, How to choose the best connectivity network for your Project, 2017, https://publisher.opensensors.io/connectivity; ETSI (European Telecommunications Standards Institute), SmartM2M; IoT standards landscape and future evolutions, ETSI TR 103 375 V1.1.1 (2016-10), 2016, http://www.etsi.org/deliver/etsi_tr/103300_103399/103375/01.01.01_60/tr_103375v010101p.pdf.

Note: M2M, machine-to-machine; LF, low frequency; HF, high frequency; UHF, ultra-high frequency.

TABLE 1.3
Characteristics of Some of the Currently Available Single-Board Computers for IoT

	Models	CPU	RAM	Operating System	Price ($)	Connectivity	Embedded Sensors
Arduino	20+ models	ATmega, ATSAM, AR9331, etc.	0.5 kB–16 MB	Linux	30	Ethernet, Wi-Fi, extension boards	Extensions
Beagle board	BeagleBone Black	AM335x, 1 GHz; ARM Cortex A8	512 MB–1 GB	Linux	50	Ethernet	Extensions
Intel	Edison, Joule, Galileo	Intel Atom, Intel Quark	256 MB–4 GB	Linux	20–60	Wi-Fi, Bluetooth	
Pine 64	Pine A64, Pine A64+	ARM Cortex A53, 1.2 GHz	512 MB–2 GB	Linux, Android, Windows IoT	15–29	Ethernet	
Raspberry Pi	Zero, RPi 1 A+, RPi 2 B+, RPi 3 B	ARM1176, 1 GHz; ARM1176; ARM Cortex A7; ARMv8, 1.2 GHz	512 MB–1 GB	Linux, Windows IoT	20–40	Wi-Fi, BLE, Ethernet, extensions	SenseHAT, other

1.5 APPLICATION AREAS: AN OVERVIEW

A symbiosis between platforms, applications, devices, and services gives the ability to improve citizens' well-being and quality of life. The great potentialities offered by the IoT make the development of a huge number of applications possible, while it also plays a crucial role in the so-called fourth Industrial Revolution (I4.0). The IIC* was created with the goal of transforming industry through intelligent, interconnected objects that may improve performance, lower costs, and increase reliability. This consortium considers that industry involves the areas of energy, healthcare, manufacturing, smart cities, and transportation.

Nevertheless, the areas of application cover various sectors of society and are grouped in diverse ways in the literature. There are at least two (Atzori et al., 2010; AIOTI, 2015) fairly complete and interesting classification schemes of the application domains. Considering the most relevant research made on applications, but essentially these two sources, an overview of the main application domains is shown in Figure 1.7.

As the figure depicts, the main areas of application are smart cities, healthcare, smart homes and buildings, mobility and transportation, energy, industry, agriculture, and the environment/planet. Note that some applications, like environmental monitoring, can fit into different groups, like smart city, smart buildings, the environment, and industry. It is worth mentioning that in the context of IoT, it is not possible to develop a "one-size-fits-all" solution. For example, a solution for home environmental monitoring may not be adequate for industrial ambient monitoring due to different types of physical conditions and relevant parameters. An industrial setting requires different levels of accuracy, security, and robustness for the deployed sensors and software, while a home environment usually does not impose such restrictions. Some examples of these applications are briefly introduced below.

1.5.1 Smart Cities

Finding ways to use technology to improve the quality of life in a city has become one of the most popular research topics in the area of IoT applications (Yin et al., 2015; Zanella et al., 2014; Vlacheas et al., 2013; AT&T, 2017). Smart city solutions include several areas, ranging from water and waste management (like smart dumpsters) (Hong et al., 2014; Phithakkitnukoon et al., 2013; Smartup Cities, 2017), lighting control (Castro et al., 2013), energy (Kyriazis et al., 2013), transportation, traffic, and parking management, to building efficiency, services, and safety (Perera et al., 2014).

Even though smart buildings, transportation and mobility, and energy are sometimes included under the smart cities umbrella, Figure 1.5 dedicates a separate branch to each, due to the importance that these areas are receiving.

Some examples of real-life implementations occur in the cities of London† and Greenwich in the United Kingdom; Santander (Sanchez et al., 2011), Barcelona, and Murcia in Spain; Amsterdam in Holland; Aarhus in Denmark; Oulu (Gil-Castineira et al., 2011) in Finland; several towns in Korea; and Bordeaux in France. AT&T is helping cities to deploy integrated smart city solutions and will deliver solutions in some cities in the United States. The CITYkeys‡ project, funded by the European Commission's Horizon 2020 program, aims at developing an evaluation framework to compare smart city solutions across European cities.

1.5.2 Healthcare

IoT technologies may bring significant benefits to the healthcare domain, namely, in two areas: clinical care and remote monitoring. Basically, the use of small-sized, low-cost, and low-power wearable

* http://www.iiconsortium.org/about-industrial-internet.htm.
† http://www.organicity.eu.
‡ http://www.citykeys-project.eu.

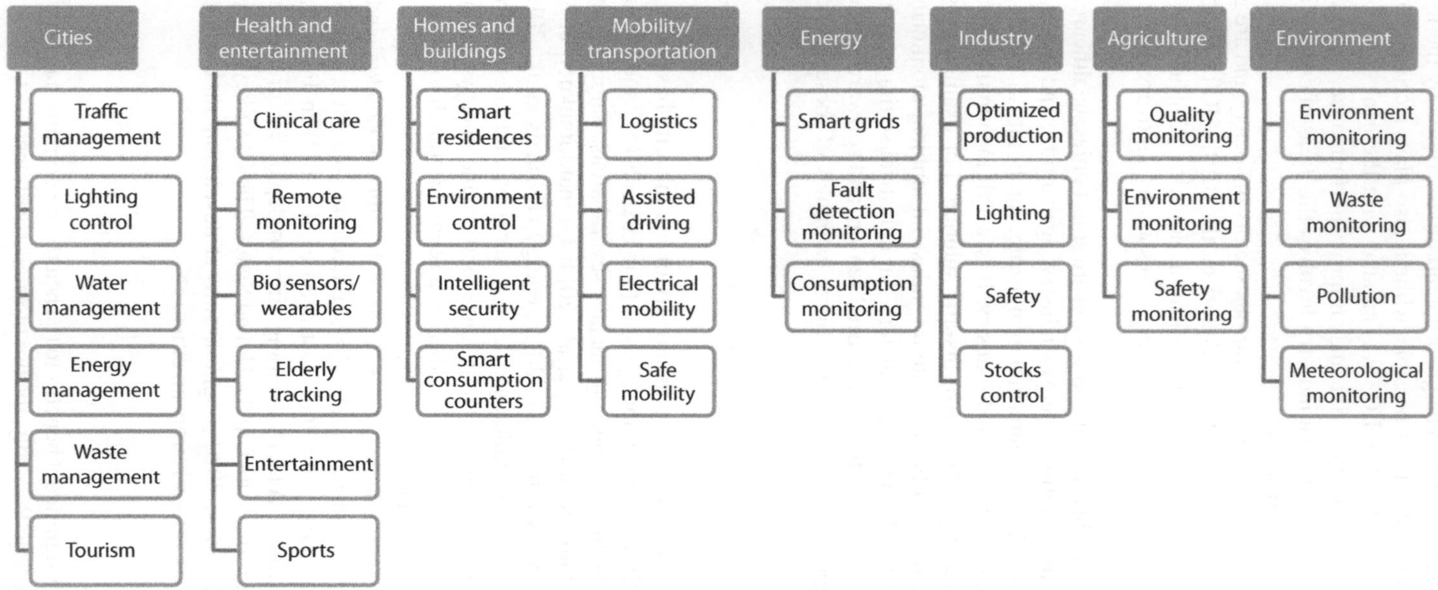

FIGURE 1.7 Applications grouped in domains. (Based on Atzori, L., et al., *Computer Networks*, 54 (15), 2787–2805, 2010; AIOTI, Internet of things applications, 2015, https://aioti.eu/aioti-wg01-report-on-internet-of-things-applications/.)

biosensors can enhance the quality of life of people suffering from chronic diseases or even during emergencies, either inside (Domingo, 2012; Doukas and Maglogiannis, 2012; Yang et al., 2014; Bui and Zorzi, 2011; Dohr et al., 2010; X. Li et al., 2011) or outside their homes (Domingo, 2012; Doukas and Maglogiannis, 2012). Elderly tracking or ambient assisted living (AAL) encompasses technical systems to support elderly people in their daily routine to allow an independent and safe lifestyle as long as possible. This is a special case of healthcare that has gained increased importance due to the problem of population aging (Yang et al., 2014; Dohr et al., 2010). Another interesting example is using IoT in noninvasive glucose-level sensing for diabetes management (Istepanian et al., 2011) or using an IoT platform to improve the life of people with disabilities (Domingo, 2012).

1.5.3 Smart Homes and Smart Buildings

Smart home applications can range from elderly monitoring (Yang et al., 2014) and home automation, in either air conditioning control and monitoring or lighting control, to consumption monitoring and energy saving (Wei and Li, 2011; Kelly et al., 2013), or even solar greenhouse production (Wang et al., 2004).

The focal point of IoT applications for "smart buildings" or "intelligent buildings" is essentially cost savings, providing the building with some intelligence, through building automation. These applications mainly focus on air conditioning and lighting monitoring and control, and on consumption monitoring and energy saving (Wei and Li, 2011; Moreno et al., 2014; Brad and Murar, 2014; Ji et al., 2014). Security is also an important issue; therefore, fire and intrusion monitoring are also crucial (Ryu, 2015; Li et al. 2013; Li, 2013). Intel, Telit, IBM, and many others have several solutions for smart building implementations (Zhong and Dong, 2011).

1.5.4 Mobility and Transportation

All types of vehicles in a city (cars, trains, buses, and bicycles) are becoming more equipped with sensors and/or actuators, resulting in a network composed of a set of mobile sensors. Both roads and rails, as well as transported goods, are also equipped with tags and sensors that send important information to traffic control sites. This not only allows monitoring of the status of the transported goods, but also allows the creation of innovative solutions, allowing transportation vehicles to better route the traffic or providing the tourist with appropriate transportation information.

Moreover, modern cars are also equipped with several sensors, forming a kind of in-vehicle network, which provides kinematics information, automotive diagnostic services, and so forth. Cars can be further equipped with external sensing devices to monitor specific physical parameters, such as pollution, humidity, and temperature. Thus, the concept of "smart vehicles" emerges.

If properly collected and delivered, such data can contribute to make the road transport greener, smarter, and safer (Campolo et al., 2012; Zouganeli and Svinnset, 2009). For example, driving recommendations that aim at eco-efficiency for public transportation and reducing fuel consumption and emission can be provided (Tielert et al., 2010; Kyriazis et al., 2013). Mobile applications, such as Google Traffic or Waze,* rely on user contributed data to monitor traffic conditions. Smart traffic light infrastructures can be used to improve the life of drivers or make cycling or driving in cities safer and smoother. For example, combining data from smartphones carried by cyclists and traffic data gathered from different kinds of sensors deployed in the traffic light infrastructure of a city may allow for an intelligent traffic light orchestration, letting cyclists drive smoothly without unnecessary stopping at each crossroad (Anagnostopoulos et al., 2016). Another specific area of application is modern logistics, which refers to monitoring the whole process of the physical movement of goods from suppliers to demanders, in order to ensure their quality (Zhengxia and Laisheng, 2010; Zhang et al., 2011).

* https://www.waze.com/.

1.5.5 Energy

The smart grid is a recent kind of intelligent power system that can improve energy efficiency, reduce environmental impact, improve the safety and reliability of the electricity supply, and reduce the electricity transmission of the grid. The integration of IoT technology in smart grids can help to implement fault detection and monitoring, as well as consumption monitoring, through the installation of energy sensors (Yun and Yuxin, 2010; L. Li et al. 2011; Liu et al., 2006; Bui et al., 2012; Ou et al., 2012).

Other groups of related solutions envision the heat and energy management in homes and buildings to accomplish an energy savings purpose (Kyriazis et al., 2013; Sundramoorthy et al., 2010). Using IoT technology to collect data on energy consumption can also help to improve the energy efficiency and competitiveness of manufacturing companies at the energy production level (Shrouf and Miragliotta, 2015).

1.5.6 Smart Manufacturing

The design and operation of a manufacturing system needs numerous types of decision making at various levels of its activities. Therefore, IoT can be applied to develop modern manufacturing enterprises characterized by dynamic and distributed environments (Da Xu et al., 2014). In these environments, IoT technology can be used to serve a variety of purposes (Fantana et al., 2013; Tao et al., 2014; Bi et al., 2014): from environment control, lighting control, and safety, to production optimization, error detection and correction, and automatic control of stocks.

1.5.7 Smart Agriculture

Modern agriculture has a different set of requirements than traditional agriculture. It must be high yield, high quality, efficient, safe, and ecological (Shifeng et al., 2011). IoT technology has contributed to agriculture modernization and improvement (Shifeng et al., 2011; Bo and Wang, 2011). WSNs, for example, have been successfully deployed for irrigation control, fertilization, pest control, and animal monitoring, as well as for greenhouse monitoring, viticulture, and horticulture (Rehman et al., 2014; Zhang et al., 2018).

1.5.8 Environment/Smart Planet

IoT can be used to allow for the development and management of sustainable cities, covering issues like environmental monitoring (Fang et al., 2014), pollution control (Du et al., 2013; Fang et al., 2014), meteorological monitoring (Du et al., 2013), disaster monitoring, or waste management (Hong et al., 2014).

IoT technology can be used to tackle rapid urbanization and related environmental problems, allowing us to study the environment, planning, and construction issues, in order to increase understanding of how to integrate urban development and ecological processes for sustainable city construction.

1.6 CHALLENGES

Currently, IoT is one of the main accelerators of technological innovation, being one of the areas with greater potential of the transformation of society and the economy. As such, all the involved stakeholders, ranging from technologists to developers, companies, and users, face several challenges that remain to be tackled. Experience from areas such as distributed systems, networks, mobile and ubiquitous computing, context awareness, and WSN could be considered a good starting point for seeking appropriate solutions for issues such as interoperability, openness, security, scalability, and failure handling in the scope of the IoT systems.

TABLE 1.4
Overview of IoT Challenges

Challenges	References
Heterogeneity and interoperability	Ortiz et al. (2014), Serbanati et al. (2011), Al-Fuqaha et al. (2015), Rose et al. (2015), Borgia (2014), Miorandi et al. (2012), Atzori et al. (2010), Alur et al. (2016)
Openness	Perera et al. (2014), Alur et al. (2016)
Security, privacy, and trust	Perera et al. (2014), Serbanati et al. (2011), Ortiz et al. (2014), Al-Fuqaha et al. (2015), Borgia (2014), Rose et al. (2015), Miorandi et al. (2012), Whitmore et al. (2014), Atzori et al. (2010), Taivalsaari and Mikkonen (2017), Alur et al. (2016)
Scalability	Alur et al. (2016), Al-Fuqaha et al. (2015), Borgia (2014), Miorandi et al. (2012)
Failure handling	Alur et al. (2016), Ortiz et al. (2014), Taivalsaari and Mikkonen (2017), and Miorandi et al. (2012)

Based on a nonexhaustive literature survey, a set of challenges is shortly discussed in the next sections. Researched publications were grouped into the five previously mentioned categories, as shown in Table 1.4.

1.6.1 Interoperability

Heterogeneity has been a great challenge in distributed systems, as a variety of networks, hardware, different operating systems, and programming languages started to coexist within the same system. IoT systems of the future will be composed of humans, machines, things, and groups of them. To accomplish the functioning of such a network, seamless communication and cooperation among all the components is crucial (Ortiz et al., 2014). Developers and programmers also need to be prepared to cope with multidevice, always-on, highly dynamic, and distributed systems and adapt and update their programming skills to this new paradigm (Taivalsaari and Mikkonen, 2017).

Even though there are other challenges for the IoT, interoperability remains one of the most challenging goals for IoT systems, unless an RA and a set of standards are developed (Atzori et al., 2010; Rose et al., 2015; Serbanati et al., 2011). There has been an effort by the ISO/IEC to create an RA, IoT RA, which is currently under development. This architecture gathers consensus from several organizations (Yoo, 2015).

Furthermore, to interoperate, IoT components must be identifiable and discoverable by other components (Atzori et al., 2010; Borgia, 2014; Miorandi et al., 2012; Ortiz et al., 2014; Serbanati et al., 2011), as discussed in Section 1.1. After the discovery and identification process, components must be able to somehow communicate (Al-Fuqaha et al., 2015; Alur et al., 2016; Borgia, 2014; Miorandi et al., 2012; Ortiz et al., 2014). Section 1.2 gave an overview of the communication protocols and patterns that are currently being used and developed. It is imperative that in the near future, existing and new wireless technologies, such as NB-IoT, LoRaWAN, and Sigfox, be thoroughly tested and further developed, to achieve steps in the direction of having standards for connectivity among IoT devices.*

1.6.2 Openness

The openness of a system is the degree to which it can be extended and reimplemented in new ways. IoT systems must be prepared to share their data and resources with other systems in sometimes unpredictable ways (Alur et al., 2016; Perera et al., 2014). In order to make sustainable IoT systems,

* https://datafloq.com/read/7-trends-of-internet-of-things-in-2017/2530.

"openness must provide a correct balance between access to functionality, human interaction, and privacy and security" (Alur et al., 2016).

1.6.3 Security, Privacy, and Trust

In addition to the current already complex security and privacy landscape, IoT introduces considerably more data security and privacy issues. Often, IoT systems rely on wireless communications that intrinsically pose security problems. Additionally, the large amount of data generated raises new concerns not only about managing, processing, and analyzing such an amount of data, but also in how to ensure data confidentiality. IoT systems, especially those that collect sensitive data (e.g., healthcare systems), need to be secured at all layers, from the physical to the application layer (Perera et al., 2014; Rose et al., 2015). Existing IoT-enabled devices and deployed systems have been shown to be particularly vulnerable to denial of service attacks.* Only with adequate security and data protection mechanisms in place can the IoT systems expect to gain trust from the users (Ortiz et al., 2014; Rose et al., 2015). Security and privacy issues should be considered from the very beginning of the system design (Miorandi et al., 2012).

Caution is advised in the way data is processed, particularly taking care to keep it anonymous, until official data protection authorities make formal recommendations (Serbanati et al., 2011). In fact, this is an area of ongoing work and more research is needed to be able to overcome these issues. Many end-user companies and customers point to security as the main reason not to have embraced the IoT concept yet (ixia, 2016), while others find that data protection and privacy are the greatest barriers to the development of the IoT (Foster, 2017). An assessment of the extent to which existing data protection regulations fully address the IoT needs to be carried out, in order to establish what actions still need to be taken (Foster, 2017). The key legal issues that might delay the IoT also need to be discussed, as well as the ways that might help to overcome these issues. Recently, the EU published a revised version of the regulations and directives† regarding personal data processing, which becomes effective in 2018 in all member states.

Nevertheless, so far, some security frameworks have been proposed to provide confidentiality, integrity, and authentication (Serbanati et al., 2011). However, these frameworks add some communication and processing overhead to achieve their goal. Security requirements for IoT systems may be grouped into four sets (Borgia, 2014): secure authentication and authorization, secure configuration and data transmission, secure data storage, and secure access to data. For all these requirements, it is necessary to keep in mind that many IoT devices are low power and resource constrained. As such, some of the conventional techniques may not be adequate and new ones must be developed (e.g., lightweight cryptography).‡

1.6.4 Scalability

A scalable system continues to work effectively even when the amount of resources and the number of users are increased significantly. For the IoT, it is predicted that 20 billion devices will be connected to the Internet by 2020. Many of these devices will be mobile and will have low power and an unstable connection. As such, the current solutions may not be enough to guarantee proper functioning of the networks (Alur et al., 2016). At least two levels of scalability may be considered in the scope of IoT: network scalability and data scalability (Borgia, 2014). As the network of interconnected objects grows, interoperability must be guaranteed, as well as data security and privacy. Issues related to energy consumption also need to be tackled (Miorandi et al., 2012).

---

* https://datafloq.com/read/7-trends-of-internet-of-things-in-2017/2530.

† http://eur-lex.europa.eu/legal-content/EN/TXT/?qid=1493140462765&uri=CELEX:32016R0679.

‡ http://nvlpubs.nist.gov/nistpubs/ir/2017/NIST.IR.8114.pdf.

The volume and diversity of data that can and will potentially be generated by IoT is overwhelming (Cisco, 2014). Additionally, with so many different types of devices generating data, there are many reasons why this data may be stored in disparate data storage: there might be too much data to put in one place; transferring data into a database might consume too much processing power, so it is recommended that data is retrieved separately from the data generation process; devices might be geographically separated, and it is advisable to process data locally in order to obtain some processing optimization; there might be the need to separate raw data from data that represents an event; and finally, different kinds of data processing might be required. For these reasons, the data abstraction level must process many different things, which include integrating multiple data formats from different sources, for which purpose ensuring consistent semantics of data across various sources is of extreme relevance. Finally, confirming that data is complete to the higher-level application is also important (Cisco, 2014).

1.6.5 Failure Handling

Failure handling includes detecting, masking, and tolerating failures; recovering from failures; and redundancy. In such a complex, dynamic, and heterogeneous environment as IoT is expected to be, systems are required to be able to self-configure, self-diagnose, and autorepair (Alur et al., 2016). Gateways, as more resourceful components of the IoT systems, may be the right place to implement self-management fault, configuration, accounting, performance, and security (FCAPS) features (Al-Fuqaha et al., 2015).

Systems that successfully meet failure handling requirements are going to gain trust more rapidly among their users (Ortiz et al., 2014). However, this will make the task of the developers far more difficult, as they will have to find the right balance between application logic and error handling (Taivalsaari and Mikkonen, 2017).

1.7 CONCLUSION

By 2020, there will be more than 20 billion interconnected IoT devices, and its market size may reach $1.5 trillion (IDC, 2017). According to ETSI (Antipolis, 2016), each person is expected to have an average of four connected devices. Despite its growth, the IoT ecosystem is a complex market, with multiple layers and hundreds of players, including device vendors, communications and IT service providers, platform providers, and software vendors.

This chapter introduced the main topics on IoT, namely, definitions, RMs and RAs, enabling technologies, standards, main application domains, and challenges. As explained previously, international bodies, enterprises, academia, and industry are working together on a common definition and an RA for the IoT, as well as on standards that will enable interoperability among systems. Solutions to tackle other challenges, such as security and privacy, openness, scalability, and failure handling, are also being explored. Many are already on the table, and new ones are still emerging. It is expected that of around 300 IoT software platforms that are available on the market today, in the long term, only 5–7 of them will be consolidated (Skerrett, 2016).

As a network of devices that communicate autonomously, without human intervention, continuously connected to the Internet, the IoT has several social, individual, economic, and environmental implications. It is expected that IoT technologies will have a positive impact on several areas of society, as discussed in Section 1.5, where an overview of application areas was given. The development of smart cities, for example, in terms of infrastructure, transport, and buildings, has had a significant societal impact by improving the efficiency and sustainability of a whole range of urban services. IoT also plays a crucial role in I4.0, in which industrial sites are being transformed through intelligent, interconnected objects that may improve performance, lower costs, and increase reliability.

According to Tech Republic, artificial intelligence, augmented reality, virtual reality, healthcare IoT, Industrial IoT, and wearables are some of the currently emerging trends for the IoT

(Maddox, 2017). The potential of the area of IoT associated with these trends points toward a promising future. Soon, the results of the collaborative work that has been done by the different groups composed by industry, academia, and SDO representatives are expected to become evident.

REFERENCES

GPP. 2016. Standardization of NB-IOT completed. June 22. http://www.3gpp.org/news-events/3gpp-news/1785-nb_iot_complete.

Adolphs, P., H. Bedenbender, M. Ehlich, U. Epple, M. Hankel, R. Heidel, M. Hoffmeister, et al. 2015. Reference Architecture Model for Industrie 4.0 (RAMI 4.0). https://www.zvei.org/fileadmin/user_upload/Presse_und_Medien/Publikationen/2016/januar/GMA_Status_Report__Reference_Archtitecture_Model_Industrie_4.0__RAMI_4.0_/GMA-Status-Report-RAMI-40-July-2015.pdf.

AIOTI (Alliance for the Internet of Things). 2015. Internet of things applications. https://aioti.eu/aioti-wg01-report-on-internet-of-things-applications/.

Al-Fuqaha, A., M. Guizani, M. Mohammadi, M. Aledhari, and M. Ayyash. 2015. Internet of things: A survey on enabling technologies, protocols, and applications. *IEEE Communications Surveys Tutorials* 17 (4): 2347–76. doi:10.1109/COMST.2015.2444095.

Alur, R., E. Berger, A. W. Drobnis, L. Fix, K. Fu, G. D. Hager, D. Lopresti, et al. 2016. Systems computing challenges in the Internet of things. ArXiv Preprint ArXiv:1604.02980. http://arxiv.org/abs/1604.02980.

Anagnostopoulos, T., D. Ferreira, A. Samodelkin, M. Ahmed, and V. Kostakos. 2016. Cyclist-aware traffic lights through distributed smartphone sensing. *Pervasive and Mobile Computing* 31 (C): 22–36. doi:10.1016/j.pmcj.2016.01.012.

Antipolis, S. 2016. ETSI IoT-M2M Workshop: Approaching a smarter world. November 25. http://www.etsi.org/index.php/news-events/news/1144-2016-11-news-etsi-iot-m2m-workshop-approaching-a-smarter-world.

Aqeel-ur-Rehman, A. Z. A., N. Islam, and Z. Ahmed Shaikh. 2014. A review of wireless sensors and networks' applications in agriculture. *Computer Standards and Interfaces* 36 (2): 263–270.

Ashton, K. 2009. That 'Internet of things' thing. *RFiD Journal* 22 (7): 97–114.

AT&T. 2017. Smart cities. Smart Cities—Internet of Things Newsroom|AT&T. Accessed January 13. http://about.att.com/sites/internet-of-things/smart_cities.

Atzori, L., A. Iera, and G. Morabito. 2010. The Internet of things: A survey. *Computer Networks* 54 (15): 2787–2805.

Bassi, A., M. Bauer, M. Fiedler, T. Kramp, R. Van Kranenburg, S. Lange, and S. Meissner. 2013. *Enabling Things to Talk: Designing IoT Solutions with the IoT Architectural Reference Model*, 163–211. Berlin: Springer.

Bauer, M., N. Bui, J. De Loof, C. Magerkurth, A. Nettsträter, J. Stefa, and J. W. Walewski. 2013. IoT reference model. In *Enabling Things to Talk: Designing IoT Solutions with the IoT Architectural Reference Model*, edited by A. Bassi et al., 113–162. Berlin: Springer.

Bi, Z., L. Da Xu, and C. Wang. 2014. Internet of things for enterprise systems of modern manufacturing. *IEEE Transactions on Industrial Informatics* 10 (2): 1537–1546.

Bo, Y. and H. Wang. 2011. The application of cloud computing and the Internet of things in agriculture and forestry. In 2011 *International Joint Conference on Service Sciences (IJCSS)*, Taipei, Taiwan, 168–172.

Borgia, E. 2014. The Internet of things vision: Key features, applications and open issues. *Computer Communications* 54: 1–31.

Brad, S. and M. Murar. 2014. Smart buildings using IoT technologies. *Construction of Unique Buildings and Structures* 5 (20): 15–27.

Bui, N., A. P. Castellani, P. Casari, and M. Zorzi. 2012. The Internet of energy: A web-enabled smart grid system. *IEEE Network* 26 (4): 39–45.

Bui, N. and M. Zorzi. 2011. Health care applications: A solution based on the Internet of things. In *Proceedings of the 4th International Symposium on Applied Sciences in Biomedical and Communication Technologies*, Barcelona, 131.

Campolo, C., A. Iera, A. Molinaro, S. Yuri Paratore, and G. Ruggeri. 2012. SMeaRTCaR: An integrated smartphone-based platform to support traffic management applications. In 2012 *First International Workshop on Vehicular Traffic Management for Smart Cities (VTM)*, Dublin, 1–6.

Castro, M., A. J. Jara, and A. F.G. Skarmeta. 2013. Smart lighting solutions for smart cities. In 2013 *27th International Conference on Advanced Information Networking and Applications Workshops (WAINA)*, Barcelona, 1374–1379.

Cisco. 2014. The Internet of things reference model. http://cdn.iotwf.com/resources/71/IoT_Reference_Model_White_Paper_June_4_2014.pdf.

Commission of the European Communities. 2009. Internet of things—An action plan for Europe. COM (2009) 278 final. Brussels: European Union. http://eur-lex.europa.eu/LexUriServ/LexUriServ.do?uri=COM:2009:0278:FIN:EN:PDF.

Dahmen-Lhuissier, S. 2016. Internet of things. Valbonne, France: ETSI. Accessed October 18. http://www.etsi.org/technologies-clusters/technologies/internet-of-things.

Da Xu, L., W. He, and S. Li. 2014. Internet of things in industries: A survey. *IEEE Transactions on Industrial Informatics* 10 (4): 2233–2243.

Dohr, A., R. Modre-Osprian, M. Drobics, D. Hayn, and G. Schreier. 2010. The Internet of things for ambient assisted living. *ITNG* 10: 804–809.

Domingo, M. C. 2012. An overview of the Internet of things for people with disabilities. *Journal of Network and Computer Applications* 35 (2): 584–596.

Doukas, C. and I. Maglogiannis. 2012. Bringing IoT and cloud computing towards pervasive healthcare. In *2012 Sixth International Conference on Innovative Mobile and Internet Services in Ubiquitous Computing (IMIS)*, Palermo, Italy, 922–926.

Du, K., C. Mu, J. Deng, and F. Yuan. 2013. Study on atmospheric visibility variations and the impacts of meteorological parameters using high temporal resolution data: An application of environmental Internet of things in China. *International Journal of Sustainable Development & World Ecology* 20 (3): 238–247.

ETSI (European Telecommunications Standards Institute). 2016. SmartM2M; IoT standards landscape and future evolutions. ETSI TR 103 375 V1.1.1 (2016-10). http://www.etsi.org/deliver/etsi_tr/103300_103399/103375/01.01.01_60/tr_103375v010101p.pdf.

Fang, S., L. Da Xu, Y. Zhu, J. Ahati, H. Pei, J. Yan, and Z. Liu. 2014. An integrated system for regional environmental monitoring and management based on Internet of things. *IEEE Transactions on Industrial Informatics* 10 (2): 1596–1605.

Fantana, N., T. Riedel, J. Schlick, S. Ferber, J. Hupp, S. Miles, F. Michahelles, and S. Svensson. 2013. IoT Applications—Value Creation for Industry, 153. River Publishers Series in Communications. Aalborg, Denmark: River Publishers.

Foster, T. 2017. Regulation of the Internet of things. Accessed January 17. http://www.scl.org/site.aspx?i=ed47967.

Gartner, Inc. 2013. Gartener says the Internet of things installed base will grow to 26 billion units by 2020. December 12. http://www.gartner.com/newsroom/id/2636073.

Gershenfeld, N. 1999. *When Things Start to Think*. New York: Henry Holt and Co.

Gil-Castineira, F., E. Costa-Montenegro, F. J. Gonzalez-Castano, C. López-Bravo, T. Ojala, and R. Bose. 2011. Experiences inside the ubiquitous Oulu smart city. *Computer* 44 (6): 48–55.

GS1. 2016. GS1 and the Internet of things. Final. http://www.gs1.org/sites/default/files/images/standards/internet-of-things/gs1-and-the-internet-of-things-iot.pdf.

GSMA (GSM Association). 2016. Annual report 2016. http://www.gsma.com/aboutus/wp-content/uploads/2016/09/GSMA_AnnualReport_2016_FINAL.pdf.

Gubbi, J., R. Buyya, S. Marusic, and M. Palaniswami. 2013. Internet of things (IoT): A vision, architectural elements, and future directions. *Future Generation Computer Systems* 29 (7): 1645–1660.

Guinard, D. and V. Trifa. 2016. *Building the Web of Things*. Shelter Island, NY: Manning Publications.

Hong, I., S. Park, B. Lee, J. Lee, D. Jeong, and S. Park. 2014. IoT-based smart garbage system for efficient food waste management. *Scientific World Journal* 2014 (2014): 646953.

IDC. 2015. IDC predicts the emergence of 'the DX economy' in a critical period of widespread digital transformation and massive scale up of 3rd platform technologies in every industry. November 4. http://www.businesswire.com/news/home/20151104005180/en/IDC-Predicts-Emergence-DX-Economy-Critical-Period.

IDC. 2017. Internet of things ecosystem and trends. Accessed January 18. http://www.idc.com/getdoc.jsp?containerId=IDC_P24793.

IEEE (Institute of Electrical and Electronics Engineers). 2015. Towards a definition of the Internet of things (IoT). http://iot.ieee.org/images/files/pdf/IEEE_IoT_Towards_Definition_Internet_of_Things_Revision1_27MAY15.pdf.

i-SCOOP. 2015. Digital transformation: Online guide to digital transformation. February 1. http://www.i-scoop.eu/digital-transformation/.

ISO (International Organization for Standardization). 2017. ISO/IEC JTC 1—Information technology. Accessed January 15. http://www.iso.org/iso/home/standards_development/list_of_iso_technical_committees/jtc1_home.htm.

ISO/IEC (International Organization for Standardization/International Electrotechnical Commission). 2014. Study report on IoT reference architectures/frameworks. https://www.itu.int/md/T13-SG17-150408-TD-PLEN-1688/_page.print.

ISO/IEC (International Organization for Standardization/International Electrotechnical Commission). 2016. Information technology—Data structure—Unique identification for the Internet of things. ISO/IEC 29161:2016. August. https://www.iso.org/obp/ui/#iso:std:iso-iec:29161:ed-1:v1:en.

ISO/IEC JTC1. 2015. Internet of things (IoT) preliminary report 2014. ISO/IEC. http://www.iso.org/iso/internet_of_things_report-jtc1.pdf.

ISO/IEC JTC1. 2016. Information technology—Internet of Things Reference Architecture (IoT RA). ISO/IEC CD 30141:20160910 (E). https://www.w3.org/WoT/IG/wiki/images/9/9a/10N0536_CD_text_of_ISO_IEC_30141.pdf.

Istepanian, R. S. H., S. Hu, N. Y. Philip, and A. Sungoor. 2011. The potential of Internet of M-health things 'm-IoT' for non-invasive glucose level sensing. In *2011 Annual International Conference of the IEEE Engineering in Medicine and Biology Society*, Boston, MA, 5264–5266.

ITU-T (International Telecommunications Union Telecommunication Standardization Sector). 2005. ITU Internet reports, the Internet of things. Geneva: ITU-T.

ITU-T (International Telecommunications Union Telecommunication Standardization Sector). 2012. Overview of the Internet of things. Y.2060. https://www.itu.int/rec/dologin_pub.asp?lang=e&id=T-REC-Y.2060-201206-I!!PDF-E&type=items.

ixia. 2016. Key building blocks of Internet of things (IOT). https://www.ixiacom.com/resources/key-building-blocks-internet-things-iot.

Ji, S. W., H. Yun Teng, and J. Feng Su. 2014. The application and development of the Internet of things in intelligent buildings. *Advanced Materials Research* 834: 1854–1857.

Karzel, D., H. Marginean, and T-S. Tran. 2016. A reference architecture for the Internet of things. January 29. https://www.infoq.com/articles/internet-of-things-reference-architecture.

Kelly, S. D. T., N. Kumar Suryadevara, and S. Chandra Mukhopadhyay. 2013. Towards the implementation of IoT for environmental condition monitoring in homes. *IEEE Sensors Journal* 13 (10): 3846–3853.

Kyriazis, D., T. Varvarigou, D. White, A. Rossi, and J. Cooper. 2013. Sustainable smart city IoT applications: Heat and electricity management & eco-conscious cruise control for public transportation. In *2013 IEEE 14th International Symposium and Workshops on a World of Wireless, Mobile and Multimedia Networks (WoWMoM)*, Madrid, 1–5.

Li, L., H. Xiaoguang, C. Ke, and H. Ketai. 2011. The applications of Wifi-based wireless sensor network in Internet of things and smart grid. In 2011 *6th IEEE Conference on Industrial Electronics and Applications*, Beijing, 789–793.

Li, X., R. Lu, X. Liang, X. Shen, J. Chen, and X. Lin. 2011. Smart community: An Internet of things application. *IEEE Communications Magazine* 49 (11): 68–75.

Li, X. 2013. Multi-day and multi-stay travel planning using geo-tagged photos. In *Proceedings of the Second ACM SIGSPATIAL International Workshop on Crowdsourced and Volunteered Geographic Information* (GEOCROWD '13), Orlando, FL, 1–8.

Li, Z., T. Wang, Z. Gong, and N. Li. 2013. Forewarning technology and application for monitoring low temperature disaster in solar greenhouses based on Internet of things. *Transactions of the Chinese Society of Agricultural Engineering* 29 (4): 229–236.

Liu, X., M. D. Ceorner, and P. Shenoy. 2006. Ferret: RFID localization for pervasive multimedia. In *UbiComp 2006: Ubiquitous Computing*, Orange County, CA, 422–440.

Maddox, T. 2017. 9 IoT Global Trends for 2017. *TechRepublic*. January 2. http://www.techrepublic.com/article/9-iot-global-trends-for-2017/.

Minoli, D. 2013. *Building the Internet of things with IPv6 and MIPv6: The Evolving World of M2M Communications*. Hoboken, NJ: Wiley.

Miorandi, D., S. Sicari, F. De Pellegrini, and I. Chlamtac. 2012. Internet of things: Vision, applications and research challenges. *Ad Hoc Networks* 10 (7): 1497–1516.

Moreno, M. V., M. A. Zamora, and A. F. Skarmeta. 2014. User-centric smart buildings for energy sustainable smart cities. *Transactions on Emerging Telecommunications Technologies* 25 (1): 41–55.

Opensensors. 2017. How to choose the best connectivity network for your project. Accessed January 15. https://publisher.opensensors.io/connectivity.

Ortiz, A. M., D. Hussein, S. Park, S. N. Han, and N. Crespi. 2014. The cluster between Internet of things and social networks: Review and research challenges. *IEEE Internet of Things Journal* 1 (3): 206–215.

Ou, Q., Y. Zhen, X. Li, Y. Zhang, and L. Zeng. 2012. Application of Internet of things in smart grid power transmission. In 2012 *Third FTRA International Conference on Mobile, Ubiquitous, and Intelligent Computing (MUSIC)*, Vancouver, BC, 96–100.

Pang, Z. 2013. Technologies and architectures of the Internet-of-Things (IoT) for health and well-being. Doctoral thesis, KTH, School of Information and Communication Technology (ICT), Electronic Systems, Kista, Sweden.

Perera, C., A. Zaslavsky, P. Christen, and D. Georgakopoulos. 2014. Context aware computing for the Internet of things: A survey. *IEEE Communications Surveys and Tutorials* 16 (1): 414–454.

Phithakkitnukoon, S., M. I. Wolf, D. Offenhuber, D. Lee, A. Biderman, and C. Ratti. 2013. Tracking trash. *IEEE Pervasive Computing* 12 (2): 38–48.

Postscapes. 2017. IoT technology guidebook. IoT Technology|2017 Overview Guide on Protocols, Software, Hardware and Network Trends. Accessed January 13. https://www.postscapes.com// internet-of-things-technologies.

Rose, K., S. Eldridge, and L. Chapin. 2015. The Internet of things: An overview. Reston, VA: Internet Society. https://pdfs.semanticscholar.org/6d12/bda69e8fcbbf1e9a10471b54e57b15cb07f6.pdf.

Ryu, C-S. 2015. IoT-based intelligent for fire emergency response systems. *International Journal of Smart Home* 9 (3): 161–168.

Sanchez, L., J. Antonio Galache, V. Gutierrez, J. Manuel Hernandez, J. Bernat, A. Gluhak, and T. Garcia. 2011. Smartsantander: The meeting point between future Internet research and experimentation and the smart cities. In *Future Network and Mobile Summit (FutureNetw)*, 2011, Warsaw, 1–8.

Serbanati, A., C. Maria Medaglia, and U. Biader Ceipidor. 2011. *Building Blocks of the Internet of Things: State of the Art and Beyond*. Rijeka, Croatia: INTECH Open Access Publisher. http://cdn.intechweb.org/pdfs/17872.pdf.

Shifeng, Y., F. Chungui, H. Yuanyuan, and Z. Shiping. 2011. Application of IOT in agriculture. *Journal of Agricultural Mechanization Research* 7: 190–193.

Shrouf, F. and G. Miragliotta. 2015. Energy management based on Internet of things: Practices and framework for adoption in production management. *Journal of Cleaner Production* 100: 235–246.

Skerrett, I. 2016. IoT trends to watch in 2017. December 19. https://ianskerrett.wordpress.com/2016/12/19/iot-trends-to-watch-in-2017/.

Smartup Cities. 2017. Smart waste containers with ultrasonic fill-level sensors. SmartUp Cities Smart City IoT Solutions for the Future Cities. Accessed January 17. http://www.smartupcities.com/smart-waste-containers/.

Sundramoorthy, V., Q. Liu, G. Cooper, N. Linge, and J. Cooper. 2010. DEHEMS: A user-driven domestic energy monitoring seystem. In *Internet of Things (IOT), 2010*, Tokyo, 1–8.

Taivalsaari, A. and T. Mikkonen. 2017. A roadmap to the programmable world: Software challenges in the IoT era. *IEEE Software* 34 (1): 72–80.

Tao, F., Y. Zuo, L. Da Xu, and L. Zhang. 2014. IoT-based intelligent perception and access of manufacturing resource toward cloud manufacturing. *IEEE Transactions on Industrial Informatics* 10 (2): 1547–1557.

Tielert, T., M. Killat, H. Hartenstein, R. Luz, S. Hausberger, and T. Benz. 2010. The impact of traffic-light-to-vehicle communication on fuel consumption and emissions. In *Interenet of Things (IOT), 2010*, Tokyo, 1–8.

Vlacheas, P., R. Giaffreda, V. Stavroulaki, D. Kelaidonis, V. Foteinos, G. Poulios, P. Demestichas, A. Somov, A. Rahim Biswas, and K. Moessner. 2013. Enabling smart cities through a cognitive management framework for the Internet of Things. *IEEE Communications Magazine* 51 (6): 102–111.

Vodafone Group. 2016. Vodafone completes the world's first trial of standardised NB-IoT on a live commercial network. September 20. https://www.vodafone.com/content/index/what/technology-blog/nbiot-commercial.html#.

Wang, X., J. Song Dong, C. Chin, S. Ravipriya Hettiarachchi, and D. Zhang. 2004. Semantic space: An infrastructure for smart apaces. *IEEE Pervasive Computing* 3 (3): 32–39.

Wei, C. and Y. Li. 2011. Design of energy consumption monitoring and energy-saving management system of intelligent building based on the Internet of things. In *2011 International Conference on Electronics, Communications and Control (ICECC)*, Ningbo, China, 3650–3652.

Weiser, M. 1991. The computer for the 21st century. *Scientific American* 265 (3): 94–104.

Weyrich, M. and C. Ebert. 2016. Reference architectures for the Internet of things. *IEEE Software* 33 (1): 112–16.

Whitmore, A., A. Agarwal, and L. Da Xu. 2014. The Internet of things—A survey of topics and trends. *Information Systems Frontiers* 17 (2): 261–274.

Yang, G., L. Xie, M. Mäntysalo, X. Zhou, Z. Pang, L. Da Xu, S. Kao-Walter, Q. Chen, and L-R. Zheng. 2014. A health-IoT platform based on the integration of intelligent packaging, unobtrusive bio-sensor, and intelligent medicine box. *IEEE Transactions on Industrial Informatics* 10 (4): 2180–2191.

Yin, C. T., Z. Xiong, H. Chen, J. Yuan Wang, D. Cooper, and B. David. 2015. A literature survey on smart cities. *Science China Information Sciences* 58 (10): 1–18.

Yoo, S. 2015. ISO/IEC JTC1/WG 10 Working Group on Internet of Things. June 16. http://iot-week.eu/wp-content/uploads/2015/06/07-JTC-1-WG-10-Introduction.pdf.

Yun, M., and B. Yuxin. 2010. Research on the architecture and key technology of Internet of things (IoT) applied on smart grid. In *2010 International Conference on Advances in Energy Engineering (ICAEE)*, Beijing, 69–72.

Zakon, R. H. 2016. Hobbes' Internet Timeline 23. The Definitive ARPAnet & Internet History. https://www.zakon.org/robert/internet/timeline/#Growth.

Zanella, A., N. Bui, A. Castellani, L. Vangelista, and M. Zorzi. 2014. Internet of things for smart cities. *IEEE Internet of Things Journal* 1 (1): 22–32.

Zhang, Y., B. Chen, and X. Lu. 2011. Intelligent monitoring system on refrigerator trucks based on the Internet of things. In *International Conference on Wireless Communications and Applications*, Sanya, China, 201–206.

Zhang, L., K. Ibibia, W. Dabipi, and L. Brown. 2018. Internet of Things Applications in Agriculture. In *Internet of Things A to Z*, edited by Q. F. Hassan. John Wiley and Sons Inc.

Zhengxia, W. and X. Laisheng. 2010. Modern logistics monitoring platform based on the Internet of things. In *2010 International Conference on Intelligent Computation Technology and Automation (ICICTA)*, Changsha, China, 2: 726–731.

Zhong, Y. and Y-T Dong. 2011. Application of the Internet of things to security automation system for intelligent buildings. *Internet of Things Technologies* 4: 041.

Zouganeli, E. and I. Einar Svinnset. 2009. Connected objects and the Internet of things—A paradigm shift. In *2009 International Conference on Photonics in Switching*, Pisa, 1–4.

2 Organizational Implementation and Management Challenges in the Internet of Things

Marta Vos

CONTENTS

2.1 Introduction 33
2.2 IoT in Organizations 34
2.3 Managing IoT Systems 35
 2.3.1 Interoperability 35
 2.3.2 Standards 36
 2.3.3 Privacy 37
 2.3.4 Security 38
 2.3.5 Trust 39
 2.3.6 Data Management 41
 2.3.7 Legislation and Governance 42
2.4 Building the Blocks into the IoT 43
2.5 Conclusion 43
References 44

2.1 INTRODUCTION

The Internet of Things (IoT) is becoming one of the most heavily researched and hyped computing concepts today. At the most basic level, the IoT concept describes how "things" can be connected to the Internet, and each other, giving them the potential to act without the mediation of humans. This connection allows for the creation of new and novel applications with interconnected devices, organizations, and users. Theoretically, any real or virtual thing could be included in the IoT, with the limits to what IoT networks can do being bounded only by the imagination of developers and users. In reality, however, there is no one seamless IoT in which devices communicate with each other and their users. Instead, almost all "IoT" networks exist within single organizations, or limited organizational collaborations, which could be considered to be "intranets of things" (Santucci, 2010). These siloed networks and the devices they include are capable of connecting to both the Internet and each other, as well as other organizational IoT networks, but currently only connect within their silos to the Internet. Thus, these individual networks powered by radio-frequency identification (RFID), wireless sensors, or mobile technology are IoT capable, but not actually networked together. It can therefore be seen that the IoT is conceptual rather than being current reality. This is recognized in the term *Future Internet of Things* (FIoT) (Tsai et al., 2014), which acknowledges the gap between the IoT concept and reality, and looks to integrate today's intranets of things to form one IoT (Zorzi et al., 2010).

While IoT systems still exist in silos or intranets, the organizational implications of joining up these systems, within or between organizations, are not as well explored as the technical aspects of IoT systems operation. From 2010 to 2015, a sample of research literature across three commonly used databases yielded more than 8000 publications in the IoT field. Of these, the majority were concerned with technology-related questions, such as solving problems with networks, wireless technologies (e.g., RFID, wireless sensor network [WSN], and wireless body area network [WBAN]), the cloud, and security issues. Publications with respect to the management of IoT systems were relatively lacking, as was qualitative or social research (Vos, 2016). The fragmentary research, the disparate nature of IoT devices, and the complexity and size of the networks they form mean that many organizations, researchers, and users do not grasp its full potential or scale.

The IoT will likely reach into all corners of our existence, from home security systems to organizational supply chains and healthcare. Users often perceive their devices to be connected to the Internet, and therefore the IoT, but many technological, organizational, and social issues lie behind this apparent integration. Organizations themselves often do not consider issues beyond the boundaries of their own IoT implementations, and are unsure about how to deal with challenges that arise when they attempt to join with other organizational IoT networks (Vos, 2014). Because of the scale of the IoT, it is not possible to present an overview of its technology, organizational, and social issues in a single chapter. This chapter therefore focuses on the issues faced by organizations considering integrating their own IoT systems (or intranets of things) into the wider IoT environment. The term *organizations* is used broadly in this chapter to apply to the full range of organizational types and sizes. The issues discussed here would also apply in the home environment to individuals who wanted to participate more fully in the IoT experience, but the literature in this field is sparse, so IoT in the home environment is not considered in detail. Initially, the chapter discusses the IoT in an organizational context, and then the need for integration and technology standards is discussed. Following this, the chapter considers the application and adoption of the IoT in organizations. Challenges specific to the organizational management of IoT systems are then considered, including privacy, security, data management, trust, legislation, and governance. The chapter concludes with an overview of how the building blocks of the IoT might be integrated into one uniform network and a discussion of the current state of IoT research, with some suggestions for future research.

2.2 IoT IN ORGANIZATIONS

While there is an enormous amount of research dedicated to developing new technologies, and solving issues related to the IoT, organizational and social research has somewhat lagged (Vos, 2016). Partly, this is due to the difficulty of studying IoT systems, as huge distributed systems make it difficult and time-consuming for research to be undertaken, and the methods for studying such systems are not well developed (Vos, 2014). Most organizational research with respect to IoT systems deals with RFID technology, as this technology formed the initial basis of IoT implementations, while applications research and case studies focus on the most common applications, particularly in the smart city, healthcare, defense, and supply chain sectors. Potential uses for IoT systems are limited only by the imagination, but recorded implementations range from animal tracking (Vlad et al., 2012) to preventing cheating in Mahjong (Tang, 2013). One of the most heavily studied areas of IoT and WSN implementation is the healthcare sector. Healthcare organizations have been found to benefit from IoT technology in a number of ways, ranging from inventory management and time savings (Wamba and Ngai, 2011) through to improvements in healthcare practices as basic as handwashing (Shi et al., 2012).

The IoT market is predicted to continue growing, possibly reaching a potential value of $11 trillion per year by 2025 (Manyika et al., n.d.). A number of factors are driving this growth, including the reduction in costs for tags and sensors, increasing efforts in standardization assisting with the interoperability of systems, improving IoT infrastructure, and competitive pressures (Wyld, 2005). However, there are also inhibitors to adoption. Ongoing difficulties with privacy, security, and standardization (Hossain and Quaddus, 2011) are hindering systems adoption, as are problems with cost, benefits realization, and technology complexity (Bose et al., 2009; Hossain and Quaddus, 2011; Ilie-Zudor et al., 2011). From the management side, trust between organizations (Spekman and Sweeney, 2006), accountability issues (Ilie-Zudor et al., 2011), organizational readiness (Hossain and Quaddus, 2011), training (Kopalchick and Monk, 2005), satisfaction with current technology solutions (Kros et al., 2011), change resistance (Carr et al., 2010), organizational size (Matta et al., 2012), and health concerns (Curtin et al., 2007) are inhibiting the adoption of IoT technologies.

The adoption and implementation of IoT systems has received some research interest, as well as the barriers to implementation and adoption, along with research based on specific applications of IoT technology. From an organizational management perspective, the amount of knowledge an organization has about the technology, along with the presence of a knowledgeable technology champion, is a predictor of IoT adoption (C.-P. Lee and Shim, 2007). Similarly, in systems shared between public and private sectors, the amount of knowledge an organization has about technology systems is important in the management of such systems, along with the expected mediators of privacy, security, cost, data management, and benefits realization (Vos et al., 2012). The extent to which organizations are transformed by systems adoption can be predicted to some extent by the benefit derived from such systems (Wamba and Chatfield, 2010). However, the degree to which organizations achieve the benefits they thought they would from their IoT implementations has also been questioned, with benefits derived not always being those expected (Vos et al., 2012). Technology reliability, placement, and data management and interpretation also continue to be ongoing challenges in systems design and implementation (Bardaki et al., 2012).

2.3 MANAGING IoT SYSTEMS

Many of the technical issues faced with implementing and operating an IoT system are also mirrored within the management of these systems. While the technical side of interoperability, privacy, and security management, for example, is subject to ongoing technical research, the organizational issues with respect to these topics are less discussed, but not less complex. Between organizations, seven closely related issues are highlighted from a business perspective as being essential to the organizational integration of IoT systems: interoperability, standards, privacy, security, trust, data management, and legislation and governance (Vos, 2014). Privacy and security are often considered to be elements of trust, as they contribute to ensuring that the members of the trusted relationship understand exactly how they should act, and how the other partners will act (Rousseau et al., 1998). In other instances, they can be considered separately, where issues of trust might not arise, for example, where security is compromised by a malicious attack. For ease of discussion, each of these will be considered separately, with organizational and technology perspectives.

2.3.1 INTEROPERABILITY

There are many technical and organizational considerations when implementing IoT-based systems, including reliability, scalability, heterogeneity, and data use, not to mention how technical solutions might be used to solve privacy or security problems. However, when considering the issue of how organizations might implement their IoT-based systems in such a way that they might join the IoT either immediately or in the future, issues around ensuring that systems can be integrated with other IoT systems rise to the fore.

The ever-increasing range of devices that can be connected to the IoT presents enormous challenges to IoT systems, as each different device must be able to connect with the IoT architecture, transmit its data, and be understood. Each system or even device may have a different hardware manufacturer, along with different circuits, data formats, legacy systems, and carrier demands (Vermesan et al., 2011). Even the selection of radio frequency for the transmission of data presents difficulties, with different countries having different frequency ranges available for the transmission of IoT data, which cannot be readily changed (European Commission, 2013). A great deal of research effort, and expense, is being aimed at developing middleware and systems architectures to allow for integration and interoperability of IoT systems. Interoperability of IoT hardware and software allows for information to flow between different devices, networks, and IoT systems. This ability to share information between systems is a crucial component of a seamlessly integrated IoT, with standardization of infrastructure, data formats, and communications protocols being a cornerstone of interoperability (Kopalchick and Monk, 2005).

At the most basic level, there is still no agreement as to what interoperability actually means, and to what degree it is required (European Commission, 2013). For example, would it be enough for users to be happy with their devices being interoperable, and thus unaware of any delay between their devices and the services they require, or does everything need to communicate with everything else? The benefits of IoT interoperability have yet to be fully explored, but apart from seamless communication, benefits could include such things as the ability for users and organizations to create unique combinations of devices and applications in "mashups."

2.3.2 Standards

Central to the effort to ensure interoperability in IoT systems is the issue of standardization. Standards allow devices to communicate with each other as the devices would all "speak the same language." They also allow new devices to be introduced to established IoT systems with the guarantee that those devices will work seamlessly. Standards are required across the full range of IoT architecture, from naming standards for device identification through to data standards ensuring that data can be processed and interpreted without difficulties in integrating different data formats.

Hardware naming standards are generally specific to the hardware type; for example, RFID standards apply to RFID tags, and WSN standards to WSN nodes. A number of standards exist for naming (numbering) RFID tags in a way that is unique and allows the tag to be identified. The IP for Smart Objects Alliance and the Ubiquitous ID Center (Zorzi et al., 2010) develop RFID naming standards, and proprietary naming schemes are developed by some organizations for their own RFID implementations. However, the most common and widely implemented RFID naming standards center on the Electronic Product Code (EPC). EPCglobal drives development of these identification standards, along with standards for RFID architecture and hardware, in an effort to allow seamless integration of RFID-enabled devices into RFID systems (EPCglobal, 2010). The EPC Information Service (EPCIS) is a permission-based service allowing information to be shared between applications, and between applications and users, through mapping RFID identification codes to the relevant information (Glover and Bhatt, 2006). These standards are RFID specific and do not include other IoT devices. Other IoT devices also have some naming standards, although these are not as well developed as those relating to RFID. Mobile devices are identified by their Media Access Control Identification (MAC-ID), the conventions for which are managed by the Institute of Electrical and Electronics Engineers (IEEE), while Sensor Web Enablement (SWE) standards identify sensors to the Internet.

A number of organizations, including the Internet Engineering Task Force (IETF) and IEEE, are developing standards with respect to how the various IoT devices communicate with the Internet, including Internet protocol version 6 (IPv6) and the Constrained Application Protocol (CoAP). For devices with low power resources, the IETF is promoting the 6LoWPAN standard. Other communications standards, particularly for WSN, include those developed by the IEEE, ZigBee, and WirelessHART. At the architecture level GS1, the IEEE, the European Union (EU), and the Internet of Things Architecture (IoT-A) Project, among others, are working toward a standardized architectural model for the IoT (Bertot and Choi, 2013). There are few data standards that have been developed particularly for the IoT, and as the number of connected devices increases, the need for an ontology-based semantic standard is becoming more urgent to ensure that data can be exchanged effectively between different devices (Vermesan et al., 2011).

Standards are the driving force behind IoT interoperability, ensuring that devices can connect with each other, if necessary, and allowing users to achieve the seamless IoT experience promised (Bertot and Choi, 2013). However, caution needs to be exercised to ensure that any standards developed do not constrain innovation by forcing a particular standard architecture, communications protocol, or hardware on the IoT, a situation that would inhibit innovation and growth.

2.3.3 Privacy

Privacy in the IoT context is difficult to define. Separate from the need to secure IoT infrastructure, privacy refers to the need to ensure that data relating to individuals remains confidential. IoT systems gather a great deal of data, and at times are invisible to individuals who may not even be aware they are carrying digitalized devices, or RFID tags. Data collected from these devices could be used to track movements or gather other information, without the knowledge of the owner. Similarly, data collected from WSN in the home environment could reveal a lot about the personal habits of the user, from his or her preferred air temperature to how many times he or she exercises (Peppet, 2014). Anonymization of data is difficult, and de-anonymizing is also possible, leading to ongoing concerns about how IoT databases are used and secured (Peppet, 2014). In the ubiquitous IoT environment, theoretically anything could reveal everything.

It is important to note that there are at least three types of information available from IoT systems:

1. Information relating to individual humans, which is considered *private*, especially when it is individually identifiable
2. Information relating to organizations, which is considered *confidential* and secured against unauthorized access
3. Information that is neither *private* nor *confidential*, but could be shared without causing individuals to be identified or businesses to be disadvantaged

Some of the data related to IoT-related systems can be shared; for example, sensor information related to weather conditions is commonly made public. However, the nature of data generated by IoT systems must be understood by organizations, and secured appropriately. The Organisation for Economic Co-operation and Development (OECD) updated its guidelines with respect to data transmitted across national borders in 2013, in part to take account of the increasing amounts of cross-border data transmission caused by the proliferation of IoT devices. In response, many OECD member states have implemented or updated legislation and privacy guidelines to ensure that cross-border data flows respect the need for the privacy and confidentiality of individuals and organizations, while still allowing for these data flows to continue (OECD, 2013). Further, jurisdictional issues are of concern in the IoT environment, as cloud servers are not always located in the country of origin of the data. Even where organizations share IoT systems, they may be spread across different countries with different privacy requirements. The European Commission on Internet of Things Governance recommends that organizations take a "privacy by design" approach to IoT systems, advising organizations to ensure that they consider privacy from the earliest development phase, and that the data collected, and its possible impact on people, is well understood (European Commission, 2013). The European Commission has also recommended that individuals have the capability to "be invisible" to IoT systems, something that is difficult to ensure with the increasing ubiquity of IoT technologies (Krotov, 2008). However, not all IoT systems require high levels of privacy protection. There is a substantial difference between data that could identify individuals and data that, for example, deals with the movement of inventory items (Vos, 2014). The EU and OECD regulations are concerned with data that can be identified to individuals, but much less concerned with organizational information.

The nature of IoT hardware also presents difficulties in ensuring both privacy and security. Passive tags by their nature are seldom secured, and the heterogeneity of devices connected to the

IoT makes one single privacy or security solution impossible (Miorandi et al., 2012). Solutions such as "kill tags," where tags are deactivated at the point of sale, have been proposed. But these solutions disable a number of the attractive features of IoT-based items, including product support and ease of item return (Ohkubo et al., 2005). Other possible solutions, such as physical shielding of tags, is only possible in limited applications, such as passports, where the individual can both identify the tag and take action to shield it. A range of technical solutions are being explored to assist in improving privacy in the IoT (Sicari et al., 2015), including tagging IoT data with privacy properties (Evans and Eyers, 2012), anonymous authentication protocols (Alcaide et al., 2013), encryption (Wang et al., 2014), and privacy protective Domain Name Systems (DNSs) that would recognize the user's identity (Wang and Wen, 2011), among other approaches.

Identity management is another area of research with respect to improving IoT privacy for individuals. Using an identity management approach, individuals could manage how they interact with various IoT devices. Such strategies could include the use of pseudonyms to obscure the identity of an individual, and in a theoretical future, individuals could entirely dictate how they interacted with computing devices (De Hert, 2008). For example, an individual could chose to suppress his or her identity when moving through a building, and any IoT devices he or she were carrying would not be identified to the owners of the building (Cas, 2005). This is the type of privacy-preserving technology recommended by the European Commission (2013).

2.3.4 Security

Without securing the IoT environment, privacy is not possible. The IoT environment presents many security challenges, including

- The heterogeneous nature of IoT devices
- The large attack surface offered by numerous IoT devices
- The fact that the simpler building blocks of the IoT, such as passive RFID tags, are not easily secured

Security solutions, such as the Trusted Platform Module (TPM), are expensive and thus limited to high-end devices, with the energy and processing power to accommodate such units. From the data protection point of view, the gold standard for data protection is the use of public key encryption, but as with TPM, the obvious drawback of encryption is the high cost involved, making it economic only in limited applications (Mykletun et al., 2006).

The most basic devices are generally secured through software-based methods that leave the devices themselves unprotected (Abera et al., 2016). Further, engineering of security into cheap consumer goods is seldom a priority for their manufacturers (Peppet, 2014). Thousands of identical modules sold to consumers, who seldom bother with updates or secure passwords, can all be attacked in exactly the same way, leading to the kind of Distributed Denial of Service (DDoS) attacks seen in recent years (Ackerman et al., 2012). With billions of online IoT devices predicted for 2020 (Nordrum, 2016) (although more generous predictions go as high as 1 trillion devices [Iwata, 2012]), the security of these devices is a major concern. Security protection for the IoT will need to consider the nature of the infrastructure, communications, and data, as well as align with social and legal expectations of privacy.

Security issues in the IoT environment have two perspectives. One is securing the IoT devices and infrastructure against attacks directed at the organization or owner; the other perspective is securing against unauthorized use of the device itself. In 2016, IoT-capable devices were used in a DDoS attack against the Dyn domain name servers, which shut down Twitter, Netflix, and Spotify, among other popular websites (Ingraham, 2016). The attack used the Mirai malware, which exploits a weakness in IoT-capable devices where factory default passwords are not reset by the users, and could not have been reset in many cases because the passwords were hardwired into the devices. In this particular case, DVRs and IP cameras were the primary devices affected (Walker, 2016). This is not the first time a large-scale attack of this type has been carried out. In 2013, more than 100,000 IoT devices were involved in a spam and phishing attack; the services were compromised in the same way as in the 2016 attack, through factory-set passwords not being reset (Proofprint, 2014).

Other types of attack are also possible; for example, home NEST thermostats can be hacked, using ransomware to demand that the thermostat be reset (Mayer, 2016). Researchers have demonstrated how simple it is to locate vulnerable IoT devices, with an October 6, 2016, Internet scan locating more than 515,000 nonsecured IoT devices (Wikholm, 2016).

The obvious vulnerability of IoT devices has led to the European Commission commencing the development of cybersecurity requirements for IoT devices, including those used in the October 2016 attack, as part of a general overhaul of EU telecommunications law (Stupp, 2016). Similarly, the U.S. Federal Trade Commission has released best practice security guidelines for IoT devices, and has prosecuted at least 50 companies for not having sufficiently secured networks or products (Mayer, 2016).

The heterogeneous nature of the IoT, as well as the complexity and lack of standardization of IoT architecture, makes securing infrastructure difficult. Security (and privacy) is frequently cited as an inhibitor of IoT adoption, and with recent highly publicized attacks, the need to secure IoT devices is becoming urgent.

2.3.5 Trust

Trust is a concept supported by a vast amount of literature in the psychological and business fields, but one that is not always considered in technology implementations. Two features are common within trust definitions: the first recognizes that a risk must be present that gives rise to the need for trust; the other is that the different parties must be interdependent or have cause to rely on each other (Rousseau et al., 1998). In technology systems, such as supply chains or the IoT, trust is seen to be an essential part of the infrastructure of systems where credentials are exchanged between infrastructure elements before services are provided (Mahinderjit-Singh and Li, 2010). The exchange of data also needs a level of trust, as organizations sharing information need to be reassured that the information will be used in appropriate ways (Eurich et al., 2010). Contracts between organizations play a role in mitigating this risk and ensuring a trusted relationship, as organizations can be sure of the behavior of their partners (Blomqvist et al., 2008). Similarly, the presence of regulations and legislation is considered one of the most effective ways of increasing trust, most likely related to the consequences of violating legislation (Luhmann, 1979). In some cases, the mitigation of risk through sufficient knowledge of partner organizations (Laeequddin et al., 2012) and consistent policies between organizations (Treglia and Park, 2009) is considered to be a possible replacement for a trusted relationship, which could then develop over time.

Trust can also emerge or be strengthened as a result of the successful implementation and use of technology systems, and where organizations have been involved in intense collaboration, it can emerge quite quickly (Blomqvist et al., 2008). However, in a ubiquitous technology environment, such as that of the IoT, trust is complicated by the large number of different participants and stakeholders, their differing needs and perspectives, and the speed with which the technology itself changes. Consumer trust is especially important in this type of environment, as consumers tend to distrust new technologies. The presence of consistent and strong privacy policies in particular has been found to assist with consumer acceptance of such technologies (Lee et al., 2007). The associated use of entity authentication and appropriate controls on access also assists users in trusting technology systems (Grandison and Sloman, 2000).

Between organizations, trust models are many and varied. Some include the characteristics of individuals, logical economic elements, and institutional trust, which includes regulation (Laeequddin et al., 2012). Other models consider trust elements, including the anticipation of certain responses, expectations of particular behaviors based on shared goals, and the knowledge of consequences for noncompliance (Tejpal et al., 2013). Across national boundaries, different national policies, the number of agencies involved in collaborative efforts, and national culture (including attitudes toward technology) add extra challenges to trusted relationships (Navarrete et al., 2009).

When considering technology systems, security and privacy are often considered to be dimensions of trust. But the situation is more complicated than simply ensuring systems security or data privacy. The nature of trust is such that the technology itself must also be trusted. Users must trust that the technology will operate without interruption, that it will be available when wanted, and that it will not transmit incorrect or malicious data (Roman et al., 2013). Similarly, each element of the IoT network is built with the assumption (or trust) that needed components will be available when they are required, even if those components are owned by other organizations in other jurisdictions (Yan et al., 2014).

From the technological standpoint, we can also observe the necessity for trust in the IoT context. Each architectural layer relies on the other layers operating as expected, and these layers rely on the overall operation of the whole system (Yan et al., 2014). The extensive nature of IoT systems also raises challenges that may not be seen in less widespread technology systems. The nature of the IoT itself is to collect vast amounts of data; therefore, data collection trust needs to be considered. If data collected is not trustworthy, through either collection error, technical problems, or outside interference, then the integrity of the whole system is compromised. Similarly, the quality of the data collected reflects directly on the services that can be provided, with poor quality or inaccurate data producing poor-quality and nontrustworthy services. Users are also expected to disclose or share data through the IoT devices. If they do not trust IoT services, they will not want to disclose data (Yan et al., 2014).

Technology-based trust management has not been extensively studied in IoT systems, while the elements of trust, such as data handling and access rights (Sicari et al., 2015) privacy, security, data perception and transmission, quality of service, identity, and systems reliability (Yan et al., 2014), are generally considered separately from the technology basis of the IoT. However, some research has taken a broader approach considering trust management in the IoT. Recent work considers building trusted networks of devices where nodes and devices can rate each other, and also exchange information with respect to other devices in a recommendation-type system (Bao and Chen, 2012). A similar approach is taken in Social IoT (SIoT) systems where reputation-based trust systems can repel some types of attacks through denying entry to nontrusted nodes, with services and information being obtained only from trusted sources (Nitti et al., 2012). Each node or device can calculate how trustworthy its associates and other nodes are, based on the ratings of its "friends" and on its own experience, choosing the most trustworthy node to transact with, similar to a trust-based model proposed for peer-to-peer (P2P) networks (Sicari et al., 2015). Beyond such trust-based calculations, trusted communities of nodes and devices can be formed using P2P principles. Each node or device, and each community, has its own identity within the IoT network, and is trusted within the network according to its behavior. This forms a trust chain within the community with parameters such as past history, proximity, consistency, availability, common warrants and goals, place within the hierarchy, and fulfillment of requests considered. Figure 2.1 summarizes this process, illustrating how a device could come to be trusted based on its behavior within the IoT network.

Security is established by considering the nodes crossed when users access this network, while initial trust is established either by the user or, where there is no user, by the manufacturer or organization controlling the device (Lacuesta et al., 2012). These trusted chains can form communities with unique identities that allow members to access services. Such distributed trust-based models go some way toward answering concerns around the difficulties traditional access control models have in securing highly dynamic and decentralized IoT networks (Mahalle et al., 2013).

Fuzzy methods can also be used to calculate trust scores, which are used to control access through sets of permissions and credentials. This fuzzy trust-based access control (FTBAC) method

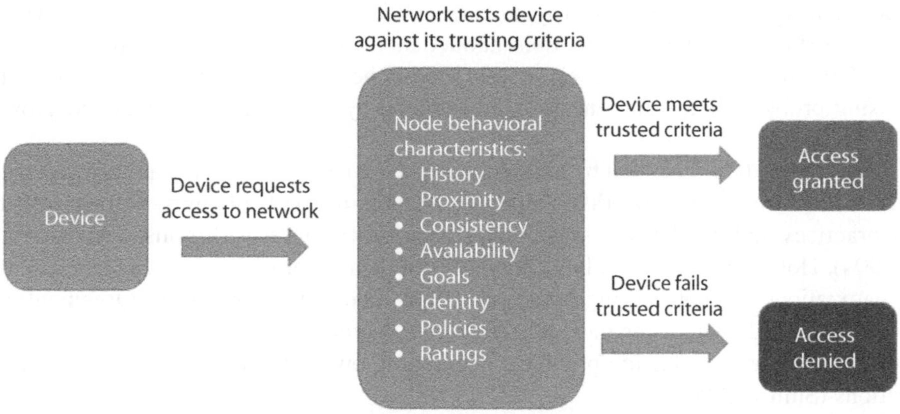

FIGURE 2.1 Demonstration of how devices come to be trusted in a trusted network.

considers three layers, a device (or sensor) layer, a request layer (which collects the trust values of knowledge, experience, and recommendations), and an access control layer. In such models, a device can access the IoT system only if its credentials meet preset policies related to the device's trust value (Wang et al., 2014). The FTBAC method has been shown to be scalable and flexible, as well as energy efficient (Bao and Chen, 2012; Mahalle et al., 2013). Other trust management mechanisms involve assessing the behavior of various nodes and calculating trust values based on prior behavior (Saied et al., 2013), and identity-based key negotiation in WSNs that recognizes suspicious nodes and reduces involvement with them (Liu et al., 2014).

Owner-defined policies are also used to verify nodes attempting to join the trusted network (Lize et al., 2014), and to control access to nonpublic information (Martinez-Julia and Skarmeta, 2013), although the lack of a common semantic language to verify differing policies still hinders this type of development. Other difficulties with policy-based management currently include determining how to manage negotiation between inconsistent and difficult-to-understand policies, and the presence of errors in security policies (Wu and Wang, 2011). Despite the difficulties with policy management, if it works correctly, it carries the advantage of allowing device owners to dictate how they interact with the IoT, and this type of trust management is preferred by the EU and similar regulatory bodies (European Union, 2009).

2.3.6 Data Management

As already discussed, IoT systems produce huge databases, up to zettabytes in size (Chen et al., 2015), which can be of considerable value to organizations (Thiesse et al., 2009). The heterogeneous nature of the data collected, along with its volume and complexity, presents challenges to effective IoT data usage (Zhou et al., 2016). NoSQL data query languages and cloud (or fog) storage are helping to reduce costs of managing big databases, and assisting in data utilization. Because of its volume, IoT data is inevitably stored in cloud databases, thus leading to problems with bandwidth, and delays in data processing and provision, or data latency (Botta et al., 2014). Fog computing (or edge computing) is an attempt to address these problems by storing data on local devices, thus removing the need for distributed data centers and improving speed of access, as well as allowing for greater mobility and real-time interaction (Marr, 2016).

From the management perspective, the ability to share information within and between organizations is enhanced by common data standards, as well as trusted networks and interoperable technical infrastructure (Yang and Pardo, 2011). Clear policies and regulations also aid data sharing within organizations or in systems where such polices were negotiated (Treglia and Park, 2009), while between jurisdictions with different regulatory structures or expectations, regulations can hinder data sharing as organizations struggle to meet disparate regulatory requirements (Ilie-Zudor et al., 2011). Further, the impact of questions about data ownership and data storage in a cloud environment, where jurisdiction is not always clear, has not been well explored, and will only become an increasing problem as the amount of data generated by IoT systems continues to grow (Chow et al., 2009).

The quantities of data generated by IoT systems also give rise to concerns about "data tsunamis" (Breur, 2015) and information overload, although ongoing research and improvement in data management practices and middleware systems has seen these concerns diminish (McKnight, 2007; Sarma, 2004). However, despite the huge quantities of data generated, or perhaps because of them, many organizations are not making optimum use of their data (Vos, 2014). Organizations have been found to struggle with using data generated by IoT-type systems to inform business decisions (Alvarez, 2004), as well as having problems negotiating ownership of data assets shared between organizations (Smith, 2006).

The massive databases derived from IoT systems have led to concerns for possible privacy or confidentiality infringements. Data collected from location-aware devices, such as RFID tags, or Bluetooth-enabled smart devices carried by individuals could theoretically be used to track the device's owner. Further, data collected by IoT systems could be aggregated and used to identify individuals, while health-related data could be compromised, leading to concerns that individual privacy rights could be violated. Solutions to these problems rely on either forming trusted relationships, allowing individuals to control their own interactions with the IoT, or regulation of IoT interactions, ensuring that organizations act in an appropriate way.

2.3.7 Legislation and Governance

The interaction between technology and regulation, or government, has always been a difficult one. Most governments are reluctant to legislate on technology matters, arguing that legislation constrains innovation (European Commission, 2013), and preferring technology-agnostic regulation that does not require changes with every new advance (Santucci, 2010). Further, the pace and scale of technology change make passing legislation difficult, with most jurisdictions focusing on technology-neutral regulations.

The scale of the IoT, and the potentially sensitive nature of the data collected by IoT systems, has attracted attention from various regulating bodies. The EU in particular has implemented a variety of regulations, and issued guidance for organizations and member states with respect to the use and implementation of IoT systems, especially in cross-border systems (European Commission, 2013). Standards bodies such as the IEEE, International Organization for Standardization (ISO), and GS1 are also heavily involved in promoting the standardization of infrastructure and other IoT components.

The need for the standardization of infrastructure and communications protocols, the traffic of data across national boundaries, and the associated cloud-based databases will require some form of governance, even if it is just to make sure the entire IoT infrastructure can function. Further, the need to ensure individual privacy often is poorly balanced with the need of organizations to use data collected from IoT systems (Friedewald and Raabe, 2011). It is likely that the DDoS attacks of recent times will spur regulators to action, if the industry does not move toward self-regulation (Wong, 2016). Governance of IoT systems is seldom discussed in literature, but the necessity to address problems integrating disparate IoT systems owned by competing organizations, adhere to national regulation, ensure the privacy of individuals and confidentiality of business data, and secure IoT infrastructure against attack is likely to require a great deal of research and organizational resources. No one can predict the unintended benefits or consequences of the IoT, and responsibility for managing the system is unclear. A multistakeholder governance model is the most likely given the large number of interested parties. It is very unlikely that a single organization or governmental group will succeed in controlling the IoT.

2.4 BUILDING THE BLOCKS INTO THE IoT

As the use of RFID and WSN technologies has become more common, the emergence of a true IoT has become a possibility, with predictions of billions of devices connected to the Internet by 2020 (Nordrum, 2016). However, challenges still remain with the implementation, connectivity, and use of IoT technology. The majority of IoT systems sit within a closed ecosystem containing the organization, its application partners, and its users. Tags and sensors from other IoT systems can be detected and read by the organizations readers, but they cannot be identified. So, while it might be possible to network these systems together into one continuous IoT, this cannot be done yet. These are the siloed "intranets of things" discussed earlier (Zorzi et al., 2010). In order to implement an effective IoT, it is essential that the various IoT components can be identified and communicate with the network or each other, be they RFID, WSN, Bluetooth devices, or something else. As new devices and protocols attempt to join the IoT, each must also be integrated and secured in some way. In order to seamlessly interact, an integrated communications protocol or a method of ensuring interoperability is required (Yoo, 2010). Connecting these IoT systems or "intranets" together presents many challenges apart from interoperability, including standards, data processing and management, privacy, security, trust, and governance, which have already been discussed.

Smart cities provide a test bed for IoT systems implementation on a larger scale. Cities such as New Songdo in Korea or the SmartSantander EU project allow for IoT systems to be explored and refined in real-world environments (Hernández-Muñoz et al., 2011). Smart city initiatives highlight the ways that daily life can be made easier through, for example, smart motorways and better use of public resources, as well as the issues and challenges facing smart environments (European Commission, 2015). Although smart city implementations offer many advantages, including more efficient management, better availability of information, improved transportation services, and opportunities to peruse creative ideas (Boulos and Al-Shorbaji, 2014), there are also challenges. Some of these might be expected in an IoT context, including heterogeneity, scalability, and integration (Hernández-Muñoz et al., 2011). Some solutions have been offered with respect to technical challenges, including the use of cognitive management frameworks based on virtual objects, which considers the value of IoT devices within the smart city, as well as how, why, and when to connect devices (Vlacheas et al., 2013). Other challenges focus more on the governance of the cities themselves, with questions being raised about the nature of agreements with technology providers, and whether those providers have too great a role in shaping the nature of the technology deployments, potentially causing technology lock-in (Greenfield, 2013). There is no doubt that FIoT research will focus on the issues presented by the smart city environments as researchers and technicians continue to address the problems of integrating siloed intranets of things into one cohesive IoT.

2.5 CONCLUSION

The heterogeneous nature of the IoT, as well as the enormous scale of IoT systems, means that grasping the full range of technologies and issues presented by the concept is challenging. This chapter has presented a high-level introduction to the issues faced in integrating today's organizational IoT systems into the wider IoT environment, from the more technical need to ensure the use of appropriate standards to streamline integration, through to privacy, security, trust, and legislative considerations. Despite the work presented here, there is still an ongoing need to consider how the increasing numbers of IoT-capable devices can be integrated into the IoT, and to find ways to unpick the complex social, organizational, and technical arrangements they bring.

In terms of the technology forming the basis of the IoT, which is not much discussed in this chapter, miniaturization (Vermesan et al., 2011), power supply (Atzori et al., 2010), the use of fog and cloud storage (Bonomi et al., 2012), and device self-management (Theodoridis et al., 2013) will continue to present research challenges. The role of the IoT in sustainability, recycling, and green

IT (Bojanova et al., 2015), and concerns about the disposal of IoT devices (Gubbi et al., 2013) are also areas for future research.

When discussing the IoT, people have the tendency to use adjectives such as *enormous*, *massive*, and *tsunami*. However, the scale and complexity of the IoT warrant these terms. No one knows, or has presented, a cohesive vision with respect to how a fully networked IoT will be realized, or even considered, if fully integrating things into everyday existence is a good idea. The IoT concept has been criticized as being too "vague," but it would be more accurate to say that it is difficult to comprehend how such a vision would play out in the real world. The potential for new kinds of human–organization–device interactions is only just beginning to be explored. In the future, it is likely that the IoT will be even more closely embedded in human social structures. Both the design of human interfaces and the question of how people could choose to interact (or not interact) in an IoT world is unsolved (Peppet, 2014).

The ongoing development of IoT systems, along with the continual integration of new and novel devices and applications, will only fuel the need for research into new ways of supporting and securing IoT systems, while organizations and individuals continue discovering how they might use IoT systems to enhance everyday experience.

REFERENCES

Abera, T., N. Asokan, L. Davi, F. Koushanfar, A. Paverd, A.-R. Sadeghi, and G. Tsudik. 2016. Invited-things, trouble, trust: On building trust in IoT systems. In *Proceedings of the 53rd Annual Design Automation Conference*, Austin, TX, 121. Retrieved from http://dl.acm.org/citation.cfm?id=2905020.

Ackerman, S. L., K. Tebb, J. C. Stein, B. W. Frazee, G. W. Hendey, L. A. Schmidt, and R. Gonzales. 2012. Benefit or burden? A sociotechnical analysis of diagnostic computer kiosks in four California hospital emergency departments. *Social Science and Medicine* 75 (12): 2378–2385.

Alcaide, A., E. Palomar, J. Montero-Castillo, and A. Ribagorda. 2013. Anonymous authentication for privacy-preserving IoT target-driven applications. *Computers and Security* 37: 111–123.

Alvarez, G. 2004. What's missing from RFID tests. *Information Week*. Retrieved from http://www.informationweek.com/news/global-cio/showArticle.jhtml?articleID=52500193.

Alvarez, G. 2004. What's missing from RFID tests? RFIDInsights. *Information Week*. November 8

Atzori, L., A. Iera, and G. Morabito. 2010. The Internet of things: A survey. *Computer Networks* 54 (15): 2787–2805.

Bao, F. and I.-R. Chen. 2012. Dynamic trust management for Internet of things applications. In *Proceedings of the 2012 International Workshop on Self-Aware Internet of Things*, San Jose, CA, 1–6. Retrieved from http://dl.acm.org/citation.cfm?id=2378025.

Bardaki, C., P. Kourouthanassis, and K. Pramatari. 2012. Deploying RFID-enabled services in the retail supply chain: Lessons learned toward the Internet of things. *Information Systems Management* 29 (3): 233–245.

Bertot, J. C. and H. Choi. 2013. Big data and e-government: Issues, policies, and recommendations. In *Proceedings of the 14th Annual International Conference on Digital Government Research*, Quebec, 1–10. Retrieved from https://doi.org/10.1145/2479724.2479730.

Blomqvist, K., P. Hurmelinna-Laukkanen, N. Nummela, and S. Saarenketo. 2008. The role of trust and contracts in the internationalization of technology-intensive born globals. *Journal of Engineering and Technology Management* 25 (1): 123–135.

Bojanova, I., J. Voas, and G. Hurlburt. 2015. The Internet of anything and sustainability. *IT Pro* 17 (3): 14–16.

Bonomi, F., R. Milito, J. Zhu, and S. Addepalli. 2012. Fog computing and its role in the Internet of things. In *Proceedings of the First Edition of the MCC Workshop on Mobile Cloud Computing*, Helsinki, 13–16. Retrieved from http://dl.acm.org/citation.cfm?id=2342513.

Bose, I., E. W. T. Ngai, T. S. H. Teo, and S. Spiekermann. 2009. Managing RFID projects in organizations. *European Journal of Information Systems* 18 (6): 534–540.

Botta, A., W. De Donato, V. Persico, and A. Pescapé. 2014. On the integration of cloud computing and Internet of things. In *2014 International Conference on Future Internet of Things and Cloud (FiCloud)*, Barcelona, 23–30. Retrieved from http://ieeexplore.ieee.org/xpls/abs_all.jsp?arnumber=6984170.

Boulos, M. N. K. and Al-Shorbaji, N. M. 2014. On the Internet of things, smart cities and the WHO healthy cities. *International Journal of Health Geographics* 13 (1): 10.

Breur, T. 2015. Big data and the Internet of things. *Journal of Marketing Analytics* 3 (1): 1–4.

Carr, A. S., M. Zhang, I. Klopping, and H. Min. 2010. RFID technology: Implications for healthcare organizations. *American Journal of Business* 25 (2): 25–41.

Cas, J. 2005. Privacy in pervasive computing environments—A contradiction in terms? *IEEE Technology and Society Magazine* 24 (1): 24–33.

Chen, F., P. Deng, J. Wan, D. Zhang, A. V. Vasilakos, and X. Rong. 2015. Data mining for the Internet of things: Literature review and challenges. *International Journal of Distributed Sensor Networks* 2015: 12.

Chow, R., P. Golle, M. Jakobsson, E. Shi, J. Staddon, R. Masuoka, and J. Molina. 2009. Controlling data in the cloud: Outsourcing computation without outsourcing control. In *Proceedings of the 2009 ACM Workshop on Cloud Computing Security*, Chicago, 85–90. Retrieved from https://doi.org/10.1145/1655008.1655020.

Curtin, J., R. Kauffman, and F. J. Riggins. 2007. Making the "MOST" out of RFID technology: A research agenda for the study of the adoption, usage and impact of RFID. *Information Technology and Management* 8 (2): 87–110.

De Hert, P. 2008. Identity management of e-ID, privacy and security in Europe. A human rights view. *Information Security Technical Report* 13 (2): 71–75.

EPCglobal. 2010. Retrieved November 19, from https://www.gs1.org/about.

Eurich, M., N. Oertel, and R. Boutellier. 2010. The impact of perceived privacy risks on organizations' willingness to share item-level event data across the supply chain. *Electronic Commerce Research* 10 (3–4): 423–440.

European Commission. 2013. Report on the public consultation on IoT governance. Brussels: European Commission. Retrieved from https://ec.europa.eu/digital-agenda/en/news/conclusions-internet-things-public-consultation.

European Commission. 2015. Smart cities: Digital agenda for Europe. Brussels: European Commission. Retrieved May 29, from https://ec.europa.eu/digital-agenda/en/smart-cities.

European Union. 2009. Internet of things—An action plan for Europe (no. 278). Brussels: European Commission. Retrieved from http://ec.europa.eu/information_society/policy/rfid/documents/commiot2009.pdf.

Evans, D. and D. M. Eyers. 2012. Efficient data tagging for managing privacy in the Internet of things. In *2012 IEEE International Conference on Green Computing and Communications (GreenCom)*, Besancon, France, 244–248. Retrieved from http://ieeexplore.ieee.org/xpls/abs_all.jsp?arnumber=6468320.

Friedewald, M. and O. Raabe. 2011. Ubiquitous computing: An overview of technology impacts. *Telematics and Informatics* 28 (2): 55–65.

Glover, B. and H. Bhatt. 2006. *RFID Essentials*. Sebastopol, CA: O'Reilly.

Grandison, T. and M. Sloman. 2000. A survey of trust in Internet applications. *IEEE Communications Surveys and Tutorials* 3 (4): 2–16.

Greenfield, A. 2013. *Against the Smart City*. New York: Do Projects.

Gubbi, J., R. Buyya, S. Marusic, and M. Palaniswami. 2013. Internet of things (IoT): A vision, architectural elements, and future directions. *Future Generation Computer Systems* 29 (7): 1645–1660.

Hernández-Muñoz, J. M., J. B. Vercher, L. Muñoz, J. A. Galache, M. Presser, L. A. H. Gómez, and J. Pettersson. 2011. Smart cities at the forefront of the future Internet. In *The Future Internet Assembly*, edited by Domingue, J., A. Galis, A. Gavras, T. Zahariadis, D. Lambert, F. Cleary, et al., 447–462. Berlin: Springer. Retrieved from http://link.springer.com/chapter/10.1007/978-3-642-20898-0_32.

Hossain, M. A. and M. Quaddus. 2011. The adoption and continued usage intention of RFID: An integrated framework. *Information Technology and People*, 24, 236–256.

Ilie-Zudor, E., Z. Kemény, F. van Blommestein, L. Monostori, and A. van der Meulen. 2011. A survey of applications and requirements of unique identification systems and RFID techniques. *Computers in Industry* 62 (3): 227.

Ingraham, N. 2016. Some of the biggest sites on the Internet were shut down this morning (update: Down again). Retrieved December 6, 2016, from https://www.engadget.com/2016/10/21/some-of-the-biggest-sites-on-the-internet-were-shut-down-this-mo/.

Iwata, J. 2012. Making markets: Smarter planet. *Presented at the IBM Investor Briefing*. Copenhagen, Denmark.

Kopalchick, J. III and C. Monk. 2005. RFID risk management. *Internal Auditor* 62 (2): 66–73.

Kros, J. F., R. G. Richey, Jr., H. Chen, and S. S. Nadler. 2011. Technology emergence between mandate and acceptance: An exploratory examination of RFID. *International Journal of Physical Distribution and Logistics Management* 41 (7): 697–716.

Krotov, V. 2008, January 22. RFID: Thinking outside of the supply chain. Retrieved December 6, 2010, from http://www.cio.com/article/174108/RFID_Thinking_Outside_of_the_Supply_Chain.

Lacuesta, R., G. Palacios-Navarro, C. Cetina, L. Peñalver, and J. Lloret. 2012. Internet of things: Where to be is to trust. *EURASIP Journal on Wireless Communications and Networking* 2012 (1): 203.

Laeequddin, M., B. Sahay, V. Sahay, and K. A. Waheed. 2012. Trust building in supply chain partners relationship: An integrated conceptual model. *Journal of Management Development* 31 (6): 550–564.

Lee, C.-P. and J. P. Shim. 2007. An exploratory study of radio frequency identification (RFID) adoption in the healthcare industry. *European Journal of Information Systems* 16 (6): 712–724.

Lee, S. M., S.-H. Park, S. N. Yoon, and S.-J. Yeon. 2007. RFID based ubiquitous commerce and consumer trust. *Industrial Management and Data Systems* 107 (5): 605–617.

Liu, Y. B., X. H. Gong, and Y. F. Feng. 2014. Trust system based on node behavior detection in Internet of things, Tongxin Xuebao. *Journal of* Communication 35 (5): 8–15.

Lize, G., W. Jingpei, and S. Bin. 2014. Trust management mechanism for Internet of things. *China Communications* 11 (2): 148–156.

Luhmann, N. 1979. *Trust; and, Power*. Chichester, UK: Wiley.

Mahalle, P. N., P. A. Thakre, N. R. Prasad, and R. Prasad. 2013. A fuzzy approach to trust based access control in Internet of things. In *2013 3rd International Conference on Wireless Communications, Vehicular Technology, Information Theory and Aerospace and Electronic Systems (VITAE)*, Atlantic City, NJ, 1–5. Retrieved from http://ieeexplore.ieee.org/abstract/document/6617083/.

Mahinderjit-Singh, M. and X. Li. 2010. Trust in RFID-enabled supply-chain management. *International Journal of Security and Networks* 5 (2): 96–105.

Manyika, J., M. Chui, P. Bisson, J. Woetzel, R. Dobbs, J. Bughin, and D. Aharon. n.d. Unlocking the potential of the Internet of things. McKinsey & Company. Retrieved December 6, 2016, from http://www.mckinsey.com/business-functions/digital-mckinsey/our-insights/the-internet-of-things-the-value-of-digitizing-the-physical-world.

Marr, B. 2016, October 14. What is fog computing? And why it matters in our big data and IoT world. *Forbes*. Retrieved December 5, 2016, from http://www.forbes.com/sites/bernardmarr/2016/10/14/what-is-fog-computing-and-why-it-matters-in-our-big-data-and-iot-world/.

Martinez-Julia, P. and A. F. Skarmeta. 2013. Beyond the separation of identifier and locator: Building an identity-based overlay network architecture for the future Internet. *Computer Networks* 57 (10): 2280–2300.

Matta, V., D. Koonce, and A. Jeyaraj. 2012. Initiation, experimentation, implementation of innovations: The case for radio frequency identification systems. *International Journal of Information Management* 32 (2): 164–174.

Mayer, C. 2016. Hack the planet: The state of IoT security. Retrieved December 6, 2016, from https://futuristech.info/posts/hack-the-planet-the-state-of-iot-security.

McKnight, W. 2007. The four pillars of RFID. *DM Review* 17 (10): 18.

Miorandi, D., S. Sicari, F. De Pellegrini, and I. Chlamtac. 2012. Internet of things: Vision, applications and research challenges. *Ad Hoc Networks* 10 (7): 1497–1516.

Mykletun, E., J. Girao, and D. Westhoff. 2006. Public key based cryptoschemes for data concealment in wireless sensor networks. In *2006 IEEE International Conference on Communications*, Istanbul, vol. 5, 2288–2295. Retrieved from http://ieeexplore.ieee.org/xpls/abs_all.jsp?arnumber=4024506.

Navarrete, A. C., S. Mellouli, T. A. Pardo, and J. R. Gil-Garcia. 2009. Information sharing at national borders: Extending the utility of border theory. In *42nd Hawaii International Conference on System Sciences, 2009 (HICSS '09)*, Waikoloa, HI, 1–10.

Nitti, M., R. Girau, L. Atzori, A. Iera, and G. Morabito. 2012. A subjective model for trustworthiness evaluation in the social Internet of things. In *2012 IEEE 23rd International Symposium on Personal Indoor and Mobile Radio Communications (PIMRC)*, Sydney, 18–23. Retrieved from http://ieeexplore.ieee.org/abstract/document/6362662/.

Nordrum, A. 2016, August 18. Popular Internet of Things forecast of 50 billion devices by 2020 is outdated. Retrieved December 11, 2016, from http://spectrum.ieee.org/tech-talk/telecom/internet/popular-internet-of-things-forecast-of-50-billion-devices-by-2020-is-outdated.

OECD (Organisation for Economic Co-operation and Development). 2013. OECD guidelines on the protection of privacy and transborder flows of personal data. Retrieved from https://www.oecd.org/internet/ieconomy/oecdguidelinesontheprotectionofprivacyandtransborderflowsofpersonaldata.htm.

Ohkubo, M., K. Suzuki, and S. Kinoshita. 2005. RFID privacy issues and technical challenges. *Communications of the ACM* 48 (9): 66.

Peppet, S. R. 2014. Regulating the Internet of things: First steps toward managing discrimination, privacy, security and consent. *Texas Law Review* 93: 85.

Proofprint. 2014. Proofpoint uncovers Internet of things (IoT) cyberattack. Retrieved December 10, 2016, from https://perma.cc/M78W-VELZ.

Roman, R., J. Zhou, and J. Lopez. 2013. On the features and challenges of security and privacy in distributed Internet of things. *Computer Networks* 57 (10): 2266–2279.

Rousseau, D. M., S. B. Sitkin, R. S. Burt, and C. Camerer. 1998. Not so different after all: A cross-discipline view of trust. *Academy of Management Review* 23 (3): 393–404.

Saied, Y. B., A. Olivereau, D. Zeghlache, and M. Laurent. 2013. Trust management system design for the Internet of things: A context-aware and multi-service approach. *Computers and Security* 39: 351–365.

Santucci, G. 2010. The Internet of things: Between the revolution of the Internet and the metamorphosis of objects. Brussels: European Commission. Retrieved from http://www.theinternetofthings.eu/content/g%C3%A9rald-santucci-internet-things-window-our-future.

Sarma, S. 2004. Integrating RFID. *Queue* 2 (7): 50–57.

Shi, J., L. Xiong, S. Li, and H. Tian. 2012. Exploration on intelligent control of the hospital infection—The intelligent reminding and administration of hand hygiene based on the technologies of Internet of things. *Journal of Translational Medicine* 10 (2): 1.

Sicari, S., A. Rizzardi, L. A. Grieco, and A. Coen-Porisini. 2015. Security, privacy and trust in Internet of things: The road ahead. *Computer Networks* 76: 146–164.

Smith, L. S. 2006. RFID and other embedded technologies: Who owns the data? *Santa Clara Computer and High Technology Law Journal* 22: 695–755.

Spekman, R. E. and P. J. Sweeney II. 2006. RFID: From concept to implementation. *International Journal of Physical Distribution and Logistics Management* 36 (10): 736.

Stupp, C. 2016, October 5. Commission plans cybersecurity rules for Internet-connected machines. Retrieved December 9, 2016, from https://www.euractiv.com/section/innovation-industry/news/commission-plans-cybersecurity-rules-for-internet-connected-machines/.

Tang, J. 2013. Designing an anti-swindle Mahjong leisure prototype system using RFID and ontology theory. *Journal of Network and Computer Applications* 39: 292–301.

Tejpal, G., R. K. Garg, and A. Sachdeva. 2013. Trust among supply chain partners: A review. *Measuring Business Excellence* 17 (1): 51–71.

Theodoridis, E., G. Mylonas, and I. Chatzigiannakis. 2013. Developing an IoT smart city framework. In *Information, Intelligence, Systems and Applications (IISA)*, Piraeus, Greece, 1–6. Retrieved from https://www.researchgate.net/profile/Evangelos_Theodoridis2/publication/261463912_Developing_an_IoT_Smart_City_framework/links/56533b9608aeafc2aabb1748.pdf.

Thiesse, F., J. Al-Kassab, and E. Fleisch. 2009. Understanding the value of integrated RFID systems: A case study from apparel retail. *European Journal of Information Systems* 18 (6): 592–614.

Treglia, J. V. and J. S. Park. 2009. Towards trusted intelligence information sharing. In *Proceedings of the ACM SIGKDD Workshop on CyberSecurity and Intelligence Informatics*, Paris, 45–52. Retrieved from https://doi.org/10.1145/1599272.1599283.

Tsai, C.-W., C.-F. Lai, and A. V. Vasilakos. 2014. Future Internet of things: Open issues and challenges. *Wireless Networks* 20 (8): 2201–2217.

Vermesan, O., P. Friess, P. Guillemin, S. Gusmeroli, H. Sundmaeker, A. Bassi, et al. 2011. Internet of things strategic research roadmap. *Internet of Things: Global Technological and Societal Trends* 1: 9–52.

Vlacheas, P., R. Giaffreda, V. Stavroulaki, D. Kelaidonis, V. Foteinos, G. Poulios, P. Demestichas, A. Somov, A. R. Biswas, and K. Moessner. 2013. Enabling smart cities through a cognitive management framework for the Internet of things. *IEEE Communications Magazine* 51 (6): 102–111.

Vlad, M., R. Parvulet, M. Vlad, and C. Pivoda. 2012. A survey of livestock identification systems. In *Proceedings of the 13th WSEAS International Conference on Automation and Information (ICAI12)*, Iasi, Romania, 165–170: WSEAS Press.

Vos, M. 2014. RFID on the boundary between the public and private sectors: An ANT/institutional theory investigation. Victoria University of Wellington. Retrieved from http://researcharchive.vuw.ac.nz/handle/10063/3431.

Vos, M. 2016. Maturity of the Internet of things research field: Or why choose rigorous keywords. *Presented at Proceedings of the Australasian Conference of Information Systems*, Adelaide. Retrieved from http://arxiv.org/abs/1606.01452.

Vos, M., R. Cullen, and J. Cranefield. 2012. RFID in the public and private sector: Key implementation considerations. *Presented at AMCIS 2012 Proceedings*, Seattle. Retrieved from http://aisel.aisnet.org/amcis2012/proceedings/EBusiness/18.

Walker, S. 2016. Hacked cameras, DVRs powered today's massive Internet outage—Krebs on security. Retrieved from https://krebsonsecurity.com/2016/10/hacked-cameras-dvrs-powered-todays-massive-internet-outage/.

Wamba, S. F. and A. T. Chatfield. 2010. The impact of RFID technology on warehouse process innovation: A pilot project in the TPL industry. *Information Systems Frontiers* 13(5): 693–706.

Wamba, S. F. and E. W. T. Ngai. 2011. Unveiling the potential of RFID-enabled intelligent patient management: Results of a Delphi study. In *2011 44th Hawaii International Conference on System Sciences (HICSS)*, Kauai, HI, 1–10. Retrieved from http://ieeexplore.ieee.org/xpls/abs_all.jsp?arnumber=5718928.

Wang, X., J. Zhang, E. M. Schooler, and M. Ion. 2014. Performance evaluation of attribute-based encryption: Toward data privacy in the IoT. In *2014 IEEE International Conference on Communications (ICC)*, Sydney, 725–730. Retrieved from http://ieeexplore.ieee.org/xpls/abs_all.jsp?arnumber=6883405.

Wang, Y. and Q. Wen. 2011. A privacy enhanced DNS scheme for the Internet of things. In *IET International Conference on Communication Technology and Application (ICCTA 2011)*, Beijing, 699–702. Retrieved from http://ieeexplore.ieee.org/xpls/abs_all.jsp?arnumber=6192955.

Wikholm, Z. 2016. July 10. Flashpoint—When vulnerabilities travel downstream. Retrieved from https://www.flashpoint-intel.com/when-vulnerabilities-travel-downstream/.

Wong, J. I. 2016. The Internet of things is totally unregulated and that might have to change. Retrieved December 5, 2016, from http://qz.com/817516/dyn-ddos-attack-the-internet-of-things-is-totally-unregulated-and-that-might-have-to-change/.

Wu, Z. and L. Wang. 2011. An innovative simulation environment for cross-domain policy enforcement. *Simulation Modelling Practice and Theory* 19 (7): 1558–1583.

Wyld, D. C. 2005. RFID: The Right Frequency for Government. Washington, DC: IBM Center for the Business of Government. Retrieved from http://www.businessofgovernment.org/report/rfid-right-frequency-government.

Yan, Z., P. Zhang, and A. V. Vasilakos. 2014. A survey on trust management for Internet of things. *Journal of Network and Computer Applications* 42: 120–134.

Yang, T.-M. and T. A. Pardo. 2011. How is information shared across boundaries? In *Proceedings of the 44th Hawaii International Conference on System Sciences*, Kauai, HI, 1–10.

Yoo, Y. 2010. Computing in everyday life: A call for research on experiential computing. *MIS Quarterly* 34 (2): 213–231.

Zhou, Z., K.-F. Tsang, Z. Zhao, and W. Gaaloul. 2016. Data intelligence on the Internet of things. *Personal and Ubiquitous Computing* 20 (3): 277–281.

Zorzi, M., A. Gluhak, S. Lange, and A. Bassi. 2010. From today's intranet of things to a future Internet of things: A wireless-and mobility-related view. *IEEE Wireless Communications* 17 (6): 44–51.

Part II

Technological Advances and Implementation Considerations

3 Cooperative Networking Techniques in the IoT Age

Luigi Alfredo Grieco, Giuseppe Piro, Gennaro Boggia, and Domenico Striccoli

CONTENTS

3.1 Introduction 51
3.2 Cooperative Approaches to Cellular Systems 52
  3.2.1 Cooperative Approaches in 4G Networks 52
  3.2.2 Cooperative Approaches in 5G Networks 54
  3.2.3 Device-to-Device Communications 55
3.3 Cooperative Approaches to WLANs and WSNs 57
  3.3.1 Cooperative Communications in WLANs 57
  3.3.2 Cooperative Communications in WSNs 58
  3.3.3 Crowd-Sourcing Systems 60
3.4 Cooperative Approaches to VANETs 60
3.5 Cooperative Approaches to Wireless Networks with Energy Harvesting Capabilities 64
3.6 Conclusions 66
Acknowledgments 67
References 67

3.1 INTRODUCTION

Today, given the heterogeneous nature of connected objects, many communication technologies are emerging in the IoT age. The most promising ones include 4G and 5G cellular systems, wireless local area networks (WLANs) and wireless sensor networks (WSNs), and vehicular ad hoc networks (VANETs), to name a few. By offering wireless connectivity with different data rates, coverage, and resilience capabilities, they generally enable heterogeneous IoT and machine-to-machine (M2M) applications (Palattella et al., 2016).

For speeding up the diffusion of IoT applications and services in the market, researchers and industries are nowadays demonstrating a growing interest to improve the efficiency of these communication systems. From the literature, it emerges that cooperative networking could be exploited for reaching this goal. In summary, cooperation is a multifolded term embracing many different technologies, protocols, and algorithms, all sharing the common ambition to improve the efficiency of communication systems thanks to some form of interaction among network nodes. At the lowest layers of the protocol stack (i.e., physical and logical link control), cooperation identifies those techniques that allow multiple sending nodes to transmit data concurrently to the same destination in order to magnify the signal-to-noise ratio (SNR), such as in virtual multiple input multiple output (VMIMO); take advantage of the channel state information (CSI) reported from several nodes in a wireless system for enabling the optimization of modulation schemas and transmission power levels; and extend the coverage of a wireless transmitter thanks to relaying operations. On the other extreme, network cooperation can be fruitfully used at the application layer to orchestrate services, enforce trust management techniques in social networks, develop crowd-sensing platforms, optimize intelligent transportation systems (ITSs), and manage harvested energy in green networks.

Of course, in the middle, cooperation can play a key role in routing and forwarding strategies that have to choose the next hop or the subset of cooperating nodes, based on collective feedback from the rest (of part) of the network. Moreover, moving from the theoretical floor to the market, the term *cooperation* is further enriched by the wide set of contexts it applies to.

In this chapter, the many facets of cooperation are explored in the most promising networking scenarios behind the corner: from 4G/5G cellular systems to cognitive radio networks, from WLANs to WSNs, and from VANETs to energy harvesting networks. Not only classic cooperation schemas are considered, based on the usage of VMIMO approaches, but also higher-level forms of cooperation are described to provide the reader with a broader picture of this complex and fascinating topic.

The rest of the chapter is organized as follows. Section 3.2 describes cooperative approaches recently formulated for offering powerful services and efficiently supporting pervasive IoT and M2M applications in cellular systems. Section 3.3 summarizes the most promising cooperative diversity techniques able to mitigate the impact of fading and reduce the reliability of communication for WLANs and WSNs. Section 3.4 focuses on VANETs and explores cooperative approaches for supporting the dissemination of multimedia contents and other kinds of data related to traffic management systems. Section 3.5 provides further considerations on cooperative techniques by investigating specific solutions for wireless networks with energy harvesting capabilities. Finally, Section 3.6 provides the conclusion.

3.2 COOPERATIVE APPROACHES TO CELLULAR SYSTEMS

In the context of 4G and 5G cellular systems, many cooperative approaches have been formulated for increasing network capabilities, offering powerful services to mobile users, and efficiently supporting pervasive IoT and M2M applications.

3.2.1 Cooperative Approaches in 4G Networks

Long Term Evolution—Advanced (LTE-A) is the communication technology devised for the fourth generation of cellular systems, simply referred to as 4G (Dahlman et al., 2011). Among its many features, LTE-A leverages the availability of relay stations to improve service coverage and system throughput. In particular, a relay station forwards the data of neighboring mobile terminals to a local base station, and vice versa, thus extending the service coverage and enhancing the spectral efficiency of a wireless communication system (Figure 3.1). The presence of relay stations requires the usage of cooperative communications techniques. Moreover, the actual performance gain contributed by the relay station strongly depends on the collaborative strategy, which includes the selection of relay types (i.e., amplify and forwarding, selective decode and forward, and demodulation and forward) and relay partners (i.e., to decide when, how, and with whom to collaborate), as well

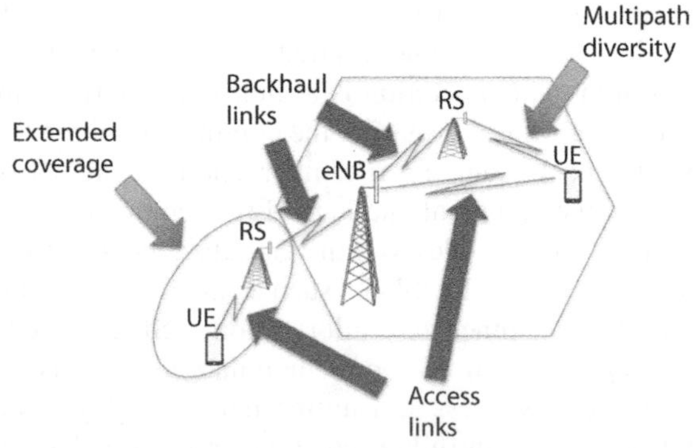

FIGURE 3.1 Different kinds of relay stations in the 4G system. evolved Node B (eNB), Relay Station (RS), User Equipment (UE).

as a sustainable business model that capitalizes the performance gain contributed by relay stations to compensate the increased costs of deployment.

Relay stations pave the way to new scenarios and business opportunities, including cooperative spectrum leasing, VANETs, and IoT applications. At the same time, they require novel design methodologies and business models. Some promising use cases are outlined below:

- *Cooperative spectrum leasing*: A primary base station leases a quota of the spectrum to a third-party unlicensed secondary relay station as a reward for the increased spectral efficiency (Gomez-Cuba et al., 2014). This kind of deployment can be particularly beneficial in IoT systems with limited mobility since the gain contributed by a third-party relay could be significantly impaired by handover operations. As described in Figure 3.2, a similar approach can also be extended to enable coexistence between the primary cellular network and secondary ad hoc networks (Zhai et al., 2014).
- *Vehicular relay stations*: In cellular systems, capacity limitations soon become evident in dense urban areas, such as downtown quarters and major events, where users might experience degraded performance. Under these circumstances, vehicles could become ideal candidates to the role of relay stations (Feteiha and Hassanein, 2015). Also, a vehicle is equipped with many different resources, including sensing, processing, storage, and communication subsystems, thus becoming an ideal mobile relay station (see Figure 3.3). The adoption of vehicular relay stations lowers the costs of infrastructure deployment, but at the same time, it requires incentive strategies to motivate vehicle owners (Zhang et al., 2013). Moreover, new physical layer designs are needed to cope with an extremely dynamic topology and frequent connectivity issues induced by mobility.
- *Interplay of social, mobile, and wireless network*: In currently available cellular networks, the underlying architecture is very centralized. Each device connects to a base station in order to gain data connectivity. Actually, mobile phones are equipped with several wireless

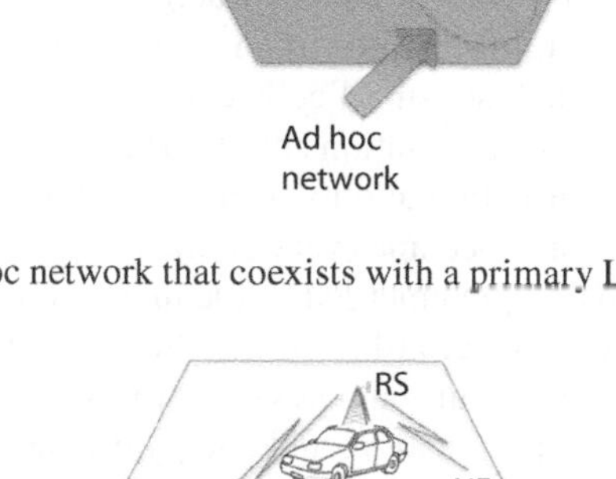

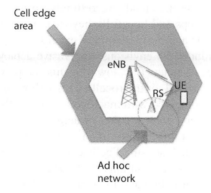

FIGURE 3.2 A secondary ad hoc network that coexists with a primary LTE-A cell.

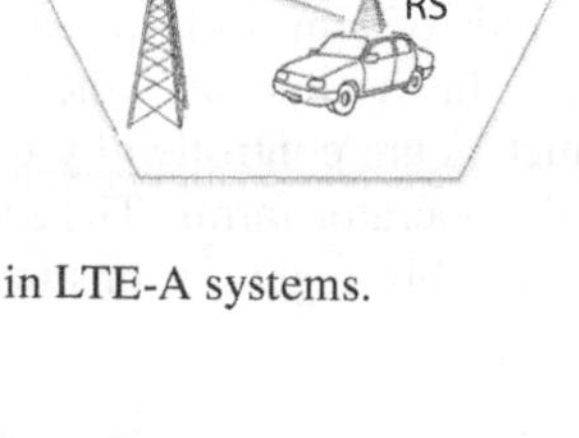

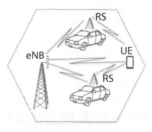

FIGURE 3.3 Vehicular relay stations in LTE-A systems.

network interfaces, based on different technologies (e.g., Wi-Fi capabilities). Therefore, Internet connections can also be set up through personal area and local area networks. This broad availability of access networks could be leveraged by network operators to face the ever-increasing needs of high-data-rate mobile applications without incurring the high costs of providing a 4G infrastructure (i.e., offloading strategies). The most relevant reasons that hinder a massive usage of Wi-Fi hotspots lie at the security and legal issues floor. In some countries, in fact, the owner of an access point is responsible for the transported traffic, so that even altruistic people are compelled to restrict the access to their hotspots. In this case, the lack of trust toward other users hinders the actual potential of cooperative offloading strategies. Luckily, trust issues can be effectively faced using social networks. If users are allowed to create trusted communities in a social network, the usage of a Wi-Fi hotspot can be allowed subject to these communities. In this way, the owner of the hotspot can identify the party responsible for any illegal use of the network and safely share its access point. Nevertheless, with respect to cellular systems up to 3G, 4G systems can deploy smaller cells that make the problem of coexistence with additional access networks (such as Wi-Fi) even more challenging (Wang et al., 2013). The good news is that this work can be seamlessly executed by the same users of the network in a distributed cooperative way, exploiting data connectivity and a variety of sensors available in the different devices. This social approach to wireless sensing can help to cut down costs, as network operators could limit their own measurement campaigns to suburban areas, where data may be insufficient (Katz et al., 2014).

3.2.2 Cooperative Approaches in 5G Networks

Upcoming 5G networks (Gupta and Jha, 2015) will be in charge of supporting the exponential rise of data traffic due to the increased number of devices that are connected to the Internet (including M2M systems), the increased processing power of smartphones, the higher pervasiveness of bandwidth-hungry services in many areas of life and industry, and the more frequent usage of smartphones as gateways to access the cloud. Accordingly, to magnify the capacity of wireless access networks (e.g., up to 10 Gbps and hopefully beyond), increase the area spectral efficiency and energy efficiency, and provide a uniform quality of experience (QoE) regardless of the position and capabilities of mobile terminals, 5G envisions a pervasive deployment of base stations with a coverage radius up to a few tens of meters (i.e., low transmission power), each one serving a small cell (e.g., femtocell, picocell, microcell, or metrocell). This will bring to a hyperdense deployment of small cells (Xu et al., 2014). Compared with macrocells, femtocells are significantly more dense and less organized, and have a very small number of active users to serve. Moreover, when hundreds of femtocells are deployed within a macrocell, cross-tier effects should be accounted for (e.g., interference and vertical handover). Also, small cells can be turned on and off by the end user or according to the traffic variation; the base station should be able to select or reselect a feasible carrier with minimal exchange of information and its own channel measurements. Thus, the management and maintenance of 5G networks becomes very challenging and expensive. In this context, the self-organizing network (SON) paradigm emerged as the most promising cooperative methodology to address these issues (Wang and Zhang, 2014). It introduces self-healing, self-configuration, and self-optimization capabilities to fulfill four main system targets: coverage expansion, capacity optimization, quality of service (QoS) optimization, and energy efficiency.

Three possible architectures for femtocell networks can be considered:

1. *Centralized architecture:* Two kinds of femtocell deployments are envisaged for 5G networks: residential and enterprise. In the first one, the base stations are installed by the users while some system parameters are controlled by the operation, administration, and management (OAM) server in the operator farms. The second type of deployment targets large enterprises, public places, and big organizations in which femtocells are set up and

fully controlled by operators for what concerns OAM. In both cases, an operator is allowed to make an OAM server to enable centralized self-healing operations. This OAM server is fed by measurements (e.g., received signal strength) coming from all femtocells, and it can monitor the entire network and suddenly detect an outage as soon as an abnormal change in such measurements is sensed. Upon outage detection, the OAM can use the global knowledge of the network to replan everything from scratch, so to drive the system toward a new optimal working point. Using a centralized architecture will always result in an optimal setting, but it also presents several subtle shortcomings related to network scalability and stability. Scalability issues arise from the communication and processing overhead that a multipoint-to-point feedback control system entails (the OAM server may suddenly become a bottleneck). Network stability, instead, can be compromised because each time an outage is detected, the settings of the entire network should be retuned. In addition, once the causes that triggered the outage disappear, the 5G system reverts back to the original settings, which were already validated and optimized on the field.

2. *Distributed architecture*: If no backhaul cooperation among femtocells is used, self-healing functionalities can be provided on a distributed basis. In this case, each femtocell runs an independent self-healing algorithm at its base station, which monitors the network environment and tunes its parameters when a nearby outage is discovered. Once an outage is detected, the base station can increase its transmission power to fill the coverage gap in its surroundings. The pros of this architecture are the very limited complexity of outage detection and compensation, which immediately translates to a high scalability. Its cons are mainly related to the sparsity of 5G femtocells, each one serving a few users, and thus becoming too sensitive to single-user habits. Also, the settling time of a distributed architecture is almost unpredictable under time-varying changes in the environment because once a femtocell changes its settings, this will propagate to the next cells, thus affecting the convergence speed of the self-healing algorithm.
3. *Local cooperative architecture*: To capitalize the advantages of centralized and distributed approaches, while counteracting their downsides, a local cooperative architecture can be considered. In this architecture, after an outage is detected based on the measurements of surrounding femtocells, a proper set of neighbor base stations retune their parameters to compensate the outage. The cooperation here is limited to neighboring cells so that the signaling overhead is limited with respect to centralized architectures. At the same time, the accuracy of outage detection schemas could be impaired because of a lack of a global view of the network. Finally, a local outage compensation should improve network stability but cannot grant global optimality. In any case, by tuning the degree of cooperation among a cluster of femtocells and the scope of the cluster itself, it is possible to achieve the desired trade-off between stability, optimality, and scalability.

Note that each of the aforementioned approaches has pros and cons, so that the actual choice mainly depends on the specific context to face. In particular, three dimensions should be considered to characterize a self-healing outage management algorithm: overhead, accuracy, and stability. The higher the signaling overhead, the higher the accuracy in outage detection and the optimality of compensation actions. At the same time, optimal settings usually entail global reconfigurations, which impair the stability. As a consequence, the trade-off among these three dimensions should be carefully tuned based on the predominant needs of the target scenario.

3.2.3 Device-to-Device Communications

Device-to-device (D2D) communications allow two devices in a cell to exchange data with each other (in the licensed spectrum) without the aid of the base station or with the partial involvement of the base station (Tehrani et al., 2014). They pave the way to a new generation of IoT scenarios and services in 5G systems:

- *Device relaying*: This can be implemented as a very promising technique that can accelerate pervasive cooperation in 5G networks beyond the fixed relaying schemes of 4G systems. In this vision, any user device with cellular connectivity (e.g., tablet, smartphone, or laptop) can act as a transmission relay for other equipment in the network, thus enabling the creation of a massive ad hoc mesh network.
- *Context-aware services*: Several IoT applications try to customize the type of service offered to the end user according to its preferences, its current location, or any other information taken from the environment (like temperature or time). These applications, which are quickly gaining in popularity, require location discovery and communication with neighboring devices, which could greatly benefit from the adoption of D2D functionalities. D2D communications, in fact, can be used to exchange key information on the context, thus enabling and customizing specific IoT services.
- *Mobile cloud computing*: Thanks to the increment of computational, energy, and memory capabilities of mobile devices, current research is proposing to extend conventional cloud computing services offered by remote data centers to mobile devices. The mobile cloud computing intends to exploit mobile devices to execute tasks related to applications running on other devices in the neighborhood, thus enabling the possibility to support, locally, more complex services (such as games, image and video processing, e-commerce, and online social networks) without exhausting their limited resources (Piro et al., 2016). In this context, D2D communications can facilitate effective sharing of resources (spectrum, computational power, applications, social contents, etc.) for users who are spatially close to each other.
- *Offload strategies*: Service providers can take advantage of D2D functionality to drain some load off of the network by allowing direct transmission among cell phones and other devices in a local area. For instance, if two users in a stadium would share multimedia content, they can exchange it directly, for example, without needing to pass through the base station.
- *Disaster recovery*: When an adverse event happens (e.g., earthquake or hurricane), it is likely that the communication infrastructure falls, as well as the electrical one. Under these circumstances, to coordinate emergence management operations, a communication network can be set up using D2D functionality in a short time, replacing the damaged communication network and Internet infrastructure. As a result, available devices may still disseminate information without requiring the coordination of a base station deployed by a given mobile operator.

However, several technical challenges need to be afforded to transform the D2D vision into reality. To reach a full integration of this new technology with 5G broadband systems, security, trust, interference management, resource discovery, and pricing issues need to be properly faced. Security and trust issues can be solved starting from the approaches devised for both IoT and M2M scenarios. In this context, in fact, the literature already provides valid solutions for addressing peer authentication, access control, data confidentiality, user privacy, and trust management services (Sicari et al., 2015). Interference management needs to account for both base station-to-device and D2D interactions, which are very difficult to control in fully distributed schemas (i.e., when the base station has no control on the setup of D2D links). Resource discovery could leverage the potential of cooperation through social networks (Katz et al., 2014). From the pricing perspective, devices that act as relays for other users use their own resources, such as battery, data storage, and bandwidth, and therefore, they should receive some incentive. Furthermore, in direct D2D communication, the devices need to have a secure environment for the process of selling and buying resources among themselves. The operator can control and create a secure environment (e.g., like an app store) for easing this kind of process. Therefore, it can expect some payment from the devices for the security and QoS in D2D communication.

3.3 COOPERATIVE APPROACHES TO WLANs AND WSNs

From a communication perspective, in WLANs (Gast, 2005) and WSNs (Atzori et al., 2010), cooperative diversity techniques can mitigate the impact of fading and improve the reliability of communications (Khan and Karl, 2014). In particular, when a transmitter–receiver pair start exchanging data, neighboring nodes can overhear the packets and replicate them over different fading channels, thus allowing the receiver to combine the different replicas and strengthen the robustness of the system. The media access control (MAC) protocol handles the coordination of all the underlying activities between the transmitter, receiver, and relays. Thus, it becomes the playmaker of the system and can also interact with the routing protocols by suggesting the nodes to select, based on the CSI. The resulting VMIMO scheme entails different degrees of freedom that refer to the dynamic selection of relay nodes, scheduling operations, and forwarding decisions, which affect the overall effectiveness of the approach and require joint optimization strategies. These concepts, as expected, are translated in different ways, depending on whether a WLAN or a WSN is considered. In addition, the perimeter of the terms *communication* and *cooperation* can be magnified when WLANs and WSNs are considered from the application point of view. Indeed, in this context, novel coworking paradigms, such as crowd sourcing and crowd sensing, can open new perspectives to cooperation among humans, the environment, and sensors in any combination (Ganti et al., 2011).

3.3.1 Cooperative Communications in WLANs

So far, many cooperative approaches to WLANs have been proposed. They mainly differ from each other based on the number of relay nodes to involve to assist communications between a couple of nodes S and D, the relay selection mechanism, the degree of compliance to the IEEE 802.11 standard, the way CSI is used during the cooperation, the initiation scheme, and the target key performance indicator to maximize (i.e., throughput or delay). In what follows, some noticeable examples are reported. A summary flowchart is reported in Figure 3.4.

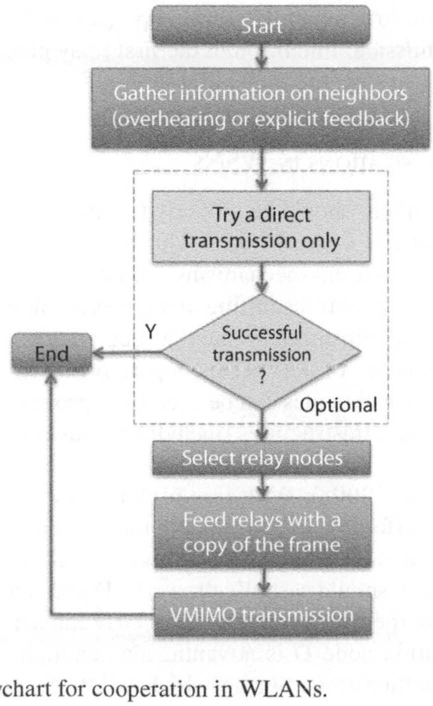

FIGURE 3.4 A summary flowchart for cooperation in WLANs.

- *CoopMAC* (Liu et al., 2007): In this protocol, a relay R can be chosen by a transmitter node S when the bitrate to the destination node D is too low and a performance gain can be pursued thanks to the higher bitrate of R. In other words, if communications from S to R and from R to D can be executed at a higher bitrate than from S to D, the relay node R is asked to take part in the transmission. In order to execute the CoopMAC protocol, each node overhears ongoing transmissions in the WLAN and takes note of the bitrate between any couple of nodes and the number of failed transmission attempts. Moreover, a new signaling packet is added to the request-to-send (RTS)–clear-to-send (CTS) handshake to check whether relay node R is available to help in the communication between S and D.
- *Selection decode and forward* (Valentin et al., 2008): A new cooperative RTS packet is added to the IEEE 802.11 standard to enforce cooperation (namely, cRTS). Moreover, cooperation is also used to strengthen the transmission of signaling frames. In particular, when a data frame has to be transmitted, the cRTS packet is sent by S, which expects to receive a CTS from D or R (or both) before sending the data frame. The cRTS packets (and the corresponding CTS from D) are relayed by R. The receiving station D sends an acknowledgment (ACK) whenever at least one correct copy of the frame originally sent by S is received. This ACK frame is relayed by R too.
- *Persistent relay carrier sensing multiple access (PRCSMA)* (Alonso-Zarate et al., 2008): In this protocol, cooperation is activated on demand. When S sends a packet to D, all neighboring nodes overhear and buffer the message in order to execute retransmissions as soon as D asks for cooperation. Only if this happens do neighboring nodes act as relays and keep retransmitting the message until D acknowledges the correct reception of the frame or the maximum number of retransmissions is reached.
- *Cooperative diversity MAC (CD-MAC)* (Moh et al., 2007): In this protocol, cooperation is activated only if the direct transmission of a packet fails. In particular, if no response is received by S to an RTS packet, S and R simultaneously transmit a copy of the frame (including signaling messages), thus forming a VMIMO system. R is selected based on link quality estimates derived by S from overheard packets.
- *2rcMAC protocol* (Khalid et al., 2011): This makes use of a second relay node to be used as a backup in order to improve reliability, throughput, and delays. The two relay nodes are chosen so that the transmission time through the first relay plus the second relay (used for backup transmission) is less than the transmission time from S to D.

3.3.2 Cooperative Communications in WSNs

In a WSN, multihop communications should be properly managed by jointly considering the limited energy and computing capabilities of network nodes. The different proposals conceived so far can be differentiated based on the way wake-up mechanisms are integrated within the relay selection and VMIMO transmissions, the number of transmitting and receiving nodes at each hop, the reactive or proactive activation of cooperative transmissions, and the degree of compliance to the IEEE 802.15.4 standard. In what follows, a summary of noticeable cooperative MAC protocols for WSNs proposed so far is reported. Note that these techniques can be used to improve the performance of IoT systems properly deployed for monitoring infrastructures (including industrial plants and smart buildings).

- *WSC-MAC* (Mainaud et al., 2008): With this protocol, the problem of selecting a relay within a neighborhood is afforded, by introducing the concept of group identifier (GID). In particular, each different node in a neighborhood is assigned to a GID. Whenever a node S needs to transmit a frame, it should casually draw a GID and send this outcome to all nodes in its neighborhood. Only the nodes with the same GID can act as relays provided the link state toward the destination node D is advantageous enough. This mechanism requires a distributed preconfiguration stage during which GIDs are assigned to WSN nodes. In addition, it cannot provide any guarantees about relayed transmissions due the random nature of its operations.

- *Cooperative Preamble Sampling MAC (CPS-MAC)* (Khan and Karl, 2014): This protocol aims to integrate cooperative communications and wake-up mechanisms in order to address energy efficiency issues too. The rationale of CPS-MAC is that any form of cooperation would fail if the destination node D is not awake to receive the data sent by S. To this end, with CPS-MAC, nodes switch between sleep and awake states, and as soon as S needs to transmit a packet, it wakes up a partner node by sending a strobe of preamble frames (long enough to cover the duration of the sleep state). After the partner is awake, it can repeat the strobe to wake up the next hop destination. Finally, transmissions from the original sender and its partner can be combined according to the cooperative paradigm. This mechanism is repeated at each hop, until the sink of the WSN is reached.
- *Generalized poor man's SIMO system (gPMSS)* (Ilyas et al., 2011): With this protocol, the source S sends a packet to the destination D and relays. If D acknowledges a correct reception, no retransmission is needed; otherwise, the frame is retransmitted by relay nodes. In particular, this operation is executed by one of the relays that correctly received the frame from S or by all relay nodes (if none of them received a correct copy of the frame). Then, D applies combination schemas and sends an ACK if it is able to decode the frame. Otherwise, S retransmits the frame and all the steps discussed above are repeated.
- *Cooperative collision resolution* (Lin and Petropulu, 2005): With this protocol, collided frames are not discarded but combined with later retransmissions. In particular, after a collision occurs in a given slot, the next slots are used by relays to transmit (one by one) the signals received during the collided slot. The receiver, combining these replicas, can reconstruct the frame. In this way, it becomes possible to capitalize on the strengths of both the ALOHA algorithm and VMIMO systems.
- *Cooperative cross-layer MAC (CC-MAC)* (Zhou et al., 2010): This is based on two transmitters—two receiver models—and combines adaptive modulation with a truncated automatic repeat request (ARQ) in a cross-layer way. Accordingly, both S and D have their own partners to capitalize the advantages brought by cooperation. Differently from other cooperative techniques, which require a first stage for a direct transmission from S to D and a second stage to enforce cooperation, the partner of S receives the copy of the message to forward from the previous hop (as S) so that it can cooperatively transmit with S at the next hop without any need for the first stage (Figure 3.5). In other words, the receiving partner of the previous hop becomes the transmitting partner of the next hop. Partners are chosen based on the link qualities at each hop, and the modulation is adaptively set at each hop based on the smallest SNR on the links S-D and partner-D.
- *Cooperative low-power MAC (CL-MAC)* (Ben Nacef et al., 2011): This jointly implements low-power listening and cooperative communications. Two variants of CL-MAC exist based on the way the relay node is chosen: either proactively (i.e., before data transmission by the source) or reactively (i.e., after data transmission). In both cases, nodes alternate fixed length sleep and activity periods. When a source S has to transmit a frame, it wakes up its neighbors by sending several short preamble packets, spaced by listing periods. The preamble indicates the destination D and the time instant the transmission has to be executed. At that time instant, the neighbors wake up to cooperatively transmit the frame. In the proactive version, the relay is selected after the neighbors are awake based on the SNR of the channel to the destination. In the reactive one, instead, the relay is selected only when the direct transmission from S to D fails.
- *Cross-layer design for multihop VMIMO* (Yuan et al., 2006): The target of this protocol is to leverage the VMIMO potential to reduce energy consumptions while providing end-to-end QoS guarantees. To this end, its design embraces all the main facets of a WSN, including radio link models, multihop routing hop-by-hop recovery, and end-to-end QoS.

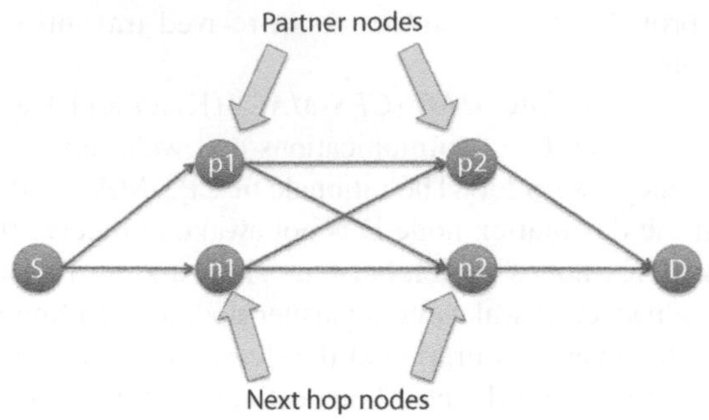

FIGURE 3.5 Forwarding and cooperation in CC-MAC.

In particular, the topology is arranged in clusters and cluster heads (CHs) are connected by a multihop backbone. The communications between any couple of CHs is strengthened using a hop-by-hop recovery scheme and VMIMO communications (cooperating nodes are chosen based on channel quality). Finally, the transmission parameters at each link are set in order to provide the expected end-to-end delay with the minimum energy consumption.

3.3.3 Crowd-Sourcing Systems

In the current IoT age, embedded devices and sensors are deployed almost everywhere and can enable a new generation of distributed and cooperative applications (Ganti et al., 2011). With reference to sensing systems, what mainly distinguishes the old WSN paradigm from the wider IoT one is that sensors and monitoring applications (i.e., data producers and consumers) can be integrated in any kind of device: from a classic mote to a smartphone, from a tablet to a workstation, from vehicles to environmental monitoring stations, and so forth. Many of these devices can execute sophisticated (in-network) processing operations on the gathered data, thus further broadening the scope and capabilities of personal and community sensing applications. Personal sensing (e.g., monitoring running or walking exercises) pertains to phenomena that refer to a single individual and, as a consequence, does not require any kind of cooperation. Social sensing, instead, includes monitoring operations that cannot be fulfilled by a single node (e.g., to map in real time the traffic condition of a city). In this case, cooperation refers to the integration of different measurements collected in very different time instants and broadly extended geographical areas. The degree of involvement of users could span from explicitly providing multimedia acquisitions (e.g., to take and deliver a photo) to providing sensed data through continuous sampling or in response to some event of interest. A general architecture of crowd-sensing applications is reported in Figure 3.6.

Mobile crowd-sensing applications can be classified based on the kind of phenomenon to monitor. In this way, environmental, infrastructure, and social applications can be distinguished. Environmental ones include measuring pollution levels in a city, measuring water levels in rivers, and monitoring wildlife habitats. Infrastructure applications include measuring traffic congestion, road conditions, parking availability, outages of fire hydrants or broken traffic lights, and real-time transit tracking. Social applications allow users to share the outcomes of their actions (e.g., daily exercises). In this kind of applications, the main technical challenges to face are related to energy efficiency, trust, and privacy (Christin et al., 2012).

3.4 COOPERATIVE APPROACHES TO VANETs

Video streaming is one of the most challenging applications in VANETs (Hartenstein and Laberteaux, 2008): in fact, fast topology dynamics could severely hinder a fluid video playout and discourage users from using multimedia systems in their vehicles. Cooperation is the key to counteract VANET

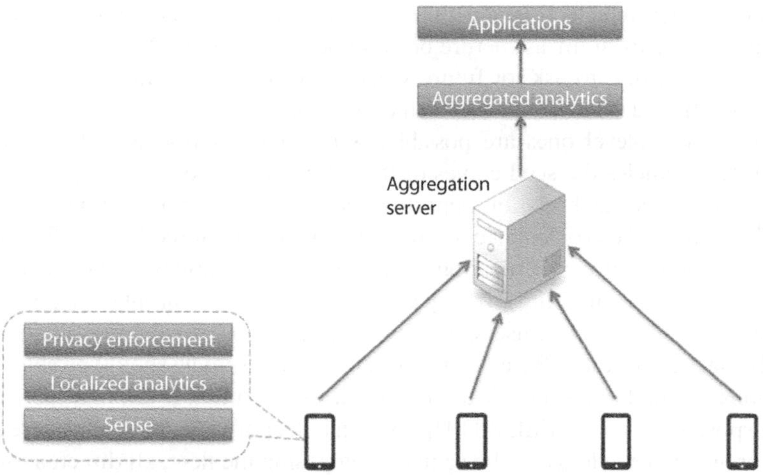

FIGURE 3.6 A general architecture for crowd-sensing application.

inefficiencies, and many proposals have been formulated so far in order to profit from the joint adoption of wide area (e.g., 3G or 4G) connectivity and local area IEEE 802.11p communications to enable seamless service provisioning across these two widely available technologies (Xu et al., 2013; Yaacoub et al., 2015; Huang et al., 2016). From one side, 4G coverage is wider and less prone to disconnection than IEEE 802.11p coverage. From the other side, IEEE 802.11p can enable short-range and high-rate data transfer among neighbor vehicles. By jointly exploring the two technologies through a cooperative approach, it becomes possible to capitalize their strengths while rejecting their weaknesses. Figure 3.7, for instance, shows a general architecture of a multihomed VANET.

In addition, scalable video encoding techniques can be used to adapt the quality of the video signals to the actual bandwidth availability of the VANET. At the same time, the setup of a cooperative system embracing different VANET technologies and encoding schemas is not straightforward because it is required to define (1) the protocols that rule the adoption of the 4G or IEEE

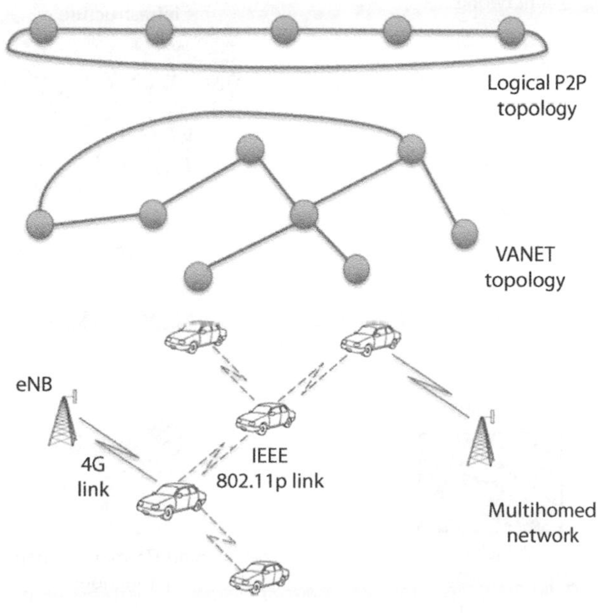

FIGURE 3.7 Multihomed VANET.

802.11p interface at each vehicle; (2) the network architecture, which could be clustered or multihop and arranged according to an infrastructure or mesh peer-to-peer (P2P) overlay; (3) the way available bandwidth is used (i.e., to ask for future video chunks or for an increased quality of chucks closer to the playout); and (4) the level at which cooperation is pursued: both application-level P2P approaches and network-level ones are possible. A possible solution would consist of distributing across different vehicles the seed copies of the different chunks of most popular contents and enabling prefetching schemas based on opportunistic interactions between neighbor vehicles (Xu et al., 2013). To simplify the problem, a clustered topology can be forced, with CH chosen based on their closeness to base stations and executing multicast communications toward nearby nodes (using IEEE 802.11p) (Yaacoub et al., 2015). In any case, an optimization problem should be formulated to maximize the QoE perceived by users, which is a function of the signal resolution, the encoding scheme, and the packet loss ratio. Based on this problem, each vehicle should be able to drive in real time its decisions on the kind of chunk to download, its resolution level, and the technology best fitted to the purpose (i.e., 4G or IEEE 802.11p). When a fleet of vehicles is considered, instead, it is possible to leverage the knowledge of the vehicles composing the fleet, all directed toward the same destination, to implement more structured forms of cooperation (Huang et al., 2016).

VANETs natively support ITS services (Dimitrakopoulos and Demestichas, 2010). By fostering cooperation between vehicles, WSNs, and traffic management systems, in fact, it becomes possible to pursue real-time sensing capabilities on the transportation infrastructure that enable quick reactions to unexpected events and traffic congestions while improving the overall safety and efficiency of the system. In fact, a vehicle could, from the one end, sense the environment and share the outcomes of its measurements and data acquisitions and, on the other hand, benefit from the information collected from other vehicles to choose the closest parking area, avoid an accident, change the path, or switch to another transportation means (i.e., multimodal systems). The degree of cooperation among vehicles can be narrowed or widened depending on the target service: while in collision avoidance systems only a local interaction among neighbor vehicles is required, in global rerouting operations the broadest level of cooperation would be necessary. A big picture of an ITS is depicted in Figure 3.8.

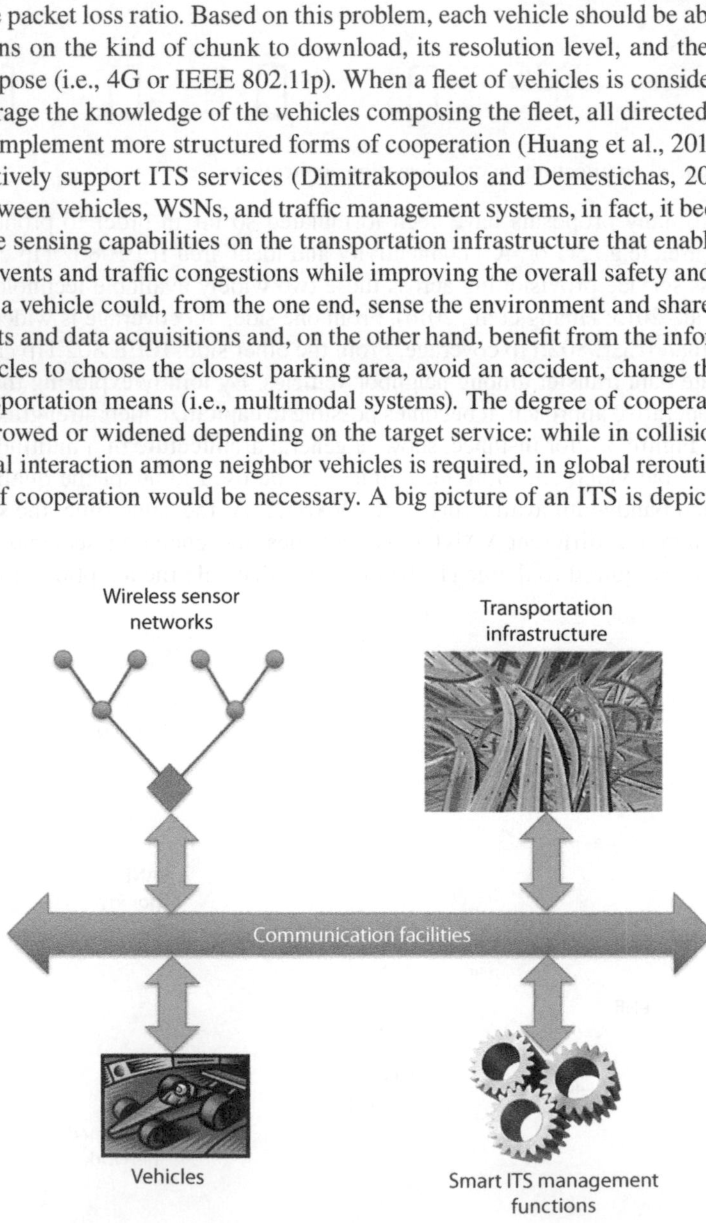

FIGURE 3.8 Key components of an ITS.

It could be adopted in our cities to alleviate the consequences of the increased mobility we are facing in the current epoch.

The main requirements of an ITS are (1) context awareness, to enable adaptation of routes and help vehicles—it is required to recognize and characterize the current context; (2) personalization, in the sense that vehicles should also be assisted based on the profile of drivers; (3) seamless support of heterogeneous communication systems; and (4) scalability. These requirements are fulfilled thanks to a pervasive exchange of information between the different actors of the system (Figure 3.9). In particular, the following data flows can be identified in a cooperative ITS (Dimitrakopoulos and Demestichas, 2010): vehicle to vehicle (raw data useful to local cooperation), vehicle to WSN (information exchange between a vehicle and the surrounding environment to enable local optimization), vehicle/WSN to smart management system (information useful to enable global optimization functionalities), and smart management system to transportation infrastructure (actuation signals to execute the global optimization strategy).

These data flows carry heterogeneous types of information, including multimedia acquisition from cameras, pollution levels, position–speed pairs from sample vehicles, information from or to travelers, modifications of transportation schedules from multimodal means, and so forth. An example of sensors and their usage in an ITS is shown in Figure 3.10.

Only thanks to such a richness of data can an ITS become a very articulated and flexible architecture able to cover advanced transportation management, advanced traveler information, advanced vehicle control, business vehicle management, advanced public transportation, and advanced urban transportation systems.

On the management side, that is, beyond communication issues, ITSs are complex systems with many static and dynamic interacting units. Moreover, the analysis of ITSs is difficult to accomplish without using holistic approaches due to the pervasive degree of cooperation among the actors of the system (Wang, 2010). In other words, although an ITS can be split in different subsystems, it is hard to predict its behavior by resorting to the superposition principle and analyzing the single components on their own. On the other side, modeling an entire ITS with a holistic approach would inflate the complexity of the management and hinder the fulfillment of requirements. Usually, to address these issues, parallel traffic management systems are made of five intertwined components: *actual transportation system* (the real infrastructure made of roads, signaling systems, public

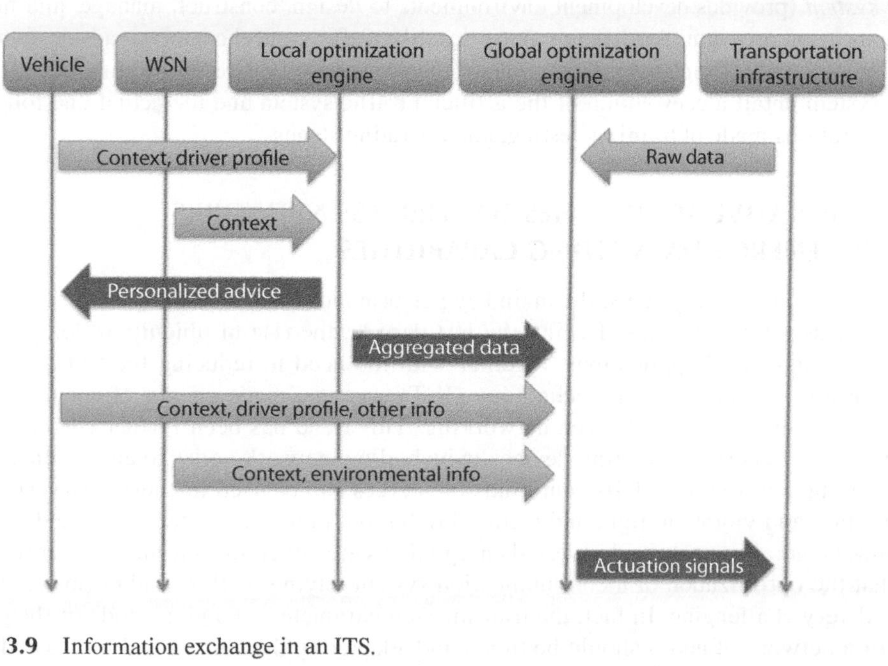

FIGURE 3.9 Information exchange in an ITS.

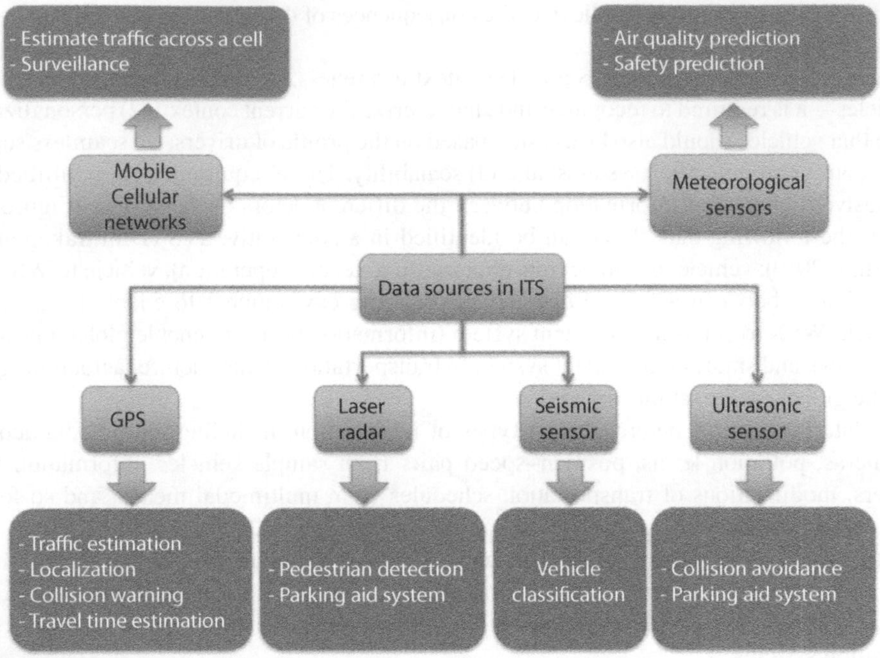

FIGURE 3.10 Usage of sensors in ITSs.

transportation, vehicles, etc.), *artificial transportation system* (simulated model of the actual transportation system used to forecast and control ITS functionalities), *traffic operator and administrator training system* (used to speed up the test and evaluation of traffic operational procedures and regulations), *decision evaluation and validation system* (eases the estimation of traffic conditions and the test of traffic control and management operations, information dissemination strategies, and decision support to traffic operators and individual drivers), and *traffic sensing, control, and management system* (provides development environments to design, construct, manage, and maintain autonomous agent programs for different purposes, like traffic control centers, roadside controllers, sensing devices, and information systems). Thus, the processes of the resulting parallel traffic management system entail a coevolution of the artificial traffic system and the actual one following a cycle of operations made of training, testing, and operating stages.

3.5 COOPERATIVE APPROACHES TO WIRELESS NETWORKS WITH ENERGY HARVESTING CAPABILITIES

In classic communication systems, the main key performance indicators to optimize are throughput, packet loss ratio, and delay. During the last decade, the rise of ubiquitous computing and pervasive IoT and M2M applications, together with the need to reducing the CO_2 footprint of information and communications technology (ICT) systems, made energy efficiency a fundamental requirement too, toward green networking. This trend has been further corroborated by the availability of energy harvesting devices, which allow network nodes to absorb energy from the surrounding environment. Different kinds of sources can be used in energy harvesting: they include motion and vibration, light and infrared radiation, radio waves, temperature differences, and airflow. Generally speaking, harvested energy flows are either not deterministic or time varying, so that the optimization of a communication system relying on this kind of power supply is considered very challenging. In fact, the transmission parameters at a given node or the protocol settings of a network of nodes should be tuned and adapted to the current and forecasted energy flows, subject to the classic requirements on QoS. Using cooperative techniques, it is possible to enable information exchange among nodes, thus improving the ability to (1) estimate the current status of the network, including channel status, battery level, and forecasted energy flows; (2) allow a node with an almost empty battery to be replaced by a set of nodes with a larger availability of energy; and (3) optimize power allocation and modulation to maximize the throughput thanks to VMIMO communications. With reference to cellular networks, energy harvesting techniques represent a possible solution to face the never-ending rise of energy consumptions of base stations. In 2011, more than 4 million base stations had been deployed to provide services for mobile users, causing an extremely high energy consumption of 25 MWh per year, on average. The deployment of self-powered cellular networks becomes economically convenient as the size of cells decreases (Piro et al., 2013), so that the current trends of cooperating 4G and 5G systems toward femtocells and D2D communications will magnify the relevance of energy harvesting techniques and green protocols in general. WSNs represent another killer scenario for cooperative energy harvesting systems. In fact, not only do WSN nodes have limited energy resources, but also their batteries are difficult to replace. In this case, it is necessary to schedule the activity of nodes based on their current availability of energy and the forecasted provisioning from harvesting modules.

In general, the problem is manyfold and in-progress research efforts are shedding some light on the different angles of the cooperation between wireless nodes equipped with energy harvesting devices. To provide an outlook on the current developments in this field, some relevant and recent contributions are summarized below.

- Minasian et al. (2014) afford the problem of throughput maximization in an energy harvesting two-hop amplify-and-forward relay network. First, an offline setting is considered in which channel states and the harvested energy profile are a priori known. Then, a more realist case is studied and a Markov decision process–based optimization is proposed, assuming a casual knowledge of channel states and harvested energy. Since the latter approach can be computationally demanding, a simpler heuristic is proposed too (as usually done in this kind of studies) to lower the complexity of the Markov decision process–based formulation.
- Zhang et al. (2014) consider cooperative underlay D2D communications in a green cellular network, in which the base station is fed also by renewable energy sources and it supports D2D data exchange by acting as a decode-and-forward relay. In particular, the problem of power allocation is faced in order to maximize the network throughput under different constraints, while avoiding outages. To this end, the charging and discharging process contributed by renewable energy sources is approximated as a G/D/1 queue.
- In Ding et al. (2014), a wireless cooperative network made of one energy harvesting relay and different couples of interacting nodes is studied. In particular, several strategies are proposed to distribute the energy accumulated at the relay among different transmitting nodes, including equal, opportunistic, and auction-based power allocations. The first strategy can help transmitters with poor channel conditions, by providing them with a richer power allocation. The second one, instead, serves users with better channel conditions before, according to a sequential water-filling scheme, thus maximizing the number of successful destinations, and (surprisingly) minimizing the worst user outage probability. The third approach, instead, achieves almost the same performance as the second one, without requiring the knowledge of CSI.
- Also in Mekikis et al. (2014), the problem of wireless energy harvesting in cooperative networks is studied, with a major emphasis on large-scale and network coding–aided scenarios. Using stochastic geometry, the lower bound of the probability of successful data exchange is derived, along with the lifetime gain thanks to the usage of wireless energy

harvesting at the relays. All in all, it has been shown that the lifetime of the network can be increased up to 70% and that, in low-noise environments, increasing the relay density improves the lifetime of the network, without compromising the QoS.

- In Nordio et al. (2014), a multihop cooperative WSN with linear topology is considered, supporting converge cast traffic. This kind of scenario is commonplace in street monitoring and video surveillance systems. The cut-set upper bound to the achievable rate for this scenario is derived using information theory arguments, also accounting for energy constraints. Also, different relaying strategies are proposed, able to perform close to the cut-set bounds, targeting the maximization of the data rate, subject to some constraints on fairness and energy consumption (typical of energy harvesting devices).
- In Misra et al. (2014), a Green Wireless Body Area Nanonetwork (GBAN) is considered, formed by nanosensors arranged in a multihop topology and conveying their data to a nanosink, which is then interfaced to a more powerful outer device. Communications between nanodevices are handled using both electromagnetic waves and molecular-based information transfers. Nanodevices are assumed to be able to harvest energy from their surrounding environment through biomechanical-to-electrical or biochemical-to-electrical energy conversion. This kind of processes does not provide sharp guarantees on the lifetime of each single node or of the networks, so that energy management should be accomplished on a cooperative basis. Accordingly, a cooperative Nash bargaining game is formulated: nanodevices bargain with one another in terms of their available energy, so that the QoS of the system is kept at an acceptable level. The resulting solution provides a unique optimal agreement or operational point while enforcing fairness and efficient use of resources.
- In Nasir et al. (2013), an amplify-and-forward wireless cooperative network is considered, in which the relay node can harvest energy from radio waves. Two different relaying protocols are investigated in the presence of delay-tolerant and delay-limited applications. The achievable throughput is derived in all those cases, and the parameters of the two energy harvesting protocols are optimized. Finally, future applications of these results are drawn, such as the analysis of finite alphabet modulation, the presence of a minimum power level to enable harvesting operations, the usage of ARQ schemas, and the availability of CSI at the relay node.
- In Li et al. (2011), the problem of scheduling cooperative communications in a WSN is considered. A time-slotted scheme is assumed, and all sensors are equipped with energy harvesting capabilities. The scheduler should decide, at every transmission, whether a relay should be used in cooperation with the transmitter to strengthen the QoS (at the expense of depleting the energy of the relay node). The objective is to maximize the long-term ratio of the data that is successfully delivered. This technique can be extremely useful in monitoring infrastructure and industrial plants, where offered services require strict constraints in terms of delivery ratio and communication latency.

3.6 CONCLUSIONS

This chapter explored the different facets of cooperation, spanning a broad class of applications, implications, and technologies. In fact, it has been shown that cooperative techniques can be extended well beyond pure VMIMO approaches, and that the potential of VMIMO can be significantly magnified by leveraging cooperation at higher layers of the protocol stack. The ways in which this happens strongly depend on the underlying technology. To this end, cooperative 4G and 5G, WLANs, WSNs, VANETs, and energy harvesting wireless systems have been described, trying to highlight common features and main differences. Future research will still work on innovative cooperative networking techniques for all the considered communication technologies. Despite the presence of a very broad background in this context, it will always be necessary to customize and/or improve existing techniques to the wide range of applications and their heterogeneous requirements.

ACKNOWLEDGMENTS

This work is supported by the project BONVOYAGE, which receives funding from the European Union's Horizon 2020 research and innovation program under grant agreement 635867; by the project symbIoTe, which receives funding from the European Union's Horizon 2020 research and innovation program under grant agreement 688156; and by the FANTASTIC-5G project, which receives funding from the European Union's Horizon 2020 research and innovation program under grant agreement ICT-671660.

REFERENCES

Alonso-Zarate, J., E. Kartsakli, C. Verikoukis, and L. Alonso. 2008. Persistent RCSMA: A MAC protocol for a distributed cooperative ARQ scheme in wireless networks. *EURASIP Journal on Advances in Signal Processing*: 817401.

Atzori, L., A. Iera, and G. Morabito. 2010. The Internet of things: A survey. *Computer Networks* 54 (15): 2787–2805.

Ben Nacef, A., S. Senouci, Y. Ghamri-Doudane, and A. L. Beylot. 2011. A cooperative low power MAC protocol for wireless sensor networks. In *IEEE International Conference on Communications*, Kyoto, 1–6.

Christin, D., C. Rokopfa, M. Hollicka, L. Martucci, and S. Kanherec. 2012. Incognisense: An anonymity-preserving reputation framework for participatory sensing applications. *IEEE International Conference on Pervasive Computing and Communications*, Lugano, 135–143.

Dahlman, E., S. Parkvall, and J. Skold. 2011. *4G: Lte/Lte-Advanced for Mobile Broadband*, 1st ed., San Diego: Academic Press.

Dimitrakopoulos, G. and P. Demestichas. 2010. Intelligent transportation systems. *IEEE Vehicular Technology Magazine* 5 (1): 77–84, March 2010.

Ding, Z., S. Perlaza, I. Esnaola, and H. Poor. 2014. Power allocation strategies in energy harvesting wireless cooperative networks. *IEEE Transactions on Wireless Communications* 13 (2): 846–860.

Feteiha, M. and H. Hassanein. 2015. Enabling cooperative relaying VANET clouds over LTE-A networks. *IEEE Transactions on Vehicular Technology* 64: 4.

Ganti, R., F. Ye, and H. Lei. 2011. Mobile crowdsensing: Current state and future challenges. *IEEE Communications Magazine* 49 (11): 32–39.

Gast, M. S. 2005. *802.11 Wireless Networks: The Definitive Guide*, 2nd ed., O'Reilly Media: Sebastopol, CA.

Gomez-Cuba, F., F. Gonzalez-Castano, and J. Munoz-Castaner. 2014. Is cooperative spectrum leasing by third-party relays advantageous in next-generation cellular networks? In *Proceedings of IEEE European Wireless Conference*, Barcelona, Spain, 1–7.

Gupta, A. and R. K. Jha. 2015. A survey of 5G network: Architecture and emerging technologies. *IEEE Access* 3: 1206–1232.

Hartenstein, H. and L. P. Laberteaux. 2008. A tutorial survey on vehicular ad hoc networks. *IEEE Communications Magazine* 46 (6): 164–171.

Huang, C., C. Yang, and Y. Lin. 2016. An adaptive video streaming system over a cooperative fleet of vehicles using the mobile bandwidth aggregation approach. *IEEE Systems Journal* 10 (2): 568–579.

Ilyas, M. U., M. Kim, and H. Radha. 2011. On enabling cooperative communication and diversity combination in IEEE 802.15.4 wireless networks using off-the-shelf sensor motes. *Wireless Networks* 17 (5): 1173–1189.

Katz, M., F. Fitzek, D. Roetter, and P. Seeling. 2014. Sharing resources locally and widely: Mobile clouds as the building blocks of shareconomy. *IEEE Vehicular Technology Magazine* 9 (3): 63–71.

Khalid, M., Y. Wang, I. Ra, and R. Sankar. 2011. Two-relay-based cooperative MAC protocol for wireless ad hoc networks. *IEEE Transactions on Vehicular Technology* 60 (7): 3361–3373.

Khan, R. and H. Karl. 2014. MAC protocols for cooperative diversity in wireless LANs and wireless sensor networks. *IEEE Communications Surveys and Tutorials* 16 (1): 46–63, First Quarter 2014.

Li, H., N. Jaggi, and B. Sikdar. 2011. Relay scheduling for cooperative communications in sensor networks with energy harvesting. *IEEE Transactions on Wireless Communications* 10 (9): 2918–2928.

Lin, R. and A. Petropulu. 2005. A new wireless network medium access protocol based on cooperation. *IEEE Transactions on Signal Processing* 53 (12): 4675–4684.

Liu, P., Z. Tao, S. Narayanan, T. Korakis, and S. Panwar. 2007. CoopMAC: A cooperative MAC for wireless LANs. *IEEE Journal on Selected Areas in Communications* 25 (2): 340–354.

Mainaud, B., V. Gauthier, and H. Afifi. 2008. Cooperative communication for wireless sensors network: A MAC protocol solution. In *Proceedings of IEEE Wireless Days*, Dubai, 1–5.

Mekikis, P. V., A. Lalos, A. Antonopoulos, L. Alonso, and C. Verikoukis. 2014. Wireless energy harvesting in two-way network coded cooperative communications: A stochastic approach for large scale networks. *IEEE Communications Letters* 18 (6): 1011–1014.

Minasian, A., S. Shahbazpanahi, and R. Adve. 2014. Energy harvesting cooperative communication systems. *IEEE Transactions on Wireless Communications* 13 (11): 6118–6131.

Misra, S., N. Islam, J. Mahapatro, and J. Rodrigues. 2014. Green wireless body area nanonetworks: Energy management and the game of survival. *IEEE Journal of Biomedical and Health Informatics* 18 (2): 467–475.

Moh, S., C. Yu, S. Park, H. N. Kim, and J. Park. 2007. CD-MAC: Cooperative diversity MAC for robust communication in wireless ad hoc networks. In *IEEE International Conference on Communications*, Glasgow, 3636–3641.

Nasir, A. A., X. Zhou, S. Durrani, and R. A. Kennedy. 2013. Relaying protocols for wireless energy harvesting and information processing. *IEEE Transactions on Wireless Communications* 12 (7): 3622–3636.

Nordio, A., C. F. Chiasserini, and A. Tarable. 2014. Bounds to fair rate allocation and communication strategies in source/relay wireless networks. *IEEE Transactions on Wireless Communications* 18 (2): 467–475.

Palattella, M. R., M. Dohler, L. A. Grieco, G. Rizzo, J. Torsner, and T. Engel. 2016. Internet of things in the 5G era: Enablers, architecture and business models. *IEEE Journal on Selected Areas in Communications* 34 (3).

Piro, G., M. Amadeo, G. Boggia, C. Campolo, L. A. Grieco, A. Molinaro, and G. Ruggeri. 2016. Gazing into the crystal ball: When the future Internet meets the mobile clouds. *IEEE Transactions on Cloud Computing* 99: 1–1.

Piro, G., M. Miozzo, G. Forte, N. Baldo, L. A. Grieco, G. Boggia, and P. Dini. 2013. HetNets Powered by Renewable Energy Sources: Sustainable Next-Generation Cellular Networks. *IEEE Internet Computing* 17 (1): 32–39.

Sicari, S., A. Rizzardi, L. A. Grieco, and A. Coen-Porisini. 2015. Security, privacy and trust in Internet of things: The road ahead. *Computer Networks* 76: 146–164.

Tehrani, M., M. Uysal, and H. Yanikomeroglu. 2014. Device-to-device communication in 5G cellular networks: Challenges, solutions, and future directions. *IEEE Communications Magazine* 52 (5): 86–92.

Valentin, S., H. Lichte, D. Warneke, T. Biermann, R. Funke, and H. Karl. 2008. Mobile cooperative WLANs—MAC and transceiver design, prototyping, and field measurements. In *Proceedings of IEEE Vehicular Technology Conference*, Calgary, BC, 1–5.

Wang, F. Y. 2010. Parallel control and management for intelligent transportation systems: Concepts, architectures, and applications. *IEEE Transactions on Intelligent Transportation Systems* 11 (3): 630–638.

Wang, W. and Q. Zhang. 2014. Local cooperation architecture for self-healing femtocell networks. *IEEE Wireless Communications* 21 (2): 42–49.

Wang, W., J. Zhang, and Q. Zhang. 2013. Cooperative cell outage detection in self-organizing femtocell networks. In *Proceedings of IEEE INFOCOM*, Turin, 782–790.

Xu, J., J. Wang, Y. Zhu, Y. Yang, X. Zheng, S. Wang, L. Liu, K. Horneman, and Y. Teng. 2014. Cooperative distributed optimization for the hyper-dense small cell deployment. *IEEE Communications Magazine* 52 (5): 61–67.

Xu, C., F. Zhao, J. Guan, H. Zhang, and G. M. Muntean. 2013. QoE-driven user-centric VOD services in urban multihomed P2P-based vehicular networks. *IEEE Transactions on Vehicular Technology* 62 (5): 2273–2289.

Yaacoub, E., F. Filali, and A. Abu-Dayya. 2015. QoE enhancement of SVC video streaming over vehicular networks using cooperative LTE/802.11p communications. *IEEE Journal of Selected Topics in Signal Processing* 9 (1): 37–49.

Yuan, Y., Z. He, and M. Chen. 2006. Virtual MIMO-based cross-layer design for wireless sensor networks. *IEEE Transactions on Vehicular Technology* 55 (3): 856–864.

Zhai, C., W. Zhang, and G. Mao. 2014. Cooperative spectrum sharing between cellular and ad-hoc networks. *IEEE Transactions on Wireless Communications* 13: 7.

Zhang, G., K. Yang, P. Liu, and X. Feng. 2013. Incentive mechanism for multiuser cooperative relaying in wireless ad hoc networks: A resource-exchange based approach. *Wireless Personal Communications* 73 (3): 697–715.

Zhang, X., Z. Zheng, Q. Shen, J. Liu, X. Shen, and L. L. Xie. 2014. Optimizing network sustainability and efficiency in green cellular networks. *IEEE Transactions on Wireless Communications* 13 (2): 1129–1139.

Zhou, Y., J. Liu, C. Zhai, and L. Zheng. 2010. Two-transmitter two-receiver cooperative MAC protocol: Cross-layer design and performance analysis. *IEEE Transactions on Vehicular Technology* 59 (8): 4116–4127.

4 Exploring Methods of Authentication for the Internet of Things

Fatemeh Tehranipoor, Nima Karimian, Paul A. Wortman, Asad Haque, Jim Fahrny, and John A. Chandy

CONTENTS

4.1 Introduction 71
4.2 Authentication Taxonomy 72
4.3 Shared Secrets 73
4.4 One-Time Password 74
4.4.1 Time-Based One-Time Password 74
4.4.2 Challenge–Response-Based OTP 75
4.4.3 Out-of-Band Transmission-Based OTP 75
4.4.4 Lockstep-Based OTP 76
4.5 Tokens 76
4.5.1 Software Tokens 77
4.5.1.1 SSL or Certificate Exchange 77
4.5.1.2 Key Exchange 78
4.5.1.3 Third Party 79
4.5.2 Hardware Tokens 80
4.5.2.1 Connected Tokens 81
4.5.2.2 Contactless Tokens 82
4.5.2.3 Disconnected Tokens 82
4.6 Intrinsic Authentication 83
4.6.1 Human Properties 83
4.6.2 Silicon Properties 84
4.7 Behavioral 84
4.7.1 Localization and Metadata 85
4.8 Next-Generation Authentication Techniques 85
4.8.1 Fast IDentity Online 85
4.8.2 CryptoPhoto 86
4.8.3 Blockchain 86
4.9 Conclusions 87
References 88

4.1 INTRODUCTION

The IoT is a large and diverse landscape of embedded systems, ranging from sensors and radio-frequency identification (RFID) tags to larger phones and tablets. Due to the integration of these IoT devices, different mechanisms and implementations are constantly being developed to meet a variety of scenarios at various scales. Applications of these IoT systems include medical, banking, governmental, home use, and infrastructure. These IoT systems are inclusive of physical objects that feature an Internet protocol (IP) as part of an ever-growing network for Internet connectivity. Communication between them and any other Internet-enabled systems is for the purpose of collecting and exchanging data. The expected growth of the IoT is 50 billion smart devices interconnected by 2020, as estimated by Cisco Evans (2011). With the exponential growth and adoption of IoT components, there are fundamental security concerns with the interaction of potentially unsecure devices. The reason for authentication is that IoT devices are often part of critical infrastructure, and one should trust they are talking to the expected IoT component. The problem with this fundamental desire for greater interconnectivity of digital lives is that the expansion of IoT will broaden the potential attack surface for cyber-criminals and hackers. Due to the increasing sophistication of these malicious individuals, new methods of authentication need to be developed in order to establish safe and secure communication or exchange of sensitive data over the IoT.

Devices in IoT environments are connected to the Internet (or other local networks) for intercommunication; therefore, they can be exposed to hacker's attacks. These attacks can manifest as not only a client authenticating to an IoT component but also as an IoT ensuring that the client accessing the device is also verified. Concerns of security of data transmissions can originate from physical aspects of the IoT's design, from protocol-specific implementations of security policies, or even from the initial assumptions of trustworthiness. The majority of security challenges arise from protocol-specific assumptions and the establishment of trustworthiness of communicating devices to authenticate remote users and other embedded devices.

During the past few years, there has been a boon in the development and implementation of a variety of authentication techniques and schemes for IoT systems. Establishing trust is essential for IoT environments for the processing and handling of data in compliance with security and privacy standards. Central to the authentication of IoT devices is the establishment of trust for these elements as they connect to a network. As an unknown user or device attempts to communicate with, or request resources from, various IoT elements, the nature of this exchange must be determined. As the application of IoT devices increases, the need to better secure these systems grows as well. Furthermore, technology improvement also lend to new and creative methods for authenticating and verifying identities. However, each technique brings their own limitations, advantages, and Challenges to the interconnected systems. This chapter provides investigative studies of the existing authentication approaches to present a comprehensive taxonomy based on each approach's framework, implementation, and application. The limitations of each IoT system dictate its security capabilities (e.g., authentication methods and communication protocols) within a network. The advantages and disadvantages of these methodologies are examined with respect to their applicability to IoT in Sections 4.3 through 4.8. Through this chapter, it was found that a handful of the current methods in literature are infeasible to be implemented for IoT authentication due to factors of resource constraints (Wortman et al., 2017), functional limitations, the novelty of new technologies, and overhead cost.

4.2 AUTHENTICATION TAXONOMY

In IoT, one of the key roles of security is safeguarding the integrity and confidentiality of machine-to-machine (M2M), machine-to-human, and machine-to-environment authentication of each device placed in any network. It should be noted that M2M is used to describe any network-enabled technology device used to exchange information and perform tasks without any manual assistance of human beings. In order to establish end-to-end data confidentiality, through encryption, one must first authenticate. Authentication in IoT is required to ensure safe and secure exchange of information, as well as nonmalicious alterations to operational function, as needed. For authenticating one's identity, a user (i.e., client device or application) is required to present some unique identifiable piece of information (i.e., factor) in order to validate it. Single-factor authentication (SFA) is a process whereby a device presents a set of credentials in response to a request by a user to authenticate them. Two-factor authentication (2FA) is a process where a user must present two

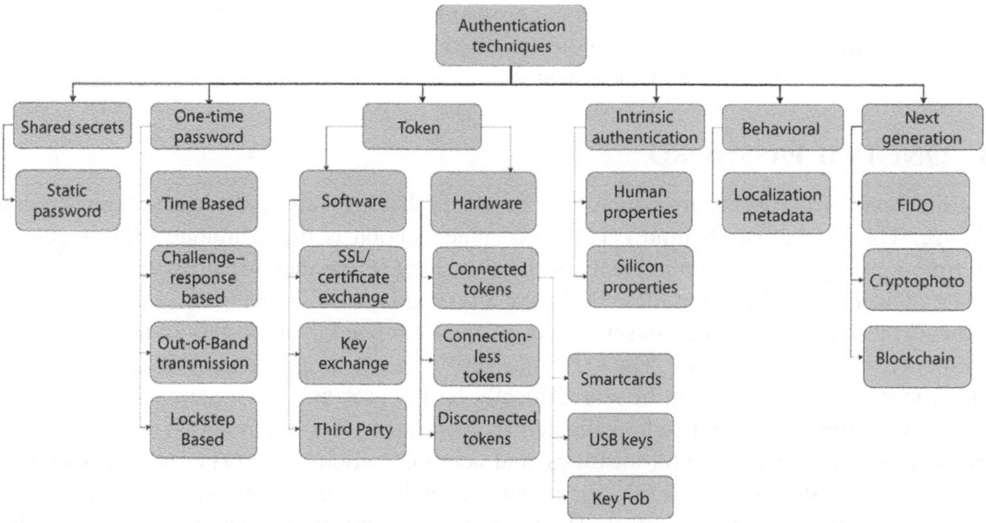

FIGURE 4.1 Taxonomy of authentication techniques in IoT.

forms of identification: a password (something they know) and a generated code (something they have). Multifactor authentication (MFA) is a method of access control whereby a user is authenticated only once they have provided several separate pieces of information to the authenticating server; currently, at least two of these are chosen: something they know, something they have, and something they are. The categories of factor-based authentication techniques lead to subcategories (SFA, 2FA, and MFA) that are too expansive to be meaningful in terms of the technologies applied. The broad categories of factor authentication and their implementation techniques are presented in Figure 4.1. These classifications of authentication technology are the most accommodating for the various implementations in the IoT area. The details of these various techniques and methodologies will be discussed in subsequent sections.

4.3 SHARED SECRETS

Shared secret (static password) is a specific sequence of data that is set once and left unchanged. A static authentication protocol means that two devices exchange a predetermined static password to authenticate each other. This type of authentication does not require any computing, just the comparison of the presented values. As a result, these shared secrets (or hashes thereof) must be stored in local memory on the IoT device, typically accomplished with a small, cheap, simple, nonvolatile storage device (e.g., smartcard). Generally, a user authenticates via an unsecure open channel in an IoT environment. If the static passwords are sent unprotected as clear texts, then an adversary can easily intercept and steal the shared secrets. Other than impersonating a legitimate user, if the secret is used to authenticate other accounts, then an adversary can access the user's private data within other network systems. Furthermore, if the password is protected via encryption, an attacker can use common methods (offline techniques), such as dictionary or brute-force attacks, to crack the unchanged static password. In addition, static passwords have small entropy. Recent work by Barreto et al. in 2015 uses the static password authentication technique in the IoT (Barreto et al., 2015). They proposed an architecture model and use cases for IoT cloud authentication. The model allows for single-sign-on (SSO) authentication and works for single-log-out (SLO) tasks. They found that designing and developing authentication schemes in emerging IoT cloud scenarios is not trivial at all due to the current technological limitations. The advantage is the multiple methods of authentication, depending on the client. But at the same time, this advantage is problematic in case the client chooses an authentication with lots of communication and steps to the process. While static passwords are lightweight solutions to authenticating IoT devices, the vulnerabilities of this method make it inappropriate for an IoT environment.

4.4 ONE-TIME PASSWORD

One-time passwords (OTPs) are temporary passwords that are valid only in one login session or exchange. The mechanism behind OTP-based authentication is that a unique OTP is generated each time a user presents login credentials. This unique OTP is then input by the user to further validate their identity. Note that these OTPs are a second layer of authenticating the user in a 2FA scheme. OTPs can be generated in several ways, and each one has trade-offs in terms of security, convenience, cost, and accuracy. The OTP mechanism can be implemented either through a randomly generated list, which can be stored locally by the user and the system, or on demand by the user every time they want to be validated by the system. These temporary OTPs can protect network access, end users' digital identities, and communications. The OTP-based authentication scheme addresses the limitations of static passwords by incorporating an additional security layer that helps protect against replay attack. However, this authentication solution is still vulnerable to session hijacking due to the possibility of using untrusted terminals. The OTP authentication model is categorized into four different groups, namely, time based, challenge–response based, out-of-band transmission based, and lockstep based. It is worth noting that while OTPs and tokens are often combined into a single solution, tokens require a physical element for storage. Each of these categories has its own advantages and disadvantages in an IoT environment. Further examination of the subcategories is made in the following subsections.

4.4.1 Time-Based One-Time Password

A time-based OTP (TOTP) is an algorithmic extension of the OTP authentication model that is generated through a function of a preset random key, known as the seed, and a current time window. The key is determined at the fabrication of the device, and the time window can be any arbitrary period (every second, minute, etc.). During this window, A older and B newer passwords are valid for this 2FA. The reason for this overlap is to accommodate the time taken by a user to read and enter the OTP. As shown in Figure 4.2, the function of creating a new OTP can be implemented into a (software or hardware) token device that has an internal clock, as can be seen in the patent in Chan et al. (2016). A challenge of this is the time synchronization between the server and a client. Due to this challenge, there is the possibility of the OTP being reused. If an attacker snoops the credentials and attached TOTP during a given time window, the adversary can authenticate as a user during this entire time period. A flavor of TOTP implementation in IoT is done by Shivraj et al. (2015). They considered a two-factor OTP authentication scheme using lightweight identity-based encryption elliptic curve cryptography (IBE-ECC), which does not require key storage. The benefit of their scheme is a robust and scalable OTP scheme using the principles of IBE-ECC. They also demonstrated that their scheme performs on par with the existing OTP schemes without compromising the security level. Unfortunately, the scheme needs more computation time for the calculation of new

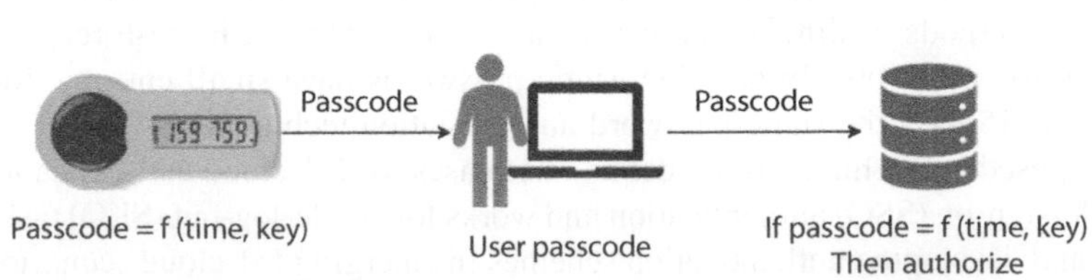

FIGURE 4.2 TOTP authentication.

keys (e.g., constant updating). In the context of TOTPs for large IoT networks, the need for constant time synchronization and communication makes this technique unfavorable.

Note that a user can set a personal identification number (PIN) to unlock the token in order to enter the challenge. Additionally, the token and the authentication server should have synchronized clocks. Unlike static passwords, dynamic passwords are convenient since they do not need to be remembered. This technique has the advantage of supporting a time-restricted authentication window for IoT systems. However, in human interaction scenarios, this time limit may produce frustration in being unable to authenticate. Furthermore, end users can easily download an application on their mobile device to generate a dynamic password, thus making any OTP technique more encouraging for human usage.

4.4.2 Challenge–Response-Based OTP

Challenge–response-based OTP (dynamic password) is the use of a predefined password that does not remain constant and changes after every login (M'Raihi et al., 2011). It is a machine-generated, random string that acts as a second layer of security and is only used once per user login attempt. As can be seen in Figure 4.3, this mechanism for generating unique OTPs is fulfilled by a token device. Each time a user presents their credentials, the authenticating server will send a challenge that the user must input to their token device. The token calculates a response that the user submits as a reply to the server. During this challenge–response exchange between a user and a server, the only data that an adversary can eavesdrop is the challenge and the encrypted result. As noted in the previous subsection, additional layers of security can be added to its IoT environment implementation to make it more favorable for human use. Despite these advantages, this method is expensive in terms of computation and communication in the IoT where resources are limited.

4.4.3 Out-of-Band Transmission-Based OTP

Out-of-band transmission is an approach of the OTP-based authentication technique where an additional level of authentication security (i.e., second-layer authentication) is sent by a server in the form of an e-mail, an SMS (Short Message Service), a call, or a fax to a user's personal device. When a user attempts to authenticate with a server, the system will present the client with an out-of-band transmission OTP, which must be input to finalize the authentication process. Since no seeds are required in this method, the passwords are completely random and impossible to predict. Furthermore, due to the use of different mediums, when an adversary eavesdrops on the original transaction, they will not be able to sniff the OTP being sent. The drawback of this OTP-based

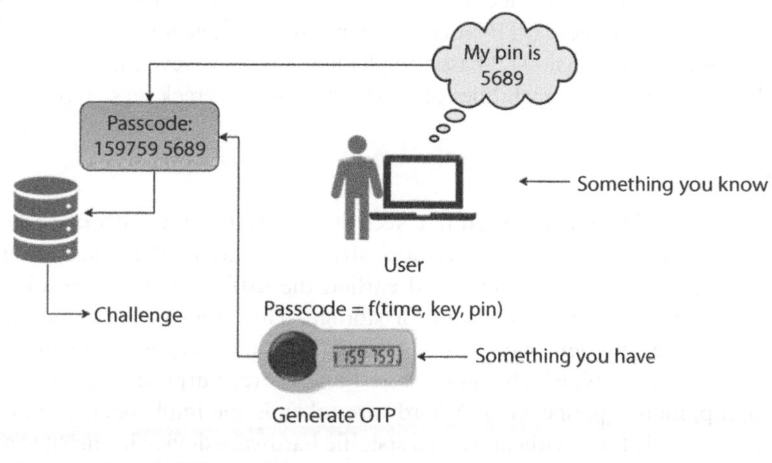

FIGURE 4.3 Dynamic-based OTP authentication.

authentication technique is that if a user loses their personal device (e.g., smartphone), the client will be unable to authenticate themselves until the device is replaced. While this is an effective method for most human authentication purposes, in the context of IoT, the cost of additional communication hardware (e.g., SMS or fax) and data transmission power consumption is a major weakness.

4.4.4 Lockstep-Based OTP

Lockstep-based OTP authentication consists of two internal counters to track the exchange of passwords between a client and a server (Aboba et al., 2004). Each time a new password is generated, a client increments its own internal counter. Once a user authenticates with a given server successfully, then the server increments its own counter. This method allows for the generation of future acceptable passwords where if the two counters become desynchronized, then the server can automatically resynchronize its counter based on the client's password. The benefit of this method is that maintaining synchronous communication is simple. However, an issue related to synchronization is its difficulty between multiple devices. If the synchronization between multiple devices is required, then one must either remember to use the same password on all devices to keep them in sync, or use various counters to track each device. Another issue is that with some devices (e.g., smartphone apps), the counter can be reset to the initial value, causing the server to resynchronize, and reuse previous passwords. If an attacker has been eavesdropping the communication over a long period of time, they can notice the reset of the counters and predict future passwords. The limitation of this method, in terms of using it in IoT devices (e.g., sensors, actuators, and cell phones), is that because these devices are resource constrained, maintaining the necessary synchronization counters is expensive and complicated.

In general, the OTP-based authentication technique is not practical for IoT systems due to the resource requirements for the generation, transmission, tracking, and synchronization of passwords. The IoT systems that could use OTPs are those that implement the production and maintenance of passwords using non-IoT components. One of the concerns with OTPs is that a malicious individual can sniff password communication. Other than that, an overall issue with OTP-based authentication implementations, except out-of-band transmission OTPs, is that they are vulnerable to a reliance on an initial seed. Traditionally, true random number generators (TRNGs) are used to produce a secure version of this initial seed value. Many different TRNG models have been developed in an attempt to better secure this generated output (seeds) (Tehranipoor et al., 2016, 2017a; Eckert et al., 2017). If the authentication server is compromised, the seeds may be accessible, allowing an attacker to easily predict passwords. With OTP authentication for M2M communication, due to the requirement of securely passing an OTP for finalizing authentication, there is overhead from both the hardware and the software standpoint to allow a machine to receive the OTP on a separate communication channel (separate interface). Unless the IoT device in question is already designed to communicate using out-of-band transmissions, none of these OTP-based solutions are favorable for the IoT systems. Initial IoT systems made use of the traditional OTP-based implementations; however, nowadays this technique does not work due to the growing capabilities of cyber-criminals to crack these types of passwords.

4.5 TOKENS

Tokens (security tokens) are generally used in a security environment to generate additional layers of authentication over an existing scheme. Traditionally, tokens are implemented as either software (soft) or hardware (hard) tokens. As mentioned earlier, the difference between tokens and OTPs are that tokens require a physical component for storage. This storage space is used to store any OTPs that are, and will be, implemented by the token device. The purpose of a software token is to generate a single-use login PIN, which can be stored on general-purpose electronic devices (desktop computer, laptop, mobile phone, etc.). A hardware token is the implementation of a token (e.g., public key infrastructure [PKI] certificate) and a specific hardware device in such a way that the two are not separated, thus keeping the stored information secure. The major difference between soft

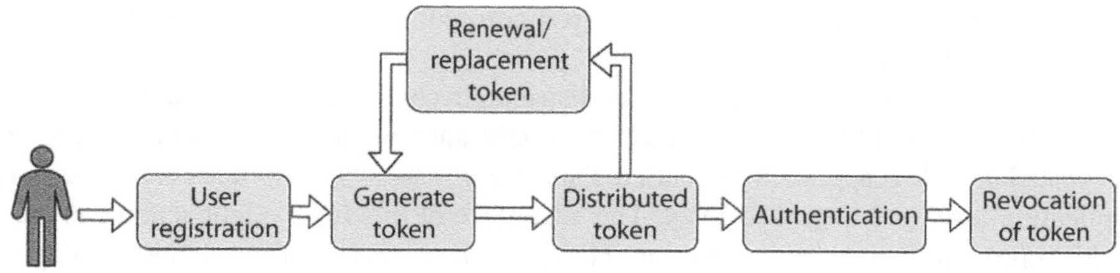

FIGURE 4.4 Token-based authentication scheme.

and hard tokens is that soft tokens can be easily duplicated because there is no dedicated hardware device storing credentials secretly. Furthermore, hard tokens are invulnerable to malicious applications, easily provide 2FA, and are self-contained devices that reduce the impact of human error. However, soft tokens do have the advantage over hard tokens of being flexible and inexpensive. They are ideal for smart device users because there is no physical token that a user must keep track of, and they do not contain batteries that will run out. Despite these advantages, the issues with software tokens are that they are vulnerable to malicious applications, keylogger, and spoofing attacks, and rely heavily on the capabilities of the general-purpose electronic devices. Figure 4.4 demonstrates a generic token-based authentication scheme for authenticating an unknown user. First, a user needs to register an ID and any devices that would use this form of authentication. Once the electronic list of registered devices and matching user IDs has been verified by a server, then these data can be used to generate the necessary authentication token for each registered device. Each token is then distributed to its assigned device (e.g., phone, PDA, or flash drive) and can be used to authenticate within the larger network. If any of the tokens are stolen, a renewal or replacement token is sent to the device, ending use of the compromised token and forcing generation of a replacement. Figure 4.4 illustrates the steps needed to generate tokens, as well as where and how the distributed tokens are renewed, replaced, or revoked.

4.5.1 Software Tokens

In the context of IoT, software tokens can provide a flexible, easy-to-change, and relatively lightweight authentication solution. Unfortunately, due to the resource limitation of these embedded devices, the implementation of software tokens can be detrimental to IoT systems. For example, if a user is accessing sensitive information from a human interactable IoT device (e.g., phone or tablet), each device has its own soft token that it uses to authenticate. Complications can arise based on how these tokens are maintained, verified, and updated.

Since software tokens are so flexible, there have been a variety of ways that they have been implemented, which can be roughly classified into the following subcategories.

4.5.1.1 SSL or Certificate Exchange

Certificate exchange is a method where an electronic document is presented and used to validate the ownership of a given public key. As long as the signature is valid, an individual that is examining the certificate will trust its origin and be confident that they have the correct public key for communicating with the certificate's owner (authentication). To further supplement trust in a presented certificate, the signer can be a third party (i.e., certificate authority [CA]), which is the model used by the Secure Sockets Layer (SSL) and central to the PKI. The advantage of using the certificate exchange authentication method is that a user can feel secure that their communication with a server is not being eavesdropped since the client has validated the server's identity. A secondary benefit to this authentication method is that the issuing of certificates and their maintenance are automated. However, if an attacker installs a fraudulent certificate, it might be taken as legitimate and trusted. Using the certificate exchange authentication method in IoT devices allows for broad and simple establishment of trust. Unfortunately, the required communication has computational needs that add overhead to the IoT systems.

Wireless sensor networks (WSNs) pose a distributed challenge to certificate-based authentication. Porambage et al. (2014) proposed a lightweight authentication and keying mechanism for WSNs in distributed IoT applications (called PauthKey), based on implicit certificates, that provides application-level end-to-end security. The advantage of this lightweight scheme is the mechanisms used to provide mutual authentication. However, the registration phase adds storage cost for obtaining cryptographic credentials. Two implementations of certificate exchange tokens in datagram transport layer security (DTLS) are proposed by the following groups. Kothmayr et al. (2013) presented a two-way mutual authentication scheme using DTLS and the Trusted Platform Module (TPM) to support RSA cryptography. The benefits of this scheme are that it can support unreliable network authentication and use, as well as lowering costs of energy, latency, and memory overhead. Unfortunately, the certificate-based DTLS handshake requires significant overhead for time synchronization and certificate status verification functionality. Following this work, Markmann et al. (2015) presented an identity-based cryptography for improving end-to-end authentication using traditional PKI with DTLS, making use of the Internet protocol version 6 (IPv6) address as the ID for devices. The advantage of this work is the use of IPv6 for communication and identification. However, this technique relies on a "border gateway" to perform authentication (or at least partial authentication) prior to communicating the public key to the IoT device. One approach that arises in the literature is to redesign the framework that implements certificate-based authentication. Initial work focused on the redesign of the security certificate (Wen et al., 2012) but was found to be highly vulnerable to all types of replay attacks, timing attacks, and node capture (Mahmoud et al., 2015). In 2013, Altolini et al. (2013) examined the implementation and performance evaluation of security at the link layer (IEEE 802.15.4) to perform certificate exchange between sensor nodes and end users in order to provide mutual authentication. The advantage of their scheme is less energy consumption and computational overhead, but it is vulnerable to capture attacks. In the following year, Chze and Leong (2014) proposed a secure multihop routing protocol (SMRP), which merges the routing and authentication processes for forming a secured IoT network without incurring significant overhead. Their scheme is not a context-aware protocol and does not result in longer network lifetime, and depending on the implementation of nodes, the memory requirement can be high.

A certificate exchange software token can be used to easily authenticate an IoT device with multiple IoT devices. This allows for easy establishment of trust for a source of information. However, this is not as effective at creating a secure line of communication between two IoT systems.

4.5.1.2 Key Exchange

Key exchange is a subcategory of software token authentication where two parties exchange cryptographic keys in order to establish secure communication between the sender and receiver. To further ensure that keys are exchanging correctly between two expected clients, only the public key is transmitted. This way a user can send a message that can only be decrypted by the owner of the private key. This property of asymmetric key ciphers is employed to provide authentication. Key exchange can enable a secure communication channel in the presence of an unsecure medium. But an adversary can produce their own public–private key pairs to cause a man-in-the-middle (MiTM) attack. The key exchange authentication method is not a good idea for IoT systems due to the need for generating, exchanging, and replacing public and private keys to establish safe communication between devices.

Development for improved key exchange over the radio-frequency (RF) medium was proposed by Liao and Hsiao (2014). They presented a secure ECC-based RFID authentication scheme integrated with an ID-verifier transfer protocol. The benefit of this work is the combination of ECC (using small key sizes and efficient computation) with RFID technology. Unfortunately, the big issue here is the tag's computation time and the memory requirement of the scheme, which lead to a lack of efficient performance. Furthermore, as pointed out by Peeters and Hermans (2013), this protocol scheme is highly vulnerable to misinformation attacks, such as tag masquerading, server spoofing attack, location tracking attack, and tag cloning attack. The development of anonymous authentication and communication techniques has led to a variety of potential solutions. In 2015, work by Gope and Hwang (2015) and Khemissa and Tandjaoui (2015) centralized around the use of a gateway server to collect and track identification information of IoT devices joining into various networks. The work by Gope and Hwang (2015) centered on a three-phase anonymous authentication scheme where home servers (i.e., gateways) are used to maintain the registration information of devices. The advantage of this scheme is to allow for anonymous movement of sensors with respect to eavesdropping by an adversary. However, it has a heavy reliance on IoT server head nodes (e.g., gateways) for storing information and performing all authentication tasks. The main contribution of the work by Khemissa and Tandjaoui (2015) is the use of hash-based message authentication code (HMAC) computation to identify nodes without sending their identity. This method has the advantage of less energy consumption through lightweight analysis. However, the energy cost of the proposed scheme is slightly higher when interacting with remote users. In 2015, Devi et al. (2015) introduced a mutual authentication scheme with two approaches (local and remote) that is based on the physical address of the user attempting to access or interact with some given nodes. The benefit of this work is that it is applicable for home automation. Locally, since physical addresses are not expected to change, providing authentication at the physical level takes less time.

In the past half decade, several approaches have tackled the limitations of authentication in IoT through the proposal of new schemes and techniques, which build on existing frameworks. Zhao et al. (2011) presented an asymmetric mutual authentication scheme that is achieved between the terminal node and platform using secret key crypto (SKC) systems. This scheme has the benefit of low computation and memory resource usage, but from the standpoint of SKC, the issues of key management and maximum storage space do not work well with this lightweight solution. Bonetto et al. (2012) proposed a lightweight method based on the offloading of computationally intensive tasks to a trusted and unconstrained node. The advantage of this technique is that the nodes will have a longer average lifetime due to less time performing computations because they are offloaded to the gateway. Unfortunately, reliance on the gateway node is the disadvantage of this work.

Jan et al. (2014) proposed a lightweight mutual authentication scheme for validating identities of participating devices before engaging them in communication for resource observation. Their model has the advantage of less connection overhead and prevents an attack from registering multiple times with a given server. But an adversary may disrupt ongoing operations in one or more clusters by constantly emitting jamming signals or launching a denial of service (DoS) attack.

In 2015, Pawlowski et al. (2015) presented a combination of Extensible Authentication Protocol (EAP) and Slim Extensible Authentication Protocol over Local Area Networks (SEAPOL) to the Trust Extension Protocol for authentication of new deployed objects and sensors. This work achieves significant network resource usage reduction through an authentication schema that reduces communication with the device manufacturer. This protocol can provide a 42% reduction in the number of transferred packets and a 35% reduction in transferred data.

The use of key exchange does create a more secure method of establishing encrypted communication between two IoT devices (e.g., PDA and laptop). This can be helpful in situations where protection of the bits being exchanged is of the utmost importance. However, this technique also has a much larger resource requirement and organizational overhead due to encryption requirements.

4.5.1.3 Third Party

Third-party authentication is a methodology whereby the communication between a client and a server has been concurrently validated by a trusted entity. The major addition is the requirement of this third-party service to verify the identity of both communicating parties. The necessity of this trust is to ensure and review all critical transaction communications between two parties. The trust for login (authentication) is established through e-mail (sending a link), phone (sending an SMS with a link or code), or social media (Facebook, Google+, etc.). This authentication method has the advantage of centralizing trust via a third party. This third party provides a token that validates the authentication process. However, it is inconvenient to a user to have to retrieve the links or codes. In the context of IoT, third-party authentication causes several problems. First, the amount of resources that are required is high since a third party needs to review all communications between IoT elements. Second, the communication protocol can be expensive due to the exchange of login links and codes.

The majority of work toward improving third-party authentication is centered on specific implementations of various schemes. In 2012, Liu et al. (2012) suggested a simple and efficient key establishment based on ECC and adopted a role-based access control (RBAC)–based authorization method using the thing's particular roles and applications. The benefit of this scheme is that it can secure mutual authentication using third-party authentication. But it is a complicated authentication process that requires too much communication to authenticate. In 2013, Alcaide et al. (2013) suggested a fully decentralized anonymous authentication protocol. Their proposal is based on a credential system that defines two roles for the participant nodes (users and data collectors). The advantage is that they implement privacy and access control, which is enhanced through third-party privacy protection. But there is a distinct lack of application scenarios and verification systems for an anonymous authentication protocol based on a third party. Pawlowski et al. (2014), in 2014, proposed a refinement of the Extensible Authentication Protocol over Local Area Networks (EAPOL) that slims down this authentication method by redefining how the EAPOL headers are used within the protocol. Through this work, they were able to show that their approach gives around 9% of memory savings in comparison with the original EAPOL implementation.

Some developments focused on the adaptation of different communication technologies to improve authentication. Yang et al. (2013) at first proposed a two-way authentication protocol for RFID systems and then later (Yang et al., 2013) introduced an authentication protocol based on hash function that uses third-party mutual authentications among tag, reader, and back-end databases. The advantage is that this allows IoT devices to exchange data between two objects. The problem is complicated and "chatty" communication that is not suitable for low-cost RFID systems.

The third-party authentication approach allows for the most secure method of communication between two IoT elements, at the assumption of a trusted third party that validates the identity of each IoT device. The trade-off is the overhead required to have concurrent verification of identity while also expending the resources to maintain this organizational backbone (e.g., third-party authentication server). However, this technique can be effective in a mixed IoT and nonconstrained device network when limitation of resources is not a concern. An example of this mixed IoT network would be authenticating Google e-mail access on a new IoT device (human interactable device, e.g., tablet or smartphone). In this case, Google maintains a database server backbone for verifying user identities. Since a user trusts Google as an authority, this technique is effective for authenticating across various IoT devices.

4.5.2 Hardware Tokens

Generally, in a hardware token–based authentication, a user enters a passcode into the token device, which then fabricates a response. Later, the user presents this generated result to finalize authentication. For IoT devices, hardware tokens prevent eavesdropping and replay attacks by malicious individuals. But the cost of designing and using a specialized hardware device for hardware token is high. Moreover, these hardware tokens are susceptible to being stolen and difficult to replace. The difference between OTPs with tokens and hardware tokens is that hardware tokens are OTPs with dedicated hardware. Hardware tokens can be an effective solution for IoT devices that move around from one environment to another, such as sensors and smartphones. The dedicated hardware has a specific resource cost in comparison with software token resource expenditure that can fluctuate depending on computational needs. However, the development of specialized hardware has its own overhead, as well as specialized resource requirements that may not meet the resource-constrained IoT ecosystem. In general, IoT devices do not authenticate through any form of physical contact.

4.5.2.1 Connected Tokens

Connected tokens transmit their authentication information once a physical connection has been established. This eliminates the need to manually enter authentication data, thus minimizing human error during this process. The most commons types of connected tokens are smartcards, USB keys, and key fob. In a scenario where one must physically authenticate with an IoT device without any manual assistance (e.g., sensors), these connected tokens are favorable. However, this does require the development of specialized hardware that can add unacceptable additional overhead to the system.

4.5.2.1.1 Smartcards

A smartcard is a lightweight, tamperproof computer that has an embedded microprocessor and non-volatile storage. These smartcards contain software tokens that are presented to log in and authenticate. This method is commonly implemented as a strong security authentication for SSO to shorten authentication time and decrease the chance of an adversary snooping the tokens. Note that SSO is a mechanism where a user logs in once and is authenticated to access different resources and applications without requiring further authentication. Using smartcards for authenticating IoT devices like tablets, wearables have the advantage of being inexpensive and fast without any human interaction (error). Similar to the disadvantage of general hardware tokens, if smartcards are stolen, then tokens will need to be changed and hardware should be replaced.

In 2014, Turkanović et al. (2014) proposed a hash function–based authentication and key agreement protocol for heterogeneous ad hoc WSNs for achieving energy efficiency, user anonymity, mutual authentication, and other security requirements. While this work was able to perform mutual authentication, it was left vulnerable to a large number of attacks (e.g., node capture, node spoofing, stolen smartcard, and offline password attack). The following year, Crossman and Liu (2015) studied using smartcards and physical unclonable function (PUF) for creating 2FA. While they have a reproducible test bed, their model may be vulnerable to attack of the stolen smartcards.

This technique of IoT authentication is good for situations where distributed IoT devices must act, and react, in a specific manner. IoT environments that would use smartcards are remote sensor networks (e.g., underwater or outer space). In these scenarios, the IoT elements must operate for prolonged periods of time without human interaction. The advantage of using smartcards for authentication purposes is that the IoT networks can function autonomously. However, the complication of this approach is that if the smartcard is compromised, hardware must be replaced, requiring a physical examination and troubleshooting of the device.

4.5.2.1.2 USB Keys

USB keys MeiHong and JiQiang (2009) are another type of connected hardware token where the software token is embedded into a USB device. USB keys are useful for being simple and a form of passwordless authentication method. If a USB key is lost, a user can fall back on a password-based authentication technique. Moreover, these USB keys' security codes can be regularly changed, thus decreasing the effect of duplicated keys. The advantage of using this technique in IoT is having passwordless authentication, while the disadvantage is the additional cost of hardware required for this solution. The drawback is that they are easily misplaced, left behind, or lost. In addition, they are susceptible to malware infections.

This technique would work for human interactable IoT devices where a user or technician must interface with them, passing a private key for authenticating themselves. The disadvantage of this scenario is that the IoT device will need the computational capability to verify the presented authentication. This introduces additional overhead.

4.5.2.1.3 Key Fob Tokens

A key fob is a small, programmable hardware device that provides access (authentication) to a physical object. In addition, these key fobs can be used to make 2FA and MFA. This type of token can act as a user's master key, which does not require a user to enter a PIN. As with any other hardware tokens, these key fobs might be lost, and will require replacement. In IoT applications, the use of key fob tokens is impractical since traditional IoT devices need to be robust and do not rely on specific areas of physical contact to authenticate. A simple implementation of using key fob–based authentication for IoT devices is for smart door locks. In this scenario, each individual has their own key fob that can be used to access privileged areas. This works for this specialized scenario because a user is interacting with a physical barrier. However, this technique does not adapt well to authenticating an IoT device. While physical contact may work in terms of a USB plug, having a specialized surface on the exterior of the IoT is extremely rare and not traditionally implemented because this technique is used for authenticating a human into an IoT network. This method does not work for M2M authentication purposes.

4.5.2.2 Contactless Tokens

Contactless tokens are tokens that are not passed over a physical connection but a logical one, for example, wireless communication. This lack of physical connection makes these tokens more convenient to use with connected and disconnected IoT devices. Traditionally, contactless tokens use a wireless communication medium (e.g., near-field communication [NFC], RF, and Bluetooth). The advantage of the contactless token authentication method is that it uses mediums that are more difficult to eavesdrop communication exchanged, and these techniques are highly convenient for the authentication of individual devices. Nonetheless, this technology is vulnerable to jamming attacks and injection attacks, and the longevity of these token devices is relatively short. In IoT authentication, contactless techniques are favored because they do not require a physical connection, making them more convenient than other types of tokens. However, the battery life of these tokens is relatively short.

In 2015, Lee (2015) focused on the use of NFC technology supporting peer-to-peer (P2P) transmission of user information and metadata through the implementation of authentication from a mobile phone to an IoT device. The benefit is the short operational range (less than or equal to 20 cm) that can be applied to various types of applications. But it is very tough to use for distances larger than 20 cm.

This type of authentication technique is useful in scenarios were IoT devices are acting in solitude and do not require much interaction with other devices. Example applications include medical IoT devices (e.g., pacemaker and insulin pump), electronic payment methods (e.g., contactless payment applications), and wearable devices (e.g., Fitbit). The disadvantage of contactless token–based authentication is that the communication mediums require either very short distances or specific processing. This introduces additional overhead and costs.

4.5.2.3 Disconnected Tokens

Disconnected tokens (Mercredi et al., 2007) are another type of hardware tokens that do not need any special input device, nor do they require a physical connection to the system that they are authenticating with. These tokens are the most common type of security token used since they have a built-in display for the generated data that the user manually enters to complete authentications. Disconnected token-based authentication methods are feasible solutions for remote 2FA in IoT systems. Nonetheless, this technology is focused on human use, thus making it useful only for human-to-machine communication.

These types of tokens can be used for any IoT scenario where a technician or user authenticates by presenting a generated key. This is an effective method of adding an additional layer of security to the authentication process; however, should the disconnected token be lost, then the hardware will require replacement. Unless working in a scenario where IoT devices need to be only physically accessed, such as satellites and deep sea sensors, there is not much of an advantage to this methodology.

4.6 INTRINSIC AUTHENTICATION

Intrinsic authentication is a methodology where a unique property of a person or IoT device is used to authenticate. In each case, this property comes from a trait that can be used to tell one client apart from another. This technique of IoT authentication is traditionally a simple and straightforward implementation that allows for authentication from a variety of users. This technique often requires the establishment of a back-end server framework to maintain usability.

4.6.1 Human Properties

A human property–based authentication (biometrics) refers to the automated recognition of individuals based on their biological and behavioral characteristics (e.g., fingerprint, iris, electrocardiogram [ECG], and face) that can be presented to an electronic system as a means of confirming a user's identity. Compared with other authentication approaches, human properties are more conclusive and cannot be guessed or stolen as easily. Moreover, they are often much simpler for users to use rather than remembering a password or inputting 2FA data into a device. As can be seen in Figure 4.5, a typical biometric authentication system has two stages of operation, namely, the enrollment phase and the authorization phase. In the enrollment phase, the biometric system acquires the human characteristic property of an individual, extracts a highlighted feature set from it, and stores the extracted feature set in a template, along with an identifier associating the feature set with an individual. During the authorization phase, the system once again acquires the biometric property of an individual, extracts a feature set from it, and compares this feature set against the templates in the database in order to determine a match or verify a claimed identity. User authentication via biometrics makes sure that it has higher security than before. Compared with the static password, biometric features are difficult to duplicate, distribute, forge, and destroy.

To consider which kind of human properties are suitable for user authentication in an IoT system, each biometric modality has its own advantages and disadvantages. For instance, fingerprints are widely accepted by the public; however, systems that use them are easy to attack by impostors since the attacker can easily collect fingerprints that the user may have left behind. For iris scans, although their accuracy is higher than that of other biometrics so far, they are not widely accepted on a mass scale because of cost and the strict requirements of the iris extractor. In 2016, Karimian et al. (2016, 2017a,b) proposed an Internet of biometric things (IoBT) scheme focusing on generating a session key from ECGs for implementation of IoT systems, specifically Kwikset Kevo door lock hardware. They have analyzed their solution in terms of security consideration, including the reliability of keys, entropy, and randomness. Their results indicate that the ECG is one of the easiest and convenient human properties that can be used for a robust key generation–based authentication technique in IoT systems.

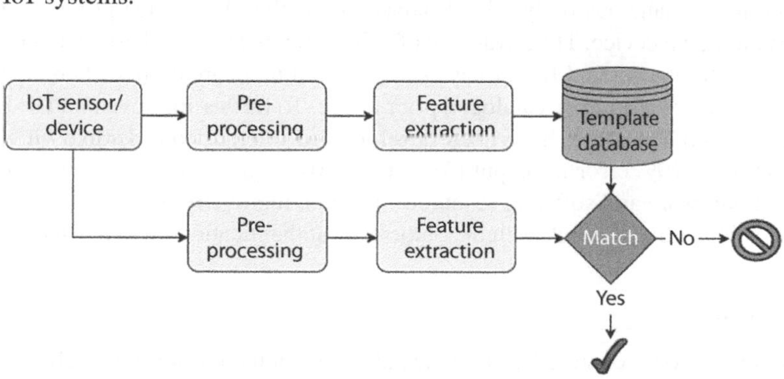

FIGURE 4.5 General human property–based authentication scheme.

For implementing a biometric or e-healthcare IoT network, this method of using human characteristics is advantageous for authenticating clients. As always, there is an overhead cost for implementing the necessary back-end server framework. Despite this additional cost, this authentication technique is cheaper to maintain than it is to implement. This technique is a lightweight solution for authenticating users in an IoT network.

4.6.2 Silicon Properties

Silicon property–based authentication (known as PUF) is the implementation of a silicon chip that uses the intrinsic device randomness caused by chip manufacturing process variations (MPVs) to generate a device-unique response for the purpose of authentication (Gassend et al., 2002). Due to their physical nature, PUF responses are traditionally unique and not perfectly reproducible (noisy), as well as not truly random. The silicon property–based authentication technique has the advantage of being reliable and is a secure way to verify PUF-based chips. Depending on the implementation of PUF design (e.g., intrinsic memory), a user can further benefit from a lightweight and low-cost solution since the existing memories, such as DRAM (Tehranipoor et al., 2015, 2017b; Anagnostopoulos et al., 2017), SRAM (Maes et al., 2009), and flash memory (Prabhu et al., 2011), can be incorporated in PUF-based authentication. For IoT devices, silicon property–based authentication is a strong technique because unlike traditional security and authentication solutions, it does not require storing secret keys; instead, it dynamically generates unique and volatile secrets for each integrated circuit (IC) for authentication purposes. Note that silicon property–based authentication is feasible for M2M authentication purposes. This feasibility comes from the fact that silicon-based authentication can be automated, removing the need for human interaction from the authentication process.

Work to improve silicon property–based authentication focuses on the use of unique properties for identification. In 2014, Cherkaoui et al. (2014) introduced a new authorization scheme for constrained resource servers, taking advantage of PUFs and embedded subscriber identity module (eSIM) features. PUF provides an inexpensive, safe, tamper-resistant key to verify constrained M2M equipment, and eSIM provides scalable mobile connectivity management, interoperability, and compliance with security protocols. Their method has the advantage of adapting current low-cost hardware solutions to current techniques and technologies to tackle problems, instead of creating a new model. Shone et al. (2015) proposed digital memories based on a two-factor user authentication mechanism for mobile devices. The advantage of their work is that the nature of digital memories is unique and unpredictable. However, the amount of data the system would have to analyze is large and unique per user. In 2016, Sharaf-Dabbagh and Saad (2016) suggested the use of device fingerprinting techniques, along with transfer learning, to effectively detect emulation attacks. In this scheme, a fingerprint allows for one to confirm that messages received by cloud are messages sent from the object itself and not a malicious party. The disadvantage is that these fingerprints are useless when dealing with an unknown device. The advantage of using silicon property–based authentication is that there is no need to store generated IDs or keys since each of them are produced on the fly. However, challenge–response pairs need to be cataloged prior to use. To further improve on PUF-based authentication, there are techniques that have been developed to tolerate considerable bit errors in PUF output, approximately 10% error in output (Yan et al., 2015, 2017). This error-correcting approach saves significant hardware and software resources, minimizing overhead for silicon property–based authentication. This method is a lightweight solution for authentication of an IoT hardware platform.

4.7 BEHAVIORAL

Behavioral authentication is centered around the habits and behavior of users or client devices (IoT). This method takes longer to implement because there is a need for knowing what regular behavior of data should look like in order to establish authentication patterns. Once this pattern is determined, it is relatively easy to track dissimilar patterns from an authenticating client.

4.7.1 Localization and Metadata

As a way to supplement traditional security authentication schemes, researchers have adopted the use of localization information and device metadata to further establish honest and legitimate authentication with a high degree of trustworthiness. In this authentication method, a user provides additional identifiable information, along with their reliable security credentials. This additional information (e.g., location information or unique device properties) is used to increase assurance that the device that has been authenticated is not being spoofed. The use of localization or metadata-based authentication is favorable in an IoT environment since each component in the IoT network should have its own traceable properties. The downside of this technique is the additional overhead in communication and computation for the authentication process.

Implementations of this method of authentication are wide and varied. From 2012 to 2013, Mahalle et al. (2012, 2013), presented the identity establishment and capability-based access control (IECAC) protocol using ECC, which protects against MiTM, replay, and DoS attacks. The advantage of their work is that it needs less memory occupation and strong resist ability against attack, but it needs more power consumption. In 2014, Shafagh and Hithnawi (2014) proposed a supplementary technique to public key cryptography to rely on ambient radio signals to infer proximity within about 1 s, and in its ability to expose impostors located several meters away. The advantage of the method is that it can be used for relatively narrow-range authentication. However, this narrow range can be problematic when performing room-level proximity detection and authentication. In 2016, Lee and Jeong (2016) proposed a scheme based on join probability in the IoT environment, which allows safe sharing of user information for users of various IoT services. This scheme improves security by assigning random variables to critical information of the IoT devices (e.g., temperature sensors, smartphones, and drones) in the information transmission and reception process. The advantage is that random variables assigned important information of IoT devices provide more convenient information exchange between people and objects. Unfortunately, they have a complex design.

This technique is advantageous for IoT systems that physically move around. Since localization and metadata-based authentication allows for tracking of behavior, this will help identify counterfeit IoT devices versus legitimate ones. Unfortunately, the requirement for behavioral log data means that additional time and resources are required to prepare implementation of this form of authentication. In a situation where an IoT device is constantly changing networks, this method could be favorable. However, this scenario rarely occurs in current IoT implementations.

4.8 NEXT-GENERATION AUTHENTICATION TECHNIQUES

Next-generation authentication techniques attempt to merge the benefits of other authentication methods while mitigating their limitations. Traditionally, this type of method supports MFA, allowing for use of a large range of authentication technologies. Next-generation authentication frameworks such as Fast IDentity Online (FIDO) (USB-based hardware MFA), CryptoPhoto (out-of-band smartphone MFA), and blockchain (Bitcoin or recording events) provide strong mutual authentication methods that are suitable for IoT devices.

4.8.1 Fast IDentity Online

FIDO aims to combine new authentication technologies in including human properties (biometric) with existing solutions and communication standards such as TPM and tokens (hardware and software tokens). As shown in Figure 4.6, FIDO has two simple steps for authentication. First, a user needs to enter their username and password in the login field of any application that supports FIDO universal two-factor (U2F) authentication. Second, they need to insert their security key in a USB port with the metallic side up; touching the metallic button on the security key can generate secure

FIGURE 4.6 FIDO two-step authentication process.

login credentials for each person. In its current implementation, it is focused on human authentication, which would be useful for remote authentication into IoT networks.

In the context of IoT, using the FIDO technique for authentication has the advantage of being able to combine hardware- and software-based authentication methods into a U2F authentication scheme. Therefore, this leads to FIDO's capability to use many forms of authentication techniques to produce a single U2F. The downside of this method is the requirement of a physical form of identification. This would cause a problem in a scenario where an IoT device is isolated and there is no physical access to it. Furthermore, if all authentication is done in software, then the FIDO technique is vulnerable to software-based attacks. Overall, this form of authentication is ideal for an IoT environment since it can incorporate any two methods of authentication together.

4.8.2 CryptoPhoto

CryptoPhoto is a 2FA framework with two-channel mutual authentication. It works by showing a user a random photo retrieved from their token device (physical card, smartphone app, etc.). This authentication technique allows a client to select photos that are unique to their token, after which a one-time authentication code is sent to complete the process. By using the CryptoPhoto authentication method, one can minimize the chance of fake or malicious data (phishing, social engineering, hijacking, MiTM, etc.) being presented to the user, and it also blocks against snooping attacks (keyloggers, viruses, Trojans, malware, etc.). While this is another human-centric authentication scheme, the current application is infeasible for IoT devices because of the resource requirements for performing image processing.

This technique does not work for an IoT environment because it would require an IoT device to process images, which is expensive. In addition, it needs special hardware to perform this task for the authentication process. Due to this limitation, the CryptoPhoto technique is an infeasible method for IoT authentication.

4.8.3 Blockchain

Blockchain is a passwordless authentication technique for M2M authentication purposes and the tracking of past operations through the use of a ledger (Herbert and Litchfield, 2015). A blockchain is a continuously growing list of ordered records (i.e., blocks) that each contain a time stamp and a link to a previous block in the chain. Each block can contain digital fingerprints, signatures, hashes of sensitive information, or a ledger for public transactions. Blockchains are traditionally implemented as a distributed database that is inherently resistant to the modification of its own data once it has been added to the blockchain. Blockchain authentication does not require third-party identity providers, because the blockchain is the directory of identities. An advantage of using blockchain-based authentication is that there are no human interactions involved since all calculations of the blockchain are independent from the user. Furthermore, it is incredibility difficult to alter previous or current transactions in a blockchain due to the heavy computational requirement for each block in the chain. Blockchain technology offers provenance in complex supply chains and authentication auditing; however, the resulting computational requirements alone are too high for even attempting this solution in a purely IoT network. Blockchain authentication can be implemented within a network of mixed IoT and non-IoT devices, such as tablets and data servers. The non-IoT elements must perform the blockchain computations because they have the resources to do so. Ideally, in order to establish M2M authentication, the IoT devices must also compute these blockchain calculations. However, since IoT systems are highly resource limited, they would not be able to perform this task. Therefore, the standard blockchain-based authentication is not feasible for IoT environment implementation because the requirements for maintaining a ledger, performing the verification computations, and using PKI are extremely resource-intensive. One attempt is to involve blockchain techniques with other known authentication practices (Guardtime, 2016). Research is currently being done to obtain this goal without intentionally weakening the blockchain ledger and verification process (Kolias et al., 2016). There has also been some work to prototype an IoT blockchain solution, but this work is inconclusive and has no results (Huh et al., 2017). In order to use blockchain techniques in IoT, one must alter the traditional blockchain method to require fewer resources for its operation.

4.9 CONCLUSIONS

This chapter reviewed several authentication approaches and their viability in an IoT environment. It was shown that the main limitations of implementing an authentication scheme in an IoT framework center around communication complexity, power distribution and consumption, computational requirements, memory storage, overhead (cost and development), and the amount of operational resources required to function. It is worth mentioning that while this chapter reviews a broad spectrum of authentication schemes, not all are feasible for IoT devices. The reason for this is that the aforementioned limitations greatly impact the implementation of these authentication frameworks. Furthermore, the scenario in which these IoT devices are used also dictates the effectiveness of each method. Due to the inherent resource-constrained nature of these embedded systems, one should favor lightweight, low-cost, reliable, and secure authentication schemes for IoT-based networks and systems.

It is believed that the most capable and beneficial authentication technologies that can meet these needs are human property–based authentication (biometrics), silicon property–based authentication (PUF), and next-generation-based authentication (FIDO). These recommended authentication solutions have the advantages of being easy to use and effective at preventing malicious attacks, minimizing the impact of human error, and being unique based on the nature of their approach. Some common-day scenarios that would benefit from this form of authentication include remote video cameras, car electronics, and home security systems. For remote video cameras, the client accessing the IoT device could be validated through the use of human property–based characteristic. This would ensure that only authorized individuals could obtain the sensitive video stream. The silicon property–based authentication technique would be ideal for overcoming authentication problems within a car's electronic systems. PUFs can be used to authenticate between hardware components minimizing malicious actions and harmful behavior. In the case of home security systems, because there are a large variety of IoT components working together, a 2FA or MFA method is preferred. Since FIDO is a merging of different authentication techniques, it would be uniquely suitable for tackling the integration of multiple IoT devices (e.g., cameras, webcams, and motion sensors) working in unison. As IoT technology grows and evolves, new approaches of implementing better authentication will continue to emerge.

REFERENCES

Aboba, B., L. Blunk, J. Vollbrecht, J. Carlson, and H. Levkowetz. 2004. Extensible authentication protocol (EAP) (No. RFC 3748). https://www.rfc-editor.org/rfc/rfc3748.txt.

Anagnostopoulos, N. A., A. Schaller, Y. Fan, W. Xiong, F. Tehranipoor, T. Arul, .. and S. Katzenbeisser. 2017. Insights into the Potential Usage of the Initial Values of DRAM Arrays of Commercial Off-the-Shelf Devices for Security Applications. 26 th Crypto-D.

Alcaide, A., E. Palomar, J. Montero-Castillo, and A. Ribagorda. 2013. Anonymous authentication for privacy-preserving IoT target-driven applications. *Computers and Security* 37: 111–123.

Altolini, D., V. Lakkundi, N. Bui, C. Tapparello, and M. Rossi. 2013. Low power link layer security for IoT: Implementation and performance analysis. In 2013 *9th International Wireless Communications and Mobile Computing Conference (IWCMC)*, Sardinia, Italy, July, 919–925.

Barreto, L., A. Celesti, M. Villari, M. Fazio, and A. Puliafito. 2015. An authentication model for IoT clouds. In *Proceedings of the 2015 IEEE/ACM International Conference on Advances in Social Networks Analysis* and *Mining 2015*, New York, August, 1032–1035.

Bonetto, R., N. Bui, V. Lakkundi, A. Olivereau, A. Serbanati, and M. Rossi. 2012. Secure communication for smart IoT objects: Protocol stacks, use cases and practical examples. In *2012 IEEE International Symposium on a World of Wireless, Mobile and Multimedia Networks (WoWMoM)*, San Francisco, CA, June, 1–7.

Chan, Y. L., M. D. Essenmacher, D. B. Lection, E. L. Masselle, and M. A. Scott. 2016. User authentication security system. U.S. Patent No. 20,160,019,382, January 21.

Cherkaoui, A., L. Bossuet, L. Seitz, G. Selander, and R. Borgaonkar. 2014. New paradigms for access control in constrained environments. In *2014 9th International Symposium on Reconfigurable and Communication-Centric Systems-on-Chip (ReCoSoC)*, Montpellier, France, May, 1–4.

Chze, P. L. R. and K. S. Leong. 2014. A secure multi-hop routing for IoT communication. *Presented at 2014 IEEE World Forum on Internet of Things (WF-IoT)*, Seoul, South Korea.

Crossman, M. A. and H. Liu. 2015. Study of authentication with IoT testbed. In *2015 IEEE International Symposium on Technologies for Homeland Security (HST)*, Waltham, MA, April, 1–7.

Devi, G. U., E. V. Balan, M. K. Priyan, and C. Gokulnath. 2015. Mutual authentication scheme for IoT application. *Indian Journal of Science and Technology*, 8 (26).

Eckert, C., F. Tehranipoor, and J. Chandy. 2017. DRNG: DRAM-based Random Number Generation using its Startup Value Behavior, *60th IEEE International Midwest Symposium on Circuits and Systems*, Boston, MA, August 2017.

Evans, D. 2011. The Internet of things: How the next evolution of Internet is changing everything. Cisco White Paper.

Gassend, B., D. Clarke, M. Van Dijk, and S. Devadas. 2002. Silicon physical random functions. In *Proceedings of the 9th ACM Conference on Computer and Communications Security*, Washington, DC, November, 148–160.

Gope, P. and T. Hwang. 2015. Untraceable sensor movement in distributed IoT infrastructure. *IEEE Sensors Journal* 15 (9): 5340–5348.

Guardtime. 2016. Internet of Things authentication: A blockchain solution using SRAM physical unclonable function. Irvine, CA: Guardtime.

Herbert, J. and A. Litchfield. 2015. A novel method for decentralised peer-to-peer software license validation using cryptocurrency blockchain technology. In *Proceedings of the 38th Australasian Computer Science Conference (ACSC 2015)*, January, vol. 27, 30.

Huh, S., S. Cho, and S. Kim. 2017. Managing IoT devices using blockchain platform. In *2017 19th International Conference on Advanced Communication Technology (ICACT)*, Bongpyeong, South Korea, February, 464–467.

Jan, M. A., P. Nanda, X. He, Z. Tan, and R. P. Liu. 2014. A robust authentication scheme for observing resources in the Internet of things environment. In *2014 IEEE 13th International Conference on Trust, Security and Privacy in Computing and Communications (TrustCom)*, Beijing, China, September, 205–211.

Karimian, N., Z. Guo, M. Tehranipoor, and D. Forte. 2017a. Highly reliable key generation from electrocardiogram (ECG). *IEEE Transactions on Biomedical Engineering* 64 (6): 1400–1411.

Karimian, N., F. Tehranipoor, Z. Guo, M. Tehranipoor, and D. Forte. 2017b. Noise assessment framework for optimizing ECG key generation. In *2017 IEEE International Symposium on Technologies for Homeland Security (HST)*, Waltham, MA, April 1–6.

Karimian, N., P. A. Wortman, and F. Tehranipoor. 2016. Evolving authentication design considerations for the Internet of biometric things (IoBT). In *2016 International Conference on Hardware/Software Codesign and System Synthesis (CODES+ ISSS)*, Pittsburgh, PA. 1–10.

Khemissa, H. and D. Tandjaoui. 2015. A lightweight authentication scheme for e-health applications in the context of Internet of things. In *2015 9th International Conference on Next Generation Mobile Applications, Services and Technologies*, Cambridge, UK, September, 90–95.

Kolias, K., A. Stavrou, I. Bojanova, J. Voas, and T. Grance. 2016. *Leveraging Blockchain-Based Protocols in IoT Systems*. Gaithersburg, MD: National Institute of Standards and Technology.

Kothmayr, T., C. Schmitt, W. Hu, M. Brnig, and G. Carle. 2013. DTLS based security and two-way authentication for the Internet of Things. *Ad Hoc Networks* 11 (8): 2710–2723.

Lee, B. M. 2015. Authorization protocol using a NFC P2P mode between IoT device and mobile phone. *Advanced Science and Technology Letter* 94: 85–88.

Lee, S. H. and Y. S. Jeong. 2016. Information authentication selection scheme of IoT devices using conditional probability. *Indian Journal of Science and Technology* 9 (24): 1–7.

Liao, Y.-P. and C.-M. Hsiao. 2014. A secure ECC-based RFID authentication scheme integrated with ID-verifier transfer protocol. *Ad Hoc Networks* 18:133–146.

Liu, J., Y. Xiao, and C. P. Chen. 2012. Authentication and access control in the Internet of things. In *2012 32nd International Conference on Distributed Computing Systems Workshops (ICDCSW)*, Macau, China, June, 588–592.

M'Raihi, D., J. Rydell, S. Bajaj, S. Machani, and D. Naccache. 2011. OCRA: OATH challenge-response algorithm (No. RFC 6287). https://www.rfc-editor.org/rfc/rfc6287.txt.

Maes, R., P. Tuyls, and I. Verbauwhede. 2009. Low-overhead implementation of a soft decision helper data algorithm for SRAM PUFs. In *Cryptographic Hardware and Embedded Systems—CHES 2009*, edited by. Clavier, C. and K. Gaj, 332–347. Berlin, Lausanne, Switzerland: Springer.

Mahalle, P. N., B. Anggorojati, N. R. Prasad, and R. Prasad. 2012. Identity establishment and capability based access control (IECAC) scheme for Internet of things. In *2012 15th International Symposium on Wireless Personal Multimedia Communications (WPMC)*, September, Taipei, Taiwan, 187–191.

Mahalle, P. N., B. Anggorojati, N. R. Prasad, and R. Prasad. 2013. Identity authentication and capability based access control (IACAC) for the Internet of things. *Journal of Cyber Security and Mobility*, 1 (4), 309–348.

Mahmoud, R., T. Yousuf, F. Aloul, and I. Zualkernan. 2015. Internet of things (IoT) security: Current status, challenges and prospective measures. In *2015 10th International Conference for Internet Technology and Secured Transactions (ICITST)*, London, UK, December, 336–341.

Markmann, T., T. C. Schmidt, and M. Wählisch. 2015. Federated end-to-end authentication for the constrained Internet of things using IBC and ECC. *ACM SIGCOMM Computer Communication Review* 45 (4): 603–604.

MeiHong, L. and L. JiQiang. 2009. USB key-based approach for software protection. In *ICIMA 2009: International Conference on Industrial Mechatronics and Automation 2009*, Chengdu, China, May, 151–153.

Mercredi, D., J. Robinson, and J. Vance. 2007. Token authentication system. U.S. Patent Application No. 11/252,040.

Pawlowski, M. P., A. J. Jara, and M. J. Ogorzalek. 2014. Extending extensible authentication protocol over IEEE 802.15. 4 networks. In *2014 Eighth International Conference on Innovative Mobile and Internet Services in Ubiquitous Computing (IMIS)*, Birmingham, UK, July, 340–345.

Pawlowski, M. P., A. J. Jara, and M. J. Ogorzalek. 2015. EAP for IoT: More efficient transport of authentication data—TEPANOM case study. In *2015 IEEE 29th International Conference on Advanced Information Networking and Applications Workshops (WAINA)*,, Gwangiu, South Korea, March, 694–699.

Peeters, R. and J. Hermans. 2013. Attack on LIAO and HSIAO's secure ECC-based RFID authentication scheme integrated with ID-verifier transfer protocol. *IACR Cryptology ePrint Archive* 2013: 399

Porambage, P., C. Schmitt, P. Kumar, A. Gurtov, and M. Ylianttila. 2014. Pauthkey: A pervasive authentication protocol and key establishment scheme for wireless sensor networks in distributed IoI applications. *International Journal of Distributed Sensor Networks* 2014: 357430.

Prabhu, P. S., A. Akel, L. M. Grupp, S. Y. Wing-Kei, G. E. Suh, E. Kan, and S. Swanson. 2011. Extracting device fingerprints from flash memory by exploiting physical variations. In *International Conference on Trust and Trustworthy Computing*, Nara, Japan, June, 188–201.

Shafagh, H. and A. Hithnawi. 2014. Poster: Come closer: Proximity-based authentication for the Internet of things. In *Proceedings of the 20th Annual International Conference on Mobile Computing and Networking*, Maui, HI, September, 421–424.

Sharaf-Dabbagh, Y. and W. Saad. 2016. On the authentication of devices in the Internet of Things. In *2016 IEEE 17th International Symposium on a World of Wireless, Mobile and Multimedia Networks (WoWMoM)*, Coimbra, Portugal, June, 1–3.

Shivraj, V. L., M. A. Rajan, M. Singh, and P. Balamuralidhar. 2015. One time password authentication scheme based on elliptic curves for Internet of things (IoT). In *2015 5th National Symposium on Information Technology: Towards New Smart World (NSITNSW)*, Riyadh, Saudi Arabia, February, 1–6.

Shone, N., C. Dobbins, W. Hurst, and Q. Shi. 2015. Digital memories based mobile user authentication for IoT. In *2015 IEEE International Conference on Computer and Information Technology; Ubiquitous Computing and Communications; Dependable, Autonomic and Secure Computing; Pervasive Intelligence and Computing (CIT/IUCC/DASC/PICOM)*, Liverpool, UK, October, 1796–1802.

Tehranipoor, F., N. Karimian, K. Xiao, and J. Chandy. 2015. DRAM based intrinsic physical unclonable functions for system level security. In *Proceedings of the 25th Edition on Great Lakes Symposium on VLSI*, May, Pittsburgh, PA, 15–20.

Tehranipoor, F., N. Karimian, W. Yan, and J. A. Chandy. 2017a. A study of power supply variation as a source of random noise. In *2017 30th International Conference on VLSI Design and 2017 16th International Conference on Embedded Systems (VLSID)*,, Hyderabad, India, January, 155–160.

Tehranipoor, F., N. Karimian, W. Yan, and J. A. Chandy. 2017b. DRAM-based intrinsic physically unclonable functions for system-level security and authentication. *In IEEE Transactions on Very Large Scale Integration (VLSI) Systems* 25 (3): 1085–1097.

Tehranipoor, F., W. Yan, and J. A. Chandy. 2016. Robust hardware true random number generators using dram remanence effects. *Presented at 2016 IEEE International Symposium on Hardware Oriented Security and Trust (HOST)*, McLean, VA.

Turkanović, M., B. Brumen, and M. Hölbl. 2014. A novel user authentication and key agreement scheme for heterogeneous ad hoc wireless sensor networks, based on the Internet of things notion. *Ad Hoc Networks* 20: 96–112.

Wen, Q., X. Dong, and R. Zhang. 2012. Application of dynamic variable cipher security certificate in Internet of things. In *IEEE 2nd International Conference on Cloud Computing and Intelligence Systems*, Hangzhou, China, vol. 3.

Wortman, P. A., F. Tehranipoor, N. Karimian, and J. A. Chandy. 2017. Proposing a modeling framework for minimizing security vulnerabilities in IoT systems in the healthcare domain. In *2017 IEEE EMBS International Conference on Biomedical and Health Informatics (BHI)*, Miami, FL, February, 185–188.

Yan, W., F. Tehranipoor, and J. A. Chandy. 2015. A novel way to authenticate untrusted integrated circuits. *Presented at Proceedings of the IEEE/ACM International Conference on Computer-Aided Design*, Austin, TX.

Yan, W., F. Tehranipoor, and J. A. Chandy. 2017. PUF-based fuzzy authentication without error correcting codes. *IEEE Transactions on Computer-Aided Design of Integrated Circuits and Systems* 36 (9): 1445–1457.

Yang, L., P. Yu, W. Bailing, Q. Yun, and Y. Xinling. 2013. A bi-direction authentication protocol for RFID based on the variable update in IOT. In *Proceedings of the 2nd International Conference on Computer and Applications ASTL*, vol. 17, 23–26.

Yang, L., P. Yu, W. Bailing, Q. Yun, B. Xuefeng, Y. Xinling, and Y. Zelong. 2013. Hash-based RFID mutual authentication protocol. *International Journal of Security and its Applications* 73: 183–194.

Zhao, G., X. Si, J. Wang, X. Long, and T. Hu. 2011. A novel mutual authentication scheme for Internet of Things. In *Proceedings of 2011 International Conference on Modelling, Identification and Control (ICMIC)*, Shanghai, China, June, 563–566.

5 Energy-Efficient Routing Protocols for Ambient Energy Harvesting in the Internet of Things

Syed Asad Hussain, Muhammad Mohsin Mehdi, and Imran Raza

CONTENTS

5.1 Introduction 91
5.2 Current Techniques for Energy-Efficient IoT 92
5.2.1 IoT Routing Protocols for Low Energy Consumption 93
5.2.2 Energy-Efficient Scheduling among IoT Devices 94
5.2.3 Minimum Energy Consumption Chain-Based Algorithm for Wireless Sensor Networks 96
5.2.4 Energy-Conserving Solutions for Area-Specific IoT 96
5.2.4.1 Energy-Conserving Solutions for WWAN-Based IoT 97
5.2.4.2 Energy-Conserving Solutions for WLAN-Based IoT 98
5.2.4.3 Energy-Conserving Solutions for WPAN-Based IoT 99
5.3 Ambient Energy Harvesting for IoT 99
5.4 Routing Challenges for IoT Powered by Ambient Energy 100
5.5 Open Research Issues for Energy-Efficient IoT 102
5.6 Conclusions 104
Key Notations 104
References 104

5.1 INTRODUCTION

The cooperation and collaboration among energy-constrained portable devices in various IoT applications has become a challenging task. The merger of heterogeneous networks and smart wireless devices requires dynamic routing algorithms, resulting in overhead on energy-constrained nodes. The smart wireless devices in most of the cases can be powered by self-reliant mechanisms like ambient energy when their main power core is depleted. The combination of energy-efficient routing protocols and ambient energy harvesting will allow devices to save energy using ambient energy as an alternative power source.

Mostly, the devices that belong to the IoT ecosystem will be portable and powered by a battery, so they must work for a long time without the need for any replacement. IoT devices must also be responsible for recharging themselves using ambient energy. Self-dependency and efficient use of energy are the major concerns for any device that joins the IoT ecosystem. Nodes in IoT can be anything from simple to complex objects. There is no need for direct connection with the public Internet, but they must be able to connect to any type of network, such as local area networks (LANs), personal area networks (PANs), and body area networks (BANs). There are tangible objects that interact and communicate with the outside environment and contain embedded technology in an IoT network. The IoT includes integrated software (software that runs the device and enables its connected capabilities), hardware (the "things" themselves), information services related to the things (which also include services dependent on the study of usage patterns and data of an actuator or a sensor), and connectivity or communication services. The solution provided by the IoT can be a product or a set of products, along with a service that can be related using one-to-one or one-to-many relations. This means that a single service can be combined either with one (set of) product(s) or with multiple (sets of) products.

In IoT, the communication system uses sensor nodes in an intelligent manner to collect and monitor the data. It also uses the wireless or wired network to transmit the data collected by sensors to the back-end server. IoT-based sensors need to be energy-efficient, and this is a major challenge due to the rapid increase in the number of devices. To achieve energy efficiency, transmission power (to the minimal necessary level) must be adjusted wisely, and distributed computing and energy harvesting techniques should be used to design an efficient communication protocol. Moreover, "activity scheduling" can be used on some nodes to turn them to a low power or sleep mode such that only a few of them remain active out of all connected nodes. The issue of energy conservation in IoT devices is also dependent on wireless technologies, for example, Bluetooth, Third Generation Partnership Project (3GPP), Long Term Evolution (LTE), LTE-Advanced, Wi-Fi, and Z-Wave. The energy conservation issues, due to wireless access technologies, can occur in many different ways, that is, how to overcome congestion or overload within the network, how to control the allocation and duty cycles of downlink or uplink radio-frequency resources in an effective manner, and so forth.

Energy harvesting is one of the best-available solutions for attaining self-dependency in IoT devices. It is a process of extracting energy from the environment to control remote-sensing capable motes. An energy harvesting unit, called a transducer, is used to transform the ambient energy, such as solar, thermal, mechanical, or piezoelectric energy, into electrical energy (Atzori et al., 2010). Energy harvesting devices capture, assemble, store, and regulate the collected energy in an efficient manner to create a sensor node–based network according to the anticipated performance. Many types of energy harvesting units, such as solar, mechanical, and thermal, are available on the market. The real challenge is the actual transducer embodiment that can be used, and in what form that transducer will provide energy, which will be converted to electrical energy (Souza and Amazonas, 2015). Energy harvesting, however, still requires added methods for the conservation of energy in IoT devices. This can be achieved by utilizing energy-efficient protocols specifically for routing data toward the base station (BS).

Energy conservation issues in IoT are closely linked with the type of wireless networks being used, as these networks use different methods for controlling overhead or congestion within the network. Major challenges and open research questions in designing routing protocols for ambient energy harvesting devices are examined in this chapter. This chapter discusses energy-efficient routing protocols and scheduling techniques for IoT devices. In particular, it describes Internet protocol version 6 (IPv6) over Low-Power Personal Area Networks (6LoWPAN) and compares its performance with various other routing protocols. Routing challenges for IoT powered by ambient energy and open research issues for energy-efficient IoT have also been presented.

This rest of this chapter is organized as follows. Section 5.2 discusses the current techniques for energy-efficient IoT devices. There are several techniques that enable efficient energy consumption. Some of the protocols might suggest a complete networking model for IoT devices to achieve efficiency. Section 5.3 includes ambient energy and energy harvesting techniques. Section 5.4 outlines routing challenges for IoT powered by ambient energy. Section 5.5 lists open research issues for energy-efficient IoT, and Section 5.6 concludes the chapter.

5.2 CURRENT TECHNIQUES FOR ENERGY-EFFICIENT IoT

Existing solutions for energy efficiency in IoT devices can be categorized based on the type of protocols and standards they adopt in different working scenarios. Notably, available methods focus on enabling either low-energy routing protocols or smart scheduling among IoT devices. Other methods may involve improving energy conservation by enhancement to physical layer concepts in IoT infrastructure. The following text classifies these techniques on the above-stated notion, highlighting the significance of each method.

5.2.1 IoT Routing Protocols for Low Energy Consumption

Ad hoc and sensor networks use various routing protocols and mechanisms, and IoT systems use the same routing mechanisms. The three main factors that may affect the routing in the IoT are consumption of energy in sensors, the node's mobility, and the kind of IoT middleware that is used. The 6LoWPAN (Kushalnagaret et al., 2007) routing protocol is used to realize an IoT ecosystem, as shown in Figure 5.1. The Internet Engineering Task Force (IETF) has identified the mechanism for routing of data for non-IP sensors in the Internet. This mechanism of routing depends on IEEE 802.15.4, which is appropriate for sensors with low power. The configuration of 6LoWPAN is composed of a series of reduced function sensors. Such sensors must be connected with full-function sensors in order to complete the topology. Furthermore, a network point is used as an entrance to another network, which means that there is specifically a gateway that acts among domains of various networks. The 6LoWPAN stack includes physical, media access control (MAC), adaption, IPv6, transport, and application layers, which are required for networking functions.

The physical layer of IEEE 802.15.4 works to effectively provide 27 channels depending on different frequencies or data rates. This medium of IEEE 802.15.4 is managed by the MAC layer through the Carrier-Sense Multiple Access with Collision Avoidance (CSMA/CA) protocol. The MAC layer also guarantees the association, disassociation, and synchronization of a device. The adaptation layer adapts various IP packets and fits them into the format appropriate for the network. Another purpose of this layer is to fragment IPv6 packets into MAC frames. It is ensured at this layer that the process of fragmentation takes place successfully. The User Datagram Protocol (UDP) is used for delay-sensitive transmission by the transport layer. The reduced function sensor starts routing (in 6LoWPAN) when it needs to transmit a packet to another IP sensor.

As data chunks are assembled at the gateway, a fully functional sensor needs to get connected to the reduced function sensors and the former is also liable for transmitting the data. The IP address is used by the gateway to find the domain of the remote IP sensor. Furthermore, it has adopted the 6LoWPAN Ad Hoc On-Demand Distance Vector Routing (LOAD) for routing. Messages for route request (RREQ) and route reply are used by LOAD. The link layer notification messages serve a function that authorizes receiving MAC messages. A mesh topology is created and runs on fully functional devices. A route can be selected by LOAD if it has fewer hops from source to destination.

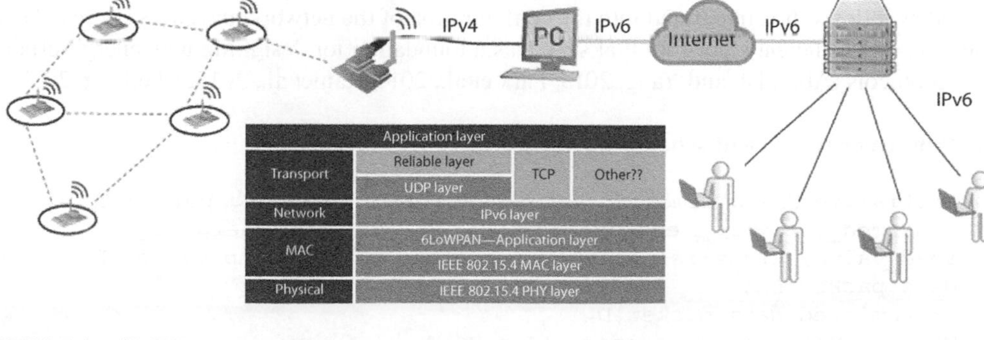

FIGURE 5.1 6LoWPAN network stack.

6LoWPAN uses a hierarchical routing (HiLow) protocol in which devices either join an existing parent or become a parent itself in the hierarchical structure.

As discussed in Tran and Thuy (2011), 6LoWPAN is suitable for networks having high processing ability. The Routing Protocol for Low-Power and Lossy Networks (RPL) (Winter et al., 2012) is developed for devices limited in power, computation, and memory capabilities. RPL is a distance vector routing protocol based on IPv6 that computes all the distances and opts for the shortest distance toward the destination. Similarly, Energy-Efficient Probabilistic Routing (EEPR) is another solution that works like AODV, but sending an RREQ packet depends on a forwarding probability determined by the residual energy and the end-of-transmission (ETX) metric.

Chen et al. (2012) have proposed a Context Awareness in Sea Computing Routing Protocol (CASCR), which generates intelligent decisions based on interactions of IoT devices at the local level. CASCR binds a state and a set of operations to every IoT device. The identified possible states can be full working, serving, single working, sleeping, and hibernating. The possible operations pertain to gathering information, transmitting information, applying information fusion, and generating a control operation. In particular, CASCR estimates the new state of every device using Markov chains by defining the new state as a function of any device's history in an IoT environment. A device having a routing request should transmit it to its first-hop neighbors, which are given a context data table specifying the network topology. Data is sent between all the devices that have a request to route and are waiting for their turns. The data is sent from a neighboring device to another through hops. The superior nodes, along with subordinates, neighbors, colleagues, and disabled devices, successfully create the network topology. Researchers in the past have discussed the performance of traditional routing protocols for ad hoc networks such as Dynamic Source Routing (DSR), Ad hoc On-Demand Distance Vector (AODV), and Optimized Link State Routing Protocol (OLSR) in an IoT environment, focusing on routing overhead, throughput, and average end-to-end delay.

5.2.2 Energy-Efficient Scheduling among IoT Devices

According to a study by Gartner (2013), the number of devices connected to the Internet will increase to 26 billion by 2020. These devices will consume a considerable amount of energy to accomplish various tasks and yield enough electronic garbage. This will result in a situation where there are power failures because control on its consumption will be difficult in the future. There is a need for adoption of new ways through which green communication can be deployed across the IoT network. Energy consumption in heterogeneous IoT devices affects the cost and availability of an IoT network. Consuming a lot of power has been an issue for a very long time, and it is likely to become more prominent with the passage of time. To overcome such issues, we need to use more energy-efficient hardware components and software to reduce overheads. Energy conservation in devices is also achieved through efficient scheduling algorithms with minimum response time, which is the time required by a processor to release a job and complete it (Albers, 2010).

Scheduling of sensors can be done by using the energy-efficient algorithm (Abedin et al., 2015), discussed as follows. It is integrated into the configuration of the network and saves energy. This is a generic energy-efficient algorithm that serves as a foundation for designing new energy-efficient routing protocols (Abdullah and Yang, 2013; Park et al., 2014; Raniet al., 2015; Chelloug, 2015).

Algorithm: Energy-efficient scheduling

1. **Initialize** Sleep_timer SL_t. Sleep_energy E_{sl} sensing period S_t, consumed_enery E_{elec} energy_reservoir e_i, maximum_energy E_{max}, read_value, transmission_value, command_value, temp_value, Received data packet D_{RK},
   Transmitted data packet D_{TK}
2. **While** stage $\neq$ current_stage && $e_i \leq E_{max}$

```
      Read stage
      Select Case of stages
        Case 1: Current stage = 'on_duty'
                 While(read_value > -1)
                        D_RK = read_value                //sensing and
                        reading sensor value
                  End of loop
                  While(transmission_value > -1)
                        D_TK = transmission_value        //transmitting
                        sensor value
                   End of loop
        Case 2:  Current stage = 'pre_off'
                   Read new_stage
                    If new_stage ≠ current_stage
                       current_stage = new_stage
                       Transmit ACK to sink node
                     Else
                       current_stage = 'off'
          Case 3: current_stage = 'off'
                  Hibernate mode with sensing capability
                  While S_t > 0
                  If temp_value = read_value
                      S_t = 0
                      Current_stage = 'pre_off'
                    Else
                      Set E_sl                                  //
                      enter sleep mode
                     While S_Lt > 0
                       Power down mode with S_Lt = 0

                      End of loop
                   End of loop
          End case
          End of loop
```

Sleep_timer SL_t is the time allotted to the specific node in the pre-off-duty state. Sleep_energy E_{sl} is the energy of the hibernating node. Sensing period S_t is the time period of the off-duty node, when it is only sensing the environment. Consumed_enery E_{elec} is the energy used by a specific node during receiving and transmitting the received data packet D_{RK} and transmitted data packet D_{TK}, respectively. Energy_reservoir e_i is the energy available to all nodes through ambient energy sources. Maximum_energy E_{max} is the amount of energy that a node can use.

In the on-duty state, a sensor node will sense, receive, and transmit the data. These sensors may behave as a relay node (RN) or sink node based on their capabilities. The pre-off-duty state follows the on-duty state whenever the device is idle for some time. This state can switch between both on-duty and off-duty whenever it needs to do so. In the pre-off-duty stage, a device only receives and transmits the required commands from the sink node. The off-duty state is constituted by hibernate, sleep, and power-off states. In the hibernate state, a device has nothing to do except sense the environment, and for this purpose, the device may use small energy resources as well. In the sleep state, the device instantaneously stops working but starts again when required to resume. The power-off state puts the device into a deep sleep (Abedin et al., 2015).

Figure 5.2 illustrates how this scheduling algorithm works. In sending a message from node 1 (sink node) to node 7, only these two nodes are required to be fully operational. For intermediate nodes 3 and 5, they are in pre-off-duty states to save power, as their job is to receive and forward.

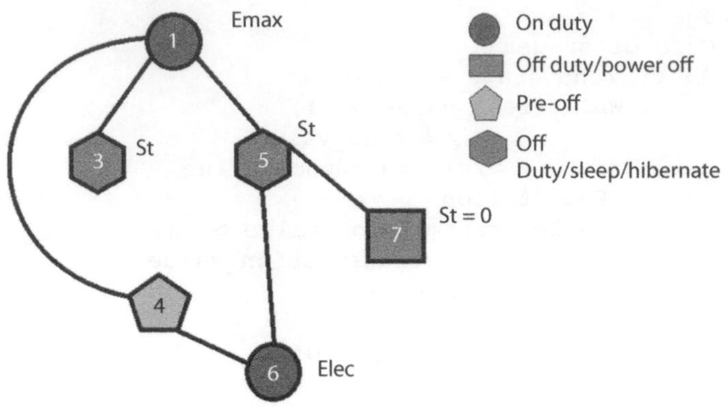

FIGURE 5.2 Generic schedule algorithm for energy efficiency.

Node 6 can be powered off, as it is not participating in communication. Node 4 is in either the hibernate or sleep mode, as it will only be used in the case of failure of node 5.

5.2.3 Minimum Energy Consumption Chain-Based Algorithm for Wireless Sensor Networks

There are many factors, such as humidity in the air, temperature, and interference, that affect the transmission of sensor nodes. These factors render the wireless sensor networks (WSNs) with dynamic routing capability pretty much unstable and useless for large-scale networking. In a dynamic routing sensor, nodes exchange data about their location. Consequently, overhead is increased, thus increasing power utilization. For the same reason, this type of routing is not feasible for IoT networks. Furthermore, components involved in the composition of IoT networks are least mobile with consistent topology; therefore, a dynamic routing configuration is of less advantage over its static counterpart.

Figure 5.3 shows several nodes that constitute a hierarchical framework according to various parameters. The nodes that are presented in this topology are static, and their routing is static as well. The lower layers are composed of normal nodes, cluster heads (CHs), cluster coordinators (CCOs), and RNs. The uppermost convergence layer includes BSs having Internet connectivity. Nodes in the lower layer sense and transmit data to their respective RNs. Afterwards, RNs pass the data to their concerned CHs. The load on CHs and CCOs is balanced by passing the data from CHs to upper-layer CCO. Afterwards, the data is handed over to the upper-layer CCO, and the same continues until the data is transmitted to the BS at the uppermost layer. On the local level, information is sent using RNs, and neighboring clusters communicate through the CHs and CCOs only. This entire deployment maintains the energy efficiency and scalability in IoT due to static routing and the simple architecture of IoT components. A lot of energy can be effectively saved by placing IoT components above this framework.

5.2.4 Energy-Conserving Solutions for Area-Specific IoT

Different types of wireless technologies, such as 3GPP, LTE, LTE-Advanced, Wi-Fi, ZigBee, and Bluetooth, can serve as vehicle to provide connectivity among IoT devices and gateways or servers. Energy-conserving issues in IoT devices are also closely linked to wireless technologies. Energy-conserving issues arise in numerous manners, depending on the category of wireless radio access technologies, for example, methods for controlling overload or congestion within the network, methods to adjust duty cycles, and allocation of uplink or downlink radio-frequency resources in an energy-efficient manner.

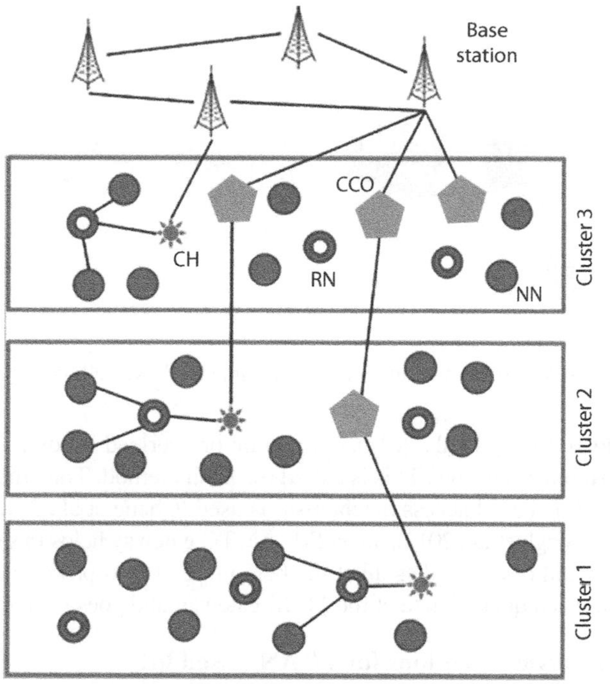

FIGURE 5.3 A multitier IoT framework. NN, normal nodes.

Wireless wide area networks (WWANs) have vast commercial usage. It is about time that they become an integral part of the generic IoT, just like WSNs. Wi-Fi-based Internet is an integral part of our lives, carrying various applications in almost every field. A BAN is the best example of a wireless personal area network (WPAN).

5.2.4.1 Energy-Conserving Solutions for WWAN-Based IoT

An important concern in adopting 3GPP LTE for the IoT is the huge number of IoT devices. The problem of overload or a congested radio access network or core network (CN) arises when both IoT devices and user devices access the network for data transfer at the same time. The problem is going to become uncontrollable as the number of devices increase in the coming years, be they human-to-human (H2H) or machine-to-machine (M2M) devices. The problem of overloading can indirectly affect IoT devices for their energy consumption. High network utilization is going to cause delay and loss of IoT data packets, and hence more battery power will be consumed in retransmission. In addition to congestion-related issues of energy consumption, different factors, such as time, frequency, and transmit power, are considered for energy-efficient allocation of radio resources to IoT devices. There is a need to control multiple devices from accessing the system at the same time because this will also reduce power consumption. Figure 5.4 shows a typical arrangement of use of a WWAN in an IoT environment.

Energy consumption is reduced through adaptive learning in fault-tolerant routing. As soon a fault is detected in the current path, the algorithm switches to the next available one with the highest goodness value (Misra et al., 2012). To conserve energy, all nodes lying on the unused path are put to sleep. Machine-type communications (MTC) are commonly used in 3GPP networks (Universal Mobile Telecommunications System [UMTS] and LTE) (Cheng et al., 2012). These are automated applications, which comprise communications between machines and devices (sensors) without human intervention. These devices generate a large amount of signaling traffic, which creates congestion and overload random access network (RAN) and CN. Although various content resolution mechanisms have been proposed in the past, none of them give satisfactory results. In RAN overload

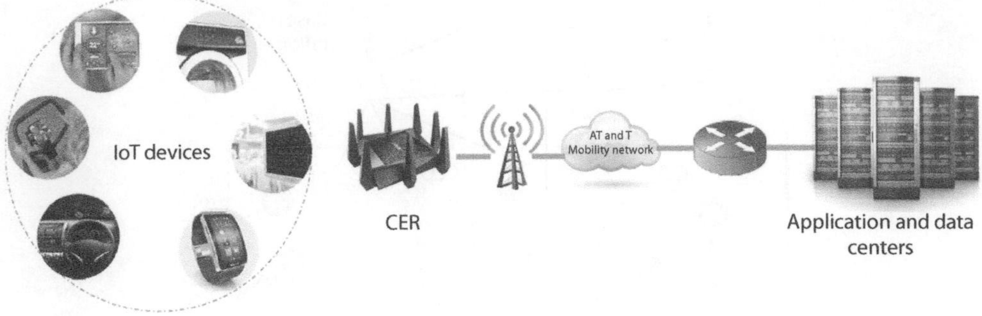

FIGURE 5.4 Typical WWAN connection in an IoT environment. CER, Cellular Enabled Router.

control method overload (Cheng et al., 2012), congestion or overload is controlled by randomly dispersing the load to different time slots. This is called the push method. The other approach is the pull method, in which a polling-based access mechanism is used (Cheng et al., 2012). In another method to prevent overloading (Singh et al., 2012), an M2M or MTC gateway helps in saving energy by queuing the data until the M2M device wakes up at the beginning of next power cycle. After waking up, the M2M device handles all queued data at the M2M gateway and goes back to the idle mode later.

5.2.4.2 Energy-Conserving Solutions for WLAN-Based IoT

A lot of power is consumed whenever an IoT device uses Wi-Fi to reach the CN of 3GPP and finally the servers due to congestion. When IoT devices use Wi-Fi to extend the area of operation, more power is consumed for multihop topologies due to severe collisions in multihop communications. Further research and standardization are necessary to evolve a mechanism for collision avoidance in multihop communication.

Figure 5.5 shows a typical example of a WLAN setup for an IoT environment. A control to stop multiple devices starts accessing the system and, at the same time, will also reduce power consumption.

The major challenge faced is congestion or overload. For this purpose, an offset listen interval algorithm was proposed in Abdullah and Yang (2013), which had the purpose to recover power loss. Further, we should be able to reduce the traffic so that the delay in the network can be reduced for the users and devices. To carry out this process, the wake-up time of the device is deferred in a random fashion, resolving congestion problems to some extent. It involves waking up the devices such that buffered data packets are transferred. All these packets will be sent to

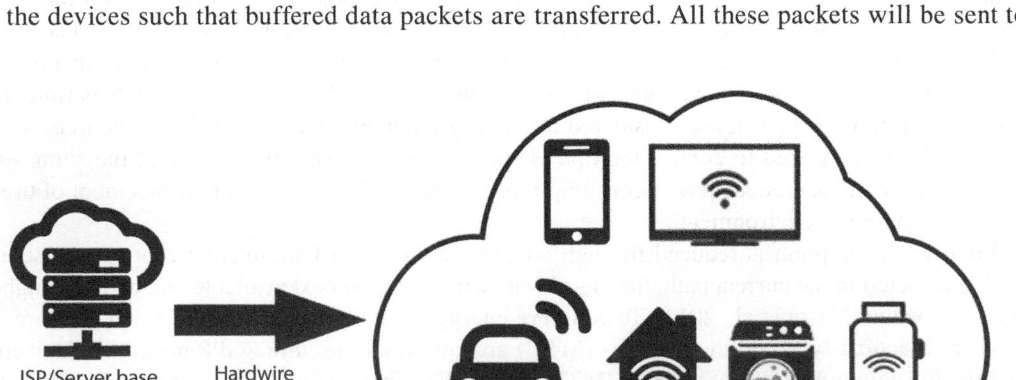

FIGURE 5.5 WLAN deployment example in IoT. ISP, Internet service provider.

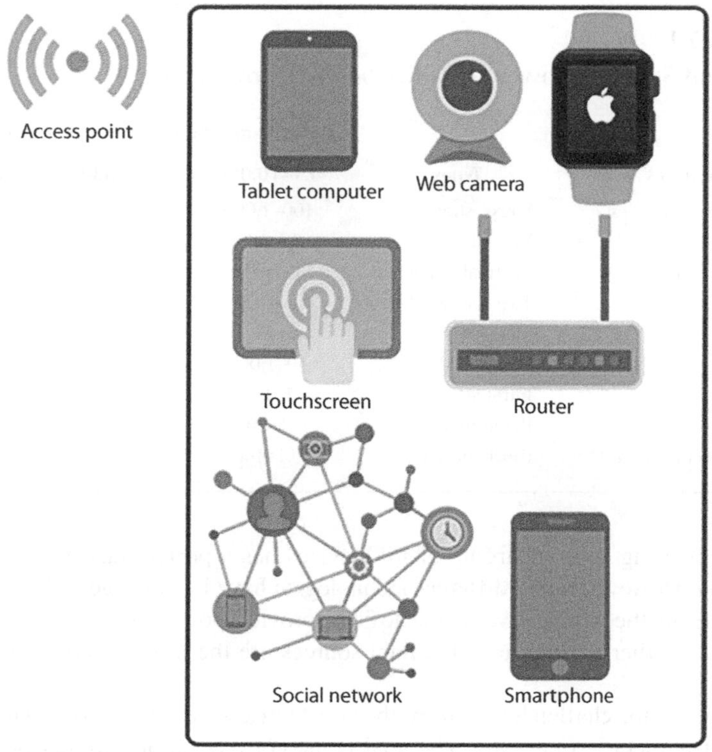

FIGURE 5.6 Basic WPAN for IoT environment.

the devices involved in the network in a timely fashion. Such mechanisms ensure efficient and effective performance of the network.

5.2.4.3 Energy-Conserving Solutions for WPAN-Based IoT

As shown in Figure 5.6, a WPAN establishes connectivity between battery-operated constrained IoT devices. Bluetooth Low Energy (BLE) is a commonly used WPAN technology for IoT setups. Different researchers (Park et al., 2014) have offered a BLE implementation and evaluated its performance as comparable to that of ZigBee/802.15.4. Energy is consumed during a master–slave discovery process, as the master and slave devices are not always in a connected mode. A master device searches for available slaves for connection simultaneously along the slave devices, which advertise their availability to the master. Energy used after establishing a connection is also considered, and parameters related to energy, such as transmission and reception, along with the inter-frame spaces, are analyzed.

A neighbor discovery mechanism multicasts a high number of messages for IPv6 over BLE, consuming higher energy in BLE-based IoT devices. A basic solution to this problem is to consider a neighbor discovery mechanism optimized to ensure that a node is removed from the neighbor cache as its lifetime is expired. To reduce the transmission of neighbor solicitation messages from other nodes, entries regarding nodes are kept in the neighbor's cache.

5.3 AMBIENT ENERGY HARVESTING FOR IoT

In any environment, ambient energy sources exist in four different forms, namely, thermal, mechanical, biochemical, and radiant. These different power sources can be differentiated based on their power densities, as shown in Table 5.1.

TABLE 5.1
Ambient Sources' Power Densities before Conversion

		Power from Ambient Sources (uW/cm^{-3})	
Power Sources	**Types**	**0.1–10,000**	**10,000–100,000**
Mechanical sources	Stress–strain	100–1,000	
	Vibrant	10–100	
Thermal sources	Thermal gradient	1–1,000	
	Thermal variation	10–100	
Radiant sources	Sun—outside		1000–100,000
	Sun—inside	10–1,000	
	Infrared	1–100	
	Radio frequency	1–10	
Biochemical sources	Biochemistry	0.1–100	

"Perpetual" powering and recharging of IoT devices has a perfect target for ambient sources' energy harvesting. However, many different technologies have been suggested for energy harvesting and offered from the perspective of the IoT. To provide power or to charge small electronic systems and devices, thermoelectric and radiant sources are the most practical solutions in terms of engineering.

At present, the major challenge faced by thermoelectric solutions is to enhance the intrinsic efficiency of thermoelectric materials. The goal is the conversion of a higher portion of the few milliwatts of thermal energy that can be harvested, and the size of the device remains small. To achieve this purpose, nano- and microtechnologies, for example, superlattices, are under consideration (Chelloug, 2015).

In the case of solar radiation, perovskite technology can boost the efficiency of solar conversions from 20%, offered by state-of-the-art cells, to more than 30% in the coming future. There are also some other advantages over photovoltaic technologies, which already exist, such as material properties by which the manufacturing process of high-performance perovskite cells is simplified. However, perovskite technology is still in its research and prototyping stage right now. The current aim of researchers is to calculate how toxic this technology is and whether lead in the cells can be replaced by some nontoxic elements (Chelloug, 2015). Meanwhile, they are also trying to get higher conversion rates.

In many practical applications, relying only on energy harvesting is obviously not good enough. Devices that require low power need 50 mW when they are in the transmission mode, and even less than that when they are in the sleep or standby mode. However, energy harvesting in a continuous active mode cannot produce this amount of energy. So, an alternate mode is also required to be integrated in devices that are powered by energy harvesting. Weather conditions also affect the performance of ambient sources; therefore, to guarantee the steady operation of a device, energy storage in it is still required. To develop advanced operations and a slim design, malleable and lightweight electronics for IoT applications, while keeping the cost low, are very important for enhancing the battery technologies.

5.4 ROUTING CHALLENGES FOR IOT POWERED BY AMBIENT ENERGY

By using an energy harvesting scheme, power restraints have been removed in a conventional IoT setup. However, it is considered the most reliable practice to overcome the energy limitations in the IoT. Along with its benefits, ambient energy–powered IoT also has some challenges that might be faced during the design process of energy harvesting circuits. Designing a routing protocol for energy harvesting IoT devices is very challenging, as it is affected by numerous factors, such as the specification of energy sources; the energy storage device, that is, the supercapacitor; power management in nodes; protocol functionality; and application requirements. The major challenges in designing routing protocols for ambient energy harvesting devices are discussed below:

- *Fault tolerance*: As sensor node charging is subjected to ambient energy obtained from the environment and has an unpredictable nature, some IoT nodes may be turned off in case of hardware equipment failure and shortage of power. Efficiency of the network should not be compromised due to this failure. A routing protocol should handle network failure by creating new routing links to reroute data. To ensure network performance, it is necessary that fault-tolerant networks support redundancy. A possible solution is a learning automata (LA)–based fault-tolerant and mixed cross-layered routing protocol for IoT (Misra et al., 2012). Even if there exist some problems between the source and destination nodes, it still ensures the successful delivery of packets. As this is related to IoT, the planned protocol must be able to scale highly and provide good performance in a diverse environment. The cross-layer and LA concepts will provide flexibility to a routing protocol so that it can be used all over the network. It shall dynamically adapt itself according to the varying network conditions. As energy is a major factor in IoT, so the protocol must provide energy-aware routing implementing fault-tolerant mechanisms. To save energy, all the nodes that are in the unused path are put in the sleep mode, and this scheduling must be adaptive and dynamic.
- *Node deployment*: The performance of a routing protocol in IoT powered by ambient energy is affected by node deployment, and it is a big challenge. Node deployment plays an important role because to charge properly, every node needs to harvest maximum energy. Node deployment is further divided in two types: randomized and deterministic. In randomized deployment, the nodes are placed randomly in an ad hoc manner. In deterministic deployment, nodes are positioned manually and a predefined path is used to route the data. In case the resulting distribution is not uniform, clustering is required for connectivity among the nodes. An ambient energy–powered IoT provides support for short durations in transmission based on unpredictable environmental factors, and that is why data must be routed in multiple hops.
- *Optimal path*: Data to the back-end server can be sent via different routes from the sender node. Some routes are lengthy and secure, while some are short. So, the routing protocol searches for the most optimal path in the network and routes data packets to the sink. A path is said to be optimal if results can be achieved by minimum overhead retransmission, resulting in less energy consumption.
- *Dynamic nature of harvesting*: An ambient energy–powered IoT has to face the unstable behavior of environmental energy because it changes with time. The wake-up and sleep units of IoT nodes are unpredictable; therefore, it cannot be anticipated that a node will be awake next time to receive the data packet (Cheng et al., 2012). As continuous and constant behavior of ambient energy sources is not certain, the node takes time to harvest energy. The routing process will be affected if at the wake-up time a node has not harvested sufficient energy. So, it can be assumed that opportunistic and broadcast methods are not good enough in ambient energy–powered IoT. A single node can ensure effective routing and communication only if it is awake all the time.
- *Topology and connectivity control*: Topology and connectivity control is dependent on power control and management. An IoT node fully charges the battery in the sleep interval. If an IoT node cannot be continuously powered due to insufficient battery, it will intentionally send the data to the sink; this action leads to a change in network topology and connectivity (Singh et al., 2012). Therefore, topology and connectivity control can be classified as a main issue in ambient energy–powered IoT.

- *Reliable data transmission*: There are two key tasks involved in forwarding data packets from an IoT node to the sink: accessing the shared wireless channel and forwarding the data packet to the next hop. In some applications, trustworthy data transmission is needed all the time. But the network faces the biggest problem whenever a node is in the sleep mode and a data packet is not forwarded to ensure reliable transmission. As the energy is harvested, there is no advantage for a node that is near the sink over one that is far, because forwarding packets have a higher priority according to opportunistic routing. The data flow must be regulated by using some transport protocol so that any source in the network can get its proper share of bandwidth despite its location, that is, regardless of the fact it is near the sink or far from it (Gomez et al., 2012).

5.5 OPEN RESEARCH ISSUES FOR ENERGY-EFFICIENT IoT

Future implementation of the IoT will encompass the insertion of different devices to sense and control the information of the physical world, with no power constraints. Power constraints in the IoT could be eliminated by replacing the battery with an energy harvesting unit, but this technology is not able to provide sustained energy to the IoT devices continuously because powering of an IoT node is highly dependent on the environmental factors, which are stochastic in nature.

Due to the stochastic nature of environmental factors, charging intervals of any IoT node are not predictable. Consequently, to design a robust, efficient, and scalable routing protocol is a challenging task because wake-up and sleep intervals of an IoT node are not certain. The selection of a harvesting unit producing the maximum amount of energy all the time is under experimentation. The following are some open research issues for energy-efficient IoT.

- *Selection of energy harvesting unit*: Mostly, a solar energy harvester is used to give power to an IoT node. The energy harvester does not give the best results by using artificial light, or in other words, the energy harvesting rate is stochastic in nature. So, the energy harvesting unit is dynamic and situation based. Under these conditions, other types of harvesters (e.g., vibrational and thermal) could also be used. The selection of the energy harvester depends on the scenarios in which it will be used, and thus it is necessary to analyze the application in which the energy harvester mote needs to be deployed. Placement of the energy harvester, at which it could harvest a great amount of energy, is also another consideration.
- *Self-configuration and reconfiguration*: If a node fails due to hardware, topology changes, or mobility, the routing protocol should have the ability to reconfigure the network and also update the current topology of the network. A routing protocol needs to perform the topology changes in a very robust manner to maintain the network functions.
- *Adaptive localization algorithms*: IoT nodes are deployed into uncertain situations. Here, the problem is to calculate the actual coordinates of any IoT node. Many existing schemes assumed that IoT nodes know their position. GPS is not applicable to IoT nodes because it works only in the outdoors and does not work well for obstructions. So, there is a need to develop an adaptive localizing routing algorithm that periodically updates the location of the IoT node or location measuring system and provides acceptable accuracy.
- *Bioinspired approaches*: In unpredictable situations, bioinspired methods are used to solve complex problems of communication and networking systems. Bioinspired approaches establish conclusions from previous situations and make intelligent decisions based on past work. For the IoT, considering several different devices, one solution is to use bioinspired ant colony optimization (ACO) for optimal path identification (OPI) (Moreira Sa de Souza et al., 2008). In target or object tracking surveillance, a vast amount of data is generated. This data is organized in a such a way as to process in real-time scenarios. The data forwarding time from the sensor node to the back-end server should be as minimum as

possible, requiring a path having the least distance and a reduced hop count. Conventional greedy algorithms like Dijkstra's and dynamic programming algorithms like the Bellman–Ford one are suitable in such cases but are computationally complex and do not guarantee finding a global optimal path (arko et al., 2014). These limitations can be overcome by using ACO optimized for using path length and reduced communication overhead.

- *Heterogeneity*: The IoT is an umbrella for uniting a wide range of application domains, such that there is a diversity of devices, protocols, network connectivity methods, and resulting application models on the market today. This problem is expected to get worse in the future due to the increased demand for low-resource IoT devices. Therefore, it is important to foresee improvements in heterogeneous power-constrained IoT devices. Cloud computing (Afergan and Wroclawski, 2004), with its high availability, elasticity (improving scalability), and low cost of computing resources, is the best choice. A cloud-based system can support heterogeneous power-constrained IoT devices by implementing an adaptation layer that provides a uniform device abstraction that hides the diversity in devices, protocols, network connectivity methods, and application models. As there are different innovations, the heterogeneity can be as far as devices and their systems administration measures. This heterogeneity includes extra many-sided qualities of the routing procedure. As the current conventions have inflexible limits, it is vital to fabricate such a routing convention that consolidates all sorts of heterogeneity within itself.
- *Scalability*: Coordinating and managing the real-time performance of devices in an IoT network poses the problem of scalability, as there is going to be a lot more data than now, originating from a variety of devices. Most of the innovations in IoT are remote, and the devices utilizing these innovations might be stationary or versatile. Cell phones may enter or leave the system, which may increment or diminish the performance measurement of the system. So, the system versatility can influence the protocol. Understanding the performance of the IoT is about the identification of patterns, trends, and anomalies, rather than just trivial ways of assessment against preestablished parameters. Even those IoT applications that are not built for instantaneous response must detect and produce near-real-time response.
- *Latency*: The information produced in the IoT may get lapsed, and because of that, it is important to convey the information to the destination inside the sought measure of time. So, it is essential to handle the latency by the routing protocols for keeping up the administration quality. The importance of the networks that relate these devices and systems together cannot be ignored. Placing and processing data closer to users whenever possible can help in reducing distance-related network latency, but applications will still be affected by bad routing decisions and network congestion. Connection latency is just one aspect of overall performance degradation. More examples are expected as automated optimization is chosen by designers for exploiting additional performance metrics, such as network transit times, server and device processing times, and response-time latency from peer services.
- *Congestion control*: Congestion is an issue in a wide range of systems. Because of the exponential increment of system movement, it has turned into a complex phenomenon. Congestion happens when the measure of activity increments past the limit of the system packet loss, causing unwanted delays. For preventing congestion at a particular node, it is vital that protocol convention do stack adjusting when the activity increments at a particular node. Because of congestion, the nodes can get to be hotspots, and if congestion perseveres for a long time, fast exhaustion of the node's vitality may happen, which may result in decline of the system's lifetime. So, it is crucial that the routing convention relay and attempt to beat the congestion promptly, and it must stop for congestion avoidance too.
- *Elimination of data redundancies*: IoT systems will create a tremendous measure of information and will send it to a destination for further handling. So, as opposed to taking care of and sending comparable information over and over again and squandering system energy, it is vital

to combine information for dispensing with information redundancies. An increased amount of data means an increase in the energy requirement for routing it to its destination. The data may be redundant most of the time, and eliminating this redundancy is going to reduce the energy required for routing, eventually causing an increase in lifetime of the network. Many data repeated techniques have already been proposed, such as data fusion or negotiation-based ones (Hassan, 2014). Redundancy can be reduced by employing data fusion techniques at the CH level, whereas in negotiation-based protocols, a communication between the source and destination can be established beforehand for eliminating redundancy in the data.

- *Multipath routing*: Multipath routing allows the communication to have multiple options to reach the destination based on the node information. Each node can also work as an intermediate node if the neighboring node is not operational. It will stop the thorough utilization of particular guardian nodes and their quick energy exhaustion; yet, it is likewise important to keep the topology reconfigurations in control. Fewer topology reconfigurations implies fewer control parcels, and fewer control bundles implies fewer energy utilizations, which at last results in expansion in system lifetime. Alongside burden adjustment, this procedure may build the adaptation to noncritical failure, dependability, and quality of service change (Varma et al., 2003).

5.6 CONCLUSIONS

Energy-efficient routing protocols for IoT devices powered by ambient energy can address a variety of challenges posed by the increase in the number of devices connected to the Internet. This will help in developing new energy conservation techniques through which green communication can be deployed across the IoT network. This chapter discussed numerous challenges for the implementation of such protocols. These challenges include fault tolerance capability, node deployment decision, optimal path search, uncertain node sleep/awake times, and topology and connectivity control. Future research should target the development of a decision support system regarding the selection of energy harvesting units as per application scenarios. This is important, as energy harvesting units are dependent on environmental factors that are stochastic in nature, affecting charging intervals of IoT devices. Other open research issues for energy-efficient IoT include the heterogeneity of devices, network connectivity, network latency, congestion control, data redundancies, and topology reconfiguration control problems.

KEY NOTATIONS

Notation	Meaning
IoT	Internet of things
3GPP	Third Generation Partnership Project
LTE	Long Term Evolution
6LoWPAN	IPv6 over Low-Power Personal Area Networks
IETF	Internet Engineering Task Force
IP	Internet protocol
AODV	Ad Hoc On-Demand Distance Vector Protocol
ETX	End of transmission
BS	Base station

REFERENCES

Abdullah, S. and K. Yang. 2013. An energy-efficient message scheduling algorithm in Internet of things environment. In *2013 9th International Wireless Communications and Mobile Computing Conference (IWCMC)*, Sardinia, Italy, July 1–5, 2013, 311–316.

Abedin, S. F., R. G. Alam, and S. C. Hong. 2015. A system model for efficient green-IoT network. In *International Conference on Information Networking (ICOIN)*, Cambodia, January 12–14, 2015, 177–182. doi:10.1109/ICOIN.2015.7057878.

Afergan, M. and J. Wroclawski. 2004. On the benefits and feasibility of incentive based routing infrastructure. In *Proceedings of the ACM SIGCOMM Workshop on Practice and Theory of Incentives in Networked Systems*, Portland, OR, August 30–September 3, 2004, 197–204.

Albers, S. 2010. Energy-efficient algorithms. *Communications of the ACM* 53 (5): 86–96.

Arko, I.P., K. Pripužić, M. Serrano, and M. Hauswirth. 2014. IoT data management methods and optimisation algorithms for mobile publish/subscribe services in cloud environments. In *Proceedings of the 2014 European Conference on Networks and Communications (EuCNC)*, Bologna, Italy, June 23–26, 2014, 1–5.

Atzori, L., A. Iera, and G. Morabito. 2010. The Internet of things: A survey. *Computer Networks* 54: 2787–2805.

Chelloug, S. 2015. Energy-efficient content-based routing in Internet of things. *Journal of Computer and Communications* 3: 9–20.

Chen, Z., H. Wang, Y. Liu, F. Bu, and Z. Wei. 2012. A context-aware routing protocol on Internet of things based on sea computing model. *Journal of Computers* 7: 96–105.

Cheng, M.-Y., G.-Y. Lin, H.-Y. Wei, and A. C.-C. Hsu. 2012. Overload control for machine-type-communications in LTE-advanced system. *IEEE Communication Magazine* 50: 38–45.

Gartner. 2013. Gartner says the Internet of things installed base will grow to 26 billion units by 2020. www.gartner.com/newsroom/id/2636073.

Gomez, C., J. Oller, and J. Paradells. 2012. Overview and evaluation of Bluetooth low energy: An emerging low-power wireless technology. *Sensors* 12: 11734–11753.

Hassan, Q. 2014. Demystifying cloud computing. *Journal of Defense Software Engineering* 1: 16–21.

Kushalnagaret, N., G. Montenegro, and C. Schumacher. 2007. IPv6 over Low-Power Wireless Personal Area Networks (6LoWPANs): Overview, assumptions, problem statement, and goals. IETF RFC 4919. Fremont, CA: Internet Engineering Task Force.

Misra, S., P. Venkata Krishna, H. Agarwal, and A. Gupta. 2012. An adaptive learning approach for fault-tolerant routing in Internet of things. In *IEEE Wireless Communications and Networking Conference (WCNC)*, Paris, April 1–4, 2012, 815–819.

Moreira Sa de Souza, L., P. Spiess, M. Koehler, D. Guinard, S. Karnouskos, and D. Savio. 2008. SOCRADES: A web service based shop floor integration infrastructure. *Lecture Notes in Computer Science* 4952: 50–67.

Park, S.-H., S. Cho, and J.-R. Lee. 2014. Energy efficient probabilistic routing algorithm for Internet of things. *Journal of Applied Mathematics*, 2014: 213106.

Rani, S., R. Talwar, J. Malhotra, S. H. Ahmed, M. Sarkar, and H. Song. 2015. A novel scheme for an energy efficient Internet of things based on wireless sensor networks. *Sensors*, 15: 28603–28626.

Singh, S., K.-L. Huang, and B.-S. P. Lin. 2012. An energy-efficient scheme for WiFi-capable M2M devices in hybrid LTE network. In *Proceedings of the IEEE International Conference on Advanced Networks and Telecommunications Systems (ANTS)*, Bangalore, India, December 16–19, 2012, 126–130.

Souza, A. M. C. and J. R. A. Amazonas. 2015. A new Internet of things architecture with cross-layer communication. In *Proceedings of the 7th International Conference on Emerging Networks and Systems Intelligence Emerging 2015*, Nice, July 19–24, 2015, 1–6.

Tran, T. and T. Thuy. 2011. Routing protocols in Internet of things. *Presented at the ESG Seminar*, Finland.

Varma, H., K. Fadaie, M. Habbane, and J. Stockhausen. 2003. Confusion in data fusion. *International Journal of Remote Sensing* 24: 627–636.

Winter, T., et al., 2012. RPL: IPv6 Routing Protocol for Low Power and Lossy Networks, RFC 6550, IETF ROLL WG, Tech. Rep.

6 IoT Hardware Development Platforms

Past, Present, and Future

Musa G. Samaila, João B. F. Sequeiros, Acácio F. P. P. Correia, Mário M. Freire, and Pedro R.M. Inácio

CONTENTS

6.1 Introduction 108
6.2 IoT Hardware Development Platforms 109
6.3 Related Work 110
6.4 IoT Hardware Development Platforms in the Past 9 Years 111
6.4.1 Processing and Memory and Storage Capabilities of IoT Hardware Platforms 112
6.4.1.1 Processing Power of IoT Hardware Development Boards 112
6.4.1.2 Memory and Storage Capacity of IoT Hardware Development Boards 114
6.4.2 Connectivity and Communication Interfaces of IoT Hardware Platforms 115
6.4.2.1 Connectivity and I/O Interfaces of Microcontroller-Based Hardware Development Boards 115
6.4.2.2 Connectivity and I/O Interfaces of Single-Board Computers 116
6.4.3 OS Support for IoT Hardware Platforms 117
6.4.3.1 OS Support for Arduino Microcontroller-Based Hardware Development Boards 117
6.4.3.2 OS Support for Single-Board Computers 117
6.4.4 Battery Life of IoT Development Boards 117
6.4.4.1 Battery Life of Arduino Microcontroller-Based Hardware Development Boards 117
6.4.4.2 Battery Life of Single-Board Computers 118
6.4.5 Size and Cost of IoT Development Boards 118
6.4.5.1 Size and Cost of Arduino Microcontroller-Based Hardware Development Boards 118
6.4.5.2 Size and Cost of Single-Board Computers 118
6.4.6 Security Features of IoT Hardware Development Platforms 118
6.4.6.1 Security Features of Arduino Microcontroller-Based Hardware Development Boards 119
6.4.6.2 Security Features of Single-Board Computers 119
6.5 Current IoT Hardware Development Platforms 119
6.5.1 Processing and Memory and Storage Power of Current IoT Hardware Platforms 122
6.5.1.1 Processing Power of Current IoT Hardware Platforms 122
6.5.1.2 Memory and Storage Capacity of Current IoT Hardware Platforms 123

6.5.2 Connectivity and Input/Output Ports of Current IoT Hardware Platforms 125
6.5.2.1 Connectivity and I/O Interfaces of Current Microcontroller Boards 125
6.5.2.2 Connectivity and I/O Interfaces of Current Single-Board Computers 125
6.5.3 OS Support for Current IoT Hardware Platforms 126
6.5.3.1 OS Support for Current Microcontroller-based Boards 126
6.5.3.2 OS Support for Current Single-Board Computers 126
6.5.4 Battery Life of Present-Day IoT Hardware Development Boards 126
6.5.4.1 Battery Life of Current Microcontroller-Based IoT Hardware Development Boards 126
6.5.4.2 Battery Life of Current Single-Board Computers 127
6.5.5 Size and Cost of Current IoT Development Boards 127
6.5.5.1 Size and Cost of Current Microcontroller-Based Development Boards 127
6.5.5.2 Size and Cost of Current Single-Board Computers 127
6.5.6 Security Features of the Present IoT Hardware Platforms 128
6.5.6.1 Security Features of Current Microcontroller-Based Boards 128
6.5.6.2 Security Features of Current Single-Board Computers 128
6.6 IoT Hardware Development Platforms in the Next 5 Years 128
6.6.1 Processing and Memory and Storage Capacity of Future IoT Hardware Platforms........131
6.6.1.1 Processing Power of Future IoT Hardware Platforms 131
6.6.1.2 Memory and Storage Capacity of Future IoT Hardware Platforms........... 131
6.6.2 Connectivity and Communication Interfaces of Future IoT Hardware Development Boards 132
6.6.2.1 Connectivity of Future IoT Hardware Development Platforms 132
6.6.2.2 Communication Interfaces of Future IoT Development Platforms 132
6.6.3 OS Support for Future IoT Development Boards 132
6.6.4 Battery Life of Future IoT Hardware Platforms 133
6.6.5 Size and Cost of Future IoT Development Boards 133
6.6.6 Security Features of Future IoT Hardware Platforms 134
6.7 Timeline of Evolution of the IoT Hardware Development Platforms 134
6.8 Conclusions 134
Acknowledgments 135
References 136

6.1 INTRODUCTION

IoT is a system of interrelated devices in which sensors and communication capabilities are embedded into everyday objects, animals, and even people, enabling them to *see*, *hear*, *think*, and *perform* some tasks by communicating with each other (Al-Fuqaha et al., 2015). As a novel paradigm that is increasingly gaining ground in recent years, the IoT is developing at a rapid pace, creating multidimensional business models and investment opportunities. This is largely due to an explosion in the availability of tiny, energy-efficient, and inexpensive IoT hardware development platforms. These platforms combine microcontrollers and processors with other components, such as memory and wireless connectivity chips in a variety of ready-to-build packages for IoT design, prototyping, and mass production (Fernandez, 2014).

IoT hardware development platforms play a vital role in the overall development of the IoT. In particular, they provide designers, developers, and researchers with prebuilt and ready-to-program kits that enable them to focus more on their projects. However, despite their critical role in the IoT development, today most discussions on the IoT are often focused on its services, such as cloud solutions, application domains, business opportunities, and software applications, as well as security. But little or nothing is mentioned in the literature about the underlying hardware platforms that are enabling the sensing, transmission, processing, sending, and receiving of data to or from the cloud and securing the whole process (Lee, 2015). While it is true that the number of connected devices and IoT cloud solutions are growing at a staggering rate, we are also seeing a similar trend with IoT hardware development boards. Currently, there are several hardware platforms that one can choose from. Arguably, knowing *where we have been*, *where we are*, and *where we are going* in terms of some hardware parameters may help developers to better understand the capabilities and limitations of various boards, and hence enable them to make the best choice.

This chapter provides an overview of a brief intellectual history of IoT development platforms, and shows how they have evolved over the years. It also predicts some features of future development boards. With respect to the aforementioned subject, the chapter focuses on some key attributes of IoT hardware platforms, which include processing and memory capacity, connectivity and communication interfaces, Operating System (OS) support, battery life, and security features. These attributes are worth considering when choosing IoT hardware, because lack of proper understanding of any of them during the design process may affect the deployment goals. Thus, a better knowledge of some of these features may enable designers and developers to make the right choice for hardware, which in turn may foster the design and development of more secure IoT devices, considering the fact that security is critically important for the successful operation of the IoT.

The remainder of this chapter is structured as follows. Section 6.2 presents some background on the study. Section 6.3 considers related works. Section 6.4 describes IoT hardware development platforms in the past 9 years. Section 6.5 considers the current IoT development platforms. Section 6.6 forecasts features of IoT development platforms in the next 5 years. Section 6.7 presents the timeline of evolution of the IoT hardware development platforms. Finally, the chapter ends in Section 6.8 with a brief summary and conclusion.

6.2 IoT HARDWARE DEVELOPMENT PLATFORMS

A few decades ago, a single computer could easily fill up an entire room (Arthur, 1996), but as semiconductor technology advanced, which led to the invention of the transistor, the size began to shrink significantly. The breakthrough in semiconductor technology has resulted in making hardware smaller in size, easier and cheaper to create, faster, and easier to integrate. This eventually led to the development of microcontroller-based boards, System-on-Chip (SoC), and Single-Board-Computers (SBCs), which in turn led to the current state of the art in the IoT hardware platforms. The days when computers were bulky are long gone. Currently, there is a computer (Michigan Micro Mote) (Vermes, 2015) that is small enough to be injected into the body.

The trend to manufacture ever smaller, lighter, and less expensive computer products and devices has sparked a revolution of computer miniaturization that has been the source of a seeming struggle between different chipmakers and technology giants worldwide. Such competitions, along with the recent technological advances in wireless communication, are enabling the production of a variety of IoT hardware development platforms.

An IoT hardware development platform is a small single electronic circuit board with limited memory and processing power that can be used by developers, hobbyists, or anyone with some level of experience to create interactive electronic objects. Many development boards have provision for Ethernet or built-in wireless connectivity. They also feature some onboard sensors or provision for connecting additional sensors. Some platforms support small Linux or Windows OSes. Hardware platforms for IoT can be programmed via an external interface to another computer, or via a web-based Integrated Development Environment (IDE).

Generally, IoT hardware development boards are classified into two main types, namely, open source and proprietary. The design details of open-source boards are made publicly available for users to study, reproduce, and modify to suit their various purposes. These boards are usually more suitable for prototyping. Most open-source boards have active and supportive communities that make building projects for the IoT easier. On the other hand, proprietary IoT boards are typically designed to be used in end applications and users are not allowed to have access to the design details. Such boards are more suited for final production. Hardware development platforms can be divided into three categories: microcontroller based, SoC, and SBCs.

A microcontroller-based IoT hardware development platform integrates a number of components, such as a microcontroller; RAM; flash memory; internal clock; various input/output (I/O) interfaces, such as Serial Peripheral Interface (SPI), Inter-Integrated Circuit (I2C), Inter-IC Sound (I2S), Universal Serial Bus (USB), and Universal Asynchronous Receiver/Transmitter (UART); and other supporting components, all built onto a single printed circuit board (PCB)* (McRoberts, 2010). This type of board provides all the necessary circuitry for some useful control tasks, and therefore they are used mainly for prototyping. A popular example is Arduino *Uno*.†

A SoC is an integrated circuit (IC) that integrates the necessary components of a computer into a single silicon chip (Saleh et al., 2006). Typically, a SoC contains a processor, a graphics processing unit (GPU), memory, and circuitry for power management, along with some I/O peripherals, such as SPI, I2C, I2S, UART, USB, and Peripheral Component Interconnect (PCI). SoC devices usually have small form factor, consume less power, and are computationally excellent. Hence, they are used for building SBCs as well as for mass production of IoT devices and other embedded systems. An example is the Intel Curie module.‡

A SBC is a complete computer on a small PCB with microprocessors, memory, power management circuitry, a GPU, real-world multimedia, and some I/O interfaces, such as USB, UART, High Definition Multimedia Interface (HDMI), and Ethernet, which allows it to function as a computer (Atwell, 2013). SBCs are usually used for IoT prototyping, for educational purposes, and for use as embedded computer controllers. One example is Raspberry Pi.§

6.3 RELATED WORK

To the best of our knowledge, there is a limited amount of reported work on this subject, and a number of those works have no direct relevance to the subject under discussion. For example, the "Gap Analysis of Internet of Things Platforms" by Mineraud et al. (2016) focused on middleware solutions that provide connectivity for sensors and actuators to the Internet. The authors defined a platform, in the context of their work, as the middleware and infrastructure that allows end users to interact with small objects. Similarly, Derhamy et al. (2015) surveyed commercial frameworks and platforms designed for developing and running IoT applications. Although their survey discussed recent developments in commercial frameworks and platforms supported by big players in the software and electronics industries, a platform in the context of their work refers to a cloud facility that can be used for integrating sensors into IoT applications, among other things.

There are, however, a few scholarly articles and books that discussed topics that have some relevance to the subject. There are also quite a few websites and blogs that discussed, in general, *the past, present, and future of the IoT*. While we focus on the devices of the IoT, particularly on the hardware development boards, some of these authors approached the subject from different perspectives, mostly focusing on different application domains of the IoT, business aspects of the IoT, or its market size.

Purcell (2016), for example, briefly highlighted the timeline of IoT evolution from 2008 to 2016, including how the term *IoT* was hyped in many popular tech websites, like Gartner. The author mentioned some industry sectors that already recorded a remarkable success story, such as energy, healthcare, and transportation. He highlighted some future challenges, which include capturing,

* https://gist.github.com/rgaidot/9132b50cdcdb455fccbe#credit-card-sized-computer.
† https://www.arduino.cc/en/Main/ArduinoBoardUno.
‡ http://eu.mouser.com/pdfdocs/intel-curie-module-fact-sheet-2.pdf.
§ https://www.raspberrypi.org/products/.

analyzing, and harnessing the vast amount of data that will be generated by the IoT. Kumar (2015) started by reviewing the number of connected devices, followed by a more detailed history of IoT development. He examined the current breakthroughs in many application areas, including smart cars, smart homes, and Google glass. The author concluded by presenting the future prospects of the IoT for consumers in the areas of smart home, medical assistance, and so forth. In the following paragraphs, we examine works that are directly or somehow related to the subject under discussion, and we also explain how our work is different from these studies.

There are several books and articles that covered specific hardware development platforms, such as Arduino, BeagleBone, and Raspberry Pi. For instance, Hughes (2016) provided a thorough description of the electrical, software, and performance aspects of the Arduino board. The author conducted an extensive study of the physical design and characteristics of different official Arduino hardware development platforms, and the internal functions of the Atmel AVR processors. More importantly, he examined the specific processors used in different Arduino boards. The author also covered a wide range of possible applications for Arduino hardware development boards. Doukas (2012) also provided ample information on building IoT with the Arduino platform. The author covered different aspects of creating IoT with Arduino, including cloud computing concepts, creating embedded projects using Arduino, connecting Arduino with the Android phone over the Internet, and reprogramming the Arduino microcontroller remotely via the cloud.

Molloy (2015) explored the BeagleBone open-source Linux SBC platform that can be used both as a general-purpose Linux computer and for building IoT or other electronics projects. The author particularly focused on BeagleBone hardware in Chapter 1, where he considered some features and different subsystems, along with the peripherals of BeagleBone Black. In another study, Maksimović et al. (2014) discussed the performance and constraints of Rasberry Pi as an IoT hardware platform. The authors presented a comparative analysis of certain features and performances of Raspberry Pi with some existing IoT hardware platforms. The results of their analysis show that despite a few limitations, Raspberry Pi remains among the most versatile and inexpensive SBCs used in IoT research applications.

Ortmeyer (2014) focused on the hardware aspect of the IoT by considering the *then* and *now* of SBCs. Although he focused specifically on SBCs, his work is more related to the subject under discussion. After presenting a brief history of SBCs, the author examined the present-day SBCs. Some of the examples he considered include Raspberry Pi, BeagleBone, and PandaBoard. He discussed several features of SBCs, such as processor speed, memory, interfaces, connectivity, and size. The author also highlighted some trends that he believes will continue into the future, which include more powerful processors, the availability of more accessories, and more miniaturization.

Compared with the work done by Ortmeyer, our idea stands out by examining both microcontroller-based development boards and the SBCs. Additionally, our study provides a more recent and detailed discussion of the features discussed by Ortmeyer using a larger number of development boards as examples. On top of these, the chapter examines other attributes of the development boards that were not considered by Ortmeyer, namely, cost and security features. Finally, it also considers all these features for development boards in the past 9 years. In addition, the chapter examines the present-day boards, and then predicts what features will be on future boards in the next 5 years.

6.4 IoT HARDWARE DEVELOPMENT PLATFORMS IN THE PAST 9 YEARS

The last few years have seen an explosion in the number of IoT hardware development boards. This can be traced back to the efforts of a team of designers in Ivrea, Italy, who developed the first easy-to-use and low-cost microcontroller-based development kit in 2005 (Ortmeyer, 2014). Their goal was to develop an open-source prototyping board that would allow people with any skill level to use microcontrollers for building their projects. This landmark achievement marked the beginning of Arduino, which paved the way for a new breed of electronics enthusiasts, popularly called hobbyists, to start building electronic projects using microcontrollers. As Arduino continued to evolve in the hands of engineers and hobbyists, its market continued to rise, which led to a dramatic drop in the cost of microprocessors and SoCs.

The next major breakthrough in the development of IoT prototyping boards occurred when BeagleBoard.org unveiled the BeagleBoard on July 28, 2008.* The board is an open-source hardware with all the functionalities of a basic computer. BeagleBoard was developed by a team of engineers at Texas Instruments in the United States. It was developed for educational purposes in order to enhance the teaching of computer science, as well as for enhancing the teaching of open-source hardware and software capabilities in schools around the world. One of the biggest developments in the rise of IoT development boards happened in February 2012, when the first Raspberry Pi was launched (Lyons, 2015), even though the decision to develop the SBC was made in 2006. The SBC was developed by a group of researchers (presently known as the Raspberry Pi Foundation) in a computer laboratory at the University of Cambridge. The purpose of developing the Raspberry Pi is quite similar to that of the BeagleBoard.

Over the years, subsequent versions of the aforementioned IoT development boards have been released, and many other boards have also been developed, which are available in a variety of shapes and sizes. The following subsections discuss different attributes of IoT hardware development platforms that have been released in the past 9 years. Since quite a number of these boards have been released within the period under review, we focus only on the following Arduino microcontroller-based boards and SBCs: *Diecimila*, *Uno*, *Duemilanove*, *Due*, and *Yún* (Arduino), and BeagleBoard, PandaBoard and the ES version, OlinuXino, Raspberry Pi 1 (A and B), Intel Galileo, and RIoTboard (specifically designed for developing IoT projects) (SBCs). Figure 6.1 shows images of the Arduino boards, while images of some of the SBCs are shown in Figure 6.2. Although this chapter classifies the devices considered in this section as those that have been used in the past, some of these hardware platforms are still being used today.

6.4.1 Processing and Memory and Storage Capabilities of IoT Hardware Platforms

Among the many underlying components that determine an IoT hardware performance are processing capacity and memory. Like other computing hardware devices, the number and type of processors and their speed determine the processing capacity of IoT hardware. Similarly, the amount and type of memory a development board possesses directly impacts its performance.

6.4.1.1 Processing Power of IoT Hardware Development Boards

Starting with the Arduino microcontroller boards, apart from Arduino *Due*, which has a different architecture (ARM), the architecture of the microcontroller units (MCUs) on the other Arduino boards considered here is AVR based. *Diecimila*, released in 2007, is based on the ATmega168 microcontroller with a 16 MHz clock speed. Its successor, Arduino *Uno*, released in 2010, is faster since it is based on ATmega328 with the same clock speed. *Duemilanove*, first released in 2008, has two versions. While the older version is based on ATmega168, the later version was made with ATmega328, and they all have a 16 MHz crystal oscillator. Arduino *Due* is based on AT91SAM3X8E microcontroller with an 84 MHz crystal oscillator and performs much better. *Yún*, released in 2013, is unique in that it is based on two processors: ATmega32U4, clocked at 16 MHz, and Atheros AR9331, with the same clock speed.

Considering the SBCs, the central processing unit (CPU) or the processor of BeagleBoard is based on ARM Cortex-A8 architecture and has a clock frequency up to 720 MHz, which makes it significantly faster than the Arduino boards. PandaBoard, launched in 2010, consists of a dual-core ARM Cortex-A9 MP CPU clocked at 1 GHz. The PandaBoard ES version, released in 2011, has the same CPU but with a 1.2 GHz clock frequency. A13-OlinuXino, released in 2012, is based

* http://beagleboard.org/beagleboard.

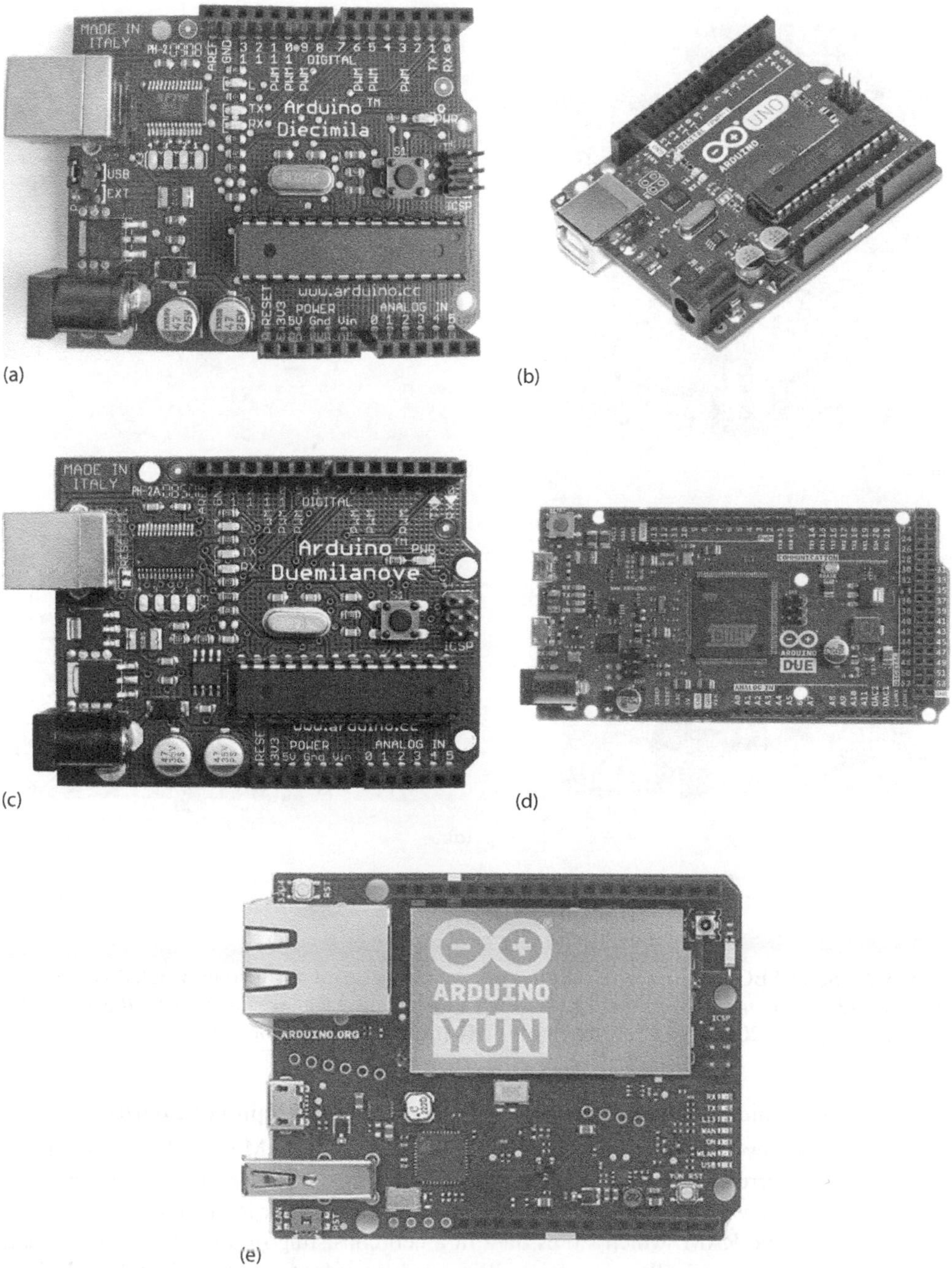

FIGURE 6.1 Some arduino boards released over the last 9 years: (a) *Diecimila*, (b) *Uno*, (c) *Duemilanove*, (d) *Due*, and (e) *Yún*. (Images CC-SA-BY from Arduino.cc.)

on a single-core A13 ARM Cortex-A8 CPU clocked at 1 GHz. The two models of Raspberry Pi 1 (A and B) are based on ARMv6 with a 700 MHz clock speed, even though Raspberry Pis are capable of increasing their clock speed to temporarily reach up to 1 GHz (Upton, 2012). Intel Galileo, released in October 2013, includes a single-core Intel Quark X1000, clocked at 400 MHz, and RIoTboard is based on ARM Cortex-A9 MPCore, clocked at 1 GHz.

(a) (b) (c) (d)

FIGURE 6.2 Some SBCs released in the last 9 years: (a) BeagleBoard rev.B (from BeagleBoard.org, 2017), (b) Olimex A13-OlinuXino (copyright © by OLIMEX Ltd. and used with permission), (c) Raspberry Pi 1 A (from Raspberrypi.org, 2017), and (d) Raspberry Pi 1 B+ (from Raspberrypi.org, 2017).

6.4.1.2 Memory and Storage Capacity of IoT Hardware Development Boards

Before proceeding, several important terms are briefly explained. RAM is random access memory for storing data in current use, and it can easily be accessed by the processor. It is very fast and not necessarily too big; however, it loses its data when power is shut off or interrupted.

DRAM is dynamic RAM, which stores data in a cell consisting of a capacitor and transistor, and must be refreshed electrically every few milliseconds in order to hold the data. SRAM is static RAM, implying that it does not have to be refreshed continuously like the DRAM. It is, however, more expensive. SDRAM is synchronous DRAM, a type of memory that synchronizes itself with the CPU timing, which allows it to run faster than the previous memory types. DDR is Double-Data-Rate SDRAM, a type of memory that achieves higher bandwidth than the SDRAM by doubling its transfer rate without increasing the clock speed. It has a prefetch buffer width of 2 bits, which is twice that of SDRAM, with a transfer rate ranging from 266 to 400 MT/s. DDR2 is DDR Two SDRAM, capable of operating the external data bus two times faster than DDR SDRAM, with a prefetch buffer width of 4 bits and a 533–800 MT/s transfer rate. DDR3 is DDR Three SDRAM; it reduces power consumption by 40% compared with DDR2. The transfer rate of DDR3 ranges from 800 to 1600MT/s, with a prefetch width of 8 bits. LPDDR is Low-Power DDR, which operates at 1.8V instead of the traditional 2.5V, and it is usually used in portable devices. Newer versions are LPDDR2, LPDDR3, and LPDDR4. ROM is read-only memory, and EEPROM is electrically erasable programmable ROM, meaning that it can be erased and reprogrammed repeatedly. Flash memory is a memory chip on the circuit board that is nonvolatile and retains data for a very long period of time, even if the device is powered off. Furthermore, it erases stored data in blocks. Finally, eMMC is embedded multimedia card, a type of flash memory that is soldered directly on the board. It is usually slower and cheaper than the Solid-State Drive (SSD), and hence used on cheap devices for data storage. Below we consider the memory and storage capacity of the devices under consideration.

Arduino *Diecimila* has a 1 KB SRAM (where the program creates and manipulates its variables) with an 8-bit Data Path Width (DPW), 512 bytes of EEPROM for storing limited long-term data, and 16 KB of flash memory for storing user programs, also called sketches (of which 2 KB is used by the boot loader). *Uno* has a 2 KB SRAM with an 8-bit DPW, 1 KB of EEPROM, and 32 KB of flash memory (of which 0.5 KB is used by the boot loader), which makes *Uno* faster than its predecessor. The earlier version of *Duemilanove* has a 1 KB SRAM, while the later version has 2 KB of SRAM, all with an 8-bit DPW. The EEPROM for the two versions are 512 bytes and 1 KB, respectively, and the flash memory for the two versions are 16 and 32 KB, respectively (of which 2 KB is used by the boot loader). *Due* has a 96 KB SRAM with a 32-bit DPW and 512 KB of flash memory completely available for the user; this makes *Due* very fast. *Yún* has a 2.5 KB SRAM with an 8-bit DPW on the ATmega32U4 processor, 1 KB of EEPROM, and 32 KB of flash memory (of which 4 KB is used by the boot loader). It also has 64 MB of DDR2 SDRAM and 2.5 KB of SRAM on the Atheros AR9331 processor with the same 8-bit DPW. It has 1 KB of EEPROM and 16 KB of flash memory and a micro-SD card reader.

The BeagleBoard SBC has 256 MB of LPDDR SDRAM and 256 MB of NAND flash memory. Both the PandaBoard and ES versions have the same 1 GB of LPDDR2 SDRAM and two caches: 32 KB for program and 32 KB for data on cache 1, and 1 MB on cache 2. A13-OlinuXino has 512 MB of DDR3 SDRAM, clocked at 408 MHz. Raspberry Pi 1 model A and model B rev. 1 have 256 MB of SDRAM, while model B rev. 2 has 512 MB of SDRAM. Intel Galileo has 256 MB of DDR3 SDRAM and 8 MB of NOR flash memory. Finally, RIoTboard has 1 GB of DDR3 SDRAM with a 32-bit DPW that is clocked at 800 MHz; it also has a 4 GB eMMC flash memory.

6.4.2 Connectivity and Communication Interfaces of IoT Hardware Platforms

In the IoT vision, *things* are expected to become part of the Internet and active participants in information, business, and social processes where every connected device is uniquely identified and accessible to the network. This underscores the need for IoT development boards to be Internet enabled and have the necessary communication interfaces that will allow them to interact and communicate with each other, as well as to sense the environment. This section examines the same IoT development boards considered in the previous section, but with respect to connectivity and I/O interfaces.

6.4.2.1 Connectivity and I/O Interfaces of Microcontroller-Based Hardware Development Boards

Among the Arduino boards considered, *Yún* is the only board that comes with a built-in Ethernet and Wi-Fi support; therefore, it can connect to the Internet directly without using Ethernet, Wi-Fi, or GSM *shield*.* The other boards rely on the use of Ethernet, Wi-Fi, or GSM *shields* to connect to the Internet.

---

* A shield refers to a compatible board that is plugged on top of an Arduino board for the purpose of extending its capabilities.

All the Arduino boards have quite a number of digital I/O pins, also known as General-Purpose I/O (GPIO), which are used as low-level peripherals. The analog pins on these boards also have all the functionality of GPIO pins. The Arduino boards under consideration also have USB ports and feature other hardware I/O and communication interfaces, such as SPI communication using the SPI library, I2C/Two Wire Interface (I2C/TWI) communications using the wire library,* and UART communication. The Arduino boards also feature the In-Circuit Serial Programming (ICSP) header.

For instance, *Diecimila*, *Uno*, and *Duemilanove* all feature a USB port, 14 GPIO digital pins (of which 6 are Pulse-Width Modulation [PWM]† output), and 6 analog input pins. They also have SPI, I2C/TWI, and UART hardware I/O and communication interfaces; each of the boards also features an ICSP header. *Due* has 2 USB ports, 54 GPIO digital pins (of which 12 are PWM output), 12 analog input pins, and 2 Digital-to-Analog Converter (DAC) analog output pins. The board also has 4 UART ports, 1 SPI header, 1 I2C, and 2 TWI headers, as well as 1 ICSP header. *Yún* features 2 USB ports, 20 GPIO digital pins (of which 7 are PWM output), and 12 analog input pins. In addition, *Yún* has 1 UART port and 1 ICSP header, and supports SPI and I2C/TWI I/O communications.

6.4.2.2 Connectivity and I/O Interfaces of Single-Board Computers

The first BeagleBoard released in 2008 has no onboard Ethernet port; however, it features some communication interfaces, including I2C, I2S, and SPI for serial communication, as well as Digital Visual Interface (DVI)-D and S-Video for video display. But BeagleBoard-xM, launched in 2010, has an onboard Ethernet jack. PandaBoard is not Internet enabled; however, the ES version has Ethernet and Wi-Fi, as well as Bluetooth connectivity. Both boards include some communication interfaces, such as DVI, HDMI, camera expansion header, audio I/O, USB, serial/RS-232, and two USB host ports, as well as a 14-pin Joint Test Action Group (JTAG) GPIO, UART, I2C, and so forth. The first version of A13-OlinuXino has a Video Graphics Array (VGA), USB, audio output, and microphone input. The board also has a Universal EXTension (UEXT) connector, which consists of power and three serial interfaces: asynchronous, I2C, and SPI, as well as other GPIO connectors with 68/74 pins. But Internet connectivity is optional, and if needed, a USB Wi-Fi modem can be used with an RTL8188CU chip that can be bought separately. However, the OLinuXino-MINI-Wi-Fi version, released in 2013, has built-in Wi-Fi connectivity.‡ Raspberry Pi 1 model A has one USB port, but no Ethernet port. The SBC features HDMI, composite video (RCA jack), a 15-pin Mobile Industry Processor Interface (MIPI) camera interface, and audio output. It also has 26 GPIO pins (of which 17 are real GPIOs) that can be used as low-level peripherals, such as UART, I2C, SPI, and I2S. Model B has an Ethernet port and two USB ports, along with the other peripherals that are on model A. Model B+ has four USB ports, an Ethernet port, and the other peripherals found on the previous versions. Intel Galileo has an Ethernet port, a USB port, and USB host ports. In addition to the traditional I/O communication interfaces, such as UART and the RS-232 serial port, it also features some industry standard I/O interfaces like Advanced Configuration and Power Interface (ACPI) and PCI Express. Intel Galileo also supports Arduino *shields*. The RIoTboard has an Ethernet port, two USB hubs, a mini-USB, HDMI, a camera interface, audio I/O, and many other peripherals, which include camera interfaces, a debug port, GPIO pins, and an expansion port.

---

* https://www.arduino.cc/en/Reference/Wire.

† Pulse-width modulation (PWM) is a technique for obtaining analog outputs from a microprocessor's digital outputs.

‡ https://www.olimex.com/Products/OLinuXino/iMX233/iMX233-OLinuXino-MICRO/resources/iMX233-OLINUXINO-MICRO.pdf.

6.4.3 OS Support for IoT Hardware Platforms

From the hardware perspective, IoT devices can be divided into two major categories depending on their performance and capability: high-end devices and low-end devices (Hahm et al., 2016). RIoTboard, Raspberry Pi, and other SBCs fall under the first category, because each of these devices has adequate resources to run lightweight versions of the traditional General-Purpose Operating System (GPOS), also known as a High-Level Operating System (HLOS), such as Linux and Android. The second category consists of devices that are resource constrained and cannot run HLOS, such as Arduino *Uno* and wireless sensor nodes like TelosB motes.*

But since precise timing and timely execution are very crucial in many IoT use cases, such as in smart healthcare and industrial automation, a Real-Time Operating System (RTOS) based on a microkernel and designed for a very small memory footprint, as well as for energy efficiency, is best suited for such devices (Hahm et al., 2016). In contrast to computers and mobile devices like smartphones, there are a wide variety of open-source and commercial RTOSs for IoT devices. RTOSs for IoT applications are usually designed for real-time performance, as well as to run efficiently on small-form-factor and low-power devices. This section discusses OS support for IoT development boards over the last 9 years.

6.4.3.1 OS Support for Arduino Microcontroller-Based Hardware Development Boards

Due to stringent resource constraints, all the Arduino boards being considered have no OS support except Arduino *Yún*. The Atheros processor on the *Yún* supports OpenWrt Linux distribution.

6.4.3.2 OS Support for Single-Board Computers

The BeagleBoard hardware supports Angstrom Linux distribution. PandaBoard runs the Linux kernel with optimized versions of Ubuntu, Android, or FirefoxOS. A13-OlinuXino supports the Linux Debian distribution and Android. The officially supported OS for Raspberry Pi is the Raspbian, which is based on Debian. The third-party OSes that can run on Pi 1 models A and B include Arch Linux and Pidora. The first generation of Intel Galileo runs the Yocto project-based Linux. Available flavors of OSes supported by the RIoTboard include Android 4.3 (Jelly Bean), Linux 3.0.35 (Ubuntu), and Linux 3.10.17 (Yocto), but the board usually comes preinstalled with Android 4.3.

6.4.4 Battery Life of IoT Development Boards

While IoT development boards are generally designed to minimize power consumption, the choice of development board is one possible contributing factor affecting battery life. One important aspect of IoT power management design is the ability to balance power consumption and device performance. Although this is not something new in embedded systems design, but due to the relatively longer battery life requirements of IoT devices, the trade-off becomes critical.

Basically, the battery life of a device can be obtained by dividing the battery capacity by the average current consumption of the device. Considering the fact that determining battery life depends on the type of project built on a particular board and what is hooked up to it, this section examines the operating voltage, the active current consumption, and the power consumption of typical bare-development boards over the last 9 years. The power consumption is presented in watts or milliwatts, as the case may be.

6.4.4.1 Battery Life of Arduino Microcontroller-Based Hardware Development Boards

The operating voltage for the *Diecimila* is 5V, the DC† current per I/O pin is 40mA, and the DC current per 3.3V pin is 50mA; therefore, the average power consumption is 200mW. The operating voltage

* https://telosbsensors.wordpress.com/.

† DC - direct current

for the *Uno* is 5V, the DC current per I/O pin is 20mA, and the DC current per 3.3V pin is 50mA; hence, the average power consumption is 250mW. The *Duemilanove* has the same specifications as the *Diecimila. The Due* has very different specifications. Its operating voltage is 3.3V, its DC current per 3.3V pin is 800mA, and its DC current per 5V pin is 800mA; therefore, its average power consumption is 2.64W. The *Yún* also has the same specifications as the *Diecimila* and the *Duemilanove.*

6.4.4.2 Battery Life of Single-Board Computers

The typical bare-board power consumption for BeagleBoard is 2W. The power consumption of PandaBoard is between 100mW and 1W (Eijndhoven, 2011). The typical bare-board power consumption for A13-OlinuXino and A13-OlinuXino-Wi-Fi are 1.56W and 2.28W, respectively. The typical power consumption for Raspberry Pi 1A and B are 1.5W and 3.5W, respectively. The average power consumption of both the first and second generations of Intel Galileo is 15W. The average input voltage and current of the RIoTboard are 5V and 1A, respectively, implying that the average power consumption is 5W.

6.4.5 Size and Cost of IoT Development Boards

The notable drop in the cost and size of IoT development boards has contributed enormously to the advancement of IoT technology. Here, we consider the physical size and cost of IoT hardware development boards over the last 9 years. We start by considering the size and cost of the Arduino boards, then followed by the size and cost of SBCs.

6.4.5.1 Size and Cost of Arduino Microcontroller-Based Hardware Development Boards

The length and width of the *Diecimila* board are 73 and 53 mm, respectively. The *Diemila* is not in the market currently; hence, its price is not specified. The *Uno* is somewhat smaller; its length and width are 68.6 and 53.4 mm, respectively, and its price in 2013 was about $30. The dimensions of the *Duemilanove* are the same as those of *Diecimila.* The price of *Duemilanov* in 2010 was about $35. The *Due* is the biggest, with a length and width of 101.52 and 53.3 mm, respectively, and it was sold for about $50 in 2013. Finally, the *Yún* has the same dimensions as the *Diecimila* and the *Duemilanove.* Its price in 2013 was about $65.

6.4.5.2 Size and Cost of Single-Board Computers

The length, width, and height of BeagleBoard are 76.2, 76.2, and 16 mm, respectively, and its price in 2008 was between $95 and $149. The length and width of PandaBoard are 114.3 and 101.6 mm, respectively, and its price in 2010 was $174. The length and width of A13-OlinuXino are 100.33 and 85.09 mm, respectively, and the cost of the board in 2012 was about $73. The length, width, and height of Rasberry Pi 1 model A are 86, 54, and 15 mm, respectively, and its price in 2012 was $25. The dimensions of Raspberry 1 model B are the same as those of 1A, except for the height, which is 17 mm in B. Its price in 2012 was $35. The dimensions of Intel Galileo in terms of its length, width, and height are 107, 74, and 23 mm, respectively, and the price of the first generation in 2013 was $70. The length and width of RIoTboard are 120 and 75 mm, respectively, and its price in 2014 ranged from $50 to $79.

6.4.6 Security Features of IoT Hardware Development Platforms

As the connectivity trend in the IoT expands, and the *chip* war among the leading industries continues, with every major company fighting for a leading position, there is a growing concern over the privacy and security of data in the IoT. Baking security into IoT hardware will help in securing smart devices by design (Swift, 2016). Fortunately, a number of chipmakers are already building security features into their hardware (Harvey, 2015). In this section, we discuss the security features of IoT hardware over the last 9 years.

6.4.6.1 Security Features of Arduino Microcontroller-Based Hardware Development Boards

Essentially, Arduino boards have no cryptographic engine, and hence are not designed with core security in mind. In addition, almost all the Arduino boards we have considered so far lack the required computational and memory capacity to run the Secure Sockets Layer (SSL) stack used for securing Hypertext Transfer Protocol (HTTP) communications. The dual architecture of *Yún*, however, makes it somehow unique. The OpenWRT-based Linux distribution on the Atheros processor provides support for running a number of protocol stacks or utilities, such as cURL, SSL as OpenSSL, or PolarSSL, which is a lightweight stack.

6.4.6.2 Security Features of Single-Board Computers

The BeagleBoard, PandaBoard, and A13-OlinuXino have no cryptographic chip. In other words, the TrustZone (Lesjak et al., 2015; Vasudevan et al., 2014) on these SBCs is disabled. The TrustZone technology* is hardware-based security capability built into SoCs by chipmakers. The technology has been integrated into different ARM-based systems, ranging from the smaller microcontrollers to high-performance processors. Although the processors on the above-named boards can perform some encryption and decryption of data, the intensive computational requirements of encryption standards like the Data Encryption Standard (DES), Advanced Encryption Standard (AES), and Elliptic-Curve Cryptography (ECC) can slow down the overall performance of the systems (Wong et al., 2015; Mossinger et al., 2016).

Both Raspberry Pi 1 (A and B) and Intel Galileo Generations 1 and 2 lack a robust hardware-based security solution, which is a basic security feature. While some level of security could be achieved in software with OpenSSL/SSH, as well as Libgcrypt (Miller, 2016), an attacker can still get remote access and take over a system, or decrypt and view sensitive data (Miller, 2016; Brumley and Boneh, 2005), if he is able to locate the data relating to security keys. One other security issue with Raspberry Pi is the fact that both user data and code reside on one SD card, which can be removed by an attacker that has physical access to the device.

On the other end of the spectrum, the hardware security features of the RIoTboard are enabled. Some of the security functions include protection against debug port attacks by blocking access to system debug features using System JTAG Controller (SJC); securing nonvolatile storage and the Real-Time Clock (RTC) using Secure Non-Volatile Storage (SNVS); and ensuring secure booting using the Advance High Assurance Boot (A-HAB) with some new embedded enhancements based on SHA-256, a 2048-bit RSA key, a version control mechanism, warm boot, and so forth.†

Tables 6.1 and 6.2 summarize the different attributes of the IoT hardware development platforms over the last 9 years. The tables present a capability comparison that shows the specifications of all the development boards featured in Section 6.4.

6.5 CURRENT IoT HARDWARE DEVELOPMENT PLATFORMS

Recent advances in microprocessor chip technology have reshaped the IoT hardware industry in profound ways (Kantoch et al., 2014). Current IoT development boards come with a wide variety of smaller, faster, and more energy-efficient processor types. Most of the boards also come with GPUs onboard, Ethernet, built-in Wi-Fi and/or Bluetooth, and built-in cryptographic engines along with compact OSes. Today there are several different IoT prototyping hardware platforms in the market, with new ones coming up on a regular basis. Although the basic functionality of these boards is very similar, each platform comes with different features, capabilities, and limitations that make it ideal for certain applications. Thus, the choice of a board depends completely on the type of project.

* http://www.arm.com/products/processors/technologies/trustzone/index.php.
† http://www.mouser.com/pdfDocs/RIoTboard_User_Manual_v1.1.pdf.

TABLE 6.1
Summary of the Features of Microcontroller-Based IoT Hardware Development Boards Released over the Last 9 Years

Specification Highlights	Diecimila	Uno	Duemilanove	Due	Yún
Processor architecture	ATmega168 AVR Core	ATmega328P AVR Core	ATmega168/328 AVR Core	AT91SAM3X8E ARM Core	ATmega32U4 and Atheros AR9331 AVR/MIPS Cores
Processor speed	16 MHz	16 MHz	16 MHz	84 MHz	16 MHz
Memory (RAM)	1 KB SRAM, 8-bit DPW	2 KB SRAM, 8-bit DPW	1/2 KB SRAM, 8-bit DPW	96 KB SRAM, 32 DPW	2.5 KB SRAM, 8 DPW on ATmega32U4 and 64 MB DDR2 and 2.5 KB SRAM on Atheros AR9331
Onboard storage	512-byte EEPROM, 16 KB flash	1 KB EEPROM, 32 KB flash	512 bytes/1 KB EEPROM, 16/32 KB flash	512 KB flash	1 KB EEPROM and 32 KB flash on ATmega32U4 and 1 KB EEPROM and 16 KB flash on Atheros AR9331
Onboard connectivity	—	—	—	—	Ethernet and Wi-Fi support
Peripheral interfaces	USB	USB	USB	USB	USB
GPIO low-level peripherals	14 digital I/O pins (6 provide PWM output), 6 analog input pins	14 digital I/O pins (6 provide PWM output), 6 PWM digital I/O pins, 6 analog input pins	14 digital I/O pins (6 provide PWM output), 6 analog input pins	54 digital I/O pins (12 provide PWM output), 12 analog input pins	ATmega32U4 provides 20 digital I/O pins, 7 PWM digital I/O pins, 12 analog input pins
OS support	—	—	—	—	Supports OpenWrt Linux distribution
Power consumption	5V/40mA 200mW	5V/50mA 250mW	5V/40mA 200mW	3.3V/800mA 2.64W	5V/40mA 200mW
Size L×W or L×W×H (mm)	73×53	68.6×53.4	73×53	101.52×53.3	73×53
Cost ($)	—	30	35	50	65
Encryption chip	—	—	—	—	—
Release date	October 22, 2007	September 24, 2010	October 19, 2008	October 22, 2012	September 10, 2013

TABLE 6.2
Summary of the Features of Single-Board Computers for IoT Development Released over the Last 9 Years

Specification Highlights	BeagleBoard	PandaBoard/ES Version	A13-OlinuXino	Raspberry Pi 1 A and B	Intel Galileo Gen. 1	RIoTboard
Processor architecture	ARM Cortex-A8	Both have dual-core ARM Cortex-A9	ARM Cortex-A8	Both have ARMv6	Intel Quark X1000	ARM Cortex-A9 MPCore
Processor speed	720 MHz	1/1.2 GHz	1 GHz	700 MHz both	400 MHz	1 GHz
Memory (RAM)	256 MB LPDDR	Both have 1 GB LPDDR2	512 MB DDR3, clocked at 408 MHz	Both A and B rev. 1 have 256 MB SDRAM, and B rev. 2 has 512 MB	256 MB DDR3	1 GB clocked at 800 MHz
Onboard storage	256 MB NAND flash	32 KB for program and 32 KB for data on cache 1 and 1 MB on cache 2	—	—	8 MB NOR flash	4 GB eMMC flash
Onboard connectivity	xM version has 10/100 Mbit/s Ethernet	ES version has 10/100 Mbit/s Ethernet and Bluetooth	MINI-Wi-Fi version has Wi-Fi	Model A has no Ethernet port, but model B has one	10/100 Mbit/s Ethernet	10/100 Mbit/s Ethernet
Peripheral interfaces	DVI-D, S-Video	DVI, HDMI, audio I/O, USB, Serial/RS-232, 2 USB host ports	VGA, USB, Audio output, microphone input	Both models have HDMI, composite video jack, 15-pin MIPI, audio output, but model A has 1 USB port, while B has 2	USB, RS-232 serial port, ACPI and PCI Express	HDMI, audio I/O, 4 USB ports and 1 mini-USB port, camera interface
GPIO low-level peripherals	I2C, I2S, SPI	14-pin JTAG GPIO, UART, I2C	UEXT connector, GPIO connector with 68/74 pins	17 GPIO pins, including UART, I2C, SPI, and I2S	GPIO pins, UART	GPIO, expansion port, debug port
OS support	Angstrom Linux	Optimized Linux: Ubuntu, Android, or Firefox OS	Linux Debian and Android	Raspbian, Arch Linux, Pidora, etc.	Yocto-based Linux	Android 4.3, Linux 3.0.35 Ubuntu, Linux 3.10.17
Power consumption	2W	1W (Max)	1.56 and 2.28W for MINI-WiFi	1.5W for A and 3.5W for B	15W	5W
Size L×W or L×W×H (mm)	76.2×76.2×16	114.3×101.6	100.33×85.09	86×54×15 for A, 86×54×17 for B	107×74×23	120×75
Cost ($)	95–149	174	73	25 for A, 35 for B	70	50–79
Encryption chips	—	—	—	—	—	SJC, ANVS, A-HAB, SHA-256, 2048-bit RSA key
Release date	July 28, 2008	October 27, 2010, ES version November 16, 2011	April 17, 2012	February 2012	October 17, 2013	January 2014

In the following sections, we discuss in more detail some attributes of a few of the countless IoT development boards that are currently available in the market. The development boards we consider are Arduino/Genuino 101,* which started shipping in January 2016; Arduino/Genuino MKR1000,† released in April 2016; Adafruit Feather M0 Wi-Fi with ATWINC1500,‡ released in March 2016; Tessel 2 (Kolker, 2015), which started shipping in April 2016; and Particle Photon§ and Electron,¶ released in March 2015 and February 2016, respectively. The others are Raspberry Pi 3,** released in February 2016; Raspberry Pi Zero Wireless (W),†† released in February 2017; BeagleBone Green (BBG),‡‡ released in mid-2015; BeagleBone Green Wireless (BBGW),§§ released in May 2016; Samsung ARTIK 520,¶¶ which started shipping in February 2016; and ARTIK 1020,*** released in mid-2016 (Samsung ARTIK 520 and ARTIK 1020 were formerly known as ARTIK 5 and ARTIK 10, respectively).

The selection of these boards is not based on performance, cost, popularity, widespread usage, or being open source, but rather on the fact that they are relatively new in the market. We classify the boards into two main categories: microcontroller-based boards and SBCs. While Arduino/Genuino 101, Arduino/Genuino MKR1000, Adafruit Feather M0 Wi-Fi, Tessel 2, and Particle Photon/Electron fall under the microcontroller-based boards, Raspberry Pi 3, Raspberry Pi Zero W, BBG, BBGW, and Samsung ARTIK 520 and ARTIK 1020 fall into the SBC category. Images of some of the current microcontroller-based development boards and SBCs are shown in Figures 6.3 and 6.4, respectively.

6.5.1 Processing and Memory and Storage Power of Current IoT Hardware Platforms

Below we discuss the computational power and memory and storage capacity of the modern-day IoT development boards.

6.5.1.1 Processing Power of Current IoT Hardware Platforms

Starting with the microcontroller-based boards, Arduino/Genuino 101 uses a low-power Intel Curie module that is based on Intel Quark SE SoC. The module has two tiny cores, x86 (Intel Quark) and a 32-bit ARC EM core architecture, which operate simultaneously and share the same memory, and all clocked at 32 MHz. Arduino/Genuino MKR1000 is based on the Atmel ATSAMW25 SoC, consisting of three main blocks. The first block is the SAMD21 Cortex-M0+ 32-bit low-power ARM MCU, clocked at 48 MHz; the other two main blocks are explained subsequently. The MKR1000 also has an onboard RTC that is clocked at 32.768 kHz; the RTC is used for coordinating auto-wake-up from the stop/standby mode. The heart of Feather M0 Wi-Fi is the ATSAMD21G18 ARM Cortex-M0 processor, clocked at 48 MHz. The design of Tessel 2 is based on Atmel SAMD21G14A-MU Cortex-M0+ MCU, which is clocked at 48 MHz (Kolker, 2015). The hardware core of both Particle Photon and Electron is the 32-bit STM32 ARM Cortex-M3 MCU, clocked at 120 MHz.

---

* https://www.arduino.cc/en/Main/ArduinoBoard101.
† https://www.arduino.cc/en/Main/ArduinoMKR1000.
‡ https://www.adafruit.com/product/3010.
§ https://docs.particle.io/datasheets/photon-datasheet/.
¶ http://staging-www2.spark.io/prototype.
** https://www.raspberrypi.org/blog/raspberry-pi-3-on-sale/.
†† https://www.raspberrypi.org/blog/raspberry-pi-zero-w-joins-family/.
‡‡ https://beagleboard.org/green.
§§ https://beagleboard.org/green-wireless.
¶¶ https://www.digikey.com/en/product-highlight/s/samsung-led/artik-520-modules-and-developer-kits.
*** https://www.digikey.com/en/product-highlight/s/samsung-led/artik-1020-modules-and-developer-kits.

FIGURE 6.3 Some current microcontroller-based IoT development boards: (a) Arduino 101 (images CC-SA-BY from Arduino.cc), (b) Arduino MKR1000 (images CC-SA-BY from Arduino.cc), (c) Particle Photon (from Particle.io, 2017), (d) Particle Electron 3G (from Particle.io, 2017), and (e) Particle Electron 2G (from Particle.io, 2017).

On the other hand, the processing capacity of the SBCs is usually much higher. For instance, the Raspberry Pi 3 SBC is powered by a 1.2 GHz 64-bit quad-core ARM Cortex-A53 CPU (MagPi, 2016), with approximately 10 times the performance of Pi 1. Raspberry Pi Zero W features a Broadcom BCM2835 SoC with a 1 GHz ARM1176JZF-S CPU core (MagPi, 2017). Similarly, both BBG and BBGW are based on the powerful AM335x ARM Cortex-A8, and clock at 1 GHz. ARTIK 520 features a dual-core Cortex-A7 ARM CPU that is clocked at 1 GHz. ARTIK 1020 features eight processing cores consisting of a quad-core Cortex-A15 ARM CPU with a 1.5 GHz clock frequency, and a quad-core Cortex-A7 ARM CPU, clocked at 1.3 GHz. When compared with the IoT development boards over the last 9 years, the processing speed of current development boards is much faster.

6.5.1.2 Memory and Storage Capacity of Current IoT Hardware Platforms

Arduino/Genuino 101 has 24 KB of SRAM and a flash memory of 196 KB. Coincidentally, Arduino/Genuino MKR1000 and Feather M0 Wi-Fi feature the same memory and storage

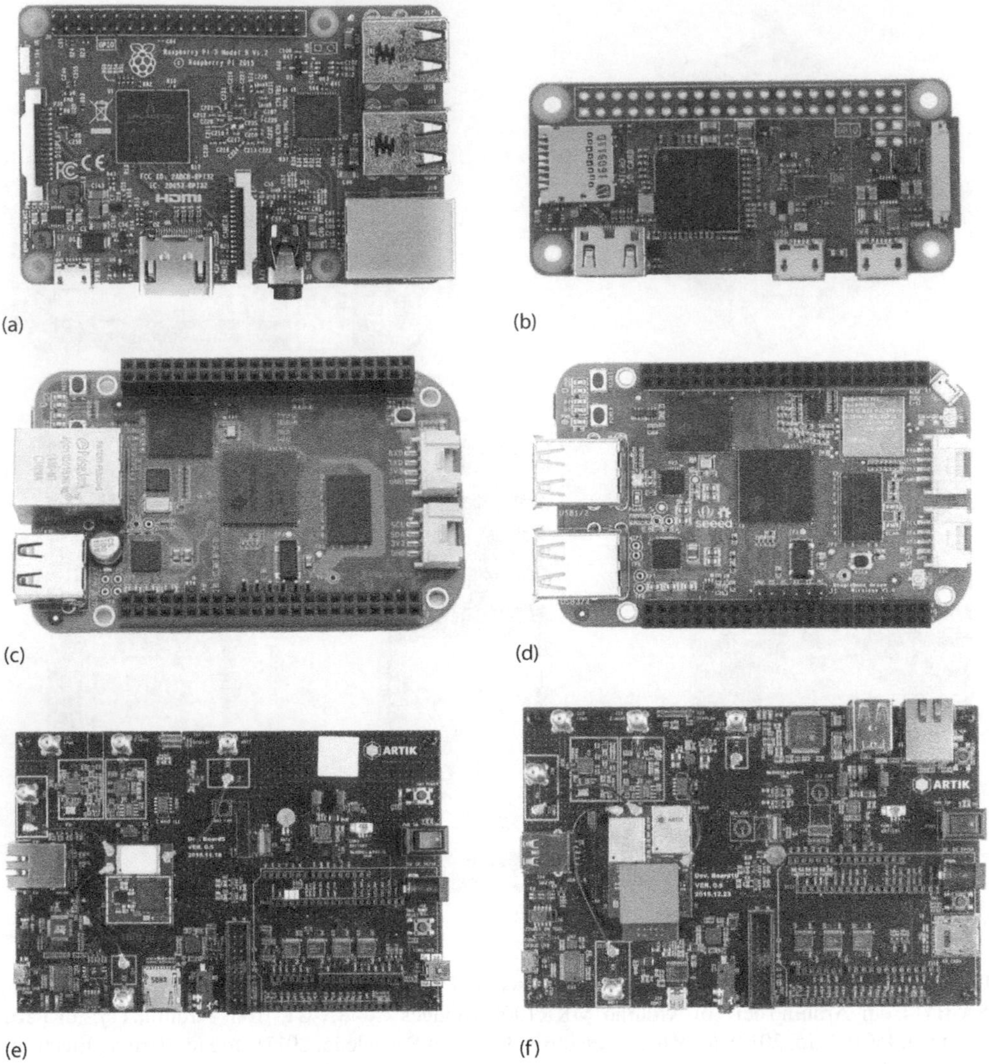

FIGURE 6.4 Some current SBCs for IoT prototyping: (a) Raspberry Pi 3 (from Raspberrypi.org, 2017), (b) Raspberry Pi Zero W (from Raspberrypi.org, 2017), (c) BBG (from source: BeagleBoard.org, 2017), (d) BBGW (from BeagleBoard.org, 2017), (e) ARTIK 520 rev. 0.5 (Copyright © Samsung Electronics Co., Ltd. Reprinted with permission. All Rights Reserved), and (f) ARTIK 1020 rev. 0.5 (Copyright © Samsung Electronics Co., Ltd. Reprinted with permission. All Rights Reserved).

specifications, 32 KB of SRAM and 256 KB of flash memory, respectively. Tessel 2 has 16 KB of SRAM, 2 KB of flash memory, 64 MB of DDR2 system memory, and 32 MB of flash memory for firmware (OpenWRT) (Kolker, 2015). Both Particle Photon and Electron also have the same specifications, which are 128 KB of RAM and 1 MB of flash memory. This shows a significant difference between the past and the current microcontroller-based IoT development boards.

Raspberry Pi 3 features a 1 GB RAM clocked at 900 MHz (MagPi, 2016). Raspberry Pi Zero W has 512 MB of LPDDR2 SDRAM. Both BBG and BBGW have 512 MB of DDR3 SDRAM and 4 GB of 8-bit eMMC onboard flash storage. Samsung ARTIK 520 has 512 MB of LPDDR3 SDRAM and 4 GB of eMMC flash storage. Finally, ARTIK 1020 has 2 GB of LPDDR3 SDRAM and 16 GB of eMMC flash storage. There are significant improvements in both memory and storage capacity if we compare the present-day IoT development boards with the ones manufactured over the last 9 years.

6.5.2 Connectivity and Input/Output Ports of Current IoT Hardware Platforms

In this section, we consider the connectivity and I/O peripherals of current microcontroller-based boards and current SBCs under two different subsections.

6.5.2.1 Connectivity and I/O Interfaces of Current Microcontroller Boards

Arduino/Genuino 101 has Bluetooth Low Energy (BLE), but cannot connect to the Internet without an Arduino Internet *shield*. Apart from a USB port, the board features 14 GPIO digital pins (of which 4 are PWM output), as well as 6 analog input pins. The board also features SPI and I2C/TWI hardware interfaces, as well as an ICSP header. The second block of ATSAMW25 on Arduino/Genuino MKR1000 is the WINC1500, a low-power 2.4 GHz IEEE 802.11b/g/n Wi-Fi that provides Internet connectivity. The MKR1000 has a micro-USB port as well as a number of GPIO low-level peripherals, including 8 digital I/O pins, 12 PWM pins, and 7 analog input pins. The remaining hardware I/O and communication interfaces are one SPI, one I2C, and one UART. Adafruit Feather M0 Wi-Fi with ATWINC1500 comes with a built-in reliable and high-speed ATWINC1500 Wi-Fi module (Ada, 2016) that enables it to connect to Internet using IEEE 802.11g or IEEE 802.11n Wi-Fi. It has a micro-USB port and 20 GPIO pins, including 8 PWM pins, 10 analog inputs, and 1 analog output. In addition, it features hardware serial, 12C, and SPI support. Tessel 2 features 10/100Mbps (megabits per second) Ethernet and IEEE 802.11b/g/n Wi-Fi with dual PCB antennas (Kolker, 2015). The board has two USB ports, one micro-USB for power and programming, and two primary sets of ports, A and B. Each port has 10 pins: 2 pins for power (3.3V and ground) and 8 GPIO pins, a total of 16 GPIOs on both ports. Particle Photon connects to the Internet via IEEE 802.11b/g/n Wi-Fi (Particle: Photon Datasheet, 2016). The board has a lot of peripherals, including 18 digital I/Os, 8 analog (analog-to-digital converter [ADC]) inputs, 2 analog (DAC) outputs, 2 SPI I/Os, 1 12S I/O, 1 12C I/O, 9 PWM output pins, and 1 micro-USB port. There are two models of Particle Electron; one model has U-blox SARA-G350 and the other has U-blox SARA-U260/U270 cellular moderns for 2G and 3G cellular connectivity, respectively (Farnham, 2016). Each model has a total of 36 I/O peripheral pins, including 28 GPIOs, TX/RX, and a micro-USB port.

6.5.2.2 Connectivity and I/O Interfaces of Current Single-Board Computers

Raspberry Pi 3 features IEEE 802.11b/g/n wireless LAN and Bluetooth 4.1, aside from the 10/100 Mbps Ethernet port (Allan, 2016). The SBC has a 40-pin GPIO header, of which 27 are accessible to the user. The GPIO pins can be configured as I2C, SPI, and UART;* there are also 3.3 and 5V sources, as well as a few grounds. Apart from the low-level peripherals, Raspberry Pi 3 also has an HDMI port, a 3.5 mm analog audio-video jack, four USB 2.0 ports, a Camera Serial Interface (CSI), and a Display Serial Interface (DSI). Like Raspberry Pi 3, the Pi Zero W features the IEEE 802.11 b/g/n wireless LAN and Bluetooth 4.1, as well as BLE; however, it does not have an Ethernet port. Raspberry Pi Zero W is also equipped with all the I/O interfaces that are on the Pi 3, although it does not have a DSI and combined 3.5 mm audio and composite video jack. In addition, the HDMI on Pi Zero W is a mini-version, and while Pi 3 has four USB 2.0 standard ports, Pi Zero W only has a mini-USB On-The-Go (OTG) port. The BBG has a 10/100Mbps RJ45 Ethernet port, a micro-USB 2.0 client for power and communication, a USB 2.0 host port, 7 analog I/O pins, 65 digital

* https://developer.microsoft.com/en-us/windows/iot/docs/pinmappingsrpi.

I/O (GPIO) pins, and 8 PWM pins.* The BBGW has a built-in 2.4 GHz IEEE 802.11b/g/n Wi-Fi module, as well as BLE 4.1 connectivity. It also features a USB host with four port hubs (Rush, 2016), plus all the other I/O peripherals that are on the BBG. Samsung's ARTIK 520 comes with IEEE 802.11a/b/g/n/ac Wi-Fi, Bluetooth BT/BLE, and IEEE 802.15.4 ZigBee connectivity. It has the following analog and digital I/O interfaces: GPIO, I2C, SPI, UART, Secure Digital I/O (SDIO), USB 2.0, JTAG, and analog input. It also has some media interfaces, such as a two-lane MIPI, an I2S audio interface, and a MIPI CSI. ARTIK 1020 has the same connectivity as ARTIK 520. In addition to the I/O interfaces of ARTIK 520, ARTIK 1020 also has an I2S low-level peripheral and USB 3.0. Its media interfaces are a four-lane MIPI DSI, a simultaneous HDMI, a one-channel pulse code modulation (PCM), and a two-channel I2S audio interface.

6.5.3 OS Support for Current IoT Hardware Platforms

This section examines the OS support for current IoT development boards. We start by examining the OS support for microcontroller-based boards, followed by OS support for SBCs.

6.5.3.1 OS Support for Current Microcontroller-Based Boards

Although Arduino/Genuino 101 is designed to run an RTOS using the Intel Curie, at the time of writing, the RTOS is still under development; however, the source code of the RTOS was released recently for hacking and studying purposes. On the other hand, Arduino/Genuino MKR1000, Adafruit Feather M0 Wi-Fi, and Tessel 2 do not support any OS. Particle Photon comes with an RTOS (FreeRTOS) that provides scheduling for the Wi-Fi connectivity code.† Users can also use the RTOS to create their own multithreaded code using the Application Programming Interface (API) that will soon be provided (Supalla, 2015). Particle Electron, however, does not support any OS.

6.5.3.2 OS Support for Current Single-Board Computers

Aside from the Raspbian, both Raspberry Pi 3 and Pi Zero W support other OSes, like Ubuntu Mate, OSMC, OpenELEC, and Windows IoT Core‡ (Klosowski, 2016). Both BBG and BBGW also support HLOSs such as Linux, Android, Debian, and Ubuntu. Samsung ARTIK 520 comes pre-installed with Linux-based Fedora OS, and ARTIK 1020 also supports Fedora, as well as Snappy Ubuntu Core and Tizen OS (Rouffineau, 2016).

6.5.4 Battery Life of Present-Day IoT Hardware Development Boards

In this section, we discuss the battery life of current IoT development boards, beginning with the microcontroller-based boards, followed by the SBCs.

6.5.4.1 Battery Life of Current Microcontroller-Based IoT Hardware Development Boards

Despite the additional functionalities and capabilities of Arduino/Genuino 101 when compared with the *Uno*, the low-power consumption of Intel Curie SoC on Arduino/Genuino 101 maintains a DC current per I/O pin of 20mA at an operating voltage of 3.3V. However, its operating current is not explicitly specified, and since it has BLE functionality, its *average power consumption* cannot be easily estimated from the given specifications. The low-power capability of Arduino/Genuino MKR1000 MCU maintains a DC current per I/O pin of 7mA only at a 3.3V operating voltage. Although the average current consumption of the MCU is about 20 mA, when operational, the Wi-Fi on MKR1000 can consume roughly 100mA or more.§ Hence, the total current consumption

* http://www.mouser.com/pdfdocs/Seeed-BBG_SRM_V3_20150804.pdf.

† https://www.seeedstudio.com/Particle-Photon-p-2527.html.

‡ https://www.raspberrypi.org/downloads/.

§ https://www.arduino.cc/en/Tutorial/MKR1000BatteryLife.

of the board for IoT applications will be about 120mA or more, implying that the average power consumption of the MKR1000 will be about 396mW. The ATWINC1500 used on the Adafruit Feather M0 Wi-Fi is an extreme low-power IEEE 802.11b/g/n IoT network controller SoC, such that the Wi-Fi draws about 130mA during transmission at an operating voltage of 3.7V. The board can be programmed to further drop the consumption to about 22mA by shutting down the unneeded parts, making the power consumption to be about 481 and 81mW, respectively. Unlike Tessel 1, the sleep modes and power control over module ports are enabled on Tessel 2 (McKay, 2015). This allows the board to save a significant amount of power; however, the power consumption or average current drawn by Tessel 2 is not specified. A typical average current consumption of Particle Photon with Wi-Fi turned on is 80mA, with 5VDC, and the deep sleep quiescent current is about 8μA; therefore, the power consumptions are 400mW and 40μW, respectively. When powered from a LiPo battery, a typical average current drawn by Particle Electron is between 180mA and 1.8A transients at 5V DC, and the deep sleep quiescent current is about 130μA; hence, Particle Electron power consumption ranges from 900mW to 9W.

6.5.4.2 Battery Life of Current Single-Board Computers

In the official Raspberry Pi magazine (MagPi, 2016), the current draws of different Raspberry Pi models are presented in two modes: standby and run. For Raspberry Pi 3, they are 0.31 and 0.58A, respectively. This implies that at 5V, the power consumption of Raspberry Pi 3 ranges from 1.55 to 2.9W. According to Klosowski (2017) and RasPi.TV (2017), the average current consumption of Raspberry Pi Zero W, when running, is about 0.18A, meaning that its average power consumption is about 0.9W. BeagleBoard.org has not specified the current or power consumption of both BBG and BBGW. Similarly, although both Samsung ARTIK 520 and ARTIK 1020 use a Power Management Integrated Circuit (PMIC) to provide power to the modules using onboard bucks and Low-DropOut (LDO) regulators, Samsung did not specify the current or power consumption of these new boards.

6.5.5 Size and Cost of Current IoT Development Boards

This section discusses the form factor and cost of current IoT hardware development boards. The size and cost are divided into two subsections: size and cost of microcontroller-based boards, and size and cost of SBCs. Note that the cost in this section refers to the price of each board at the time of writing this chapter.

6.5.5.1 Size and Cost of Current Microcontroller-Based Development Boards

The length and width of Arduino/Genuino 101 are 68.6 and 53.4 mm, respectively, and the board is currently selling for $30. The dimensions of Arduino/Genuino MKR1000 in terms of length, width, and height are 65, 25, and 6 mm, respectively, and its price is $34.99. Adafruit Feather M0 Wi-Fi measures $53.65 \times 23 \times 8$ mm^3 (without headers soldered), and it costs $34.95. The dimensions of Tessel 2 are not really specified, but it costs $35. The dimensions of Particle Photon with and without headers in terms of length, width, and height are $36.58 \times 20.32 \times 6.86$ mm^3 and $36.58 \times 20.32 \times 4.32$ mm^3, respectively. The cost of Photon with headers is $19. The dimensions of both models of Particle Electron with and without headers are nearly the same: $50.8 \times 20.32 \times 12.7$ mm^3 and $50.8 \times 20.32 \times 7.62$ mm^3. While the cost of the 2G model of Particle Electron is $49, the 3G model costs $69.

6.5.5.2 Size and Cost of Current Single-Board Computers

The length of Raspberry Pi 3 is 85 mm, and its width is 56 mm, and like its predecessor, Pi 3 is also selling for $35. However, being one of the smallest Raspberry Pis, the dimensions of Pi Zero W are $65 \times 30 \times 5.4$ mm^3, and it is selling for $10. The dimensions of BBG and BBGW are, respectively, $86.36 \times 53.34 \times 19.05$ mm^3 and $86.36 \times 53.34 \times 21.59$ mm^3, and while BBG costs $39, BBGW is selling for $44.9. Finally, the dimensions of ARTIK 520 and ARTIK 1020 are

$29 \times 25 \times 3.5$ mm^3 and $39 \times 29 \times 3.5$ mm^3, respectively. Currently, the price of ARTIK 520 is \$99.99 and the ARTIK 1020 costs \$149.99.

6.5.6 Security Features of the Present IoT Hardware Platforms

This section highlights the security features of current IoT hardware development platforms. We discuss the security features under two subsections: security features of microcontroller-based boards and security features of SBCs.

6.5.6.1 Security Features of Current Microcontroller-Based Boards

Arduino/Genuino 101 does not have any onboard encryption chip. Conversely, Arduino/Genuino MKR1000 has an encryption chip; the third block in the ATSAMW25 is ECC508, which provides crypto-authentication. The built-in Wi-Fi also has a crypto-chip for ensuring secure communication. The ATWINC1500 on Adafruit Feather M0 Wi-Fi supports Wired Equivalent Privacy (WEP), Wi-Fi Protected Access (WPA), WPA2, Transport Layer Security (TLS), and SSL encryption. On the other hand, Tessel 2 has no encryption engine onboard. All communications between every Particle device and the Particle cloud are encrypted by default using industry standard WPA2 (i.e., for Photon) or standard 3G/2G (i.e., for Electron) radio protocols. In addition, to defend against port scans, all incoming ports on Particle Photon and Electron are closed by default (Particle, 2017).

6.5.6.2 Security Features of Current Single-Board Computers

Raspberry Pi 3, Raspberry Pi Zero W, BBG, and BBGW have no onboard encryption chips. ARTIK 520 and ARTIK 1020, on the other hand, both have encryption chips based on ARM TrustZone and Trusted Execution Environment (TEE) from Trustonic, which provide *bank-level* end-to-end security. These security features ensure secure point-to-point authentication and secure data transfer.

Tables 6.3 and 6.4 provide an overview of the various specifications of the current IoT hardware development boards that were used as examples in Section 6.5.

6.6 IoT HARDWARE DEVELOPMENT PLATFORMS IN THE NEXT 5 YEARS

The IoT is presently an emerging technology that is growing at a breathtaking pace and impacting many aspects of life, as well as various industries. It is also among the top technologies on the horizon that are poised to change life as we know it today. In the space of just a decade, the number of heterogeneously connected devices is already overwhelming. In addition, millions more devices are still expected to come online over the next 5 years (Gartner, 2015), which will make the interaction of humans with technology commonplace. Considering the present technology trends, it would be no exaggeration to say that in the not too distant future, almost every *object* will be connected to the Internet, thanks to Internet Protocol version 6 (IPv6). This will usher in a new era of ubiquity and a new era of endless opportunities for consumers to enhance their living conditions, and for business owners to increase productivity, reduce cost, and save energy.

To a large extent, the future of the IoT will depend on the development of a variety of new optimized hardware platforms that will compete for the attention of the developer community. The future IoT hardware platforms are envisioned to allow developers and designers to build more efficient, smaller, and ultra-energy-efficient IoT devices, as well as dramatically cut the cost of production and reduce time to market. This will be driven mainly by chip miniaturization, resulting from a revolution in cheap sensor technology, the affordability of Bluetooth wireless technology, and the growing ubiquity of more secure and low-power Wi-Fi technologies.

This section focuses on predicting the attributes of IoT hardware development platforms in the next 5 years. Our projections are based on the latest technological breakthroughs in chip design

TABLE 6.3

Summary of the Specifications of Current Microcontroller-Based IoT Hardware Development Platforms

Specification Highlights	Arduino 101	Arduino MKR1000	Adafruit Feather M0 Wi-Fi	Tessel 2	Particle Photon	Particle Electron
Processor architecture	2 cores: × 86 (Intel Quark) core and 32-bit ARC core	SAMD21 Cortex-M0+ 32-bit low-power ARM MCU	ATSAMD21G18 ARM Cortex-M0 MCU	Atmel SAMD21G14A-MU Cortex-M0+ MCU	32-bit STM32 ARM Cortex-M3 MCU	32-bit STM32 ARM Cortex-M3 MCU
Processor speed	Both cores clocked at 32 MHz	48 MHz	48 MHz	48 MHz	120 MHz	120 MHz
Memory (RAM)	24 KB of SRAM	32 KB of SRAM	32 KB of SRAM	16 KB of SRAM	128 KB of RAM	128 KB of RAM
Onboard storage	196 KB of flash	256 KB of flash	256 KB of flash	2 KB of flash, 64 MB of DDR2 system memory, 32 MB flash for firmware	1 MB of flash memory	1 MB of flash memory
Onboard connectivity	BLE	IEEE 802.11b/g/n Wi-Fi	IEEE 802.11g or n Wi-Fi	10/100 Mbit/s Ethernet and IEEE 802.11b/g/n Wi-Fi	IEEE 802.11b/g/n Wi-Fi	2 models: U-blox SARA G350/U260 for 2G and U-blox SARA G350/U270 for 3G
Peripheral interfaces	Micro-USB	Micro-USB	Micro-USB	2 USB ports, 1 micro-USB	Micro-USB	Micro-USB
GPIO low-level peripherals	14 digital I/O pins (of which 4 are PWM), 6 analog input pins	8 digital I/O pins, 12 PWM pins, 7 analog input pins, 1 UART, 1 SPI, 1 I2C	8 PWM pins, 10 analog input pins, and 1 analog output pin, I2C, SPI	16 GPIO pins on 2 primary ports, A and B	18 digital I/O pins, 8 analog (ADC) inputs, 2 analog (DAC) outputs, 9 PWM output pins, 2 SPI, 1 I2S, 1 I2C	28 GPIOs and TX/RX
OS support	Open-source RTOS	—	—	—	FreeRTOS	—
Power consumption	—	396 mW	481 and 81mW when unneeded parts shut down	Not specified	400mW and 40 μW during deep sleep	900mW to 9W
Size L×W or L×W×H (mm)	68.6×53.4	65×25×6	53.65×23×8 (without headers soldered)	Not specified	36.58×20.32×6.86 (with headers), 36.58×20.32×4.32 (without)	50.8×20.32×12.7 (with headers), 50.8×20.32×7.62 (without)
Cost ($)	30	34.99	34.99	35	19	49 for 2G, 69 for 3G
Encryption chips	—	ECC508, WINC1500	ATWINC1500	—	On the MCU	—
Release date	January 2016	April 2016	March 2016	April 2016	March 2015	February 2016

TABLE 6.4
Summary of the Specifications of Current Single-Board Computers for IoT Development

Specification Highlights	Raspberry Pi 3	RaspberryPi Zero W	BeagleBone Green	BeagleBone Green Wireless	Samsung ARTIK 520	Samsung ARTIK 1020
Processor architecture	64-bit quad-core ARM Cortex-A53 CPU	BCM2835 SoC ARM1176JZF-S CPU core	AM335x ARM Cortex-A8 CPU	AM335x ARM Cortex-A8 CPU	Dual-core ARM Cortex-A7 CPU	8 cores: A quad-core Cortex-A15 ARM CPU and a quad-core Cortex-A7 ARM CPU
Processor speed	1.2 GHz	1 GHz	1 GHz	1 GHz	1 GHz	1.5 GH/1.3 GHz
Memory (RAM)	1 GB, clocked at 900 MHz	512 MB of LPDDR2 SDRAM	512 MB of DDR3 SDRAM	512 MB of DDR3 SDRAM	512 MB of LPDDR3 SDRAM	2 GB of LPDDR3 SDRAM
Onboard storage	—	—	4 GB 8-bit eMMC flash storage	4 GB 8-bit eMMC flash storage	4 GB of eMMC flash storage	16 GB of eMMC flash storage
Onboard connectivity	IEEE 802.11 b/g/n wireless LAN, 10/100 Mbit/s Ethernet, Bluetooth 4.1	IEEE 802.11 b/g/n wireless LAN, Bluetooth 4.1, BLE	10/100 Mbit/s Ethernet	2.4 GHz IEEE 802.11b/g/n Wi-Fi, Bluetooth 4.1 with BLE	IEEE 802.11a/b/g/n/ac Wi-Fi, Bluetooth BT/BLE, IEEE 802.15.4 ZigBee	IEEE 802.11a/b/g/n/ac Wi-Fi, Bluetooth BT/BLE, IEEE 802.15.4 ZigBee
Peripheral interfaces	HDMI, 3.5 mm analog audio/video jack, 4 USB ports, CSI, DSI	HDMI mini, CSI, mini-USB OTG port	USB port, micro-USB port	USB host with 4 port hubs	2-lane MIPI, I2S audio interface, MIPI CSI, USB port	4-lane MIPI DSI, a simultaneous HDMI, 1-channel PCM, 2-channel I2C audio interface USB
GPIO low-level peripherals	40 GPIO pins, including I2C, SPI, UART	40 GPIO pins, including I2C, SPI, UART	65 digital I/O pins, 7 analog I/O pins, 8 PWM pins	65 digital I/O pins, 7 analog I/O pins, 8 PWM pins	I2C, SPI, UART, SDIO, JTAG	I2C, I2S, SPI, UART, SDIO, JTAG
OS support	Raspbian, Ubuntu Mate, OSMC, OpenELEC, Windows IoT Core, etc.	Raspbian, Ubuntu Mate, OSMC, OpenELEC, Windows IoT Core, etc.	Supports HLOSs like Linux, Android, Debian, Ubuntu	Supports HLOSs like Linux, Android, Debian, Ubuntu	Preinstalled Linux-based Fedora	Fedora, Snappy Ubuntu Core, Tizen OS
Power consumption	1.55–2.9W	0.9W on the average	Not specified	Not specified	Not specified	Not specified
Size L×W or L×W×H (mm)	85×56	65×30×5.4	86.36×53.34×19.05	86.36×53.34×21.59	29×25×3.5	39×29×3.5
Cost ($)	35	10	39	44.9	99.99	149.99
Encryption chip	—	—	—	—	ARM TrustZone, TEE	ARM TrustZone, TEE
Release date	February 2016	February 2017	Mid-2015	May 2016	February 2016	Mid-2016

and manufacturing, and the anticipated technological developments in the chipmaking industry. We make our predictions on the same hardware features that we have considered in the previous sections.

6.6.1 Processing and Memory and Storage Capacity of Future IoT Hardware Platforms

This section looks at the processing power of future IoT hardware development platforms, as well as discusses their memory and storage capacity.

6.6.1.1 Processing Power of Future IoT Hardware Platforms

While Intel has admitted that transistors have become almost infinitesimally small (Paul, 2016), implying that the era of Moore's law may be coming to an end (Green, 2015), and is now focusing on sacrificing speed for power efficiency (Paul, 2016), other companies will just not give up. For example, in an effort to push the limits of chip technology needed to power emerging technologies, an IBM-led consortium in July 2015 announced the first 7 nm node test chips in the world (Savov, 2015). Other collaborators in the group include Samsung and SUNY Polytechnic Institute (Savov, 2015).

Processor performance requirements for connected devices depend largely on the type of sensing, processing, and communication needed for the target application (Voica, 2016). For instance, some devices are designed to perform a limited amount of processing on datasets like temperature or humidity, and others are designed to handle more complicated tasks, such as video streams or high-resolution sound. Due to the diversity of IoT applications, and especially as the IoT matures in the near future, with smart devices able to perform much more complex tasks without human intervention, there may be a need for greater diversity of chip configurations than what is found in computers and smartphones. Consequently, some chip construction models are emerging. One such model is the Multi-Chip-Module (MCM), which promises low volume, high system performance, and high reliability. Another model is the System-in-a-Package (SiP), consisting of ICs with different functionalities, and it may include passive components and/or a Micro-Electro-Mechanical System (MEMS), all assembled into one package that functions as a subsystem or system (Derhacobian, 2016).

Given the foregoing, we arguably forecast that in the next 5 years, the processors in IoT hardware platforms will have the following:

1. Different architectures depending on the application.
2. One or more 32- to 64-bit cores.
3. Clock frequency from 2 GHz and above.

6.6.1.2 Memory and Storage Capacity of Future IoT Hardware Platforms

The current exponential trends in capacity and price of memory and storage devices that have been consistent for more than 50 years are expected to continue into the future, even if miniaturization limits are reached. This is supported by the fact that in May 2016, researchers at IBM succeeded in storing 3 bits of data per cell using a new technology known as Phase-change Memory (PcM) (Future Timeline.net, 2016). Using PcM technology, a memory can provide high read and write speed, endure at least 10 million write cycles, and not lose data when powered off, unlike DRAM. This remarkable achievement can provide fast and efficient storage that can take care of the exponential growth of data from IoT devices. In addition, in 2016 Intel and Micron released a Three-Dimensional (3D) XPoint memory, also known as Optane (Mearian, 2016). The new memory uses byte addressing, can endure write cycles 1000 times more than the traditional NAND flash, and has 1000 times faster I/Os. In the long run, Optane may replace DRAM (Hruska, 2016).

Going by the above trend, we arguably predict that the RAM of IoT hardware platforms in the next 5 years will be from 2.5 GB and above, and the flash memory will be from 16 GB and above.

6.6.2 Connectivity and Communication Interfaces of Future IoT Hardware Development Boards

In this section, we try to forecast the connectivity and communication interfaces of IoT hardware development platforms in the next 5 years.

6.6.2.1 Connectivity of Future IoT Hardware Development Platforms

Reliable Internet connectivity is perhaps the most important component of the IoT. As the Internet connectivity is being extended to more and more devices through wireless mobile connectivity, and the complexity of connected devices is increasing by the day, the need for connected devices with more reliable Internet connectivity is becoming increasingly apparent. Although the Internet of space (IoS) (Raman et al., 2016), a high-data-rate suborbital-based communication network, is still on the horizon, mobile networks such as 3G and 4G, as well as other wireless technologies like Wi-Fi will continue to play crucial roles in providing the IoT with Internet connectivity, at least in the near future.

Based on the above discussion, the connectivity of IoT hardware development platforms in the next 5 years may be forecasted as follows:

1. Most development boards will feature both Wi-Fi and cellular connectivity.
2. Some hardware platforms will start using the new Wi-Fi HaLow (Sartain, 2016), which extends Wi-Fi into the 900 MHz band. HaLow is suited for small data payloads, has a better range than the traditional 2.4 and 5 GHz bands, and can penetrate physical barriers.
3. The mobile wireless connectivity will be 4G and 5G.
4. Some devices may use Wi-Fi-like technologies, such as White-Fi (IEEE 802.11af) or the IEEE 802.11ah (DeLisle, 2015). It is expected that these technologies will use the sub-1 GHz spectrum, which will provide long-range and low-power operation. The range of IEEE 802.11ah, in particular, is expected to be up to 1 km with data rates between 150 kbps and 8 Mbps (Tian et al., 2016; Park, 2015).
5. Some devices will also be connected via IoS networks.

6.6.2.2 Communication Interfaces of Future IoT Development Platforms

While many IoT applications like wireless sensor networks (WSNs) for simple environmental monitoring require only a limited number of interfaces for connecting a few sensors, there are other applications that will require quite a number of different I/O peripherals. In the near future, the situation with regards to I/O interfaces on IoT devices is very much likely to change. As the IoT takes shape and offers endless possibilities for organizations, businesses, and services, the need for a greater diversity of peripherals will exponentially rise in order to meet the ever-increasing demands. In the next 5 years, it is most likely that:

1. IoT devices will be used for services that will demand more user interaction through multimedia than ever before.
2. The speed requirements of the peripherals will be comparable to the rate at which the processor demands data or instructions from the memory.
3. Some interfaces may require very high bandwidths, and there will still be some whose requirements may be minimal.

6.6.3 OS Support for Future IoT Development Boards

An OS usually acts as a resource manager, making it essential for managing the limited resources on some resource-constrained devices. As the deployment of IoT is increasingly becoming more cost-effective (i.e., as sensors and chips become smaller and cheaper) and the network becomes more complex in terms of diversity and number of devices (Morgan, 2015; Gil et al., 2016), more data-driven services will evolve in the future that may require IoT devices to run more robust and reliable RTOSs (Gaur and Tahiliani, 2015).

We therefore forecast that in the next 5 years, all IoT hardware development boards will have OS support and the OSes would:

1. Have robust security capabilities.
2. Be scalable, such that they can be used on different devices.
3. Be modular, so that developers can choose components based on their system requirements.
4. Support most connectivity standards, such as Wi-Fi and Bluetooth.
5. Support most cellular standards, like 3G, 4G, LTE, and 5G.

6.6.4 Battery Life of Future IoT Hardware Platforms

How much battery energy is consumed by an IoT device depends significantly on the radio transmitter type, protocols used for communications, the sensors, and the processor type (Vujovic et al., 2015). Examples of batteries commonly used in small IoT devices include lithium, nickel, and alkaline batteries. Most of these battery chemistries offer very low self-discharging characteristics, making them well suited for long service intervals. Nonetheless, one of the most important things to consider when designing and deploying IoT devices for certain applications is the battery life. This is because replacing batteries in the field is not economically viable, especially if the replacement will involve thousands or even millions of devices. Achieving ultra-low-power consumption and extending the battery life of IoT devices are active areas of research (Somov and Giaffreda, 2015).

With that in mind, we arguably predict that in the next 5 years,

1. A battery life of upto 10 years and above is attainable.
2. The new Wi-Fi HaLow that is also aimed at reducing power needs will greatly lower the power consumption of the new IoT hardware platforms that will use the technology.
3. Future IoT hardware platforms will use energy harvesting schemes, such as solar and thermal gradients to charge batteries onboard.
4. Future IoT devices may also be powered by the *ambient signals from Wi-Fi routers*, a novel technology developed by engineers at the University of Washington (Langston, 2015).

6.6.5 Size and Cost of Future IoT Development Boards

In general, a relationship exists between the size and the cost of electronic devices. As devices become smaller, their prices normally increase. But when the technology matures and the process of miniaturization is fully automated, prices begin to decline (B'Far, 2005). In this section, we predict the size and cost of future IoT devices in the next 5 years.

Over the last few years, the idea of shrinking transistor sizes onto microcontrollers and computer processors to enhance performance as well as to reduce size and cost has become more complex following the twilight of Moore's law. Recently, a group of researchers in a French microelectronic laboratory has developed a new process for stacking thin layers of semiconductor material with transistors while the performance of the transistors remains intact (Morra, 2016). The result is a landmark achievement that led to the development of monolithic 3D chips, which behave like a single device, having the same size as the two-dimensional (2D) chips, consuming less power and generating less heat. Furthermore, research advances in a number of areas other than IC dies are proving to be highly promising in the continued progress toward miniaturization (Mehta et al., 2016; Pinkerton, 2002). A notable example is the use of flex PCB and High-Density Interconnect (HDI) PCBs to manufacture smaller, but yet sophisticated IoT wearable devices, ranging from medical implants and hearing aids to fitness trackers (Bahl, 2016).

The bottom line, however, is that several electronics devices have reached a near-optimal form factor already, and stretching further to get a cutting-edge miniaturization will be very expensive (B'Far, 2005). Additionally, a major trade-off in miniaturization is whether the market will be able to support the cost. There is no doubt that markets like the military, medical electronics, and aerospace can support the cost. However, in several IoT applications, where devices are expected to be disposable or too numerous to count, this will certainly be an issue, at least in the short run.

Having said that, we arguably make the following forecasts (for the next 5 years) for both the size and cost of future IoT hardware development boards, depending on whether they are microcontroller-based boards or SBCs:

- The sizes of these devices in terms of length, width, and height will be approximately $25 \times 18 \times 3$ mm and $60 \times 45 \times 15$ mm for microcontroller-based boards and SBCs, respectively.
- The prices of microcontroller-based boards and SBCs will be about \$5 and \$18, respectively.

6.6.6 Security Features of Future IoT Hardware Platforms

Virtually any *thing* that is connected to other *things* and accessible over the Internet is prone to cyber attacks. Software security alone has proven inadequate to protect devices against many known threats (Hanna, 2016), such as denial of service (DoS), distributed denial of service (DDoS), malware, espionage, tampering, and hijacking. Therefore, the future of IoT security and privacy will depend on the ability of the various hardware vendors to implement reliable security at the hardware level. This can be achieved by including an encryption chip, also known as a Security Controller (SC), in the hardware. The SC performs a defined set of cryptographic operations using cryptographic keys that are securely stored in the SC (Lesjak et al., 2015). The operations include identifying unauthorized access and detecting tampering. If tampering or microprobing is detected, the chip should cause a tamper response that will result in an immediate *zeroization* (Moritz et al., 2015). The SC should also be resistant against any side-channel attacks like Differential Power Analysis (DPA). Based on the foregoing, we can arguably predict that virtually all the IoT hardware development platforms that will be developed in the next 5 years will have encryption chips onboard.

6.7 TIMELINE OF EVOLUTION OF THE IoT HARDWARE DEVELOPMENT PLATFORMS

This section presents a graphical timeline of the IoT hardware development platforms discussed in the preceding sections, as shown in Figure 6.5. The figure depicts the evolution of the hardware development boards that were considered in this chapter, starting with the hardware development platforms that were used in the past 9 years, followed by those that are currently being used today, and it also portrays the ones that are on the horizon in the next 5 years.

6.8 CONCLUSIONS

As we enter into a new era of IoT, with humans becoming the minority on the Internet, IoT developers will require hardware platforms with more powerful features that will be able to cope with the complexities of the future IoT. In this chapter, we have explored several features of a number of IoT hardware development platforms in the past 9 years, and the features of many that are currently being used today. We have also attempted to predict the same attributes for the hardware platforms that will be developed in the next 5 years.

Although the first release of each of those boards was considered a groundbreaking achievement because of their various capabilities, from this study it is clear that given the features of the past hardware platforms, their performance is not comparable to the modern-day hardware platforms. This is becoming more obvious as the new platforms are beginning to attain the capability of the present-day

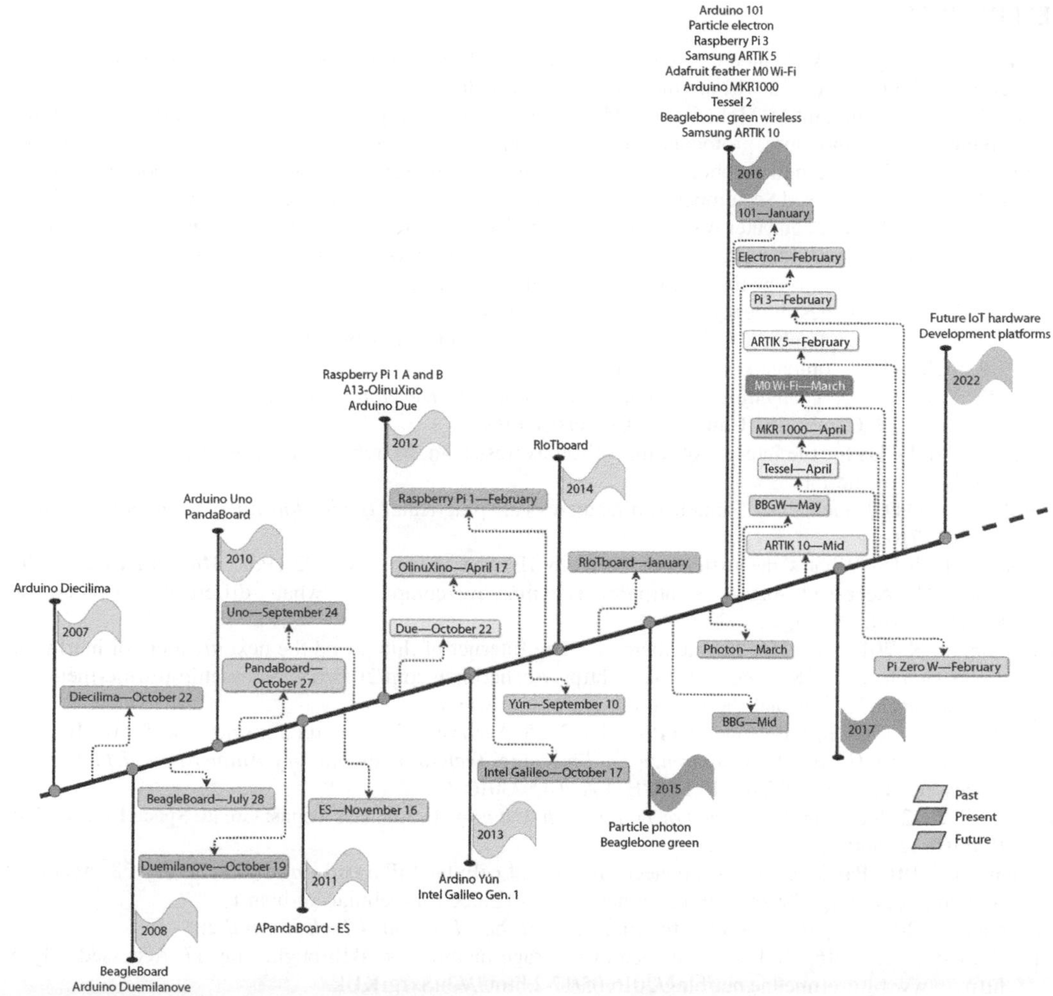

FIGURE 6.5 Graphical timeline of evolution of the IoT hardware development platforms.

smartphones and computers. Furthermore, the trends in the current technological advances and the ongoing research that we have reported in different fields show that the capabilities of the future hardware platforms will no doubt surpass the capabilities of the modern-day development boards. This is actually good news for designers and developers, because the increased capacity in terms of processing power, memory and storage capacity, and battery life, as well as the inclusion of hardware security features will translate into more robust development boards. Such boards can support the necessary security mechanisms that can be used for protecting user-sensitive data and privacy.

While designers and developers eagerly anticipate these improvements, they probably will never be satisfied. This is because, as the performance of these IoT devices improves, the complexity of our work and the type and size of real-time data we will have to deal with in the near future will also increase.

ACKNOWLEDGMENTS

The authors wish to thank the Centre for Geodesy and Geodynamics, National Space Research and Development Agency, Toro, Bauchi State, Nigeria for supporting this work. This work was supported by National Funding from the FCT—Fundação para a Ciência e a Tecnologia, through the UID/EEA/50008/2013 Project.

REFERENCES

Ada, L. 2016. Using the Wi-Fi module Adafruit. September 21. Accessed September 27. https://learn.adafruit.com/adafruit-feather-m0-wifi-atwinc1500/using-the-wifi-module.

Al-Fuqaha, A., M. Guizani, M. Mohammadi, M. Aledhari, and M. Ayyash. 2015. Internet of things: A survey on enabling technologies, protocols, and applications. *IEEE Communications Surveys Tutorials* 17: 4.

Allan, A. 2016. Meet the new raspberry Pi 3—A 64-bit Pi with built-in wireless and bluetooth LE. *Make*, February 28. Accessed September 28. http://makezine.com/2016/02/28/meet-the-new-raspberry-pi-3/.

Arthur, C. 1996. The first computer was as big as a room. Now they're the size of a full stop... and getting even maller. *Independent*, May 28. Accessed June 14. http://www.independent.co.uk/news/the-first-computer-was-as-big-as-a-room-now-theyre-the-size-of-a-full-stop-and-getting-even- smaller-1349636.html.

Atwell, C. 2013. The biggest-little revolution: 10 single-board computers for under $100. EDN Network, August 21. Accessed September 25. https://www.edn.com/design/diy/4419990/The-biggest-little-revolution--10-single-board-computers-for-under--100.

B'Far, R. 2005. *Mobile Computing Principles: Designing and Developing Mobile Applications with UML and XML*, 13. Cambridge: Cambridge University Press.

Bahl, A. 2016. PCBs for the Internet of things. Protoexpress.com, March 14. https://www.protoexpress.com/blog/pcbs-for-the-iot/.

Brumley, D. and D. Boneh. 2005. Remote timing attacks are practical. *Elsevier Journal of Computer Networks* 48: 701–716.

DeLisle, J-J. 2015. What's the difference between IEEE 802.11af and 802.11ah? *Microwaves and RF*, April 24. Accessed August 3. http://mwrf.com/active-components/what-s-difference-between-ieee-80211af-and-80211ah.

Derhacobian, N. 2016. One chio to rule them all? The Internet of things and the next great era of hardware. *TechCrunch*, May 28. Accessed July 5. https://techcrunch.com/2016/05/28/one-chip-to-rule-them-all-the-internet-of-things-and-the-next-great-era-of-hardware/.

Derhamy, H., J. Eliasson, J. Delsing, and P. Priller. 2015. A survey of commercial frameworks for the Internet of things. In *IEEE 20th Conference on Emerging Technologies Factory Automation (ETFA)*, 1–8. Accessed March 8, 2017. doi:10.1109/ETFA.2015.7301661.

Doukas, C. 2012. *Building the Internet of Things with Arduino*, 1–348. Las Vegas: Create Space Independent Publishing Platform.

Farnham, K. 2016. Particle.io ships the electron, a small cellular IoT board. *InforQ*, February 23. Accessed September 28. https://www.infoq.com/news/2016/02/electron-cellular-iot-board.

Fernandez, A. 2014. Rapid prototyping the Internet of things. *Element14 Tech Journal* 2: 8–10.

FutureTimeline.net. 2016. IBM scientists achieve storage memory breakthrough. May 17. Accessed July 5. http://www.futuretimeline.net/blog/2016/05/17-2.htm#.V3uSxrgrKUk.

Gaur, P. and M. P. Tahiliani. 2015. Operating systems for IoT devices: A critical survey. In *IEEE Region 10 Symposium (TENSYMP)*, 33–36. Accessed September 30. doi:10.1109/TENSYMP.2015.17.

Gil, D., A. Ferrández, H. Mora-Mora, and J. Peral. 2016. Internet of things: A review of surveys based on context aware intelligent services. *Sensor* 16: 1. Accessed September 30. doi:10.3390/s16071069.

Green, C. 2015. The end of Moore's law? Why the theory that computer processors will double in power every two years may be becoming obsolete. *Independent*, July 16. Accessed July 5, 2016. http://www.independent.co.uk/life-style/gadgets-and-tech/news/the-end-of-moores-law-why-the-theory-that-computer-processors-will-double-in-power-every-two-years-10394659.html.

Hahm, O., E. Baccelli, H. Petersen, and N. Tsiftes. 2016. Operating systems for low-end devices in the Internet of things: A survey. *IEEE Internet of Things Journal* 3: 1.

Hanna, S. 2016. Hardware is the foundation of IoT security. GlobalSign, July 26. Accessed October 1. https://www.globalsign.com/en/blog/iot-security-hardware/.

Harvey, L. 2015. Service providers are uniquely positioned for secure IoT. TM Forum, March 15. http://inform.tmforum.org/perspectives2015/2015/03/service-providers-are-uniquely-positioned-for-secure-iot/.

Hruska, J. 2016. Phase change memory can operate thousands of times faster than current ram. *Extremetech*, August 15. Accessed September 29. http://www.extremetech.com/computing/233691-phase-change-memory-can-operate-thousands-of-times-faster-than-current-ram.

Hughes, J. M. 2016. *Arduino: A Technical Reference*, 1–534. Sebastopol, CA: O'Reilly Media.

Kantoch, E., P. Augustyniak, M. Markiewicz, and D. Prusak. 2014. Monitoring activities of daily living based on wearable wireless body sensor network. Presented at IEEE 36th Annual International Conference on Engineering in Medicine and Biology Society. Accessed July 3, 2016. doi:10.1109/EMBC.2014.6943659.

Klosowski, T. 2016. The best operating systems for your Raspberry Pi projects. *Lifehacker*, May 5. Accessed September 28. http://lifehacker.com/the-best-operating-systems-for-your-raspberry-pi-projec-1774669829.

Klosowski, T. 2017. How much power the Raspberry Pi Zero W uses compared to other models. *Lifehacker*, March 1. Accessed March 5. http://lifehacker.com/how-much-power-the-raspberry-pi-zero-w-uses-compared-to-1792854782.

Kolker, E. 2015. Tessel 2 hardware overview. TESSEL, March 10. Accessed June 13, 2016. https://tessel.io/blog/113259439202/tessel-2-hardware-overview.

Kumar, S. 2015. Past, present and future of IoT [Internet of Things]. Tips2Secure.com, December 3. http://www.tips2secure.com/2015/08/internet-of-things-past-present-future.html.

Langston, J. 2015. Popular science names 'power over Wi-Fi' one of the year's game-changing technologies. University of Washington Today, November 18. Accessed July 7, 2016. http://www.washington.edu/news/2015/11/18/popular-science-names-power-over-wi-fi-one-of-the-years-game-changing-technologies/.

Lee, T. 2015. The hardware enablers for the Internet of things—Part I. IEEE Internet of Things Newsletter. Accessed June 27, 2016. http://iot.ieee.org/newsletter/january-2015/the-hardware-enablers-for-the-internet-of-things-part-i.html.

Lesjak, C., D. Hein, and J. Winter. 2015. Hardware-security technologies for industrial IoT: TrustZone and security controller. In *41st Annual Conference of the IEEE Industrial Electronics Society*: 002589–002595. Accessed September 27, 2016. doi:10.1109/IECON.2015.7392493.

Lyons, C. 2015. A history of the Raspberry Pi. Nova Blog, March 4. http://novadigitalmedia.com/history-raspberry-pi/.

Maksimović, M., V. Vujović, N. Davidović, V. Milošević, and B. Perišić. 2014. Raspberry Pi as Internet of things hardware: Performances and constraints. In *Proceedings of 1st International Conference on Electrical, Electronic and Computing Engineering IcETRAN*, 1–6. Accessed March 13, 2017. https://www.researchgate.net/profile/Vladimir_Vujovic/publication/280344140_ELI16_Maksimovic_Vujovic_Davidovic_Milosevic_Perisic/links/55b3368608ae9289a08594aa/ELI16-Maksimovic-Vujovic-Davidovic-Milosevic-Perisic.pdf?origin=publication_list.

MagPi. 2016. Raspberry Pi 3 is out now! Specs, benchmarks and more. *Official Raspberry Pi Magazine*, March. Accessed June 27. https://www.raspberrypi.org/magpi/raspberry-pi-3-specs-benchmarks/.

MagPi. 2017. Introducing raspberry Pi Zero W. *Official Raspberry Pi Magazine*. Accessed June 5. https://www.raspberrypi.org/magpi/pi-zero-w/.

McKay, J. 2015. How Tessel 2 compares with the original Tessel board—First look: Tessel 2 embeds node.js in your project for 35 bucks. *Make*, March 6. http://makezine.com/2016/06/30/these-custom-night-vision-goggles-dont-even-look-homemade/.

McRoberts, M. 2010. *Beginning Arduino*, 1–434. New York: Springer.

Mearian, L. 2016. These technologies will blow the lid off data storage. *ComputerWorld*, March 9. Accessed July 5. http://www.computerworld.com/article/3041947/data-storage/how-these-technologies-will-blow-the-lid-off-data-storage.html.

Mehta, Y. P., V. P. Dadhich, and P. H. Pandey. 2016. Internet of Things (IoT). *International Journal of Technical Research and Applications* 41: 20. Accessed July 8. http://www.ijtra.com/ijtra-special-issue01.php?issue=Special%20Issue%2041

Miller, S. 2016. Enhancing Raspberry Pi security Zymbit. Accessed September 26. https://zymbit.com/securing-your-iot-devices-2/.

Mineraud, J., O. Mazhelis, X. Su, and S. Tarkoma. 2016. A gap analysis of Internet of things platforms. *Elsevier Journal of Computer Communications* 89–90:5–16.

Molloy, D. 2015. *Exploring BeagleBone: Tools and Techniques for Building with Embedded Linux*, 3–22. Indianapolis: Wiley.

Morgan, L. 2015. 14 ways IoT will change big data and business forever. *InformationWeek*, December 14. Accessed September 30, 2016. http://www.informationweek.com/iot/14-ways-iot-will-change-big-data-and-business-forever/d/d-id/1323531.

Andras Moritz, C., S. Chheda, and K. Carver. 2015. Security microprocessor against information leakage and physical tampering. Google patents, June 30. Accessed October 1, 2016. https://www.google.com/patents/US9069938.

Morra, J. 2016. Forget shrinking transistors. Fuse them together in three dimensions. *Electronic Design*, April 22. Accessed July 8. http://electronicdesign.com/microprocessors/forget-shrinking-transistors-fuse-them-together-three-dimensions.

Mossinger, M., B. Petschkuhn, J. Bauer, R. C. Staudemeyer, M. Wojcik, and H. C. Pohls. 2016. Towards quantifying the cost of a secure IoT: Overhead and energy consumption of ECC signatures on an ARM-based device. In *IEEE 17th International Symposium on a World of Wireless, Mobile and Multimedia Networks (WoWMoM)*, 1–6. Accessed September 26. doi:10.1109/WoWMoM.2016.7523559.

Ortmeyer, C. 2014. Then and now: A brief history of single board computers. *Electronic Design Uncovered,* December 6. Accessed June 18, 2016. http://www.newark.com/wcsstore/ExtendedSitesCatalogAssetStore/cms/asset/pdf/americas/common/NE14-ElectronicDesignUncovered-Dec14.pdf.

Park, M. 2015. IEEE 802.11ah: Sub-1-GHz license-exempt operation for the Internet of things. *IEEE Communications Magazine* 53: 145–151.

Particle. 2017. Security checklist for the Internet of things: An essential guide to securing connected products. *Particle*, January 3. Accessed April 5. https://www.particle.io/.

Paul, I. 2016. Future Intel chip tech will sacrifice speed gains for power efficiency. *PCWorld*, February 9. Accessed July 5. http://www.pcworld.com/article/3031222/tech-events-dupe/future-intel-chip-tech-will-sacrifice-speed-gains-for-power-efficiency.html.

Pinkerton, G. 2002. Many technologies contribute to miniaturization. *Electronic Design*, December 23. Accessed July 8, 2016. http://electronicdesign.com/boards/many-technologies-contribute-miniaturization.

Purcell, L. 2016. The past, present and future of IoT. *Intel Developer Zone*, March 3. https://software.intel.com/en-us/articles/the-past-present-and-future-of-iot.

Raman, S., R. Weigel, and T. Lee. 2016. Internet of space (IoS): A future backbone for the Internet of things? *IEEE Internet of Things*. March 8. Accessed July 6. http://iot.ieee.org/newsletter/march-2016/the-internet-of-space-ios-a-future-backbone-for-the-internet-of-things.html.

RasPi.TV. 2017. How much power does Pi Zero W use? RasPi.TV, March 1. Accessed March 5. http://raspi.tv/2017/how-much-power-does-pi-zero-w-use.

Rouffineau, T. 2016. Ubuntu core now available for Samsung ARTIK 520 and 10. *Ubuntu*, May 5. Accessed September 28. https://insights.ubuntu.com/2016/05/05/ubuntu-core-now-available-for-samsung-artik-5-and-10/.

Rush, C. 2016. BeagleBone Green Wireless—802.11 b/h/n Wi-Fi and Bluetooth 4.1. *Maker.10*, May 20. Accessed September 28. https://www.maker.io/en/blogs/beaglebone-green-wireless-802-11-b-g-n-wi-fi-bluetooth-4-1/4ea7485457f240be9f914f78667bbe39.

Saleh, R., S. Wilton, S. Mirabbasi, A. Hu, M. Greenstreet, G. Lemieux, P. Pratim Pande, C. Grecu, and A. Ivanov. 2006. System-on-chip: Reuse and integration. *Proceedings of the IEEE* 94: 1050–1069.

Sartain, J. D. 2016. Hello HaLow: Your guide to the Wi-Fi alliance's new IoT spec. *Networkworld*, May 23. http://www.networkworld.com/article/3072961/internet-of-things/hello-halow-your-guide-to-the-wi-fi-alliance-s-new-iot-spec.html.

Savov, V. 2015. IBM's 7nm chip breakthrough points to smaller, faster processors. *THE VERGE*, July 9. Accessed July 5, 2016. http://www.theverge.com/2015/7/9/8919091/ibm-7nm-transistor-processor.

Gartner. 2016. Gartner says 6.4 billion connected "things" will be in use in 2016, up 30 percent from 2015. Gartner. Accessed July 4. http://www.gartner.com/newsroom/id/3165317.

Somov A. and R. Giaffreda. 2015. Powering IoT devices: Technologies and opportunities. *IEEE Internet of Things*, November 9. Accessed July 7, 2016. http://iot.ieee.org/newsletter/november-2015/powering-iot-devices-technologies-and-opportunities.html.

Supalla, Z. 2015. Photon changelog. *Particle*, May 15. Accessed June 30, 2016. https://community.particle.io/t/is-photon-open-source/12440/3.

Swift, A. 2016. How open, hardware-based IoT security can be a win-win for innovation and regulation. IEEE Computer Society, February 25. https://www.computer.org/web/computingnow.

Tian, L., J. Famaey, and S. Latré. 2016. Evaluation of the IEEE 802.11ah restricted access window mechanism for dense IoT networks. In *IEEE 17th International Symposium on a World of Wireless, Mobile and Multimedia Networks (WoWMoM)*, 1–9. Accessed August 3. doi: 10.1109/WoWMoM.2016.7523502.

Upton, E. 2012. Introducing turbo mode: Up to 50% more performance for free. Raspberry Pi Foundation, September 19. https://www.raspberrypi.org/blog/introducing-turbo-mode-up-to-50-more-performance-for-free/.

van Eijndhoven, J. 2011. Measuring power consumption of the OMAP4430 using the PandaBoard. Vector Fabrics Blog, November 17. https://www.design-reuse.com/industryexpertblogs/27827/omap4430-mobile-soc.html.

Vasudevan, A., J. M. McCune, and J. Newsome. 2014. *Trustworthy Execution on Mobile Devices*, 33. Berlin: Springer.

Vermes, K. 2015. World's smallest computer is smaller than a grain of rice, powered by light. *Digital Trends*, April 7. Accessed June 13, 2016. https://www.digitaltrends.com/computing/say-hello-worlds-tiniest-computer-michigan-micro-mote/.

Voica, A. 2016. A guide to IoT processors. Imagination Technologies Blog, June 21. https://imgtec.com/blog/a-guide-to-iot-processors/.

Vujovic, V., S. Jokic, and M. Maksimovic. 2015. Power efficiency analysis in Internet of things sensor nodes. In *Second International Electronic Conference on Sensors and Applications*, 2. Accessed July 7, 2016. doi:10.3390/ecsa-2-D005.

Wong, M. M., M. L. D. Wong, C. Zhang, and I. Hijazin. 2015. Compact and short critical path finite field inverter for cryptographic S-box. In *IEEE International Conference on Digital Signal Processing (DSP)*, 775. Accessed September 26, 2016. doi:10.1109/ICDSP.2015.7251981.

7 IoT System Development Methods

Görkem Giray, Bedir Tekinerdogan, and Eray Tüzün

CONTENTS

7.1 Introduction 141
7.2 Background 142
7.2.1 System Development Methods 142
7.2.2 IoT System Building Blocks 143
7.3 Iot SDMs in the Literature 144
7.3.1 IgniteIIoT Methodology 145
7.3.2 IoT Methodology 146
7.3.3 IoT Application Development 147
7.3.4 ELDAMeth 147
7.3.5 Software Product Line Process to Develop Agents for the IoT 148
7.3.6 General Software Engineering Methodology for IoT 150
7.4 Evaluation of IoT SDMs 151
7.4.1 Method Artifacts 152
7.4.2 Process Steps 153
7.4.3 Support for Life Cycle Activities 153
7.4.4 Coverage of IoT System Elements 156
7.4.5 Design Viewpoints 156
7.4.6 Stakeholder Concern Coverage 156
7.4.7 Metrics 156
7.4.8 Addressed Discipline 157
7.4.9 Scope 157
7.4.10 Process Paradigm 157
7.4.11 Rigidity of the Method 157
7.4.12 Maturity of the Method 157
7.4.13 Documentation of the Method 158
7.4.14 Tool Support 158
7.5 Conclusion 158
References 159

7.1 INTRODUCTION

It is generally believed that the application of methods plays an important role in developing quality systems. A development method is mainly necessary for structuring the process in producing large-scale and complex systems that involve high costs. Similar to the development of other systems, it is important for IoT systems to be developed in a systematic manner in order to achieve a proper system with respect to both the functional and nonfunctional requirements. Development methods for IoT systems are more complex than traditional software systems and possess challenges from the process perspective. So far, several IoT system development methods (SDMs) have already been proposed in the literature, but an overview and evaluation of SDMs for IoT is still missing.

In this chapter, an overview of SDMs dedicated for IoT systems is presented. For this, the key IoT SDMs (the Ignite | IoT Methodology [Slama et al., 2016], the IoT Methodology [Collins, 2017], IoT Application Development [Patel, 2014], ELDAMeth [Fortino and Russo, 2012], a Software Product Line Process to Develop Agents for the IoT [Ayala and Amor, 2012], and a General Software Engineering Methodology for IoT [Zambonelli, 2016]) found in the literature were identified. Based on these methods, a set of evaluation criteria that include necessary features for characterizing and evaluating IoT SDMs were presented.

The remainder of the chapter is organized as follows. In Section 7.2, a background on SDMs and IoT systems is presented. Section 7.3 summarizes IoT SDMs along with their process flows represented using Business Process Model and Notation (BPMN), and Section 7.4 presents the characterization of these methods using the evaluation criteria. Finally, Section 7.5 concludes the chapter.

7.2 BACKGROUND

System development is generally a complex endeavor, especially for the systems including software, hardware, and communication components. To reduce and control this complexity, many SDMs have been proposed. The use of such methods brings some benefits. IoT system development is also a complex endeavor and encompasses dealing with a diverse set of components. The objective of an IoT SDM is to guide a project team in developing and combining these components in order to be able to fulfill user requirements.

7.2.1 System Development Methods

An SDM is an approach to develop a system systematically based on directions and rules (Brinkkemper, 1996). In this chapter, SDM refers to the development of an IoT system. SDM is preferred over the "software development method" concept since an IoT system encompasses many software, hardware, and communication components. Therefore, the development of an IoT system calls for more holistic methods than software development endeavors. Since IoT systems have software components as well, SDMs can also contain and/or benefit from software development methods.

It is generally believed that the application of SDMs plays an important role in developing quality systems. SDMs provide many benefits, including

- An SDM provides engineers with a set of guidelines for developing some artifacts and their verification against the requirements defined in a problem statement.
- Since SDMs formalize certain procedures of design and externalize design thinking, they help to avoid the occurrence of overlooked issues in the development process and tend to widen the search for appropriate solutions by encouraging and enabling the engineer to think beyond the first solution that comes to mind.
- SDMs help to provide logical consistency among the different processes and phases in the development process. This is particularly important for the development of large and complex systems, which are produced by large teams of designers and developers. A development method provides a set of common standards, criteria, and goals for the team members.
- An SDM helps reduce possible errors in the development process and provides heuristic rules for evaluating design decisions.
- Mainly from the organizational point of view, an SDM helps to identify important progress milestones. This information is necessary to control and coordinate the different phases in system development.

An SDM is mainly necessary for structuring the process in producing large-scale and complex systems that involve high costs. The motivation for SDMs can thus be summarized as directing the project team, leading to better systems by considering various solution alternatives, providing consistency among different processes, reducing failures, and identifying important milestones.

7.2.2 IoT System Building Blocks

An IoT system is made up of various components interacting with each other. Figure 7.1 provides a conceptual model of the components making up a typical IoT system, including the relations among the basic IoT concepts. The model has been adopted from the Alliance for Internet of Things Innovation (AIOTI) domain model (AIOTI, 2016). The domain model represents the basic concepts and relationships in the domain at the highest level. Since the AIOTI domain model includes intuitive high-level basic concepts, it has been also referenced by some other standardization efforts, such as the IEEE P2413 Working Group, whose objective is to form some standards to support the IoT's growth.

In the model, a user interacts with a physical entity from the physical world, that is, a thing. The user can be a human person or a software agent that has a goal. To achieve this goal, an interaction with the physical environment is needed, and this interaction is performed through the mediation of an IoT system. A thing is a discrete, identifiable part of the physical environment that can be of interest to the user for the completion of his goal. Things can be any physical entities, such as humans, cars, animals, or computers.

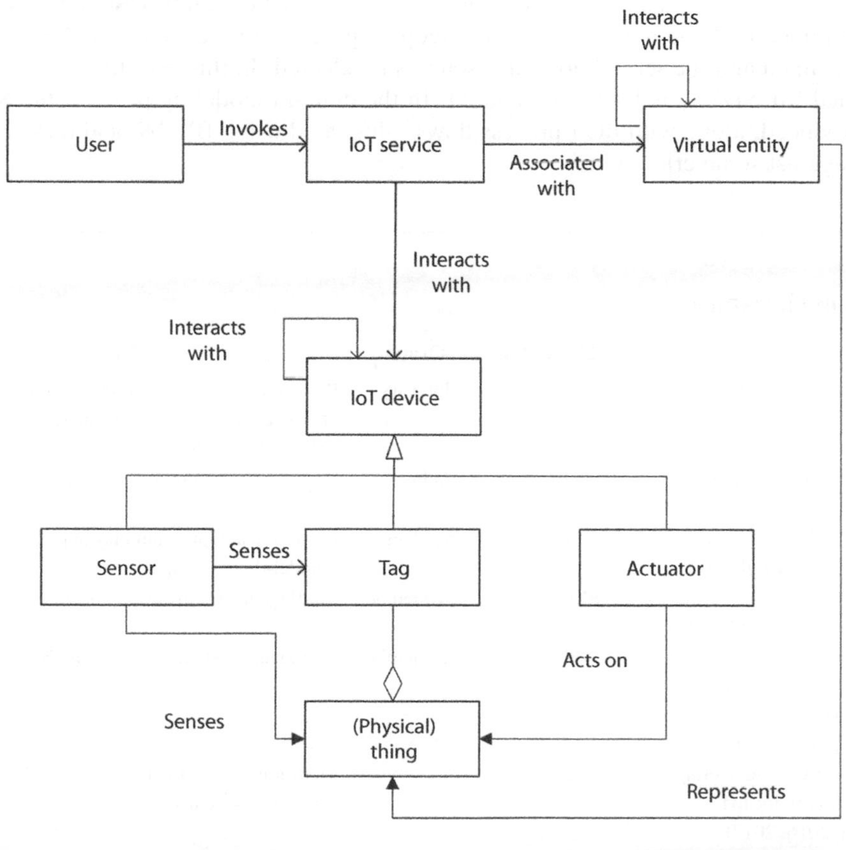

FIGURE 7.1 Conceptual model of a typical IoT system. (Adapted from Alliance for Internet of Things Innovation, High Level Architecture [HLA], Release 2.1, AIOTI WG03—IoT Standardisation, 2016.)

The interaction between a user and a thing in the IoT is mediated by a service, which constitutes the interface between a user and an IoT system. A service is associated with a virtual entity, that is, a digital representation of the physical entity. A thing can be represented in the digital world by a virtual entity. Different kinds of digital representations of things can be used, such as objects, three-dimensional models, avatars, or even social network accounts. Some virtual entities can also interact with other virtual entities to fulfill their goal.

An important aspect in IoT is that changes in the properties of a thing and its corresponding virtual entity need to be synchronized. This is usually realized by a device that is embedding into, attached to, or simply placed in close vicinity of the thing. In principle, three types of devices can be identified, namely, sensors, tags, and actuators. Sensors are used to measure the state of things they monitor. Essentially, sensors take a mechanical, optical, magnetic, or thermal signal and convert this into voltage and current. This provided data can then be processed and used to define the required action. Tags are devices used to support the identification process, typically using specialized sensors called readers. The identification process can take place in different forms, including optical, as in the case of barcodes and QR code, or radio frequency based. Actuators, on the other hand, are employed to change or affect the things.

7.3 IoT SDMS IN THE LITERATURE

To identify the different IoT SDMs, a thorough domain analysis process has been applied, in which relevant studies were searched in the literature that directly propose an IoT SDM or partially address the development of one or more components of an IoT system. As a result, six SDMs were identified. In the domain analysis, two basic activities can be distinguished, namely, domain scoping and domain modeling. In the scoping process, the scope of the domain analysis process is defined and the set of knowledge sources is selected. In this text, the scope consists of the identified IoT SDMs, as listed in Table 7.1. In the domain modeling process, the SDMs are briefly introduced, along with their process flows, illustrated using BPMN, and these SDMs are evaluated against some criteria.

TABLE 7.1
IoT SDMs in Literature

Method	Abbreviation	Origin	Base
The Ignite I IoT Methodology (Slama et al., 2016)	Ignite	Industry	Best practices from the projects in the industry; project management guidelines, such as Project Management Body of Knowledge (PMBOK)
The IoT Methodology (Collins, 2017)	IoT-Meth	Industry	Best practices from the projects in the industry
IoT Application Development (Patel and Cassou, 2015; Patel, 2014)	IoT-AD	Academia	Built on macroprogramming approach; inspired by model-driven design
ELDAMeth (Fortino and Russo 2012; Fortino et al., 2014, 2015)	ELDAMeth	Academia	Multiagent system development
A Software Product Line (SPL) Process to Develop Agents for the IoT (Ayala and Amor 2012; Ayala et al., 2012, 2014, 2015)	SPLP-IoT	Academia	Multiagent system development; SPLE
A General Software Engineering (SE) Methodology for IoT (Zambonelli, 2016, 2017)	GSEM-IoT	Academia	Traditional software development; abstractions from IoT systems

Note: SPLE, Software Product Line Engineering.

7.3.1 IGNITE|IoT METHODOLOGY

Derived from real-world IoT projects, the Ignite | IoT Methodology (Slama et al., 2016) ("Ignite" in this chapter) is aimed at various IoT stakeholders, including product managers, project managers, and solution architects. The methodology has two major groups of activities: IoT Strategy Execution and IoT Solution Delivery. IoT Strategy Execution aims to define an IoT strategy and a project portfolio supporting this strategy. IoT Solution Delivery supports IoT solution design and IoT project management, along with some artifacts, such as project templates, checklists, and solution architecture blueprints. These two groups of activities should be synchronized to keep the project portfolio in line with the strategy and revise the strategy according to the outcomes of the project portfolio. Figure 7.2 illustrates the process flow of Ignite.

IoT Strategy Execution is about business perspective and involves identifying an opportunity, developing a business model, and making decisions on how to manage these opportunities (such as internal project, external acquisition, and spin-off).

IoT Solution Delivery is about delivering a solution, which is conceptually defined during the IoT Strategy Execution phase, and has a life cycle consisting of the planning, building, and running of the system. Planning starts with project initiation, in which an initial solution design and a project organization chart are delivered. Moreover, stakeholder, environment, requirements, risk, and resource analysis should be conducted. After the initiation, the tasks are managed under seven work streams: (1) project management, (2) crosscutting tasks, (3) solution infrastructure and operations, (4) back-end services, (5) communication services, (6) on-asset components, and (7) asset preparation. Project management encompasses the activities for initiating, planning, executing, monitoring, controlling, and closing a project. Crosscutting tasks address the dependencies among subsequent work streams, such as security and testing. Solution infrastructure and operations include the

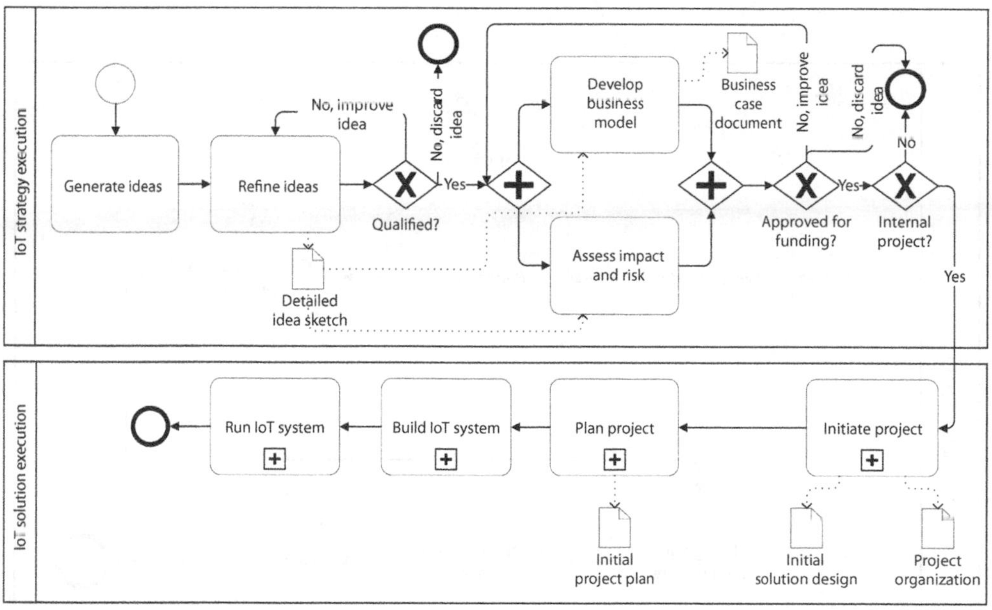

FIGURE 7.2 Ignite process flow.*

* The process flows are illustrated using BPMN, which provides a standard graphical notation for representing processes. For more information, see http://www.bpmn.org/.

X: Exclusive gateway, which means that only one of the following branches can be traversed.

+: Parallel gateway, which means that the following tasks can be performed in parallel.

installation and management of hardware and software infrastructure, on which an IoT system will be developed and operated. Back-end services refer to the IoT services typically hosted on a private or public cloud and interacting with IoT devices. Communication services encompass the installation and management of communication infrastructure. On-asset components refer to the development or procurement of software and the manufacturing or procurement of hardware and network components to be integrated with a thing in an IoT system. Asset preparation addresses the manufacturing and procurement of a thing in an IoT system.

7.3.2 IoT Methodology

The IoT Methodology (Collins, 2017) ("IoT-Meth" in this chapter) is a generic, lightweight method built on iterative prototyping and Lean start-up approaches. Figure 7.3 illustrates the steps of IoT-Meth, eliminating its iterative nature for the sake of simplicity (the steps are renamed for better understanding). IoT-Meth involves the following iterative steps (the original names of the steps are in parenthesis):

1. *Generate ideas (cocreate)*: This step involves the identification of problem areas by communicating with stakeholders, especially end users. The objective is to generate ideas on potential problems from a business perspective. Some of these ideas are elaborated with their use cases, to be refined in the next step.
2. *Refine ideas (ideate)*: Some of the ideas identified in the former step are further elaborated to be communicated with project managers, designers, and implementers. The artifact named IoT Canvas can be used in brainstorming sessions with stakeholders to identify and validate high-level requirements. IoT Canvas mainly consists of a problem statement, key actors, things in the physical environment, sensors and actuators, data models, middleware

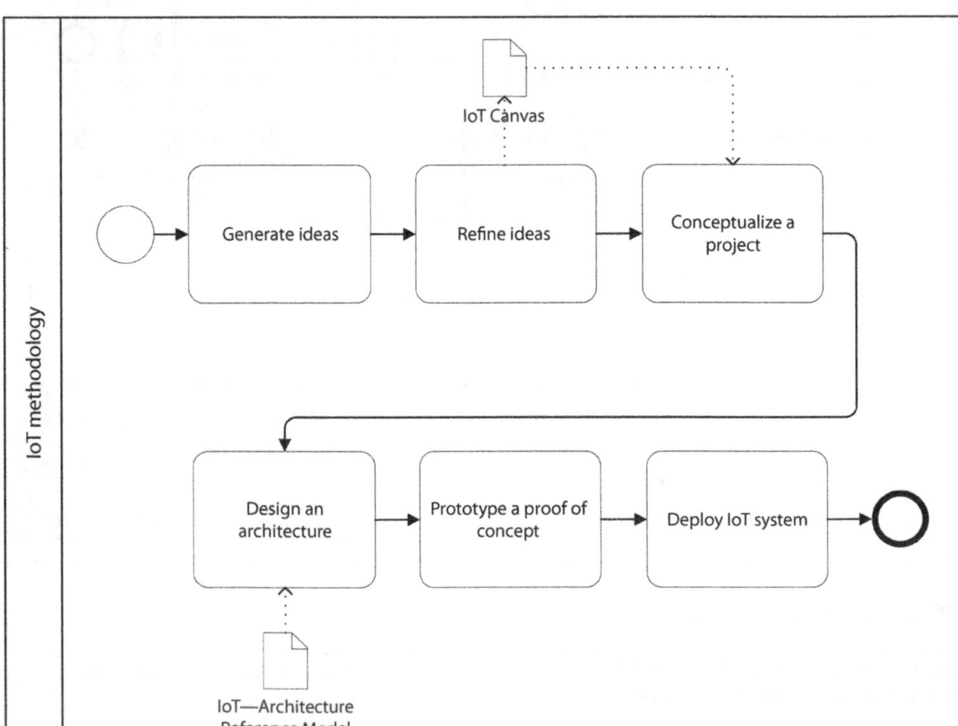

FIGURE 7.3 IoT-Meth process flow.

requirements to connect IoT services, third-party web services to be integrated, and user interface (UI) sketches.

3. *Conceptualize a project (Q&A)*: This step involves analyzing refined ideas further to close the gap between idea and implementation. The requirements are analyzed and validated; the domain is analyzed further.
4. *Design an architecture (IoT OSI)*: In this step, the requirements are mapped to an architecture and infrastructure. The artifact named IoT—Architecture Reference Model is an input for this step and basically an adaptation of the seven-layer International Organization for Standardization/Open Systems Interconnection (ISO/OSI) reference model for IoT solutions. This reference model comprises five layers: end points, connectivity, middleware, IoT services, and applications. These layers help in classifying components and hence managing complexity.
5. *Prototype a proof of concept (prototype)*: This step encompasses building prototypes and iterating toward minimal viable IoT systems. The prototypes are assessed, and forthcoming iteration plans are revised accordingly.
6. *Deploy IoT system (deploy)*: The last step is about deploying the system and closing the feedback loop. Generally, the system is improved continuously according to the feedback.

IoT-Meth does not define well-defined roles with descriptions and responsibilities. It only addresses some roles, such as end user, designer, implementer, and project manager, without any detail.

7.3.3 IoT Application Development

Patel and Cassou (2015; Patel, 2014) propose an approach to IoT application development ("IoT-AD" in this chapter) built on macroprogramming (in contrast to node-centric programming), in which the behavior of a system is specified using high-level abstractions and then compiled to node-level code (Pathak et al., 2007). IoT-AD encompasses a development methodology and a concrete development framework realizing this methodology. IoT-AD is built on the separation of concerns principle and classifies the IoT domain into four areas of concern: (1) domain, (2) functional, (3) deployment, and (4) platform. The domain concern is related to domain-specific concepts of an IoT system and mainly encompasses the identification of a domain vocabulary. The functional concern is about specifying an architecture and implementation of this architecture. The deployment concern encapsulates deployment specifications of each thing. The platform concern addresses the development of platform-specific device drivers for each type of thing. The process flow of IoT-AD organized according to the concerns is illustrated in Figure 7.4. IoT-AD provides a set of modeling languages for modeling the concepts in these areas of concern and some automation techniques for reducing development effort.

7.3.4 ELDAMeth

ELDAMeth (Event-driven Lightweight Distilled state charts based Agents Methodology) utilizes an agent-based paradigm to guide the development of software for things in an IoT system (Fortino and Russo, 2012). Agents can be defined as autonomous software components acting in a distributed, networked environment. Therefore, agents are very similar to the software components of things operating in a distributed, networked environment within an IoT system. Things are named smart objects (SOs), referring to the autonomy, which is an important feature of agents. SOs (refers to the software component of things) are treated as the fundamental building blocks of IoT systems (Fortino et al., 2015).

As illustrated in Figure 7.5, this methodology mainly encompasses three main phases (Fortino and Russo, 2012; Fortino et al., 2014): (1) The modeling phase takes high-level design models and requirements as inputs and produces a detailed design of SOs, which can be translated into

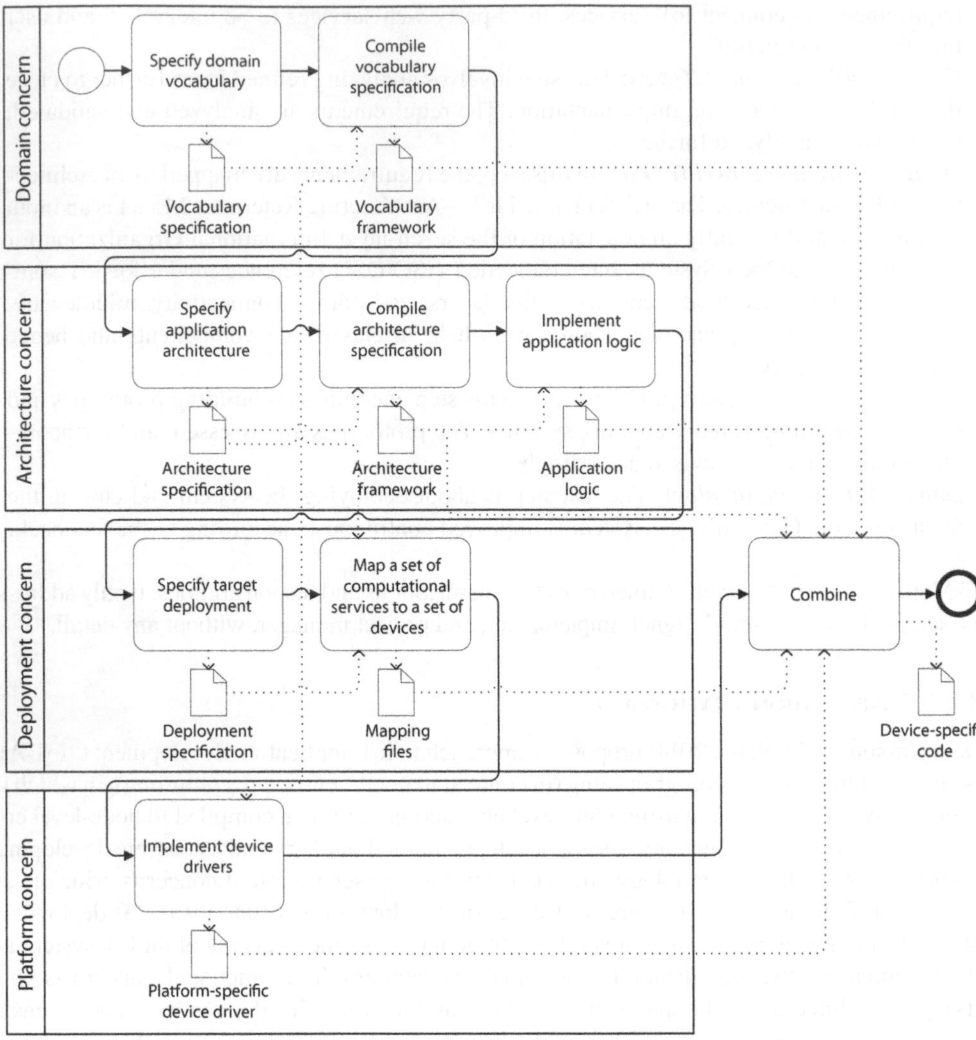

FIGURE 7.4 IoT-AD process flow.

platform-independent code. (2) The simulation phase takes requirements and platform-independent ELDA SO code as inputs and produces simulation results to be evaluated against requirements. (3) The implementation phase encompasses the development and test of platform-specific ELDA SO code.

ELDAMeth focuses on the technical aspects of developing SOs. The business aspects, as well as requirements engineering for developing IoT systems, are not addressed in this methodology.

7.3.5 Software Product Line Process to Develop Agents for the IoT

The Software Product Line Engineering (SPLE) paradigm aims to develop software by identifying commonalities and variabilities of a family of software products. Commonalities form a reusable platform, and variabilities refer to the specific features of particular software within the scope of a family of software products. SPLE separates two main processes (Pohl et al., 2005): Domain engineering is for establishing the platform with common features, and application engineering is for deriving particular applications built on top of the platform. There are several motivations for SPLE reported in the literature (Pohl et al., 2005), such as reduction of development costs, enhancement of

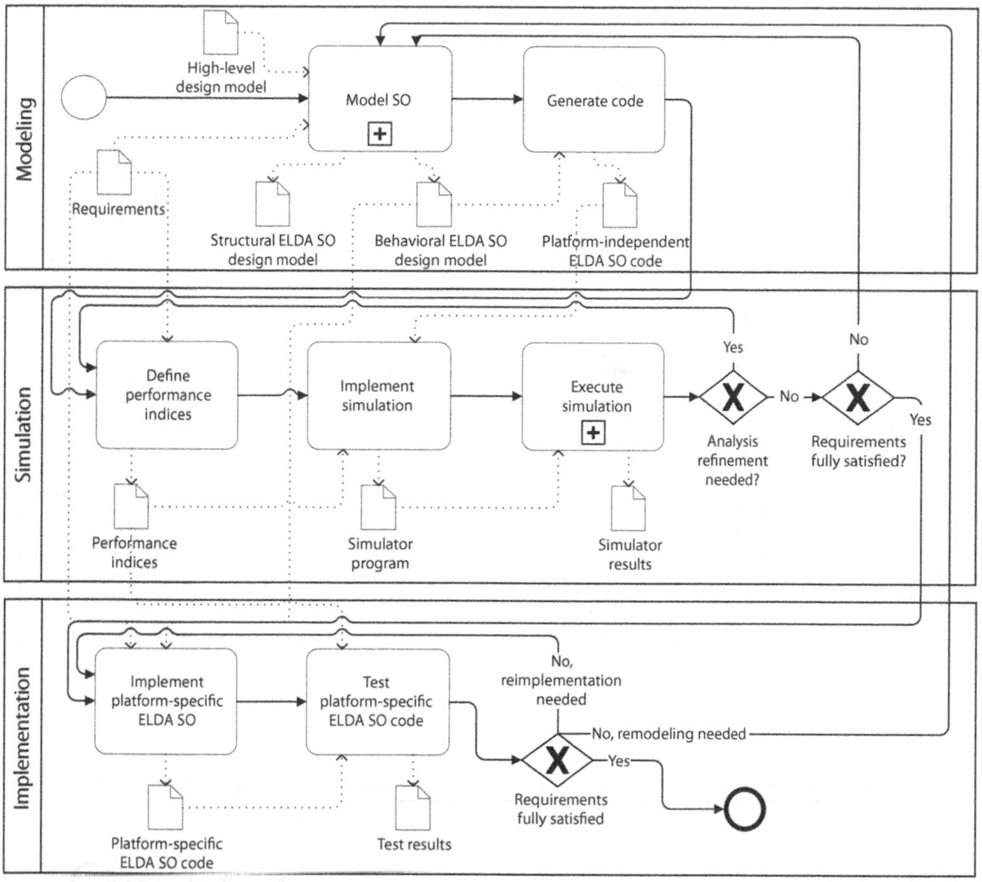

FIGURE 7.5 ELDAMeth process flow.

quality, reduction of time to market, reduction of maintenance effort, coping with evolution, coping with complexity, and improving cost estimation.

IoT systems envision a large-scale, complex, heterogeneous network of things. Therefore, agents, which are autonomous software components, can be an enabler for self-managed IoT systems (Ayala et al., 2015). Developing software components of things using an agent-based paradigm is considered due to the agents' distributed nature, context awareness, and self-adaptation (Ayala et al., 2015).

A Software Product Line Process to Develop Agents for the IoT ("SPLP-IoT" in this chapter) combines SPLE with an agent-based paradigm to utilize the benefits of both of these paradigms for developing IoT systems (Ayala et al., 2015). The aim is to identify commonalities among software agents and develop common reference architecture, and hence obtain a reduction in implementation time and cost, as well as an increase in quality.

Figure 7.6 illustrates an overview of SPLP-IoT. Domain engineering is responsible for establishing a reusable platform and thus defining commonalities and variabilities of a multiagent system. This is achieved by mining domain requirements, analyzing the IoT multiagent system domain, and specifying and realizing domain variability. Domain variability defines the variation points to be configured to obtain specific agents. Domain variability is specified by an IoT multiagent system variability model represented using Common Variability Language (CVL). CVL is a domain-independent language defined using a Meta-Object Facility (MOF)–based metamodel, which is used for specifying and resolving variability (Haugen et al., 2013). An IoT multiagent system architecture, which represents commonalities and variabilities, is produced after the variability model is defined.

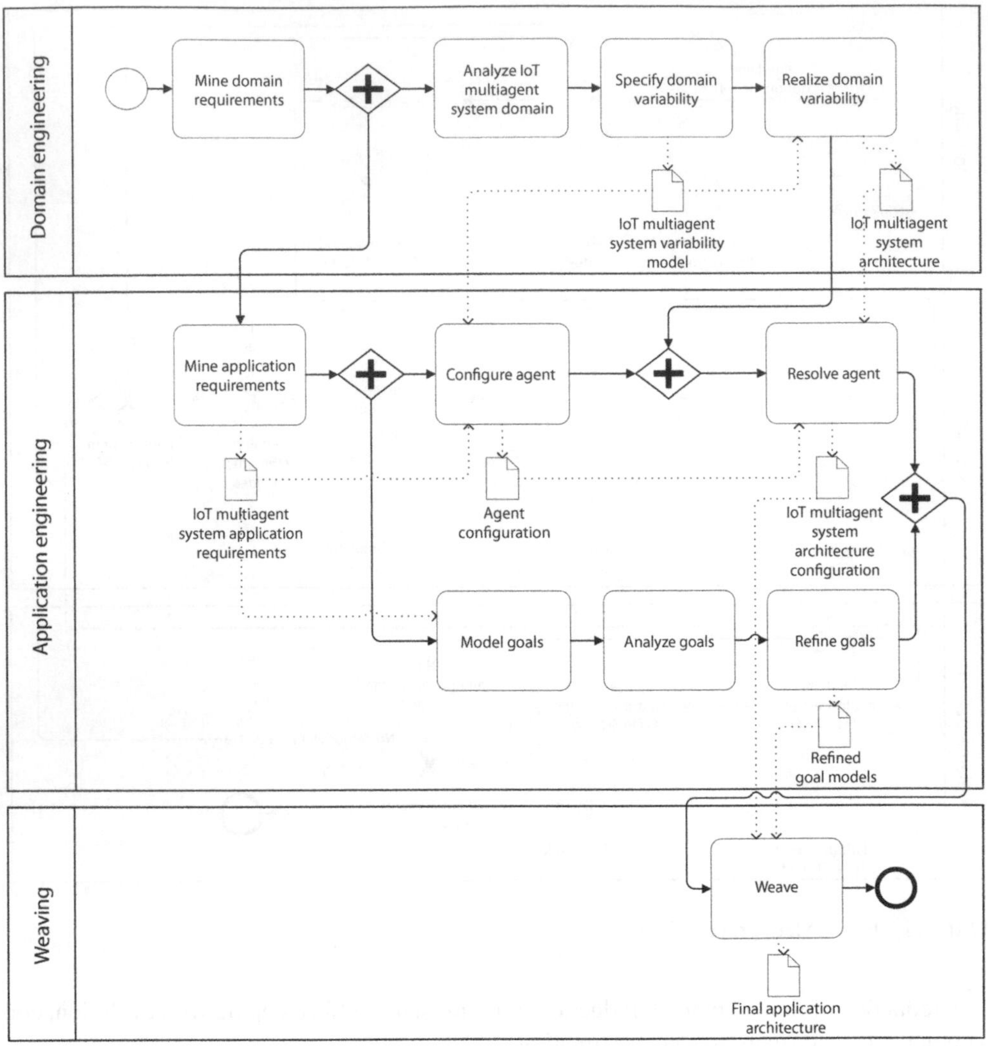

FIGURE 7.6 SPLP-IoT process flow.

Therefore, an IoT multiagent system architecture acts as a base model, forming both the common platform and the variability points to be configured for each particular agent. Application engineering encompasses building agents by exploiting the variability model and IoT multiagent system architecture. In this part, an agent is configured according to specific application requirements and an agent configuration is obtained as a result. When resolving the agent, an IoT multiagent system architecture configuration is generated by a CVL tool. Modeling and analyzing goals are for checking consistency among agent goals, context, and plans. Afterwards, the goals are refined. Refined goal models and IoT multiagent system architecture configuration are processed to produce a final application architecture. This final application architecture corresponds to a software component on a thing, and this software component is built on top of a platform (built around commonalities).

7.3.6 General Software Engineering Methodology for IoT

Zambonelli (2016, 2017) proposes some general guidelines and steps of a general software engineering methodology for developing IoT systems ("GSEM-IoT" in this chapter). GSEM-IoT consists of three phases (Zambonelli, 2016): (1) analysis phase, which includes the identification and analysis of actors, existing infrastructure, functionality, and requirements; (2) design phase, which includes the design of avatars, groups, and coalitions and identification of new infrastructural needs; (3) and implementation phase, which includes the implementation of avatars and coordinators, along with the deployment of new "things" and new middleware. Figure 7.7 illustrates the overview of GSEM-IoT, which is attributed as a first small step toward a general method for developing IoT systems.

7.4 EVALUATION OF IoT SDMS

An SDM guides a system development endeavor by providing guidance on many different aspects, such as addressing stakeholder concerns, steps to be followed, and artifacts to be produced. Figure 7.8 illustrates a conceptual model of the important concepts related to an SDM. This model is adapted to an IoT context. In essence, each IoT system will be important for a number of stakeholders that have a number of concerns. From a simplified perspective, an SDM consists of a set of life cycle phases, a number of steps, and artifacts that reflect IoT elements and design viewpoints for modeling various design perspectives. Further, an SDM is focused on a particular paradigm and is supported by documentation and tools.

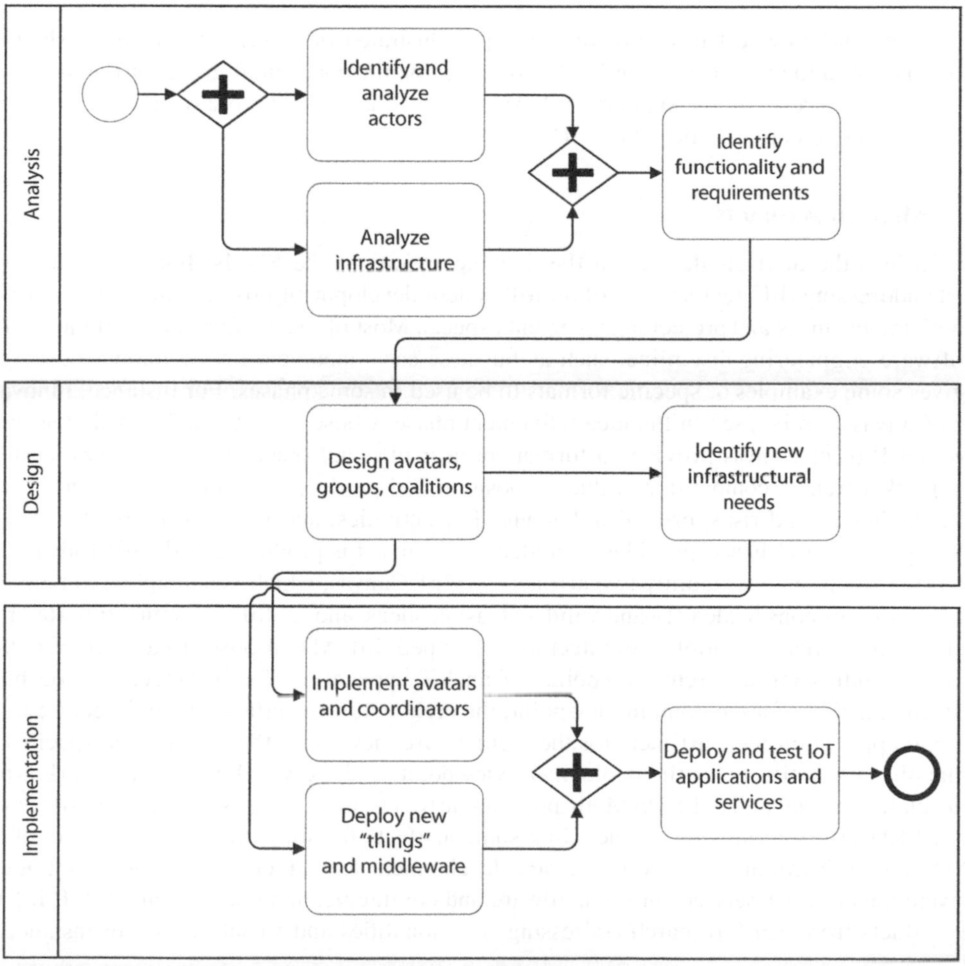

FIGURE 7.7 GSEM-IoT process flow.

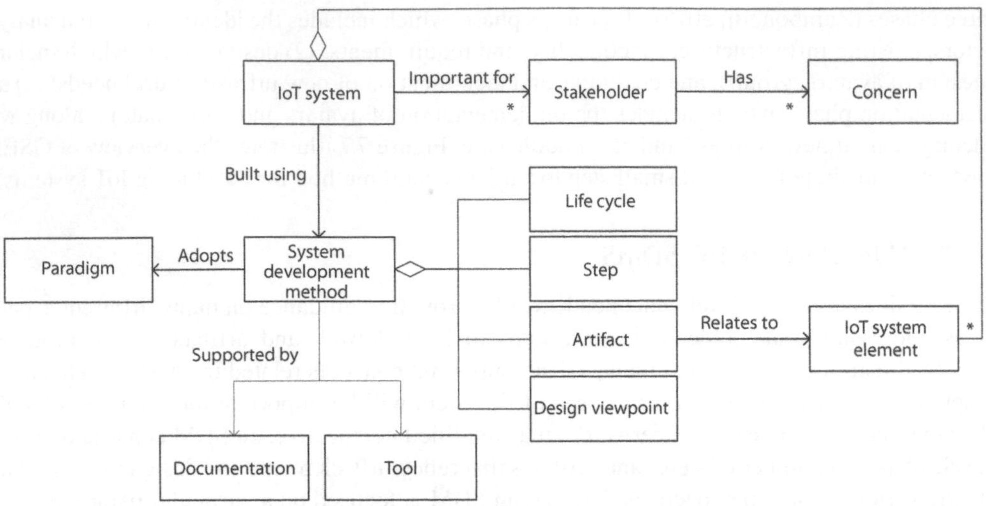

FIGURE 7.8 Conceptual model for a method and related concepts.

An SDM can be evaluated against the concepts illustrated in Figure 7.8. Table 7.2 shows the evaluation criteria to analyze existing IoT SDMs based on the conceptual model of Figure 7.8. The evaluation criteria relate to the support for SDM elements, as well as more qualitative issues with respect to the use and pragmatics of the SDM.

7.4.1 Method Artifacts

Table 7.3 lists the artifacts defined in the documentations of the SDMs. Ignite provides many artifacts addressing different aspects of an IoT system development project. Most of the artifacts deal with the business and project management aspects. Most of the artifacts are well known from the software engineering discipline, such as business case document and domain model. Ignite also gives some examples of specific formats to be used in some phases. For instance, Innovation Project Canvas* can be used in the idea refinement phase, whose output is a detailed idea sketch. Innovation Project Canvas provides a format (project title and objectives, customers, customer needs, market trends, competition, value proposition and product description, solution, business model, challenges and risks, critical unknowns, key activities, and issues to be treated first) for specifying ideas, and hence provides a predefined format for producing a detailed idea sketch artifact. IoT-Meth includes mainly two artifacts: (1) IoT Canvas, which is a template to be used in brainstorming sessions to identify and validate feasible ideas, and (2) IoT—Architecture Reference Model, against which a solution's architecture is mapped. IoT-AD proposes to use different kinds of artifacts addressing different viewpoints of an IoT application, for instance, the vocabulary specification artifact for the domain viewpoint; the architecture specification, architecture framework, and application logic artifacts for the architecture viewpoint; the deployment specification and mapping file artifacts for the deployment viewpoint; and the vocabulary framework artifact for the platform viewpoint. ELDAMeth provides artifacts to develop software for IoT devices (SOs in ELDAMeth terminology), including some artifacts for simulation. The artifacts of both IoT-AD and ELDAMeth only cover software development for IoT devices; they do not address the development of IoT services or the hardware and communication components. SPLP-IoT uses some artifacts from SPLE research addressing commonalities and variabilities. For instance, the

* Innovation Project Canvas website, www.innovationprojectcanvas.com.

TABLE 7.2
Evaluation Criteria for Analyzing IoT SDMs

Criteria	Description
Method artifacts	What are the method artifacts in the overall process?
Process steps	What are the process steps?
Support for life cycle activities	Which life cycle activities are supported by the method? (feasibility analysis, requirements definition, design, development, testing, deployment, project management, maintenance)
Coverage of IoT system elements	Is the process related to all the IoT system elements? (e.g., sensors and actuators)
Design viewpoints	Does the method include different design viewpoints?
Stakeholder concern coverage	Does the method support the required stakeholder concerns (expectations from the IoT system, expectations from the project realizing the IoT system)?
Metrics	Does the method provide any metrics? (such as requirements quality metrics, code quality metrics, project management metrics, test metrics, etc.)
Addressed discipline	What is the addressed engineering discipline? (system, software, mechanical, etc.)
Scope	What is the scope of the method? General purpose vs. domain specific?
Process paradigm	What is the adopted process paradigm? (plan-driven vs. agile)
Rigidity of the method	Is the method extensible? (process, rules, artifacts)
Maturity of the method	Has the method been validated?
Documentation of the method	How well is the method documented?
Tool support	Does the method have tool support? If so, which?

IoT multiagent system variability model represents the possible variability points for each agent (software component of IoT device in this case). Moreover, the IoT multiagent system architecture includes a blueprint of a common platform (besides variability points) to establish a software product line. No artifact has been defined for GSEM-IoT yet, as the method is still under development (Zambonelli, 2016).

7.4.2 Process Steps

Both Ignite and IoT-Meth define a high-level process flow addressing the development of an IoT system from idea to a running system (Figures 7.2 and 7.3). IoT-AD, ELDAMeth, and SPLP-IoT include a process flow for developing software to be deployed on IoT devices (Figures 7.4 through 7.6). ELDAMeth has some steps for simulation, which aims to validate the solution design before implementation. SPLP-IoT addresses commonality and variability analysis for SPL with some steps. GSEM-IoT provides a process flow for systematic development of IoT systems (Figure 7.7). GSEM-IoT tries to enrich the traditional software development process with key concepts and abstractions from the IoT domain.

7.4.3 Support for Life Cycle Activities

Ignite addresses the whole life cycle of developing an IoT system, focusing more on project management, feasibility analysis, requirements engineering, and analysis and design. Ignite also includes software development and testing activities; on the other hand, it does not provide any technical details on how to develop and test software for IoT systems. IoT-Meth partially addresses feasibility analysis, requirements engineering, analysis and design, and deployment. IoT-Meth does not provide any details on these life cycle activities. IoT-AD, ELDAMeth, SPLP-IoT, and GSEM-IoT focus on requirements engineering, analysis and design, and development at varying levels of detail. In addition, IoT-AD

TABLE 7.3
Artifacts Defined in the Identified SDMs

SDM	Artifact	Description
Ignite	Detailed idea sketch	Describes the key elements of an idea to be elaborated for further evaluation
	Business case document	Refinement and validation of a detailed idea sketch; evaluated for funding
	Project organization	Defines the teams involved and how these teams are structured to achieve the objective of a project
	Initial project plan	Formed after funding decision, based on the idea and requirements
	Problem statement	Contains a short description of the problem domain and vision for the IoT solution
	Stakeholder analysis	Artifact including an analysis of the stakeholders by their interest and influence on the project
	Site survey	A document covering all aspects of an IoT device (hardware) to be manufactured or procured
	Solution sketch	Narrows down the solution scope and creates a basis for the communication between business and technical stakeholders
	Project dimensions	Capture all important aspects of an IoT solution and made up of five main dimensions: (1) assets and devices, (2) communications and connectivity, (3) back-end services, (4) standards and regulatory compliance, and (5) project environment
	Quantity structure	Includes projections for possible changes on the numbers of users, assets, etc.
	Milestone plan	Defines the key milestones of the project
	Process maps/use cases	Demonstrate how an IoT solution addresses customer's problem
	UI mock-ups	Visualizations of key UIs; provide a basis for discussing ideas and validating requirements with end users and business stakeholders
	Domain model	Encompasses a business-oriented, consolidated view of the key data entities of an IoT solution
	Asset integration architecture	Describes the relationships between assets and devices (things) and the back-end
	SOA landscape	Describes the key software components and their main business functions; different than asset integration architecture in being technology agnostic and focusing on business functions
	Software architecture	Defines the key software components and their relationships
	Technical infrastructure	High-level view of an IoT solution and its environment; contains assets, communication infrastructure, etc.
	Hardware design	Addresses the main components of an asset on an IoT device, such as CPU, memory, power supply, digital I/O, and communication modules
IoT-Meth	IoT Canvas	Defines high-level characteristics of an IoT solution for providing a basis for further assessment (including funding); can include a problem statement, things in the physical environment, key actors, etc.
	IoT—Architecture Reference Model	Defines the layers identifying an important aspect of an IoT solution; such layers can include connectivity, middleware, IoT services, and applications
IoT-AD	Vocabulary specification	Defines resources, which are conceptual representations of sensors, actuators, storages, and UIs
	Vocabulary framework	Contains concrete classes and interfaces corresponding to resources defined in a vocabulary specification
	Architecture specification	Guides software developers in developing software components of IoT devices
	Architecture framework	Contains abstract classes that hide interaction details with other software components and allow software developers to focus on the application logic of a particular software component

(Continued)

TABLE 7.3 (CONTINUED)
Artifacts Defined in the Identified SDMs

SDM	Artifact	Description
	Application logic	Represents a concrete implementation of abstract classes defined in an architectural framework in line with the architecture specification
	Deployment specification	Includes the details of each IoT device, such as resources (sensor, actuator, storage, and UI) hosted by devices and types of devices
	Mapping files	Include the mapping information between software components and IoT devices
	Platform-specific device driver	Concrete implementation of some functionality specific to an IoT device, such as reading a barcode using an IoT device with an Android operating system
	Device-specific code	Executed on each particular IoT device
ELDAMeth	Requirements	Define functional and nonfunctional requirements of a multiagent system under development
	High-level design models	Represent a solution without any details based on requirements
	Structural ELDA SO design models	Class diagrams representing the interaction relationships among agents and roles
	Behavioral ELDA SO design models	State charts representing agent behaviors and/or roles
	Platform-independent ELDA SO code	Part of the code that is independent of any specific technology
	Platform-specific ELDA SO code	Part of the code that is dependent on a specific technology
	Performance indices	Define the criteria against which the results of a simulation will be evaluated
	Simulator program	Enables execution of the simulation
	Simulation results	Findings obtained from execution of the simulation
	Test results	Document the findings obtained from tests, also considering the performance indices
SPLP-IoT	IoT multiagent system variability model	Defines variation points, which can be configured differently when implementing a particular agent
	IoT multiagent system architecture	Defines the main components of a system, along with their relationships
	IoT multiagent system application requirements	Refer to the specific expectations from a particular software component running on an IoT device
	Agent configuration	Refers to the implementation of a particular IoT device by adjusting variation points defined in a variability model
	IoT multiagent system architecture configuration	Contains the set of components and connections that leads to the realization of the final application architecture
	Refined goal models	Contain a set of goals and plans and the context of the agent that are consistent and whose conflicts are detected; these goals define how these agents (IoT devices in this case) will behave
	Final application architecture	Derived from IoT multiagent system architecture configuration and refined goal models
GSEM-IoT	No artifact defined	

covers deployment by proposing to produce a deployment specification, which includes the details of each IoT device. Moreover, IoT-AD claims that its approach supports the maintenance phase. The rationale behind this is the separation of different concerns (domain, architecture, deployment, and platform) and automation techniques provided in the method. ELDAMeth includes a testing activity in the implementation phase and proposes to produce an artifact, including test results.

7.4.4 Coverage of IoT System Elements

Ignite provides guidance for developing IoT services and software components for IoT devices from a business and project management perspective; however, it does not provide a low-level, technical guidance. Ignite excludes the manufacturing of assets (things). IoT-Meth addresses IoT services and software components for IoT devices at an architectural level; however, it does not include any detailed subprocesses for developing each of these elements. IoT-AD, ELDAMeth, and SPLP-IoT only cover the development of software components, which run on IoT devices (things). GSEM-IoT mainly provides a conceptual view (without providing any technical details) on the development of software components of IoT devices.

7.4.5 Design Viewpoints

A design viewpoint examines a system from a particular perspective. A design viewpoint of a system can be represented by one or more artifacts. These artifacts should address the concerns of that particular viewpoint. For instance, a functional viewpoint of a system can focus on the functionalities to be provided by a system, whereas a deployment viewpoint can treat the concerns on the deployment of a system.

Ignite addresses business, usage, functional, and implementation viewpoints. It defines certain artifacts to address these viewpoints:

1. Business: Business case document, site survey, project dimensions, milestone plan
2. Usage: Process maps and use cases
3. Functional: UI mock-ups and domain model
4. Implementation: Software architecture and technical infrastructure

IoT-AD specifies four viewpoints concerning domain, architecture, deployment, and platform (Patel, 2014). The domain viewpoint addresses the specification of a domain-specific vocabulary for an IoT application. The architecture viewpoint encompasses specifying application architecture, compiling architecture specification, and implementing application logic. The deployment viewpoint is about describing deployment specifications for devices and mapping a set of computational services to a set of devices. The platform viewpoint is for implementing platform-specific device drivers. IoT-Meth, ELDAMeth, SPLP-IoT, and GSEM-IoT do not explicitly address different viewpoints.

7.4.6 Stakeholder Concern Coverage

Ignite proposes to identify stakeholders and analyze their expectations when initiating a project to address their concerns. IoT-Meth discusses communication with stakeholders during the idea generation step, with no further detail. GSEM-IoT has a step for identifying and analyzing actors. IoT-AD, ELDAMeth, and SPLP-IoT do not address stakeholders explicitly.

7.4.7 Metrics

None of the methods offer a metric set to be used during IoT system development for tracking different aspects of the process and product. Ignite states that IoT-specific metrics should be created in an organization if that organization considers IoT systems as a strategic component (Slama et al., 2016). IoT-AD has been evaluated using some well-known metrics measuring development effort, success of reusability, and code quality (Patel, 2014). On the other hand, IoT-AD does not contain any specific metric to be used when applying the method to a specific project. IoT-Meth, ELDAMeth, SPLP-IoT, and GSEM-IoT do not include any information on metrics.

7.4.8 Addressed Discipline

Ignite provides a comprehensive view by handling IoT devices and IoT services, ending up with a system engineering perspective to an IoT system development project. Although IoT-Meth mentions IoT system elements, it has a very superficial system engineering perspective. IoT-AD, ELDAMeth, SPLP-IoT, and GSEM-IoT mainly focus on the software engineering aspect of IoT systems. GSEM-IoT partially addresses back-end services from a software engineering perspective as well.

7.4.9 Scope

All the methods are designed for general-purposes usage and are not specialized for a specific domain (e.g., agriculture or transportation).

7.4.10 Process Paradigm

Ignite, according to its documentation (Slama et al., 2016), is compatible with both plan-driven and agile paradigms. The creators of Ignite claim that an agile paradigm can be applied in IoT system development, but on the other hand, they emphasize some issues, which are potentially challenging: (1) scaling agile to large, distributed project organizations; (2) potential cultural differences between hardware and software engineers; (3) long-term planning needed for hardware (IoT device) design, implementation, and testing; and (4) challenging release management due to distributed nature of IoT devices. IoT-Meth does not specifically define a process paradigm, but on the other hand, the method itself uses a terminology close to that of the agile paradigm. It favors iterative development and producing prototypes rapidly. Moreover, it uses concepts favored in the agile paradigm, such as continuous deployment, a feedback loop, and a minimum viable product. IoT-AD claims that it supports iterative development through automation in different phases of IoT application development. The documentation of ELDAMeth covers iterative development with no discussion of a particular paradigm. The documentation on SPLP-IoT and GSEM-IoT does not include any discussion on a process paradigm, not even iterative development.

7.4.11 Rigidity of the Method

Although Ignite's documentation includes many references to other practices and artifacts that can be used with Ignite, it does not include any information on the flexibility of tailoring the process. Ayala et al. (2015) discuss the ability to customize the development process within the scope of SPLP-IoT, but they do not give any information on how to do this. The documentations of IoT-Meth, IoT-AD, ELDAMeth, and GSEM-IoT do not cover process tailoring, which might be needed to meet project-specific needs.

7.4.12 Maturity of the Method

Ignite was developed by analyzing best practices and real-world projects. Since Ignite was not used in these projects, a project has been selected to validate it. The project team found Ignite useful in providing a high-level roadmap for implementing an IoT system realizing a business idea. They think that Ignite can be improved further, for instance, to include an approach for keeping track of artifacts during a project. IoT-AD has been evaluated via two case studies in a lab environment. It is reported that IoT-AD, along with IoTSuite, enables us to generate a significant percentage of total application code, resulting in a reduction in development effort, especially for IoT applications involving a large number of devices. It is reported that ELDAMeth has been validated in many case studies from different application domains, such as distributed information retrieval, mobile e-marketplaces, content delivery infrastructures, and wireless sensor network–based systems (Fortino and Russo, 2012). Some case studies have been conducted to validate SPLP-IoT, and the results have shown that SPLP-IoT leads to autonomous agent systems for IoT systems (Ayala and Amor, 2012; Ayala et al., 2012, 2014). No evaluation or validation cases were reported for IoT-Meth and GSEM-IoT.

7.4.13 Documentation of the Method

Ignite is documented via a book (Slama et al., 2016), which defines the method by providing information from real-world projects. IoT-Meth is only documented through a website (Collins, 2017), which includes a high-level process flow with a presentation giving some information on the method. The main document for IoT-AD is a PhD thesis (Patel, 2014), along with a journal article (Patel and Cassou, 2015). ELDAMeth is a result of a research project, whose outputs are presented on the website* and in some academic articles (Fortino and Russo, 2012; Fortino et al., 2014, 2015). SPLP-IoT is documented by some journal articles (Ayala and Amor, 2012; Ayala et al., 2015) and conference papers (Ayala et al., 2012, 2014). Some conference papers (Zambonelli, 2016, 2017) have been published on GSEM-IoT.

7.4.14 Tool Support

Ignite does not propose the use of any specific tool. However, it gives some examples of tools to support development, testing, and so forth, from real-world projects. IoT-AD is supported by an open-source tool called IoTSuite,† which aims to make IoT application development easier by supporting the separation of concerns, high-level modeling, and automation. ELDAMeth is supported by a CASE tool called ELDATool, which provides support to developers during the modeling, simulation, and implementation phases for developing ELDA-based distributed agent systems (Fortino and Russo, 2012). The tool was implemented as Eclipse plug-ins. SPLP-IoT discusses some tools that can be used separately to support some steps (configure agent, resolve agent, and model goals illustrated in Figure 7.6) of the process. A software architect can use a tool to check dependencies while configuring an agent to avoid errors that can result from performing this task manually. Moreover, a CVL tool can be used to resolve an agent (combining an agent configuration with an IoT multiagent system architecture to obtain an IoT multiagent system architecture configuration) and produce an IoT multiagent system architecture configuration. Another tool can be used when modeling goals to check the consistency and detect conflicts among agent goals. IoT-Meth and GSEM-IoT do not discuss any specific tool.

7.5 CONCLUSION

As several opportunities are afforded by IoT systems, developing such systems efficiently becomes more important. Generally, developing IoT systems is nontrivial and more complex than traditional software systems, since they encompass many software, hardware, and communication components. Therefore, efficient development of IoT systems requires systematic approaches that come into existence mainly in the form of SDMs. This chapter presents a brief overview of IoT SDMs in the literature and their evaluation based on 14 criteria.

Some IoT SDMs are built based on experience from real-world projects. Ignite and IoT-Meth are two examples of such SDMs emerged from the industry. The SDMs established by the researchers from academia are mainly grounded on previous research in various areas. IoT-AD is built on macroprogramming and inspired by model-driven design. ELDAMeth is based on multiagent system development. SPLP-IoT utilizes the SPLE paradigm, as well as multiagent system development. GSEM-IoT adapts some concepts from traditional software development to the IoT domain.

* ELDAMeth website, http://eldameth.deis.unical.it/.

† IoTSuite website, https://github.com/pankeshlinux/IoTSuite/wiki.

None of the identified IoT SDMs can be considered a complete method that covers all the important phases necessary for developing IoT systems. Ignite presents a more holistic view for developing IoT systems than the rest of the SDMs. It roughly defines a process from an idea to a running IoT system and provides some artifacts. IoT-Meth includes six main steps, which are far from having sufficient detail for guiding the development of an IoT system. IoT-AD, ELDAMeth, SPLP-IoT, and GSEM-IoT focus only the software components of an IoT system.

Validation of these SDMs is an important topic that needs to be addressed. Only Ignite has been validated by one real-world project, but is this enough? Validating these SDMs in real-world projects would allow us to not only assess them, but also to improve them. Moreover, the SDMs should provide adequate documentation covering some basic topics of a method description, such as activities, artifacts, roles, and phases.

REFERENCES

AIOTI [Alliance for Internet of Things Innovation]. (2016). High Level Architecture (HLA). Release 2.1. AIOTI WG03—IoT Standardisation.

Ayala, I., and Amor, M. (2012). Self-configuring agents for ambient assisted living applications. *Ubiquitous Computing* 17(6):1159–1169.

Ayala, I., Amor, M., and Fuentes, L. (2012). An agent platform for self-configuring agents in the Internet of things. In *3rd International Workshop on Infrastructures and Tools for Multi-Agent Systems, ITMAS*, Valencia, Spain, pp. 65–78.

Ayala, I., Amor , M., and Fuentes, L. (2014). Towards a CVL process to develop agents for the IoT. In *8th International Conference, UCAmI 2014*, Belfast, UK, pp. 304–311.

Ayala, I., Amor, M., and Fuentes, L. (2015). A software product line process to develop agents for the IoT. *Sensors* 15(7):15640–15660.

Brinkkemper, S. (1996). Method engineering: Engineering of information systems development methods and tools. *Information and Software Technology* 38(4):275–280.

Collins, T.A. (2017). Methodology for building the Internet of Things. http://www.iotmethodology.com/ (accessed January 16, 2017).

Fortino, G., Guerrieri, A., Russo, W., and Savaglio, C. (2015). Towards a development methodology for smart object-oriented IoT systems: A metamodel approach, pp. 1297–1302. In *IEEE International Conference on Systems, Man, and Cybernetics*, Kowloon, China.

Fortino, G., Rango, F., and Russo, W. (2014). ELDAMeth design process. In *Handbook on Agent-Oriented Design Processes*, ed. M. Cossentino, V. Hilaire, A. Molesini, and V. Seidita, 115–139. Berlin: Springer.

Fortino, G., and Russo, W. (2012). ELDAMeth: An agent-oriented methodology for simulation-based prototyping of distributed agent systems. *Information and Software Technology* 54(6):608–624.

Haugen, Ø., Wąsowski, A., and Czarnecki, K. (2013). CVL: Common Variability Language. In *17th International Software Product Line Conference (SPLC '13)*, New York, pp. 277–277.

Patel, P. (2014). Enabling high-level application development for the Internet of things. PhD dissertation, Universite Pierre et Marie Curie—Paris VI.

Patel, P., and Cassou, D. (2015). Enabling high-level application development for the Internet of things. *Journal of Systems and Software* 103(C):62–84.

Pathak, L. M., Bakshi, A., Prasanna, V., and Picco, G. (2007). A compilation framework for macroprogramming networked sensors. *Distributed Computing in Sensor Systems* 4549:189–204.

Pohl, K., Böckle, G., and van der Linden, F. (2005). *Software Product Line Engineering*. Berlin: Springer-Verlag.

Slama, D., Puhlmann, F., Morrish, J., and Bhatnagar, R. M. (2016). *Enterprise IoT Strategies and Best Practices for Connected Products and Services*. Sebastopol, CA: O'Reilly Media.

Zambonelli, F. (2016). Towards a discipline of IoT-oriented software engineering. In *17th Workshop "From Objects to Agents,"* Catania, Italy, July 29–30, pp. 1–7.

Zambonelli, F. (2017). Key abstractions for IoT-oriented software engineering. *IEEE Software* 34(1):38–45.

8 Design Considerations for Wireless Power Delivery Using RFID

Akaa Agbaeze Eteng, Sharul Kamal Abdul Rahim, and Chee Yen Leow

CONTENTS

8.1 Introduction 161
8.2 RFID Principles 162
8.2.1 Inductive Coupling RFID 163
8.2.2 Backscatter RFID 164
8.3 Wireless Power Transfer through Inductive Coupling RFID 164
8.3.1 General System Architecture 165
8.3.2 Wireless Power Transfer Link Design 167
8.3.2.1 Coil Antenna 167
8.3.2.2 Link Transfer Efficiency 169
8.3.2.3 Spatial Freedom 170
8.3.2.4 Multiobjective Link Considerations 170
8.4 RFID Radio-Frequency Energy Harvesting 172
8.4.1 General System Architecture 172
8.4.2 Design Considerations 173
8.4.2.1 RFID Rectenna Design 173
8.5 Conclusion 174
References 175

8.1 INTRODUCTION

In a most general form, radio-frequency identification (RFID) is a short-range radio technology for communicating digital information, either between a stationary location and a movable object, or between two nonstationary objects (Landt, 2005). From its shared origins with Radar systems in the 1940s, RFID has since matured into a technology used in a wide variety of applications. Typical examples of RFID usage include access control in buildings, supply chain management, library book-tracking systems, livestock management, and automated vehicle toll systems, to name a few.

The rapid and widespread adoption of RFID over the years created a need for the standardization of the technology. Specifications for the operation of RFID systems not only ensured that solutions from various vendors were interoperable, but also expanded the adoption of RFID technology from isolated niche applications to more horizontally networked usage scenarios (McFarlane, 2005).

The IoT is a vital networked application in which RFID finds contemporary relevance. The growth of the IoT is inextricably linked to a massive increase in the deployment of sensing platforms. It is projected that sensors will account for 75% of the growth of connected electronic devices from current levels to about 40.9 billion interconnected devices by 2020 (ABI Research, 2014).

Enabling technologies for the IoT can be categorized either as information acquisition, information processing, or security-enhancing technologies (Vermesan and Friess, 2013). On this basis, RFID can primarily be regarded as a technology to enable IoT objects acquire and communicate contextual information. For example, RFID can be employed to provide vital real-time information about the location and condition of specific inventory items in a supply chain. Also, an RFID-enabled wearable patient monitor could be used to provide secure real-time updates of a patient's medical condition to a doctor.

IoT objects will often contain electronics, which should be characterized by eco-friendliness and energy autonomy (Roselli et al., 2015). Eco-friendliness is largely an issue of the environmental impact of materials used in the physical realization of the technology. Energy autonomy, on the other hand, implies that the sensing electronics need not depend on conventional wired power delivery from the grid, or on external battery systems. This requirement is critical for applications in which sensing electronics are placed in inaccessible environments.

It is noteworthy that passive RFID transponders are, in principle, examples of energy-autonomous electronic devices that can be used in sensing and communication interactions. The energy transfer mechanisms of RFID schemes are adequate for the transfer of typically short bursts of data in traditional identification applications. However, emerging RFID sensing applications require the use of the technology for power delivery to support sensor electronics (Zhao et al., 2015; Sample et al., 2008).

Consequently, this chapter presents a discussion of considerations for wireless power delivery through passive RFID, as well as approaches to its actualization. First, we present an overview of the working principles behind passive RFID, with emphasis on load modulation and backscatter mechanisms. Next, design considerations to enable wireless power transfer (WPT) applications through load-modulated RFID links are discussed. Finally, we examine considerations for radio-frequency (RF) harvesting RFID power delivery.

8.2 RFID PRINCIPLES

RFID employs RF communication principles for the automatic identification of objects, locations, or individuals. It provides a mechanism for acquiring data, which could then be stored or processed as dictated by the particular application. In RFID schemes, objects can be identified by RFID readers based on unique codes contained in RFID transponders attached to or embedded within the identified objects. Consequently, the basic communication interaction in RFID systems is between readers and transponders.

Broadly, there are three variants of reader–transponder energy interactions used to facilitate RFID data transfers. The first, passive RFID, is one where the electrical power required for the operation of the transponder is wholly provided by the reader. In this case, the energy field from the reader is used for both reader-to-transponder and transponder-to-reader communications. Passive transponders transfer data to the reader by modulating the energy field emitted by the reader (Finkenzeller, 2010). The second scheme, active RFID, employs transponders with independent power sources, usually in the form of onboard batteries. These batteries provide the power required to transmit the modulated transponder signal back to the reader. Active RFID schemes are thus able to operate at greater communication ranges than passive RFID schemes. The third RFID variant employs semipassive transponders. These transponders contain batteries, but still depend on the reader energy field for transmission of a radio signal (Bolic et al., 2010).

An alternative classification of RFID schemes is based on the nature of the energy field used to facilitate communication. Differences in the communication-facilitating energy fields influence the energy-autonomous application realized with RFID systems. Near-field RFID schemes employ non-propagating field interactions between reader and transponder antennas. Far-field RFID schemes, on the other hand, are based on propagating electromagnetic waves.

Near-field RFID systems employ nonradiating electric or magnetic fields within the near-field region of the reader antenna. This means that for oscillating fields with wavelength λ, the range of these RFID systems is less than 2π/λ. Near-field RFID systems with very short ranges are known as close-coupling systems. Physically, close-coupling systems operate at ranges of less than 1 cm, using oscillating fields at frequencies of less than 30 MHz. This short range enables the transfer of relatively greater energy between coupled readers and transponders than with other RFID variants. Close-coupling systems are typically employed in contactless card applications in which security of transaction, rather than range, is premium. Remote-coupling near-field RFID systems, on the other hand, operate at ranges up to 1 m. Near-field RFID systems in which the coupling field is magnetic are commonly described as inductive coupling RFID, while capacitive coupling RFID systems employ electrical coupling fields. The vast majority of available near-field RFID systems are inductively coupled (Finkenzeller, 2010), and these are the near-field RFID links under consideration in this chapter.

Far-field RFID systems operate at ranges greater than 2π/λ, typically employing electromagnetic waves in the ultra-high-frequency (UHF) and microwave spectral ranges. Physically, this translates to ranges up to 3 m using passive transponders, and 15 m with active transponders (Finkenzeller, 2010). While the reader–transponder energy interaction in inductively coupled RFID is highly directional, the energy radiated from far-field RFID reader antennas is much less directional. The farther range and less directional nature of these RFID systems suggest a rather different wireless power delivery application for far-field RFID compared with its inductively coupled equivalent.

8.2.1 Inductive Coupling RFID

The operation of an inductive coupling RFID link is based on the oscillating magnetic field created around the coil antenna of a reader by the time-varying current flowing within it, as predicted by the Biot–Savart law. When this oscillating field intercepts a transponder's coil antenna, a voltage is induced across the antenna terminals, as predicted by Faraday's law. This voltage is rectified, and serves as the supply voltage to the data-carrying transponder microchip. Inductive coupling RFID is illustrated in Figure 8.1.

Currents flowing in the coil antennas of the transponder and reader generate magnetic fluxes. Normalizing the magnetic flux generated in the area enclosed by a coil antenna by the current flowing through it provides the self-inductance L of the coil antenna. The computation of the enclosed magnetic flux for complex coil antenna geometries is often complicated. Consequently, various

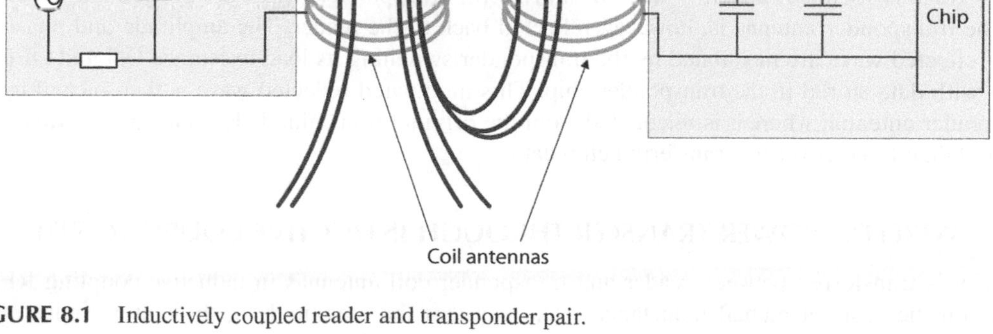

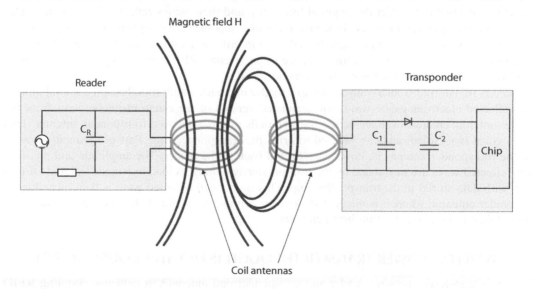

FIGURE 8.1 Inductively coupled reader and transponder pair.

handbook methods have been developed to calculate the self-inductance of various coil shapes (Grover, 1946; Mohan et al., 1999; Greenhouse, 1974). Also, coil self-inductance can be determined using computer-based electromagnetic simulators.

The inductive coupling between a reader and transponder coil antenna is modeled as a mutual inductance M. This characterizes the amount of magnetic flux generated by the reader coil antenna that passes through the transponder coil antenna. Mutual inductance depends on the magnetic properties of the medium between the coupled antennas, their relative orientation, and their geometries. Roughly, mutual inductance can be increased by employing bigger, perfectly aligned coil antenna pairs, at shorter separation distances. Semiempirical formulas have been developed to enable the calculation of mutual inductance between coils of various geometries (Grover, 1946). In the alternative, mutual inductance can also be determined using electromagnetic simulation software.

A more general characterization of inductive coupling is provided by the coupling coefficient k, which is a normalization of the mutual inductance M_{12} between a coupled pair of coil antennas to the geometric mean of coil antenna self-inductances, namely, L_1 and L_2. Consequently, for a pair of coil antennas, the coupling coefficient is defined as

$$k_{12} = \frac{M_{12}}{\sqrt{L_1 L_2}} \tag{8.1}$$

Data is transferred in an inductive coupling RFID link from a transponder to a reader by load modulation, in which circuit parameters of the transponder coil antenna resonant circuit are varied in step with the intended data stream, thereby modulating the amplitude and phase of the reflected transponder impedance. Generally, this load modulation can be achieved by varying either the transponder load resistance or its associated parallel capacitance. Transferred data is reconstructed at the reader by demodulation. Load-modulated RFID systems usually operate within the high-frequency spectral designation, with the unlicensed 13.56 MHz industrial, scientific, and medical (ISM) band being very popular.

8.2.2 Backscatter RFID

Backscatter RFID systems are typically operated at the UHFs of 868 and 915 MHz, or at the microwave frequencies of 2.5 and 5.8 GHz. These RFID systems operate by reader antennas radiating electromagnetic waves at the required frequency, and these waves reflected back to the reader sources by transponder antennas. Typically, reflecting antennas are designed to be comparable in size to the wavelength of electromagnetic radiation in order to facilitate radiation and reflection of electromagnetic waves. The operating distance of backscatter RFID links is typically greater than 0.16 times the wavelength of the electromagnetic signal.

Data is transferred from a transponder to a reader in a backscatter link through a modulation of the reflected electromagnetic wave. An attenuated version of the electromagnetic wave from the reader antenna creates a voltage across the terminals of the associated transponder antenna. This RF voltage is rectified, and can be used to drive the transponder chip. Part of the incident wave on the transponder antenna is, however, reflected back to the reader. The amplitude and phase of this reflected wave are modulated by the transponder switching its load resistance ON and OFF in step with data stored in the transponder chip. This modulated reflected wave is then picked up at the reader antenna, where it is uncoupled from the original transmitted electromagnetic wave and demodulated to recover the transferred chip data.

8.3 WIRELESS POWER TRANSFER THROUGH INDUCTIVE COUPLING RFID

Energy is transferred between reader and transponder coil antennas in inductive coupling RFID links on the basis of mutual inductances. The fact that the magnitude of the mutual inductance is strongly influenced by the relative orientation of the paired coil antennas means that inductive coupling RFID links are strongly directional. More energy is transferred between the reader and transponder coil antennas when they are perfectly aligned, and within the prescribed operating range. Consequently, wireless power delivery infrastructure using inductive RFID links requires intentional transmission of electrical energy from dedicated sources, rather than ambient energy scavenging.

Conceptually, designing an inductive coupling RFID-based WPT implementation involves harnessing the energy wirelessly transferred by an inductive RFID reader to power up a suitable load. IoT infrastructure in which inductive coupling RFID-based sensors are deployed for data gathering in scenarios that preclude wired power delivery would benefit from such WPT implementations. An example of this WPT application is found in a recent project where near-field communication (NFC), a technology based on inductive coupling RFID, has been integrated with a sensing platform to create the NFC–Wireless Identification and Sensing Platform (NFC-WISP) (Zhao et al., 2015). This programmable sensing and computing platform can be powered up and interrogated by inductive RFID readers, as well as NFC-enabled smartphones. In the cited example, the NFC-WISP was used in a data logging application, where the sensing platform is placed on milk packaging in a simulated cold-supply shipment. By integrating the sensor platform with the milk packaging, the milk container is effectively converted into a smart object. The temperature of the container, its motion, and three-dimensional (3D) orientation are recorded by the NFC-WISP. Once the shipment arrives at a destination, an RFID reader can be used to download the recorded data to a host computer for further processing. Alternatively, an NFC-enabled mobile phone could be used to download the recorded data by bringing it close to the milk container. The energy used by the sensor platform to sense temperature, 3D position, and motion is acquired during the inductive coupling interactions involving RFID readers or NFC-enabled smartphones. This energy is stored onboard in a thin-film battery, which powers the sensor electronics when the sensing platform is far from a reader. Other examples of RFID WPT to smart IoT objects, as reported in the literature, include an NFC-enabled blood glucose monitor (DeHennis et al., 2013; Tankiewicz et al., 2013) and a wireless-powered smart watch (Lin et al., 2015).

In this section, we examine general system architecture for realizing inductive coupling RFID-based WPT links, as well as some design considerations, and methods to optimize link performance.

8.3.1 General System Architecture

Figure 8.2 illustrates the principal components of an inductive coupling RFID WPT scheme.

The RFID reader serves as the power transmitter in the link. In a more general sense, the RFID reader is a transceiver, since it can also detect load-modulated signaling. This facility can be employed to provide feedback control for the wireless power delivery scheme. The RFID reader can either be battery powered or powered externally. The output of the RFID reader is essentially an oscillating magnetic field, whose frequency depends on the specific RFID protocol in use. For the widely employed ISO/IEC 14443 and ISO/IEC 15693 specifications, the frequency of oscillation is 13.56 MHz. Typically, the RFID reader would have inbuilt power amplification to drive the reader coil antenna.

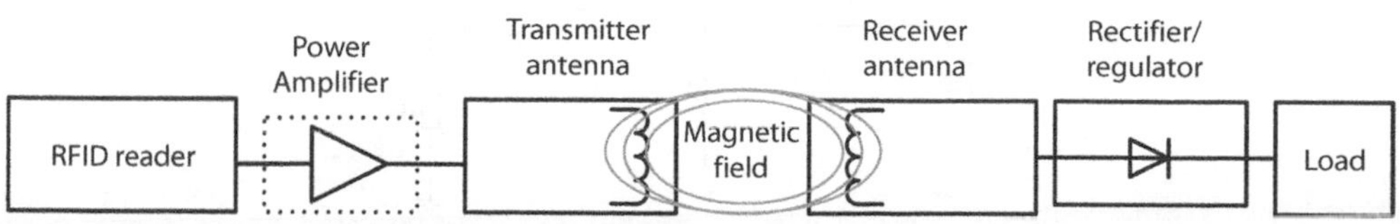

FIGURE 8.2 Inductive RFID WPT system.

More power-demanding WPT applications may require added external power amplification at the link transmitter. The power amplifier is often designed to meet the specific drive requirements of the associated power transmitter coil. For example, it could be used to increase the read range of an ISO/IEC 14443A inductive RFID system (De Mulder et al., 2009). Most power amplifier designs for inductive wireless power delivery applications are based on either Class D or Class E topologies. These are essentially switching power amplifiers. As shown in Figure 8.3, the basic Class D power amplifier uses a pair of transistors driven such that they are alternately switched ON and OFF. This arrangement realizes a two-pole switch that presents either a rectangular current or voltage waveform at the input of a tuned load circuit. The tuned load circuit filters out harmonics from the rectangular waveform, resulting in a sinusoidal output (Albulet, 2001). In a basic Class E amplifier, as shown in Figure 8.4, a single transistor serves as the switch. The arrangement is such that current and voltage waveforms do not overlap during the switching time interval. Consequently, there is virtually no power loss associated with transistor switching, resulting in very efficient power amplification (Kazimierczuk, 2008). The Class E power amplifier topology is generally preferred at higher frequencies (Pinuela et al., 2013b).

Although they are commonly known as coil antennas, the coil structures employed at the front ends of RFID reader and tag infrastructure are not true antennas, as they do not radiate and intercept electromagnetic waves. This is because the frequency of operation of inductive RFID links is such that the sizes of these coils are significantly smaller than the wavelength by some orders of magnitude. However, these coils are closed-loop structures, which can be analyzed as derivatives of magnetic dipoles (Ramo et al., 1994). In other words, currents flowing through these structures excite more significant magnetic field components than electrical field components. Thus, a coil structure at the front end of the RFID reader enables the generation of an oscillating magnetic field. Similarly, a coil structure at the front end of the power receiver enables the coupling of the magnetic energy in the reader-generated magnetic field. This coupled energy presents itself as a voltage across the terminals of the receiving coil structure. Physically, magnetic coupling can be achieved using conductor loop structures such as helices (Tak et al., 2011), spirals (Jonah et al., 2014), and solenoids

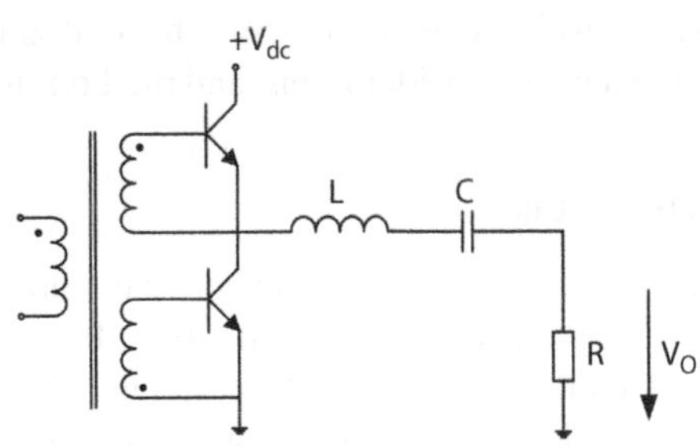

FIGURE 8.3 Basic Class D power amplifier topology.

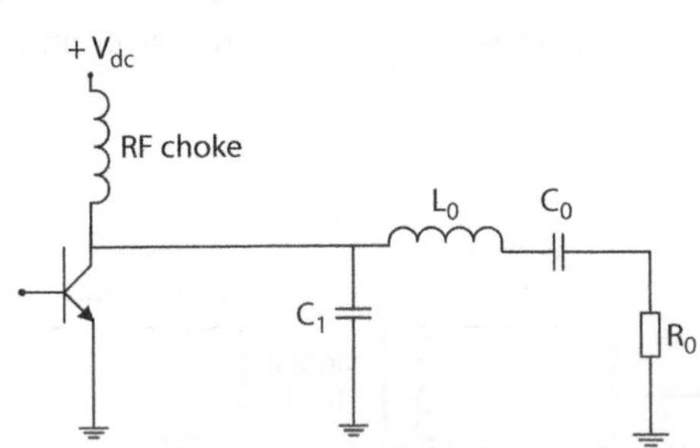

FIGURE 8.4 Basic Class E power amplifier topology.

(Lee et al., 2014), among others. The printed spiral coil (PSC), where the spiral conductors are printed on suitable substrate materials, is perhaps the most widespread implementation of planar tag spiral coil structures. To eliminate losses and power reflections, impedance matching networks are usually inserted between the power amplifier and transmitter coil, and between the rectifier and receiver coil.

The oscillating alternating current (AC) voltage across the terminals of the receiver coil structure needs to be rectified to obtain a direct current (DC) voltage. This DC voltage could be filtered and further conditioned before being used to drive a designated electrical load. Consequently, the major role of the rectifier is to provide a DC voltage from the high-frequency voltage at the terminals of the receiver coil antenna. Discrete component rectifier implementations usually employ Schottky diodes in a full-bridge configuration. Alternatively, single-chip complementary metal-oxide semiconductor (CMOS) rectifier implementations are widespread.

8.3.2 Wireless Power Transfer Link Design

Employing inductive RFID infrastructure in power-centered applications requires a shift from the more typical data- and range-centered design considerations adopted in traditional RFID systems (Sample et al., 2011). In a lot of respects, design considerations for inductive RFID-based WPT are similar to concerns for inductive power transfer systems. However, the power delivery functionality of these RFID links should not impede their traditional data transfer roles.

At the transmitter end, an AC flowing through the coil excites a magnetic field around the coil. The energy in this field is inductively coupled to the receiving coil in proximity to the transmitter coil. The link is therefore analogous to a loosely coupled transformer circuit. Figure 8.5 is a simplified circuit model of the coupling of a pair of coils.

8.3.2.1 Coil Antenna

As shown in Figure 8.5, each coil can be modeled as a parasitic capacitance in parallel with a series combination of a resistance and inductance.

The coil inductance arises from the magnetic field generated by current flowing in the coil loops. A popular formula for determining the inductance of multiturn planar spiral coils is given as (Mohan et al., 1999)

$$L = 0.5\left(\mu n^2 d_{avg} c_1\right)\left[\ln\left(c_2/\rho\right) + c_3\rho + c_4\rho^2\right] \tag{8.2}$$

The values of the coefficients $c_1 - c_4$ depend on the geometric layout of the PSC. μ refers to the permeability of the conductor material, while ρ is the spiral fill ratio, namely,

$$\rho = \frac{d_o - d_i}{d_o + d_i} \tag{8.3}$$

d_0 and d_i refer to the outer and inner diameters of the coil turns on the PSC, as illustrated in Figure 8.6.

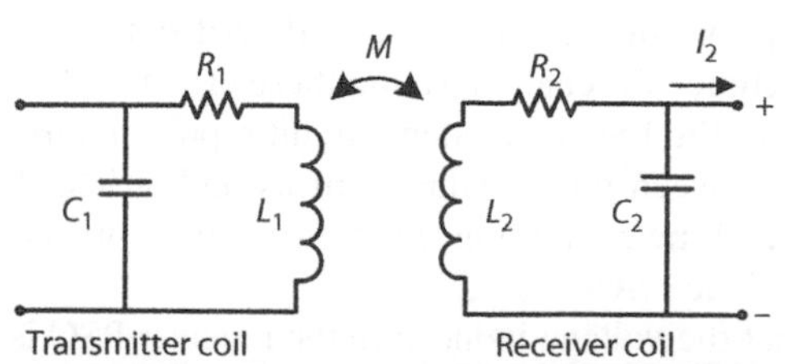

FIGURE 8.5 Inductive coupling link.

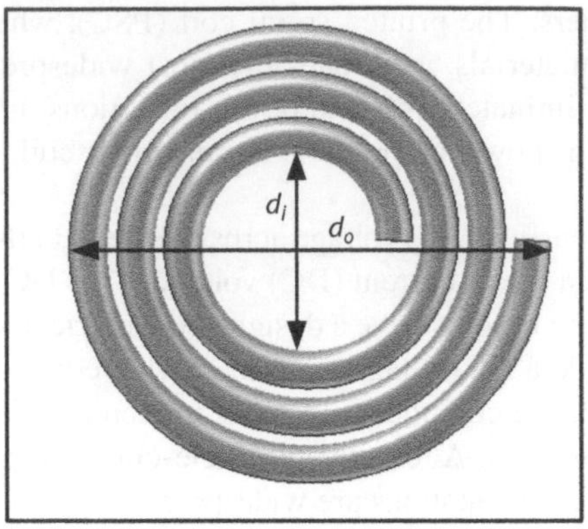

FIGURE 8.6 PSC layout.

The average diameter of the spiral trace is given by

$$d_{avg} = 0.5(d_o + d_i) \tag{8.4}$$

Analytic models for resistance in PSCs usually take skin effects into account. The resistance due to DCs flowing in a conductor is given by

$$R_{DC} = \frac{l}{A\sigma} \tag{8.5}$$

where:

- l is total the conductor length
- A is the cross-sectional area of the conductor
- σ is the conductivity of the conductor material (Reinhold et al., 2007)

The skin effect occurs as ACs flow through the conductor, with the current migrating toward the conductor surface, away from its core. This modifies the conductor resistance to

$$R_{AC} = R_{DC} \cdot \frac{t}{\delta_{skin}\left(1 - e^{-t_c/\delta_{skin}}\right)}, \quad \delta_{skin} = \sqrt{(\pi f \mu \sigma)^{-1}}, \tag{8.6}$$

where:

- δ_{skin} is the skin depth
- t_c is the conductor thickness (Jow and Ghovanloo, 2007; Reinhold et al., 2007)

In multiturn coil structures, the skin effect is further modified by proximity effects, in which magnetic fields excited by adjacent loop turns cause asymmetric current distributions. Although some studies have modeled proximity effects (Felic et al., 2013), full-wave electromagnetic simulations still remain a popular choice for characterizing these effects.

The parallel capacitance in the PSC equivalent circuit representation models the aggregate parasitic capacitance forming between coil conductor turns and strips, through free space, and the dielectric substrate. Actual analytic expressions for parasitic capacitance depend on the geometrical layout and implementation of the PSCs.

Faraday's law predicts that the voltage induced in the receiver PSC is

$$V_{ind} = j\omega M I_2 \tag{8.7}$$

where:

I_2 is the current in the receiver coil
M is the mutual inductance between the transmitter and receiver coils

As earlier noted, there are several handbook methods for computing the mutual inductance between a pair of inductively coupled coils. For instance, the mutual inductance between a pair of circular multiturn coils can be computed using (Raju et al., 2014)

$$M = \sum_{i=1}^{n_a}\sum_{j=1}^{n_b}\left(\frac{\mu_0 \pi a_i^2 b_j^2}{2\left(a_i^2 + b_j^2 + z^2\right)^{3/2}}\left(1 + \frac{15}{32}\gamma^2 + \frac{315}{1024}\gamma^4\right)\right), \quad \gamma = \frac{2a_i b_j}{a_i^2 + b_j^2 + z^2} \tag{8.8}$$

where:

n_a and n_b are the number of turns of the transmit and receive loop, respectively
a_i and b_i are the radii of the ith transmitter and jth receiver coil turns, respectively
z is the axial separation distance between both coils

Here, μ_0 refers to the permeability of the free-space medium separating the transmitter and receiver coils.

8.3.2.2 Link Transfer Efficiency

The end-to-end system efficiency of an inductive RFID WPT scheme is the product of the efficiencies of the constituent subsystems. In most WPT cases, however, the weakest link in the chain is the coupling between the interacting coil antennas (Vandevoorde and Puers, 2001). Consequently, the link transfer efficiency is a most critical determinant of the efficiency of the WPT scheme. In a general inductive WPT scenario employing a pair of biconjugate matched coil antennas, the link transfer efficiency is given by

$$n_{link} = \frac{k^2 Q_1 Q_2}{1 + \sqrt{1 + k^2 Q_1 Q_2}} \tag{8.9}$$

where:

$Q_{1,2}$ are the Q-factors of the coil antennas
k is the link coupling coefficient

Conventional inductive power transfer links typically employ high Q-factor coil structures to facilitate efficient power delivery. The Q-factor of a coil antenna is the ratio of its reactive self-impedance to its resistive self-impedance. Consequently, high Q-factor coil antennas are realized using designs with high reactance-to-resistance ratios. Unfortunately, due to the inverse relationship between coil Q-factors and their bandwidths, Q-factor enhancements could be detrimental to the data transfer capability of an inductive link. Usually, inductive RFID reader coil antennas are implemented to have as low a Q-factor as possible within the design specification (Aerts et al., 2008). Consequently, Q-factor-based transfer efficiency enhancements may not be the most viable approach to efficient inductive RFID WPT schemes.

The product of coil antenna Q-factors and the coupling coefficient is often viewed as a figure of merit (FoM) of traditional inductive WPT schemes (Bosshard et al., 2013; Inagaki, 2014). By implication, Q-factors and coupling levels equivalently determine the transfer efficiency of an inductive WPT link (Waffenschmidt and Staring, 2009). Consequently, coupling enhancements could be used as an alternative means to achieve efficient power transfer in Q-factor-constrained inductive RFID WPT applications. Coupling is a function of self-inductances of the coupled coils, and the mutual inductance between them. The mutual inductance is proportional to the magnetic flux enclosed by the receiving coil as a consequence of the magnetic field excited by current flow in the transmitter coil. Since this mutual inductance is also influenced by the geometry of the interacting coils, design undertakings to enhance coupling between paired coils focus on two areas, namely, field enhancement and coil geometry.

Magnetic field enhancement is widely employed in short-range inductive WPT links. Typically, ferrite materials are employed to redirect excited magnetic fields toward intended coupling directions (Jiseong et al., 2013). Ferrite sheets can be incorporated into the structure of inductive RFID coil antennas to alter the distribution of the magnetic field (B. Lee et al., 2014; Bauernfeind, 2013). Alternatively, coil turns can be distributed away from the coil edge, resulting in an increase in the coupling coefficient between paired coils (Zierhofer and Hochmair, 1990). This technique has been further harnessed to strengthen the excited magnetic field from a Q-factor-constrained reader coil antenna (Sharma et al., 2013a, 2013b), and to improve the link transfer efficiency in Q-factor-constrained symmetric inductive WPT links (Eteng et al., 2016).

8.3.2.3 Spatial Freedom

Misalignment between a pair of coupled coil antennas results in a reduction in the intensity of the magnetic coupling between them. Consequently, less power is transferred across misaligned coil antennas than if they were perfectly aligned. Axial misalignments have a more significant impact in design scenarios that require smaller transponder receiver coil antennas. Usually, in such cases, larger reader transmitter coil antennas are used to compensate for the small size of the receiver infrastructure. Such design scenarios make it necessary to investigate the limits of power transfer performance under various degrees of coil antenna misalignment. Compact analytical models to describe coil misalignments in inductive RFID telemetry links have been proposed (Fotopoulou and Flynn, 2011). Such models enable a designer predict the impact of lateral and angular coil antenna misalignments without resorting to lengthy electromagnetic simulations.

A unique problem arises when inductive coupled coils are at separation distances closer than the link is designed to operate at. Such scenarios lead to the appearance of multiple new resonance frequencies, a phenomenon known as frequency splitting (Kim and Ling, 2007; Inagaki, 2014). Frequency splitting is associated with link overcoupling, and is usually accompanied by a loss in transfer efficiency at the original frequency the link was designed to operate in. Efficiency loss at the link operating frequency can be prevented by proper link design. One proposal is to introduce a reverse coupling to counteract the overcoupling arising from coil antennas being brought closer together. This can be achieved through the design of the reader coil antenna with antiparallel loop turns (Lee et al., 2013), or with capacitance-controlled reverse-current flow (Lee et al., 2014).

Spatial freedom can also be achieved by incorporating control schemes in the inductive RFID link. The general system architecture described in Figure 8.2 is essentially an open-loop configuration, whose performance will be significantly impaired by changes in the physical orientation of the coupled coil structures. This can be counteracted by implementing a closed-loop RFID link (Kiani and Ghovanloo, 2010). The RFID implementation in the cited reference employs the commercially available TRF 7960 RFID reader (Texas Instruments, Dallas), and an external control unit composed of a digital potentiometer and a microcontroller. The implemented WPT link harnesses the back telemetry capability of the RFID reader to provide the feedback required by the control unit for corrective measures against voltage fluctuations arising from coupling and loading variations.

8.3.2.4 Multiobjective Link Considerations

Designers of inductive power transfer links frequently have to reach compromises between various conflicting and competing link performance parameters. These competing interests are often expressed in FoMs, which characterize the power transfer performance in terms of multiobjective criteria.

It has been demonstrated that the maximum power delivered to a load at the receiver terminal of an inductive coupling WPT link does not necessarily coincide with the maximum source-to-load power transfer efficiency (Kiani and Ghovanloo, 2013). Consequently, it is imperative for designers to strike a delicate balance between the source-to-load power transfer efficiency η and the actual power delivered to the load. This compromise is expressed in a FoM defined as (Kiani and Ghovanloo, 2013)

$$\text{FoM} = \frac{\eta^n P_L}{V_s} \tag{8.10}$$

where:

P_L is the power delivered to the load
V_s is the source voltage

The value of weighting parameter n is chosen to reflect the relative criticality of either power transfer efficiency or delivered power, as required by the target application. This FoM, measured in Siemens, describes source-to-load power transfer efficiency if $n \rightarrow \infty$, or the power delivered to the load if $n \rightarrow 0$. These two extremes allow for the determination of limits of power transfer efficiency or delivered power that can be achieved in a usage scenario.

It is necessary that power is delivered to connected loads without a prohibitive increase in the size of the inductively coupled coil antennas. Usually, the receiver coil is subject to more severe size constraints than the transmitting coil antenna, as it is embedded in the IoT object. Since coil antenna sizes have an impact on the achieved inductive coupling, the operating range of the inductive coupling link must not be compromised in the attempt to achieve small antenna form factors. These concerns can be addressed in a FoM defined as (Mirbozorgi et al., 2014)

$$\text{FoM} = \frac{\eta \times P_L \times d}{D_r} \tag{8.11}$$

where:

D_r refers to the diameter of the receiver coil
d is the distance of separation between the transmitter and receiver coils

In order to leverage the data transfer capability of inductive RFID in a power transfer scenario, it would be necessary to note the impact of data bandwidth on efficient power transfer. This impact can be characterized in a FoM that evaluates coil antenna diameters $D_{1,2}$ at both terminals, coupling range d, power transfer efficiency η, transmission bandwidth BW, and the link frequency f (Catrysse et al., 2004), namely,

$$\text{FoM} = 10Log_{10}\left(\frac{d^2 BW\eta}{D_1 D_2 f}\right), \qquad \in \left[\text{FoM} < 0\right] \tag{8.12}$$

This FoM has been further modified to include the achievable voltage gain by the telemetry link, namely,

$$\text{FoM} = 10Log_{10}\left(\frac{d^2 BW\eta}{D_1 D_2 f} G\right), \qquad \in \left[\text{FoM} < 0\right] \tag{8.13}$$

where:

G is the linear voltage gain between the transmit and receive terminals (RamRakhyani and Lazzi, 2013)

The inclusion of voltage gain in the performance assessment of the inductive coupling link allows for links to be designed for high-voltage gain, thereby enabling more cost-effective implementations using lower source voltages at the transmitter input.

8.4 RFID RADIO-FREQUENCY ENERGY HARVESTING

The less directional characteristic of backscatter UHF-RFID systems provides an opportunity to develop transponder solutions able to harvest UHF microwave energy, in addition to their basic RFID functionality. Some examples of wireless energy harvesting (WEH) implementations using UHF-RFID are reported in the literature. These implementations mostly serve to energize RFID sensor implementations embedded in or placed on smart objects in an IoT infrastructure. The RFID Augmented Module for Smart Environmental Sensing (RAMSES) is one such RFID-based module for environmental sensing (De Donno et al., 2014). The RAMSES is able to harvest RF energy transmitted by an RFID reader placed at distances up to 10 m away. The harvested energy is used to power up electronics for monitoring ambient temperature, illumination, and motion, as well as perform computations on the sensed data, and communicate the data as required. An auxiliary onboard battery is used to store energy to ensure that sensor electronics remain powered up when the module is away from a source of RF energy. The RAMSES is an improvement over the WISP (Sample et al., 2008), which was one of the earliest UHF-RFID energy harvesting implementations for sensing, computation, and communication. More recently, the Self-Powered Augmented RFID Tag for Autonomous Computing and Ubiquitous Sensing (SPARTACUS) has been demonstrated (Colella et al., 2015), which targets augmented functionalities beyond what is available with other RFID-based sensing platforms. Unlike the aforementioned modules, SPARTACUS supports two-way communication with an RFID reader, RFID-based sensing, local computing, and actuation control. Placing SPARTACUS in an object extends the functionality of such a smart object, from passive environmental data acquisition to being an active collaborative actuation controller in an IoT infrastructure. Once again, the energy for the operation of SPARTACUS is harvested from the RF energy transmitted by RFID readers during reader–tag interactions.

This section describes the main components of an RFID energy harvesting implementation, as well as general considerations for ensuring optimal power delivery performance.

8.4.1 General System Architecture

The main components of an RF energy harvesting link are shown in Figure 8.7.

The communications channel is rich in ambient electromagnetic energy from multiple RF sources. The receiver antenna intercepts electromagnetic energy at the required frequency, from where it is rectified to a DC voltage for power delivery. A matching network is inserted between the receiver antenna and the rectifier to ensure lossless power transmission to the rectifier. The combination of the antenna, its matching network, and the rectifier is commonly referred to as a rectenna. The rectifier output may be used directly to supply power to electronics. Alternatively, it could be stored in a storage capacitor, or used to charge a battery, for later use.

The time-varying received power level at the receiver antenna has a significant impact on the rectifier performance. This leads to time-dependent optimal loading conditions. A DC-DC power

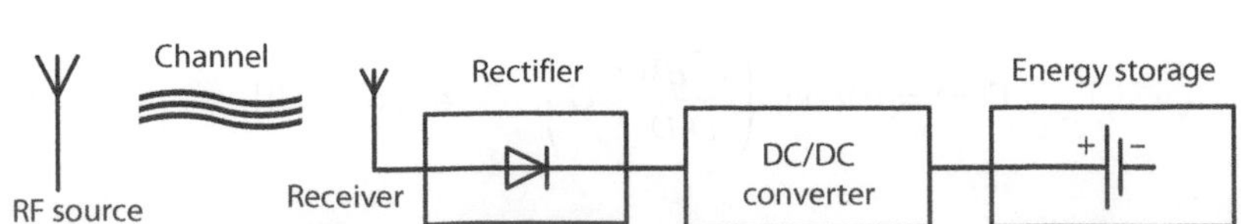

FIGURE 8.7 Major components of RF energy harvesting link.

management scheme is then needed at the rectifier output, to dynamically track the maximum power point, thereby achieving optimum rectifier performance (Piñuela et al., 2013a; Huang et al., 2013).

8.4.2 Design Considerations

An important aspect of RF energy harvester design is maximizing the scavenging of RF energy from multiple sources with unpredictable link budgets. Consequently, the RF-DC efficiency of the energy harvesting receiver is a critical factor determining the performance of the wireless power delivery scheme. The RF-DC efficiency depends on the radio-link efficiency η_{link} and the efficiency of the rectification stage n_{rect} (Costanzo et al., 2014):

$$\eta_{RF-DC} = \eta_{link}\eta_{rect} \tag{8.14}$$

Basically, the radio-link efficiency is the ratio of the received power at the receiver antenna to the transmitted power:

$$\eta_{link} = \frac{P_{rx}}{T_{tx}} \tag{8.15}$$

Given that a stated transmitter and receiver antenna are in the far-field region of each other, the power can determined by the application of the Friis transmission formula (Balanis, 2005), leading to the computation of η_{link}. In applying this approach, it is important to note that non-line-of-sight scenarios require one to account for the fading, polarization attenuation, and multipath effects, which lead to frequency-dispersive channel characteristics (Costanzo et al., 2014). To this end, the Friis formula could be adapted, or one could adopt a corresponding statistical channel characterization (Galinina et al., 2016).

η_{rect}, on the other hand, is often viewed as a FoM in rectenna design. It is the ratio of the rectifier output power P_{DC} to the received power at the receiver antenna P_{rx}:

$$\eta_{rect} = \frac{P_{DC}}{P_{rx}} \tag{8.16}$$

8.4.2.1 RFID Rectenna Design

Wireless power delivery to backscatter RFID transponders employed in sensing and location applications, in addition to the conventional identification usage, presents a usage scenario for RFID energy harvesting. In this case, designers often have to choose between tag antenna designs able to harvest RF energy over multiple frequencies and single-resonant frequency antennas. RF energy harvesting over multiple frequencies would require either wideband or narrowband antennas with multiple resonance frequencies. Some examples of single-frequency approaches include the use of fractal antenna geometries (Olgun et al., 2010), dipoles (Ladan et al., 2013; Vera et al., 2015a), and monopoles (Bito et al., 2015). Single-frequency RF energy harvesting would most likely require band-pass filtering between the receiver antenna and rectifier. The combination of the antenna and filter stages is sometimes called a filtenna (Sabran et al., 2014). Wideband antennas can be readily implemented through structural modifications to single-frequency designs, to provide nonresonant behavior over the frequency range of interest (often from 900 MHz to 2.45 GHz). Multiresonant frequency designs can be achieved by exploiting the multiple resonant modes of the same antenna radiating element, often through the inclusion of slots (Costanzo et al., 2014).

Transponder antennas, in addition, need to have high efficiencies and gains to compensate for the typically weak ambient RF energy. Further, the lack of a clearly defined angle of arrival of ambient RF energy suggests the need for omnidirectional circular polarized antenna implementations.

However, intentional transmission of RF energy for wireless power delivery does open up the possibility for directional energy transmission. In this case, RF energy can be delivered to multiple receivers using directional antennas with narrow power beams, which could be steered in multiple desired directions. Although such a scheme would be more complex to implement, the directional RF energy could provide longer operational ranges, even under conditions with significant RF attenuation (Galinina et al., 2016).

As earlier noted, the performance of the rectifier stage determines the level of usable electrical power derivable from an energy harvesting arrangement. The efficiency of the rectifier stage depends on the choice of rectification devices, the rectification topology, the input power level from the receiving antenna, and the load at the rectifier output. Schottky diodes are commonly used as rectifying devices, due to their power handling features. Zero-bias diodes (Hagerty et al., 2004), as well as CMOS devices (Yi et al., 2007) have also been employed in low-power rectification schemes. Common rectifier topologies include the single-element envelope detector, or variants of the Dickson charge pump (Dickson, 1976), commonly employed in UHF-RFID transponders (Vera et al., 2015b).

Generally, topologies employing fewer rectification devices lead to more efficient RF-DC conversion, as this would minimize the input power required to switch the various diode stages. However, due to the nonlinearity of the rectification stage, an optimum design requires a joint consideration of the antenna and rectifier stages, as well as the matching network inserted between them. This goal can be achieved by a harmonic balance simulation in combination with an imposition of a minimum desired value of η_{rect} at the operating frequency (Boaventura et al., 2013). Values of matching network components can be computed to achieve this desired efficiency level. The value of the received power at the receiving antenna is based on Thevenin or Norton equivalent circuits of the receiving antenna. This optimization approach enables the design of single-band, dual-band, or even broadband rectifiers, with optimal rectification efficiencies at the frequencies of interest.

In a scheme where RF energy can be intentionally transmitted and harvested by RFID transponders, the power delivery could be enhanced by the transmission of power-optimized waveforms, rather than traditional single-carrier continuous-wave (CW) signals (Valenta and Durgin, 2013). Demonstrations have established that waveforms with high peak-to-average power ratio (PAPR) improve the efficiency of wireless energy transfer (Collado and Georgiadis, 2014). This improvement can be attributed to the fact that these waveforms present higher power levels at the input of the rectifier stage. Consequently, the rectifying devices are more fully switched on, leading to higher rectification efficiency. Some examples of waveforms with high PAPR include ultra-wideband (UWB) signals, chaotic waveforms, white noise, and orthogonal frequency division multiplexing (OFDM) waveforms (Collado and Georgiadis, 2014).

8.5 CONCLUSION

RFID is a mature technology, widely used in asset tracking, telematics, and mobile commerce. Current developments in sensor applications have opened up prospects for nontraditional applications of RFID. One such application is the embedding of the technology in smart objects, enabling these objects to acquire data about their internal or external environments in IoT implementations. In such usage scenarios, the native wireless energy transfer features of RFID can be leveraged to deliver electrical power to the embedded sensing electronics as well. With current trends in the growth of interconnected sensors, RFID wireless power delivery is envisaged to become an increasingly more crucial aspect of the implementation of IoT infrastructure. This chapter has presented an overview of design considerations for such RFID wireless power delivery applications. Two application scenarios have been presented. The first scenario involves WPT using inductive coupling RFID. The second scenario deals with RF energy harvesting using backscatter UHF-RFID. Design trade-offs for achieving efficient energy transfer in both application scenarios have been highlighted.

The widespread adoption of RFID wireless power delivery has future implications on the realization of the IoT on a global scale. It provides a means to surmount challenges in the deployment of data gathering sensors arising from the bottleneck of limited battery life. Current trends in the development of this technology focus on improvements in power delivery functionality, and the miniaturization of implementations. A related consideration is the implementation of designs using alternative eco-friendly materials, which would reduce the bulk of electronic waste at the end of life of the powered devices.

REFERENCES

ABI Research. 2014. The Internet of things will drive wireless connected devices to 40.9 billion in 2020. https://www.abiresearch.com/press/the-internet-of-things-will-drive-wireless-connect.

Aerts, W., E. De Mulder, B. Preneel, G. A. E. Vandenbosch, and I. Verbauwhede. 2008. Dependence of RFID reader antenna design on read out distance. IEEE Transactions on Antennas and Propagation 56 (12): 3829–3837.

Albulet, M. 2001. *RF Power Amplifiers*. Atlanta, GA: Noble Publishing Corporation.

Balanis, C. A. 2005. *Antenna Theory: Analysis and Design*. 3rd ed. Hoboken, NJ: Wiley.

Bauernfeind, T. 2013. Equivalent circuit parameter extraction for controlled detuned NFC antenna systems utilizing thin ferrite foils. In *2013 12th International Conference on Telecommunications (ConTEL)*, 251–256, Zagreb, Croatia.

Bito, J., J. G. Hester, and M. M. Tentzeris. 2015. Ambient RF energy harvesting from a two-way talk radio for flexible wearable wireless sensor devices utilizing inkjet printing technologies. *IEEE Transactions on Microwave Theory and Techniques* 63 (12): 4533–4543.

Boaventura, A., A. Collado, N. B. Carvalho, and A. Georgiadis. 2013. Optimum behavior: Wireless power transmission system design through behavioral models and efficient synthesis techniques. *IEEE Microwave Magazine* 14 (2): 26–35.

Bolic, M., D. Simplot-Ryl, and I. Stojmenovic, eds. 2010. *RFID Systems: Research Trends and Challenges*. Hoboken, NJ: Wiley.

Bosshard, R., J. Muhlethaler, J. W. Kolar, and I. Stevanovic. 2013. Optimized magnetic design for inductive power transfer coils. In *Conference Proceedings—IEEE Applied Power Electronics Conference and Exposition (APEC)*, 1812–1819, Long Beach, CA.

Catrysse, M., B. Hermans, and R. Puers. 2004. An inductive power system with integrated bi-directional datatransmission. *Sensors and Actuators A: Physical* 115 (2–3): 221–229.

Colella, R., L. Tarricone, and L. Catarinucci. 2015. SPARTACUS: Self-powered augmented RFID tag for autonomous computing and ubiquitous sensing. *IEEE Transactions on Antennas and Propagation* 63 (5): 2272–22781.

Collado, A. and A. Georgiadis. 2014. Optimal waveforms for efficient wireless power transmission. *IEEE Microwave and Wireless Components Letters* 24 (5): 354–356.

Costanzo, A., D. Masotti, A. Romani, and M. Tartagni. 2014. Energy scavenging and storage for RFID systems. In Green RFID Systems, edited by L. Roselli, 39–75. Cambridge: Cambridge University Press.

De Donno, D., L. Catarinucci, and L. Tarricone. 2014. RAMSES: RFID augmented module for smart environmental sensing. *IEEE Transactions on Instrumentation and Measurement* 63 (7): 1701–1708.

DeHennis, A., M. Mailand, D. Grice, S. Getzlaff, and A. E. Colvin. 2013. A near-field-communication (NFC) enabled wireless fluorimeter for fully implantable biosensing applications. In *IEEE International Solid-State Circuits Conference Digest of Technical Papers (ISSCC)*, 298–299, San Francisco, CA.

De Mulder, E., W. Aerts, B. Preneel, I. Verbauwhede, and G. A. E. Vandenbosch. 2009. Case study: A class E power amplifier for ISO-14443A. In *2009 12th International Symposium on Design and Diagnostics of Electronic Circuits and Systems*, 20–23, Liberec, Czech Republic.

Dickson, J. F. 1976. On-chip high-voltage generation in MNOS integrated circuits using an improved voltage multiplier technique. *IEEE Journal of Solid-State Circuits* 11 (3): 374–378.

Eteng, A. A., S. S. K. Rahim, C. Y. Leow, B. W. Chew, and G. A. E. Vandenbosch. 2016. Two-stage design method for enhanced inductive energy transmission with Q-constrained planar square loops. *PLoS ONE* 11 (2): e0148808.

Felic, G. K., D. Ng, and E. Skafidas. 2013. Investigation of frequency-dependent effects in inductive coils for implantable electronics. *IEEE Transactions on Magnetics* 49 (4): 1353–1360.

Finkenzeller, K. 2010. *RFID Handbook: Fundamentals and Applications in Contactless Smart Cards, Radio Frequency Identification and Near-Field Communication*, 3rd ed. Chichester, UK: Wiley.

Fotopoulou, K. and B. W. Flynn. 2011. Wireless power transfer in loosely coupled links: Coil misalignment model. *IEEE Transactions on Magnetics* 47 (2 PART 2): 416–430.

Galinina, O., H. Tabassum, K. Mikhaylov, S. Andreev, E. Hossain, and Y. Koucheryavy. 2016. On feasibility of 5G-grade dedicated RF charging technology for wireless-powered wearables. *IEEE Wireless Communications* 23 (2): 28–37.

Greenhouse, H. M. 1974. Design of planar rectangular microelectronic inductors. *IEEE Transactions on Parts, Hybrids, and Packaging* II (2): 101–109.

Grover, F. W. 1946. *Inductance Calculations: Working Formulas and Tables*. North Chelmsford, MA: Courier Corporation.

Hagerty, J. A., F. B. Helmbrecht, W. H. McCalpin, R. Zane, and Z. B. Popovi. 2004. Recycling ambient microwave energy with broad-band rectenna arrays. *IEEE Transactions on Microwave Theory and Techniques* 52 (3): 1014–1024.

Huang, Y., N. Shinohara, and T. Mitani. 2013. A study on low power rectenna using DC-DC converter to track maximum power point. In *2013 Asia-Pacific Microwave Conference Proceedings (APMC)*, Seoul, South Korea. 83–85

Inagaki, N. 2014. Theory of image impedance matching for inductively coupled power transfer systems. *IEEE Transactions on Microwave Theory and Techniques* 62 (4): 901–908.

Jiseong, K., K. Jonghoon, K. Sunkyu, K. Hongseok, S. In-Soo, S. Nam Pyo, C. Dong-Ho, K. Joungho, and A. Seungyoung. 2013. Coil design and shielding methods for a magnetic resonant wireless power transfer system. *Proceedings of the IEEE* 101 (6): 1332–1342.

Jonah, O., A. Merwaday, S. V. Georgakopoulos, and M. M. Tentzeris. 2014. Spiral resonators for optimally efficient strongly coupled magnetic resonant systems. *Wireless Power Transfer* 1 (1): 21–26.

Jow, U.-M. and M. Ghovanloo. 2007. Design and optimization of printed spiral coils for efficient transcutaneous inductive power transmission. *IEEE Transactions on Biomedical Circuits and Systems* 1 (3): 193–202.

Kazimierczuk, M. 2008. *RF Power Amplifiers*. Chichester, UK: Wiley.

Kiani, M. and M. Ghovanloo. 2010. An RFID-based closed-loop wireless power transmission system for biomedical applications. *IEEE Transactions on Circuits and Systems II: Express Briefs* 57 (4): 260–264.

Kiani, M. and M. Ghovanloo. 2013. A figure-of-merit for designing high-performance inductive power transmission links. *IEEE Transactions on Industrial Electronics* 60 (11): 5292–5305.

Kim, Y. and H. Ling. 2007. Investigation of coupled mode behaviour of electrically small meander antennas. *Electronics Letters* 43 (23).

Ladan, S., N. Ghassemi, A. Ghiotto, and K. Wu. 2013. Compact rectenna for wireless energy harvesting application. *IEEE Microwave Magazine* 14 (1): 117–122.

Landt, J. 2005. The history of RFID. *IEEE Potentials* 24 (4): 8–11.

Lee, B., B. Kim, F. J. Harackiewicz, B. Mun, and H. Lee. 2014. NFC antenna design for low-permeability ferromagnetic material. *IEEE Antennas and Wireless Propagation Letters* 13: 59–62.

Lee, K, Z. Pantic, and S. M. Lukic. 2014. Reflexive field containment in dynamic inductive power transfer systems. *IEEE Transactions on Power Electronics* 29 (9): 4592–4602.

Lee, W.-S., K.-S. Oh, and J.-W. Yu. 2014. Distance-insensitive wireless power transfer and near-field communication using a current-controlled loop with a loaded capacitance. *IEEE Transactions on Antennas and Propagation* 62 (2): 936–940.

Lee, W.-S., W.-I. Son, K.-S. Oh, and J.-W. Yu. 2013. Contactless energy transfer systems using antiparallel resonant loops. *IEEE Transactions on Industrial Electronics* 60 (1): 350–359.

Lin, D. B., T. H. Wang, and F. J. Chen. 2015. Wireless power transfer via RFID technology for wearable device applications. In *2015 IEEE MTT-S 2015 International Microwave Workshop Series on RF and Wireless Technologies for Biomedical and Healthcare Applications (IMWS-BIO)*, Taipei, Taiwan, 210–211.

McFarlane, D. 2005. Networked RFID in industrial control: Current and future. In *Emerging Solutions for Future Manufacturing Systems*, edited by L. M. Camarinha-Matos, 3–12. Berlin: Springer.

Mirbozorgi, S. A., H. Bahrami, M. Sawan, and B. Gosselin. 2014. A smart multicoil inductively coupled array for wireless power transmission. *IEEE Transactions on Industrial Electronics* 61 (11): 6061–6070.

Mohan, S. S., M. del Mar Hershenson, S. P. Boyd, and T. H. Lee. 1999. Simple accurate expressions for planar spiral inductances. *IEEE Journal of Solid-State Circuits* 34 (10): 1419–1424.

Olgun, U., C. C. Chen, and J. L. Volakis. 2010. Wireless power harvesting with planar rectennas for 2.45 GHz RFIDs. In Symposium Digest—20th URSI International Symposium on Electromagnetic Theory, EMTS 2010, Berlin, Germany, 329–331.

Piñuela, M., P. D. Mitcheson, and S. Lucyszyn. 2013a. Ambient RF energy harvesting in urban and semi-urban environments. *IEEE Transactions on Microwave Theory and Techniques* 61 (7): 2715–2726.

Pinuela, M., D. C. Yates, S. Lucyszyn, and P. D. Mitcheson. 2013b. Maximizing DC-to-load efficiency for inductive power transfer. *IEEE Transactions on Power Electronics* 28 (5): 2437–2447.

Raju, S., R. Wu, M. Chan, and C. Patrick Yue. 2014. Modeling of mutual coupling between planar inductors in wireless power applications. *IEEE Transactions on Power Electronics* 29 (1): 481–490.

Ramo, S., J. R. Whinnery, and T. Duzer. 1994. Fields and Waves in Communication Electronics, 3rd ed. Hoboken, NJ: Wiley.

Ramrakhyani, A. K., and G. Lazzi. 2013. On the design of efficient multi-coil telemetry system for biomedical implants. *IEEE Transactions on Biomedical Circuits and Systems* 7 (1): 11–23.

Reinhold, C., P. Scholz, W. John, and U. Hilleringmann. 2007. Efficient antenna design of inductive coupled RFID-systems with high power demand. *Journal of Communications* 2 (6): 14–23.

Roselli, L., C. Mariotti, P. Mezzanotte, F. Alimenti, G. Orecchini, M. Virili, and N. B. Carvalho. 2015. Review of the present technologies concurrently contributing to the implementation of the Internet of Things (IoT) paradigm: RFID, green electronics, WPT and energy harvesting. In *2015 IEEE Topical Conference on Wireless Sensors and Sensor Networks, WiSNet,* San Diego, CA, 1–3.

Sabran, M. I., S. K. A. Rahim, T. A. Rahman, A. A. Eteng, and Y. Yamada. 2014. U-shaped harmonic rejection filtenna for compact rectenna application. In *2014 Asia-Pacific Microwave Conference*, Sendai, Japan, 1007–1009.

Sample, A. P., D. A. Meyer, and J. R. Smith. 2011. Analysis, experimental results, and range adaptation of magnetically coupled resonators for wireless power transfer. *IEEE Transactions on Industrial Electronic*, 58 (2): 544–554.

Sample, A. P., D. J. Yeager, P. S. Powledge, A. V. Mamishev, and J. R. Smith. 2008. Design of an RFID-based battery-free programmable sensing platform. *IEEE Transactions on Instrumentation and Measurement* 57 (11): 2608–2615.

Sharma, A., I. J. G. Zuazola, A. Gupta, A. Perallos, and J. C. Batchelor. 2013a. Non-uniformly distributed-turns coil antenna for enhanced H-field in HF-RFID. *IEEE Transactions on Antennas and Propagation* 61 (10): 4900–4907.

Sharma, A., I. J. G. Zuazola, A. Gupta, A. Perallos, and J. C. Batchelor. 2013b. Enhanced H-field in HF RFID systems by optimizing the loop spacing of antenna coils. *Microwave and Optical Technology Letters* 55 (4): 944–948.

Tak, Y., J. Park, and S. Nam. 2011. The optimum operating frequency for near-field coupled small antennas. *IEEE Transactions on Antennas and Propagation* 59 (3): 1027–1031.

Tankiewicz, S., J. Schaefer, and A. DeHennis. 2013. A co-planar, near field communication telemetry link for a fully-implantable glucose sensor using high permeability ferrites. In *2013 IEEE Sensors*,Baltimore, MD, 523–526.

Valenta, C. R. and G. D. Durgin. 2013. Rectenna performance under power optimized waveform excitation. In *2013 IEEE International Conference on RFID,* Penang, Malaysia, 237–244.

Vandevoorde, G. and R. Puers. 2001. Wireless energy transfer for stand-alone systems: A comparison between low and high power applicability. *Sensors and Actuators A: Physical* 92 (1–3): 305–311.

Vera, G. A., S. D. Nawale, Y. Duroc, and S. Tedjini. 2015a. Read range enhancement by harmonic energy harvesting in passive UHF RFID. *IEEE Microwave and Wireless Components Letters* 25 (9): 627–629.

Vera, G. A., Y. Duroc, and S. Tedjini. 2015b. Third harmonic exploitation in passive UHF RFID. *IEEE Transactions on Microwave Theory and Techniques* 63 (9): 2991–3004.

Vermesan, O. and P. Friess, eds. 2013. *Internet of Things: Converging Technologies for Smart Environments and Integrated Ecosystems*. Aalborg, Denmark: River Publishers.

Waffenschmidt, E., and T. Staring. 2009. Limitation of inductive power transfer for consumer applications. In *2009 EPE '09 13th European Conference on Power Electronics and Applications*, Barcelona, Spain, 1–10.

Yi, J., W. H. Ki, and C. Y. Tsui. 2007. Analysis and design strategy of UHF micro-power CMOS rectifiers for micro-sensor and RFID applications. *IEEE Transactions on Circuits and Systems I: Regular Papers* 54 (1): 153–166.

Zhao, Y., J. R. Smith, and A. Sample. 2015. NFC-WISP: A sensing and computationally enhanced near-field RFID platform. In *2015 IEEE International Conference on RFID (RFID)*, San Diego, CA, 174–181.

Zierhofer, C. M., and E. S. Hochmair. 1990. High-efficiency coupling-insensitive transcutaneous power and data transmission via an inductive link. *IEEE Transactions on Biomedical Engineering* 37 (7): 716–722.

Part III

Issues and Novel Solutions

9 Overcoming Interoperability Barriers in IoT by Utilizing a Use Case–Based Protocol Selection Framework

Supriya Mitra and Shalaka Shinde

CONTENTS

9.1 Introduction 181
9.2 Interoperability Constraints 182
9.3 Assessment of IoT-Relevant Protocols based on Open System Interconnection Topology 183
9.3.1 Link Layer Protocols 183
9.3.2 Network Layer Protocols 185
9.3.3 Transport Layer Protocols 185
9.3.4 Application Layer Protocols 186
9.4 Use Case–based Protocol Selection Framework 187
9.4.1 Use Case 1: Low-Cost IoT Application with Low-Power Device Landscape 187
9.4.2 Use Case 2: Web-Based IoT Application with Negligible Power Constraints in Landscape 188
9.4.3 Use Case 3: IoT Application Requiring Real-Time Data and Multicast Abilities from Devices 188
9.4.4 Use Case 4: IoT Application For a High-Latency and Bandwidth-Constrained Device Landscape 188
9.4.5 Use Case 5: Messaging-Oriented IoT Applications for Minimal Resource-Constrained Iot Landscape 189
9.4.6 Use Case 6: IoT Application with Large Data Volumes and Open and Interoperability Requirements 189
9.4.7 Use Case 7: IoT Application with Java-Based Language and Platform Constraints 189
9.4.8 Use Case 8: IoT Application for Low-Power-Constrained Landscape with Long-Range Connectivity Requirement 190
9.5 Conclusions 190
Acknowledgments 191
References 191

9.1 INTRODUCTION

The IoT will be vital to influencing the potential evolution of the Internet by facilitating linkages among dissimilar things, smart objects, and machines not only between themselves but also with the Internet, resulting in the creation of interoperable and value-added services and applications. The Third Generation Partnership Project (3GPP), a consortium of seven telecommunication standards development organizations (Association of Radio Industries and Businesses [ARIB], Alliance for Telecommunications Industry Solutions [ATIS], China Communications

Standards Association [CCSA], European Telecommunications Standards Institute [ETSI], Telecommunications Standards Development Society of India [TSDSI], Telecommunications Technology Association (TTA) of Korea, and the Telecommunication Technology Committee (TTC) of Japan,) and the Institute of Electrical and Electronics Engineers (IEEE), defines interoperability as "the ability of two or more systems or components to exchange data and use information" (Sutaria and Govindachari, 2013). According to a report by McKinsey, interoperability enables 40% of the total potential economic value from implementing IoT (Manyika et al., 2015). McKinsey advocates adoption of open standards and implementation of open systems and/or cross-platforms that enable different IoT systems to communicate with one another.

With future IoT projections in billions of interconnected devices, networks need to not only accommodate the growing number of devices but also handle varying traffic characteristics with respect to reliability, latency, and delay tolerance. Unfortunately, 3GPP has acknowledged the limitations of conventional wireless access networks for addressing this need and has commenced initiatives to manage the short-term needs. However, long-term needs still require identification and standardization. Government and standards organizations are also working hand in hand to mitigate this challenge—for example, the European Commission's mandates (M/411 and M/490) to European standards organizations (The European Committee for Standardization [CEN], European Committee for Electrotechnical Standardization [CENELEC], and ETSI) on establishing interoperability standards for smart meters and smart grids (Scarrone and Boswarthick, 2012).

This chapter reviews the relevant Open System Interconnection (OSI) layer IoT standards and protocols at the link layer, network layer, transport layer, and application layer. Based on the merits and demerits of each protocol, it proposes combinations of protocols across the OSI layers for usage in different use cases. The rest of the chapter is organized as follows. Section 9.2 establishes some of the challenges in enabling IoT interoperability. Section 9.3 does a deep dive into the merits and demerits of the various OSI layer protocols. Section 9.4 provides insights into the grouping of candidate sets of OSI layer protocols for differing use cases. Section 9.5 summarizes the previous sections and provides future directions for research.

9.2 INTEROPERABILITY CONSTRAINTS

Interoperability may broadly be classified into technical and syntactical categories. *Technical interoperability* is typically linked with hardware, software, systems, and platforms that enable device-to-device (D2D) communication. Communication protocols coupled with the underlying relevant infrastructure are key to enabling this form of interoperability. *Syntactical interoperability* is usually associated with data formats, including syntax and encoding (Veer and Wiles, 2006).

Interoperability is usually enabled by designing standards both within and between domains (a domain refers to a specific organization, enterprise, or industry realizing an IoT). Within a domain, standards provide long-term efficiencies of solutions, while between domains, standards encourage collaboration. The resultant value is reduced Total Cost of Ownership (TCO), faster market time, and economies of scale. However, the existing D2D-related technology topography is highly disintegrated, thus inhibiting reusability. Even within a particular domain, multiple contending standards and technologies are used and promoted (to circumvent vendor lock-in, etc.). This diversity promotes nonstandardized data formats—hence impeding integration. Future IoT networks will be characterized as heterogeneous, multivendor, multiservice, and largely dispersed. The inherent risk of noninteroperability is the unavailability of efficient IoT services related to health and emergency. Also, users and applications are likely to lose key information due to the lack of IoT interoperability.

According to the European Commission (EC) (Walewski et al., 2011), fostering a consistent, interoperable, and accessible IoT across sectors, including standardization, remains one of the biggest challenges. This is due to the following four inherent characteristics of IoT:

- Multifaceted: Due to the coexistence of multifarious systems (such as devices, sensors, and equipment) that may need to communicate. For example, vehicles have multisystems for engine control, communication, safety, and so forth. Similarly, buildings have multifarious systems for air conditioning, heating, security, lighting, and so forth.
- High diversity: Due to multiversion systems created by multiple manufacturers over time and designed for varied application domains, thus making it extremely difficult to formulate global agreements and commonly accepted specifications.
- Dynamic and nonlinear: Due to new "things" that get invented and introduced and which support new unanticipated structures and protocols.
- Complex data quality: Due to the existence of many data formats, multifarious languages, differing data models and constructs, and complex interrelationships between data. The data collected by diverse devices and sensors from the real world is dynamic and location and time dependent, and the data quality varies over different devices. This combined complexity exhibited across a substantial volume of devices creates a deluge of heterogeneous data.

A unique constraint in IoT is the abundant existence of low-powered devices, which may have minimal likelihood or accessibility for a power recharge in months or years. Added to this is the need for these devices to exchange data over "lossy" networks. Cable-powered devices and/or battery replacements are difficult and costly to deploy in remote locations. In the future, energy harvesting devices, such as solar cells, piezoelectric devices, and thermoelectric generators will replace batteries. Advances in micro-, pico-, and femtocell manufacturing will provide power and cost savings and higher throughput. As a consequence, for large-scale and self-sufficient IoT systems, lower power technologies, high-efficiency batteries, and alternate energy sources should be considered. In a recent study, it was established that the electrical efficiency of computation has roughly doubled every 18 months (also referred to as Koomey's law) (Keysight Technologies, 2016; 2015).

A recent study by the International Organization for Standardization (ISO) revealed 400+ standards that were related to IoT (2014). This plethora of standards intensifies the constant dilemma faced by technical architects and stakeholders working with IoT implementations, especially with each IoT provider self-eulogizing its own standards. A vendor's view is usually biased toward its offerings. Microcontroller vendors focus on device-level protocols, while microprocessor vendors emphasize protocols at the router level. Similarly, cloud-offering vendors focus on higher-level application protocols. IoT providers often view the market with a gold rush mentality—they assume that by leveraging their own technology, they can lock in customers and thus increase market share. However, the fact remains that there is no single leader in IoT standards. Fear of vendor lock-in is another motivator that will continue to encourage multiple standards to flourish in the market.

9.3 ASSESSMENT OF IoT-RELEVANT PROTOCOLS BASED ON OPEN SYSTEM INTERCONNECTION TOPOLOGY

To facilitate the decision-making process, this section provides a ready reckoner on the leading standards and protocols available at various layers of the OSI model, that is, at the link layer, network layer, transport layer, and application layer. Each standard has its merits and demerits—hence it is necessary to analyze requirements on a case-to-case basis before concluding on a set of protocols and standards to follow.

9.3.1 Link Layer Protocols

Link layer protocols govern and enable data packet exchanges over the network's physical layer or medium (e.g., radio wave and copper wire). The scope of the link layer is the local network connection to which the device is attached. The link layer determines how the packets are coded and signaled over the medium to which the device is attached. The following are some of the relevant link layer standards:

- IEEE 802.3 is a collection of wired Ethernet standards for the link layer. These standards provide data rates from 10 Mbps to 40 Gbps and higher. The shared medium (coaxial cable, twisted-pair wire or an optical fiber) carries the communication for all the devices on the network. Thus, data sent by one device can be received by all devices subject to transmission settings and transmitter and receiver capabilities (Ray, 2015).
- IEEE 802.11 is a collection of wireless local area network (WLAN) communication standards for the link layer. These standards provide data rates from 1 Mbps to 6.75 Gbps (Wong, 2016).
- As the original IEEE 802.11 was not effective for IoT requirements due to overhead and power consumption, IEEE 802.11 AH is a low-energy version of IEEE 802.11 that was designed to support low power and overhead requirements (Rahman, 2015).
- Bluetooth technology is based on the IEEE 802.15.1 standard and operates within the 2.4 GHz industrial, scientific, and medical (ISM) radio band. Bluetooth, which was developed by Ericsson initially, is a small-range wireless technology used for portable personal devices (Bandara, 2016). Bluetooth available in phones typically connects within 50–100 m distance with power boost.
- A new version of Bluetooth, Bluetooth Low Energy (BLE), is a subset to Bluetooth v4.0. BLE offers low power consumption and cost while providing the same range of communications as the Bluetooth 1.0. It has applications in healthcare, home automation, and security (Andersson, 2014).
- IEEE 802.15.4 is a collection of standards for low-rate wireless personal area networks (LR-WPANs). These standards provide low-cost and low-speed communication for power-constrained devices. The power constraint limits transmission distances to 10–100 m line of sight and a transfer rate from 40 to 250 Kbps (Gubbi et al., 2013). ZigBee is a low-cost wireless technology based on the IEEE 802.15.4 standard that consumes less power and offers long battery life. ZigBee uses less power than Bluetooth.
- 2G/3G/4G is a collection of mobile communication standards, such as second generation (2G, which includes the Global System for Mobile Communication [GSM] and Code Division Multiple Access [CDMA]), third generation (3G, which include Universal Mobile Telecommunications Service [UMTS], Enhanced Data for Global Evolution [EDGE] and, CDMA2000), and fourth generation (4G, including Long Term Evolution [LTE]). Data rates for these standards range from 9.6 Kbps (for 2G) to 100 Mbps (for 4G) (Bahga and Madisetti, 2015).
- IEEE 802.16 is a collection of fixed wireless broadband (also called WiMax) standards for the link layer. WiMax standards provide data rates from 1.5 Mbps to 1 Gbps.
- IEEE 802.22 is a standard for low-power long-range transmission networks (LORA). It operates within white spaces of the television spectrum in rural areas. It covers up to 32 km at a rate of 54 to 864 Mhz (i-SCOOP, 2017).
- WirelessHART is a protocol that operates on top of the IEEE 802.15.4 physical layer. It offers peer-to-peer security by using an advanced encryption mechanism to encrypt messages. It is mainly designed for industrial applications that have a self-healing mesh architecture (Al-Fuqaha et al., 2015).
- Insteon technology is another unique link layer technology mainly used for home automation, as it connects all devices using electrical power lines from building and/or radio frequency (RF). Its features a low-cost network and 38.4 kbps speed (Gazis et al., 2015). Insteon works on two protocols—the Insteon RF protocol (for facilitating communication between RF devices) and the Insteon power line protocol (for communication between power line devices).

9.3.2 Network Layer Protocols

The network layers are responsible for sending packets from the source network to the destination network. This layer performs the host addressing and packet routing. Host identification is done using IP addressing schemes such as

- Internet protocol version 4 (IPv4), which uses a 32-bit address scheme that allows a total of 2^{32} addresses. As more and more devices got connected to the Internet, these addresses were exhausted in the year 2011. The IPs establish connectivity on the packet network, but do not guarantee packet delivery. Guaranteed delivery and data integrity are handled by protocols (such as Transmission Control Protocol [TCP]) at the transport layer (Wong, 2014).
- Internet protocol version 6 (IPv6) uses 128-bit address scheme that allows a total of 2^{128} addresses.

  Over and above these two standard protocols, there exists some network layer protocols that are very specific to constrained environments. These protocols exist at the two sublayers of the network, namely, the routing sublayer, for routing packets from the source to the destination, and the encapsulation sublayer (Salman, 2016). The encapsulation sublayer leverages different data link layer frames to overcome the constraints of the standard IoT data link frame size, which cannot accommodate the long IPv6 addresses.
- Routing sublayer protocol
  - RPL (also known as the "Ripple" routing protocol) is a distance vector IPv6 routing protocol for low-power and lossy networks (LLNs). The protocol is based on computing the optimal path based on a destination-oriented directed acyclic graph (DODAG) using an objective function and a set of metrics and constraints (Vasseur et al., 2011).
- Encapsulation sublayer protocols
  - 6LoWPAN: IPv6 over Low-Power Personal Area Networks (6LoWPAN) brings the IP to low power-devices that have limited processing capability. It works in conjunction with the 802.15.4 link layer protocol (RS Components, 2015).
  - 6TISCH: Developed by the Internet Engineering Task Force (IETF) group, it allows IPv6 packets to flow in the time-slotted channel-hopping (TSCH) mode in IEEE 802.15.4e data link networks to mitigate interferences from colocated wireless systems using the same spectrum (IETF, 2017a). This is achieved using a channel distribution matrix where "frequencies allotted" are stored in columns and "time slots available" in rows. Each chunk of matrix containing time and frequency is known to all nodes connected in the network.
  - 6Lo: This is developed by the IETF group, mainly for IPv6 in resource-constrained networks. Examples include IPv6 over BLE and IPv6 over near-field communication (NFC) (IETF, 2017b).

9.3.3 Transport Layer Protocols

The transport layer provides functions such as error control, packet segmentation, flow control, and congestion control. The following are the two prevalent transport layer protocols:

- TCP is a familiar transport layer protocol used by browsers and file transfer and mail programs. TCP ensures reliable and orderly transmission of packets and error detection capability so that duplicate packets can be discarded and lost packets retransmitted (Greengard, 2015). The flow control capability of TCP ensures that the sender data rate is not too high for the receiver to process.
- User Datagram Protocol (UDP) is a transport layer protocol used for time-sensitive applications where packet dropping is preferable to delayed packets. UDP applications neither

have overhead of connection setups nor have requirements for message ordering, duplication elimination, and congestion control (Cole, 2011).

9.3.4 Application Layer Protocols

Application layer protocols work at the final layer of the OSI model and enable and define data communication between applications and IoT devices (in coordination with the lower-layer protocols). The following are some of the relevant application layer protocols used for IoT:

- Hypertext Transfer Protocol (HTTP) is request–response protocol wherein a client sends requests to a server using HTTP commands (Chen et al., 2014). HTTP is a stateless protocol wherein each HTTP request is independent of the other requests. An HTTP client can be a browser or an application running on an IoT device.
- Advanced Message Queuing Protocol (AMQP) is an open-source application protocol for business messaging. AMQP supports both point-to-point and publisher–subscriber models. AMQP brokers receive messages from publishers (e.g., devices) and route them to consumers (applications that process data) via a queuing mechanism. Either messages are delivered to the consumers that have subscribed to the queues or the consumers can pull the messages from the queues (VFabric-Team, 2016).
- Extensible Messaging and Presence Protocol (XMPP) is another application-level protocol using streaming XML data between network entities (Schneider, 2013). XMPP is a decentralized protocol using a client–server architecture. XMPP supports both client-to-server and server-to-server communication paths.
- Message Queuing Telemetry Transport (MQTT) is a publish–subscribe messaging protocol that is suitable for lightweight applications. It uses a client–server architecture where the client (devices) connects to the server (MQTT broker), which in turn forwards the messages to other subscribing clients. MQTT is well suited for constrained environments where devices have limited processing and memory resources and the network bandwidth is low (Stansberry, 2015). A recent application of MQTT is in Facebook messaging to have ensured and faster message delivery.
- Constrained Application Protocol (CoAP) is a protocol for D2D applications meant for environments with constrained devices and networks. CoAP is a web transfer protocol and uses a request–response model; however, it runs on top of UDP instead of TCP in its transport layer. The advantage of using UDP is low overhead, high transmission speed, and multicast support, allowing broadcast to multiple devices simultaneously. Hence, CoAP is the protocol of choice for LLNs where minimal overhead is desirable (Ishaq et al., 2016).
- Data Distribution Service (DDS) is a D2D standard for real-time, high-performance data exchange. DDS uses a publish–subscribe model where publishers (devices that generate data) create topics to which subscribers (devices that consume data) can subscribe. One of the key advantages of DDS is its data-centric technology since data is something that users understand. Message-centric systems like HTTP, CoAP, AMQP, MQTT, and XMPP require users to implement data sharing through the exchange of messages using complex message handling logic. DDS also uses UDP instead of TCP in its transport layer (Pardo-Castellote, 2008).
- Java Messaging Service (JMS) is a messaging protocol for Java-based platforms only. It allows the sending and receiving of messages among Java-based clients. JMS is an application layer protocol that runs on top of TCP. It is used in Java applications found in mobiles, tablets, and laptops, and also in smart grid applications (Parizo, 2014).

Since each protocol has its distinct characteristics, choosing the right set of protocols necessitates deeper analysis of the use case. In Section 9.4, a grouping of protocols across the OSI layers for adaptation in different use cases is proposed.

TABLE 9.1
Use Case–Based OSI Layer Protocol Selection Table

Use Case Number	Use Case	Application Layer	Transport Layer	Network Layer	Link Layer
1	Low-cost IoT application with low-power device landscape	COAP	UDP	6LoWPAN/RPL	IEEE 802.15.4/IEEE 802.11 AH/BLE/ Insteon
2	Web-based IoT application with negligible power constraints in landscape	HTTP	TCP	IPv4/IPv6	2G/3G/4G/IEEE 802.11/IEEE 802.16
3	IoT application requiring real-time data and multicast abilities from devices	DDS	UDP	IPv6	IEEE 802.11
4	IoT application for a high-latency and bandwidth-constrained device landscape	MQTT	TCP	6LoWPAN/6Lo/ 6TiSCH	IEEE 802.15.4 (WirelessHART)/ IEEE 802.11
5	Messaging-oriented IoT applications for minimal resource-constrained device landscape	XMPP	TCP	IPv6	IEEE 802.11
6	IoT application with large data volumes and open/ interoperability requirements	AMPQ	TCP	IPv6	IEEE 802.11
7	IoT application with Java-based language/platform constraints	JMS	TCP	IPv6	IEEE 802.11
8	IoT application for low-power-constrained landscape with long-range connectivity requirement	MQTT	TCP	6LoWPAN	IEEE 802.22 LORA/SIGFOX/ NEUL

9.4 USE CASE–BASED PROTOCOL SELECTION FRAMEWORK

In this section, a grouping of protocols across the OSI layers for different use cases are proposed. The objective is to provide guiding principles for the selection of IoT technologies. An incorrect selection of OSI level protocols can sometimes be very difficult and expensive to revise, resulting in IoT implementation failures. For example, a customer choosing a message-oriented protocol stack like XMPP in a resource-constrained IoT environment, or choosing a 2G/3G/4G/BLE link layer protocol for a long-range connectivity requirement will definitely encounter challenges. For each use case below, the relevant OSI layer protocols and real-life examples are discussed. The use case–based protocol selections are summarized in Table 9.1.

9.4.1 Use Case 1: Low-Cost IoT Application with Low-Power Device Landscape

In this scenario, CoAP is suited for the application layer protocol since it is designed for resource-constrained environments with low power, memory, and processing capabilities. Since CoAP runs on UDP only, it has low overhead (compared with TCP), which makes it ideal to use for constrained devices (Minoli, 2013). As shown in Table 9.1, 6LoWPAN can be used for the network encapsulation sublayer since it is designed for LLNs, which utilize less bandwidth and have low packet overhead and minimum power consumption. For the network routing layer, the RPL protocol is suggested since it provides a mechanism for application-specific requirements for constrained nodes. At the link layer, technologies like ZigBee (IEEE 802.15.4), BLE, and Insteon are suitable candidates.

Application examples of this use case include the smart grid, building automations, thermostats sensing the temperature of an area and sending alerts to the user, smart air sensors designed to track the amount of CO and NO_2 in a home environment, burglar alarm sensors, and coffee machine alerts to a user's phone. Domain standards like the Building Automation and Control Network (BACnet) have been built over CoAP (Jaisinghani and Maini, 2013). CoAP multicasts may be used for effective group communications, for example, between similar types of sensors in a room. CoAP has been used for controlling drones and streaming live sensor data for agriculture IoT applications (Johnson, 2017).

9.4.2 Use Case 2: Web-Based IoT Application with Negligible Power Constraints in Landscape

Use case 2 from Table 9.1 depicts a suitable stack for IoT applications with no power constraints. In such scenarios, HTTP is suitable at the application layer. HTTP is a request—response protocol commonly used in web browser application communication in client–server architectures (Minoli, 2013). Even though it is not power efficient during communication, HTTP is reliable since it runs on top of TCP. Addressing at the network layer can be IPv4/IPv6, which enables identification of different devices connected over the network and also provides scalability, expandable addresses, plug-and-play features, and security. At the link layer, 2G/3G/4G can be used in the case of mobile IoT applications, whereas WiMax (IEEE 802.16) and Wi-Fi (IEEE 802.11) are suitable candidates for fixed wireless and non-mobile applications, respectively.

There are multiple use case examples for this scenario in smart web applications, such as the Uber application (Uber Developers, 2017). The NEST thermostat is a real-time IoT application developed by Google that monitors and controls the temperature of houses (En.wikipedia.org, 2017). It can also be used for cloud-based applications and other mobile devices, such as computers, tablets, and cell phones.

9.4.3 Use Case 3: IoT Application Requiring Real-Time Data and Multicast Abilities from Devices

DDS is an application layer protocol best suited for "real-time" data transmission. It is a D2D multicast messaging protocol that runs on top of UDP (Table 9.1), which supports low overhead. It allows self-discovery that will automatically connect appropriate publishers to subscribers. DDS emerged mainly for the aerospace and defense community to address the data distribution requirements of mission-critical systems (Corsaro, 2014). At the network layer, IPv6 provides sufficient addresses to identify each device over the network. At the link layer, Wi-Fi (IEEE 802.11) enables connection to several sensors and other embedded devices within a 10 km radius with better speed.

Use case examples of this scenario are found not only in defense-related applications but also in financial trades, air traffic control, transportation, medicine, and smart grid management (Schmidt, 2016). DDS has also been used in patient monitoring systems where sensor data is sent across the nurse's station and even on a physician's mobile device (Foster, 2017).

9.4.4 Use Case 4: IoT Application For a High-Latency and Bandwidth-Constrained Device Landscape

MQTT is the proposed application layer protocol for this scenario. MQTT has a fixed header size of 2 bytes minimum, leading to low packet size and hence low overhead and latency. MQTT runs on top of TCP (Table 9.1) since MQTT needs to have a live connection in order to notify clients when a topic changes. MQTT is used to enable smaller devices to transfer the data to a higher infrastructure, like the cloud (Foster, 2015). At the network layer, MQTT supports technologies such as 6LoWPAN, 6Lo, and 6TiSCH, which are designed to work on low-power network-constrained nodes. At the link layer, possible candidates include IEEE 802.15.4 (e.g., WirelessHART) and IEEE 802.11 (Wi-Fi).

Use case examples for this scenario include monitoring a huge oil pipeline for leaks, power usage monitoring, lighting control, and intelligent gardening. Facebook uses MQTT in its mobile applications in view of its low-power and network bandwidth usage (Sovani, 2017).

9.4.5 Use Case 5: Messaging-Oriented IoT Applications for Minimal Resource-Constrained Iot Landscape

XMPP is the proposed application layer for this scenario, as it supports a small message footprint and low-latency message exchange. XMPP is easily extensible and can directly interact with other objects running XMPP. By using a push–pull mechanism, it can store contents if the receiving entity is in the sleep mode or offline (Wang et al., 2014). XMPP can be used to connect your home thermostat or any electronic device to a web server and can access information from your phone. It features addressing, security, and scalability, making it ideal for consumer-oriented IoT applications. As shown in Table 9.1, XMPP runs on top of TCP and supports IPv6 addressing. Internet connectivity can be from Wi-Fi (IEEE 802.11), which has convenient and better data rates over a given distance. XMPP is not suitable for resource-constrained devices since it consumes a lot of power. Hence, to make XMPP lightweight, it has been redesigned to run on UDP, which reduces overhead in the network (Postscapes, n.d.).

The IoT use case example includes XMPP notification services used for vehicle tracking. XMPP is also widely used in building identity and authorization services, such as OpenID and OAuth (Barrett, 2016). However, XMPP has largely been used in non-IoT applications, like instant messaging, video and voice calls, Google talk, and game applications.

9.4.6 Use Case 6: IoT Application with Large Data Volumes and Open and Interoperability Requirements

The preferred application protocol for this scenario is AMQP, which handles reliable queuing, topic-based publish-and-subscribe messaging, flexible routing, and appropriate security (Open AMQ, 2009). AMQP is a message-centric wire-level protocol that provides interoperability between different clients having implementations from different vendors. AMQP enables the transfer of large volumes of data while simultaneously receiving updates on the same communication channel. As shown in Table 9.1, AMQP typically runs on TCP and supports IPv6 addressing, using the IEEE standard 802.11 at the link layer.

Use case examples include the Aadhar project in India, which is one of the world's largest biometric databases, with 1.2 billion identity records (Sutaria and Govindachari, 2013). Another use case is the Ocean Observatories Initiative, which collects 8 terabytes of data per day from sensors around the world. An example in Manufacturing Execution System (MES) application is use of AMQP in conjunction with the Open Platform Communications (OPC) layer in supervisory control and data acquisition (SCADA) to eliminate the overhead of tag management and data integration.

9.4.7 Use Case 7: IoT Application with Java-Based Language and Platform Constraints

This scenario assumes that the base IoT systems are already using Java-based interfaces. In such scenarios, JMS is the preferred application layer protocol. JMS defines the standard messaging application program interface (API) for Java-based platforms and clients (only). It uses messages to interact with application components and also allows components to create send, receive, and read messages (DZone/Java Zone, 2016). However, JMS implementations from different vendors might not interoperate with each other. As shown in Table 9.1, JMS runs on top of TCP and is used in conjunction with Wi-Fi at the link layer.

Use case examples include building messaging systems, large enterprise applications, smart grid applications with sensors using a Java-based interface, and Java applications in mobiles and tablets enabled using JMS (Xavient Information Systems, 2016).

9.4.8 Use Case 8: IoT Application for Low-Power-Constrained Landscape with Long-Range Connectivity Requirement

A suitable link layer protocol for this scenario is IEEE 802.22 (also called LORA), which provides long-range connectivity to several nodes operating with low power. It operates within white spaces of the television frequency spectrum between 54 and 864 MHz, especially in rural areas where the spectrum usage is comparatively low (Yomas and Sebastian, 2015). It covers up to 15 km in rural areas versus 2–5 km in urban areas and is capable of eliminating interference, which helps in improving the network efficiency. DASH7, a LORA-based technology, offers long-range low-power service with ranges up to 10 km. As shown in Table 9.1, LORA can be used with 6LoWPAN, TCP, and MQTT protocols in its higher layers to meet requirements of low power consumption, as well as provide long-range connectivity.

An alternative LPWAN technology is Sigfox, which is mainly used in European countries. Sigfox can transmit data at better rates in the narrow spectrum using low power. Hence, it is suitable for IoT devices that are usually constrained in terms of power, memory, and energy. It uses technology called ultra-narrowband (UNB), which consumes about 50 μW vis-à-vis cellular technology that consumes more than 5000 μW (Radio Electronics, 2016). Since the range covered by Sigfox is 30–50 km, it is considered a better alternative than Wi-Fi, which covers only a short range and is also quite expensive to use. Sigfox provides longer battery life for the connected devices.

Similar to Sigfox, another technology—Weightless by Neul—offers low-cost, power-efficient, and long-coverage connectivity (up to 10 km). Weightless accesses the high-quality ultra-high-frequency (UHF) spectrum using white space radio, which is now accessible due to the transition from analog to digital television. It is a new wide area wireless networking technology designed for the IoT that largely competes against existing General Packet Radio Service (GPRS), CDMA, 3G, and LTE WAN technologies. Data rates can be anything from a few bits per second up to 100 kbps over the same single link, and devices can consume as little as 20–30 mA from two AA batteries, implying a battery life of 10–15 years.

Use case examples include IoT applications for automatic meter readers, GPS tracking devices, logistic applications, farming, smart meters, security applications, and smart mining (RF Wireless World, 2012).

9.5 CONCLUSIONS

Some of the abovementioned protocols have similar capabilities and hence may pose a dilemma for a decision maker. For example, both MQTT and CoAP are appropriate for low-power and network-constrained devices in lightweight environments. However, there are specific characteristics of each protocol that make them suitable for differing applications. If the end goal is to control an air conditioner from a smartphone, that is, sending commands from the smartphone to the air conditioner to trigger some functions, CoAP is the appropriate choice. However, if one has a machine-to-machine network where one wishes to control some devices by publishing messages from a sensor (based on its readings), MQTT is the appropriate choice.

To summarize, each protocol has its strengths and weaknesses. Hence, choosing the right set of protocols necessitates deeper analysis of the use case. The chapter proposed a holistic approach for choosing the appropriate protocol set for each use case by examining the strengths, weaknesses, and compatibility of protocols at each of the OSI layers. These suggested combinations may evolve and change over time as protocols are redesigned to overcome their weaknesses and compatibility limits. Also, an IoT landscape need not be limited to one set of protocols only. A mix of protocols can coexist with each other and be connected via gateways. For example, there may exist a situation where CoAP over UDP (from IoT devices) needs to connect to a cloud infrastructure that uses TCP with an existing enterprise infrastructure. A TCP-to-UDP gateway can be used at the cloud boundary to communicate with the UDP-based IoT device (Bormann et al., 2016).

While standardization should be driven by standards developing organizations (SDOs), collaboration is essential with open-source communities, special interest groups, and certification forums. One should also leverage best-practice learnings from the mobile devices industry, where interoperability was achieved not only by instituting global standards but also via the Global Certification Forum, which was a joint partnership consisting of handset manufacturers, mobile test equipment manufacturers, and network operators. Hence, certification will also play a leading role in ensuring interoperability in IoT.

Finally, a lot of emphasis is being put on protocol/communication standards at the OSI layers, but not much on standardizing IoT applications at the domain level. Noteworthy exemplars are the BACnet and KNX standards that were created for the building automation domains. Since the spate of IoT initiatives will possibly touch most domains and industries and human lives, there is a dire need to close this chasm at the earliest. A possible extension of the above research would be a detailed gap analysis and comparison of extant domain-level standards.

ACKNOWLEDGMENTS

The authors would like to thank Thomas George, Shankha Mukherjee, and Yaseen Kazi from Schneider Electric for their constant support for this project.

REFERENCES

Al-Fuqaha, A., M. Guizani, M. Mohammadi, M. Aledhari, and M. Ayyash. 2015. Internet of things: A survey on enabling technologies, protocols and applications, *IEEE Communications Surveys and Tutorials* 17(4).

Andersson, M. 2014. Use case possibilities with Bluetooth Low Energy in IoT applications, Ublox Whitepaper, 1–16. http://www.spezial.de/sites/default/files/bluetoothlowenergy-iot-applications_whitepaper_ubx-14054580.pdf.

Bahga, A. and V. Madisetti. 2015. *Internet of Things: A Hands-On Approach*, Hyderabad, India: Orient Blackswan Private Limited.

Bandara, D. 2016. Other types of networks: Bluetooth, Zigbee, & NFC. http://www.slideshare.net/DilumBandara/other-types-of-networks-2015-3.

Barrett, K. 2016. What can you do with XMPP, O'Reilly FYI Blog. http://fyi.oreilly.com/2009/05/what-can-you-do-with-xmpp.html.

Bormann, C., et al. 2016. CoAP (Constrained Application Protocol) over TCP, TLS, and WebSockets, draft-ietf-core-coap-tcp-tls-04. https://tools.ietf.org/html/draft-ietf-core-coap-tcp-tls-04.html.

Chen, S., H. Xu, D. Liu, B. Hu, and H. Wang. 2014. A vision of IoT: Applications, challenges, and opportunities with China perspective, *IEEE Internet of Things Journal* 1(4).

Cole, B. 2011. UDP and the embedded wireless Internet of things. http://www.embedded.com/electronics-blogs/cole-bin/4229531/UDP---the-embedded-wireless--Internet-of-Things.

Corsaro, A. 2014. Building the Internet of things with DDS, PrismTech White Paper, PrismTech, Woburn, MA.

DZone/Java Zone. 2016. All about JMS messages. https://dzone.com/articles/all-about-jms-messages.

En.wikipedia.org. 2017. Internet of things. Accessed January 31. https://en.wikipedia.org/wiki/Internet_of_things.

Foster, A. 2015. Messaging technologies for the industrial Internet and the Internet of things, White Paper, PrismTech, Woburn, MA, vol. 1.

Foster, A. 2017. Using DDS for scalable, high performance, real-time data sharing between medical devices in next generation healthcare systems. Accessed January 31. http://www.prismtech.com/sites/default/files/documents/DDS-Healthcare-Medical-WP-050914.pdf.

Gazis V., G. Manuel, H. Marco, L. Alessandro, M. Kostas, W. Alexander, Z. Florian, and V. Emmanouil. 2015. A survey of technologies for the Internet of things, *Presented at International Wireless Communications and Mobile Computing Conference (IWCMC)*. University of Dubrovnik, Dubrovnik, Croatia.

Greengard, S. 2015. *Internet of Things*, MIT Press, Cambridge, MA.

Gubbi, J., R. Buyya, S. Marusic, and M. Palaniswamia. 2013. Internet of Things (IoT): A vision, architectural elements, and future directions, *Future Generation Computer Systems*, 29 (7).

IETF. 2017a. IPv6 over networks of resource-constrained nodes. https://datatracker.ietf.org/wg/6lo/charter/.

IETF. 2017b. IPv6 over the TSCH mode of IEEE 802.15.4e. https://datatracker.ietf.org/wg/6tisch/about/.

i-SCOOP. 2017. LoRaWAN across the globe: LoRa Internet of things networks overview. Accessed January 31. https://www.i-scoop.eu/internet-of-things/iot-network-lora-lorawan/.

Ishaq, I., J. Hoebeke, I. Moerman, and P. Demeester. 2016, Experimental evaluation of unicast and multicast CoAP group communication, *Sensors (Basel)* 16 (7).

ISO/IEC JTC 1. 2014. Internet of things (IoT), Preliminary Report.

Jaisinghani, D. and P. Maini. 2013. CoAP: Constrained Application Protocol, April 5. https://www.iiitd.edu.in/~amarjeet/EmSys2013/CoAPv5.1.pdf.

Johnson, S. 2017. Constrained application protocol: CoAP is IoT's "modern" protocol, IoT Agenda. Accessed January 31. http://internetofthingsagenda.techtarget.com/feature/Constrained-Application-Protocol-CoAP-is-IoTs-modern-protocol.

Keysight Technologies. 2015. Battery life challenges in IoT wireless sensors and the implications for test, Keysight Technologies.

Keysight Technologies. 2016. IoT: With great power comes great challenges, Keysight Technologies.

Manyika, J., M. Chui, P. Bisson, J. Woetzel, R. Dobbs, J. Bughin, and D. Aharon. 2015. Unlocking the potential of the Internet of things, McKinsey Global Institute Report, McKinsey, Washington, DC.

Minoli, D. 2013. *Building the Internet of Things with IPv6 and MIPv6*, 1st ed., Hoboken, NJ: Wiley.

Open AMQ. 2009. Enterprise AMPQ messaging. http://www.openamq.org/doc: amqp-background.

Pardo-Castellote, G. 2008. Introduction to DDS, *Presented at OMG Real-Time Workshop*, Washington DC, July 2008.

Parizo, C. 2014. IoT technology can revive interest in JMS: Top five use cases for Java messaging, November 24 2014. Accessed January 31. http://searchmicroservices.techtarget.com/photostory/2240235284/Top-five-use-cases-for-Java-messaging/5/IoT-technology-can-revive-interest-in-JMS.

Postscapes. n.d. IoT standards and protocols. https://www.postscapes.com/internet-of-things-protocols/.

Radio Electronics. 2016. LoRa wireless for M2M and IoT. http://www.radio-electronics.com/info/wireless/lora/basics-tutorial.php.

Rahman, A. 2015. *Comparison of Internet of Things (IoT) Data Link Protocols*, 1st ed. Student Report, Last modified on November 30, 2015, https://www.cse.wustl.edu/~jain/cse570-15/ftp/iot_dlc.pdf.

Ray, A. 2015. Internet of things: Tomorrow is today, *CIO Review (India Edition)*, December 2015, https://www.cioreviewindia.com/magazine/Internet-of-Things--Tomorrow-is-Today-JHCR602293226.html.

RF Wireless World. 2012. Sigfox wireless technology basics in M2M and IoT. http://www.rfwireless-world.com/Terminology/SIGFOX-technology-basics.html.

RS Components. 2015. 11 Internet of things (IoT) protocols you need to know about, April 20, 2015. https://www.rs-online.com/designspark/eleven-internet-of-things-iot-protocols-you-need-to-know-about.

Salman, T. 2016. Internet of things protocols and standards. http://www.cse.wustl.edu/~jain/cse570-15/ftp/iot_prot/.

Scarrone, E. and D. Boswarthick. 2012. Welcome to the world of standards: Overview of ETSI TC M2M activities. https://portal.etsi.org/m2m/M2M_presentation.pdf.

Schmidt, D. C. 2016. Accelerating the industrial Internet with the OMG Data Distribution Service, White Paper. https://www.rti.com/whitepapers/OMG_DDS_Industrial_Internet.pdf.

Schneider, S. 2013. Understanding the protocols behind the Internet of things, *Electronic Design*, October 9, 2013.

Sovani, K. 2017. Which is the best protocol to use for IOT implementation: MQTT, CoAP, XMPP, SOAP, UPnP—Quora. Accessed January 31. https://www.quora.com/Which-is-the-best-protocol-to-use-for-IOT-implementation-MQTT-CoAP-XMPP-SOAP-UPnP.

Stansberry, J. 2015. MQTT and CoAP: Underlying protocols for the IoT, *Electronic Design*, October 7, 2015.

Sutaria, R. and R. Govindachari. 2013. Understanding the Internet of things, *Electronic Design*, May 1, 2013.

Uber Developers. 2017. Riders API reference. Accessed January 31. https://developer.uber.com/docs/api-overview.

Vasseur, J. P., et al. 2011. RPL: The IP routing protocol designed for low power and lossy networks, IPSO Alliance. http://www.ipso-alliance.org/wp-content/media/rpl.pdf.

Veer, H. and A. Wiles. 2006. Achieving technical interoperability: The ETSI approach, ETSI White Paper, European Telecommunications Standards Institute, Valbonne, France, vol. 3.

VFabric-Team. 2016. Choosing your messaging protocol: AMQP, MQTT, or STOMP, VMware vFabric Blog, VMware Blogs. http://blogs.vmware.com/vfabric/2013/02/choosing-your-messaging-protocol-amqp-mqtt-or-stomp.html.

Walewski, J., et al. 2011. The Internet-of-things architecture, European Commission, Brussels.

Wang, P., H. Wang, and W. Wu. 2014. Design and implementation of XMPP for wireless sensor networks based on IPv6, *Presented at International Conference on Logistics Engineering, Management and Computer Science*. Shenyang, China

Wong, B. 2016. These software trends will influence IoT strategies, *Electronic Design*, April 22, 2016.

Wong, W. 2014. IoT—The Industrial Way, O'Reilly. *Electronic Design*.

Xavient Information Systems. 2016. Introduction to messaging technologies. https://techblog.xavient.com/introduction-to-messaging/.

Yomas, A. K. J. and E. J. Sebastian. 2015. A review on IoT protocols for long distance and low power, *IRACST—Engineering Science and Technology: An International Journal (ESTIJ)* 5 (4): 2250–3498.

10 Enabling Cloud-Centric IoT with Publish/Subscribe Systems

Daniel Happ, Niels Karowski, Thomas Menzel, Vlado Handziski, and Adam Wolisz

CONTENTS

10.1 Introduction 195
10.2 A Publish/Subscribe Architecture for Cloud-Connected Things 197
10.2.1 Cloud-Based IoT Architecture 198
10.2.2 Publish/Subscribe in IoT Platforms 199
10.3 Discovery Using Metadata and Aggregates 201
10.3.1 Sensor Advertisements with Metadata 201
10.3.2 Using Sensor Aggregates as Describing Metadata 204
10.3.3 Sensor Search Based on Metadata and Aggregates 206
10.4 Exposing Application Requirements to Optimize Network Parameters 207
10.4.1 Capturing Application Requirements 207
10.4.2 Distributing Requirements over Publish/Subscribe Networks 208
10.4.3 Adaptation of Network Parameters Based on Requirements 210
10.5 Conclusions 210
References 211

10.1 INTRODUCTION

In the last decade, the vision of the IoT has become a reality: various low-cost sensors have been embedded in our environment, from cars, phones, and smart watches to homes and roads and industrial or agricultural equipment. Several studies expect the number of machine-to-machine (M2M) connections to increase further and reach billions in number by 2020 (Bradley et al., 2013; Chui et al., 2010; Gubbi et al., 2013). Connecting these manifold sensors has the potential to enable the development of new innovative applications in diverse arising fields, such as smart healthcare, smart cities, or smart grids, and to considerably enhance our daily lives. For example, sensor systems already monitor traffic conditions and suggest different routes to users based on current and predicted traffic conditions, leading to less time in traffic jams, reduced waste of fossil fuel, and less air pollution (Singh et al., 2014; Zhang et al., 2011).

The rise of the IoT paradigm is not only making a new source of information available, but also changing the way the resulting data is processed, consumed, and used. While classical sensor networks were application-specific silos (Tschofenig et al., 2015), where the network was used for a specific application, this chapter follows the vision of future systems enabling a shared economy of IoT devices by sharing sensor and actuator functionality across many diverse applications. This creates a great opportunity for advances in large-scale information analysis without the need to deploy large application-specific networks. Likewise, it is a good opportunity for network operators to help fund sensor hardware purchase and operating costs. Such a shared economy of sensors has great potential to enable novel applications while reducing operation costs, as well as the environmental footprint.

To enable such a shared economy of sensors, an interoperable and open platform is needed that provides the means for easy and efficient integration of diverse hardware and software. In the face of the heterogeneity in devices, manufacturers, access technologies, and data formats, the platform needs to ensure seamless interoperability between devices from different vendors in a plug-and-play manner.

Existing IoT platforms assume the role of a highly centralized cloud-based service that sensor data is pushed to and is only accessible via this centralized service using a dedicated application program interface (API) (Gubbi et al., 2013; Menzel et al., 2014). Cloud computing is a type of computing that enables convenient, ubiquitous, on-demand access to a shared pool of configurable computing resources (Hassan et al., 2012; Mell and Grance, 2011). Resources are accessible over the network, usually the public Internet, and usually reside in a third-party data center. Based on enablers like server virtualization, fast networking, and reliable distributed storage, cloud computing offers important benefits when used as a vehicle for realizing the vision of the IoT. The flexibility, scalability, and usage-based cost model enable elastic matching of the growing communication, computation, and storage requirements associated with IoT applications. Indeed, the approach to send sensor data to a cloud-based service for messaging and processing is widely adopted by developers (Menzel et al., 2014; Zhang et al. 2915).

Current platforms typically use a publish/subscribe messaging model, and therefore have a rigid decoupling between the cloud layer and sensor data producers and consumers. The decoupling properties that are introduced by the publish/subscribe system make some common use cases in IoT settings hard to achieve. It is, for instance, not obvious how an interested client should know the topic a publisher might choose to publish on without exchanging those topics in advance. The reason is that there is no scheme for topic discovery in most publish/subscribe systems. Likewise, the feature of exposing feedback from a subscriber to a publisher is absent, so that data producers do not perceive which subscribers, if any, are currently interested in their data. While current IoT platforms usually use standard protocols, such as Message Queue Telemetry Transport (MQTT), additional services building on these standard protocols are not standardized. For instance, there is no standard for the discovery of sensors over publish/subscribe networks or how to enable feedback from subscriber to publisher. Since those technologies are needed, individual implementations of stopgap technologies will have to be used, which may put interoperability with other vendors and systems at stake. This leads to two main implications: (1) current approaches lack a uniform way of discovering devices, as well as tracking presence, and (2) there is a lack of feedback from the cloud platform toward data producers regarding application requirements, leading to unnecessary sampling and poor sensor network parameter adaptation. These limitations pose a major obstacle for building large-scale applications on top of billions of heterogeneous sensing and actuating devices.

To cope with these shortcomings, the concept of cloud-connected IoT is extended in this work by introducing novel approaches for the discovery of sensor data producers in publish/subscribe networks. By introducing proactive advertisements from the devices containing metadata, as well as actual sensor values, the platform enables the search for sensing devices based on fixed attributes (e.g., sensor type, value range, and location), as well as measured data (e.g., the temperature measured by a particular sensor). Another challenging aspect in the shift from classical sensor networks to a cloud-connected IoT that is addressed in this chapter is the shift from networks highly optimized for one application to a general-purpose sensor and actuator hardware substrate for various applications, where application requirements are not obvious when planning and deploying the network and may change over time. The rigid decoupling of components hampers the disclosure of application layer requirements to the sensor network management plane. The publish/subscribe architecture is further extended to enable feedback from applications to the underlying sensor network by introducing additional requirement attributes to subscription requests.

The remainder of this chapter is structured as follows. The chapter starts with a publish/subscribe architecture for cloud-connected things in Section 10.2, which acts as an overview over the current state of IoT deployments and resembles the architecture used by several large cloud-based IoT platform providers. The architecture is a loosely coupled infrastructure using publish/subscribe messaging. On top of this architecture, it introduces a discovery scheme using metadata and sensor measurement aggregates in Section 10.3. The architecture is further extended by providing a preliminary version of network parameter optimization based on application requirements in Section 10.4. And finally, Section 10.5 concludes the chapter.

10.2 A PUBLISH/SUBSCRIBE ARCHITECTURE FOR CLOUD-CONNECTED THINGS

A central challenge when building cloud-centric IoT platforms is the development of a meaningful reference architecture. To derive this reference architecture of a generic IoT infrastructure, this work first analyzes the functional requirement of the system:

1. *Messaging*: The vast diversity in sensing hardware and software emphasizes the need for a unified messaging middleware or message bus that simplifies the interconnection of sensors and client applications. For the success of a cross-vendor middleware, it is essential that the platform provides a uniform and standardized way of providing and accessing data. This includes the use of a platform-independent data format, such as JavaScript Object Notation (JSON), Binary JSON (BSON), or Extensible Markup Language (XML). Additionally, to be open to anyone, protocols with royalty-free specifications and open-source implementations will have to be used. This messaging layer should support the synchronous request/response pattern to request specific sensor data points, as well as an asynchronous interface, where a client can subscribe to real-time updates in the form of push-based notifications.
2. *Discovery*: When a new device gets connected, its sensor data should become available to interested clients. Likewise, a disconnected device should be detected. The lack of well-defined schemes of sensor and actuator discovery is a key challenge that has to be overcome to enable the interoperability of different IoT data providers and producers.
3. *Storage*: While sensor data may not be needed at a certain point in time, it might provide very valuable insights in the future. Therefore, for many use cases, storage of sensor data will be a requirement of future IoT systems. Ideally, storage should be transparently distributed to achieve high fault tolerance, as well as offering high durability using revisions. Additionally, there is a need for further research on new database systems, such as time-series databases, NoSQL systems, or multimodel databases, as well as the underlying storage techniques, such as fully memory-based databases (Hassan, 2016).
4. *Processing*: Access to sensor data alone is not sufficient to gain any insight into that data, which is what makes IoT such a promising technology. The amount of data produced is one example of what is commonly called "Big Data." Sensor data has to be efficiently and effectively processed and analyzed to gain meaningful insights. A form of high-performance computing (HPC) will be needed to cope with these challenges (Hassan, 2016). In addition to the more traditional batch processing of data, commonly done with technologies like Apache Hadoop,* data will be processed continuously, which creates new data streams. Those new data streams can be seen as virtual sensors.

* Apache Software Foundation. Apache Hadoop. 2016b. http://hadoop.apache.org.

In addition, important nonfunctional requirements have to be met: (1) Due to the large number of devices expected, the system has to be scalable. (2) As data may be urgent, messaging should have low latency. This stresses the need for lightweight data formats and efficient processing. (3) The platform has to offer easy-to-use programming interfaces to enable widespread usage.

Publish/subscribe has been the dominant messaging pattern used for many-to-many communication in the IoT context, so this work builds on the assumption that such a system is used for messaging. Although not considered in this work, there is great potential to adopt ideas related to the enterprise service bus (ESB) (Chappell, 2004; Riad et al., 2010) and service-oriented architecture (SOA) (Krafzig et al., 2005; Papazoglou, 2003) approaches. While those approaches are more focused on business logic, they essentially offer solutions to some of the challenges mentioned above, such as messaging, orchestration, or complex event processing (CEP). The ESB concept can possibly be extended for the IoT, where not services but sensor data producers and consumers are loosely coupled, and independently deployed, heterogeneous components in a SOA, but this is beyond the focus of this chapter.

In the following, the chapter presents a three-layered cloud-based architecture similar to that of many deployed systems (Menzel et al., 2014). It proceeds with an introduction to publish/subscribe systems that are commonly used as a messaging middleware to interconnect the various components in the architecture and emphasizes the limitations of this messaging pattern to complete this overview of the status quo of current cloud-based IoT systems.

10.2.1 Cloud-Based IoT Architecture

Most cloud-based IoT providers use a three-layered architecture similar to the one depicted in Figure 10.1 (Gubbi et al., 2013; Menzel et al., 2014): The device tier includes sensor and actuator devices, as well as gateways connecting them to the cloud platform. The cloud tier consists of an event queuing and messaging system, which handles the messaging between all tiers, and the service tier, which offers value-added services to customers, such as CEP, storage, or data visualization. The application tier consists of applications interested in sensor data, either directly or through a value-added service. That means that in most systems today, sensor data is sent from the actual sensors over a gateway to the cloud, where it is distributed to different services and interested clients.

In this generic architecture, which is widely used, the challenge of discovering sensors is not tackled explicitly. Furthermore, the architecture does not easily allow cross-layer optimization, because tiers are decoupled using the event queuing and messaging system. In the following, this widely adopted architecture is extended to include additional modules that enable discovery as well as cross-layer optimization. Figure 10.2 shows an overview of those modules that are envisioned in the cloud tier.

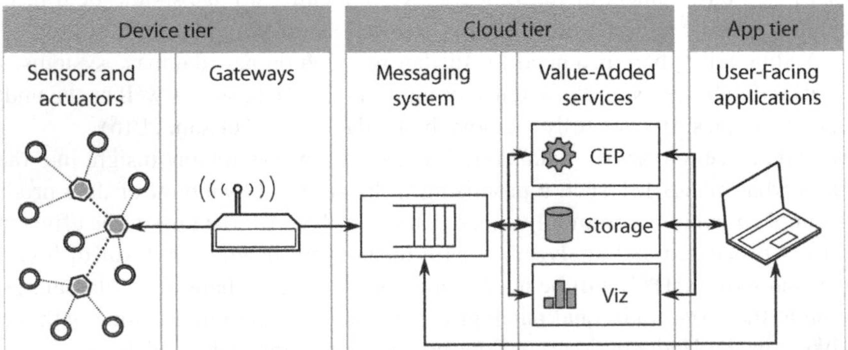

FIGURE 10.1 General architecture of a cloud-centric IoT system. Viz, visualization. (© 2016 IEEE. Reprinted with permission, from Happ, D. and Wolisz, A. Limitations of the Pub/Sub pattern for Cloud based IoT and their Implications. CIoT 2016, Paris, IEEE.)

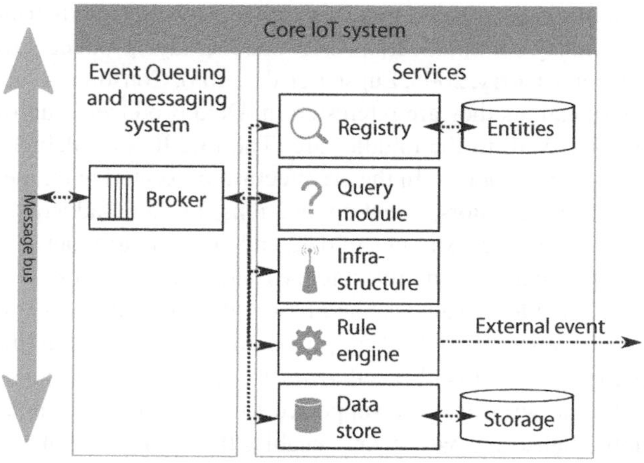

FIGURE 10.2 Building blocks of the core cloud layer of the proposed IoT architecture.

1. *Message broker*: This module offers a publish/subscribe interface for the messaging between sensors or their gateways, value-added services, and client applications. In particular, the module enables the matching of interests of subscribing clients or services and data producers.
2. *Registry service module*: This component stores metadata describing the attached data sources and enables the discovery of sensors, actuators, and value-added services, in particular the search for data producers. Newly connected devices or value-added services advertise themselves with static attributes and metadata to this service, so that interested applications or services can use those attributes to search for appropriate data sources.
3. *Querying service module*: This module handles incoming subscription queries that include user-defined requirements. It identifies overlapping requirements and issues combined subscriptions via the publish/subscribe interface. This enables adaptation of lower-level network parameters.
4. *Infrastructure management module*: This module optionally enables network parameters of lower layers according to the application requirements stored in the querying service module.

Additionally, as outlined in the requirement analysis, an IoT system would have storage and rule engine (CEP) services. In this work, the focus lies on the interconnection of sensors with the different components of IoT systems as a first step. So far, storage and processing services are seen as ordinary consumers and producers of data and are not considered in detail.

In the proof-of-concept implementation outlined here and most current systems, all those modules run on public cloud infrastructure. The loose decoupling makes it possible to run those components distributed, either on different cloud instances, cloud, and fog instances (Bonomi et al., 2012), or even fully distributed on gateway hardware. In the future, those core components would be present on each of those instances (gateway, fog, or cloud) and prepare the architecture for such a shift. Every instance, then, is connected to the other instances via a message bus and has local modules for local registry, storage, and so on.

10.2.2 Publish/Subscribe in IoT Platforms

For the loosely coupled messaging between sensors, actuators, applications, and the various cloud-based components, there is a trend to use message-oriented brokerage using the publish/subscribe pattern (Antonic et al., 2014; Menzel et al., 2014; Rowe et al., 2011). The publish/subscribe pattern is usually provided by a message-oriented middleware (MoM) offering distributed, asynchronous, loosely coupled many-to-many communications between message producers (publishers) and message consumers (subscribers) (Curry, 2005; Eugster et al., 2003). Consumers can use the middleware to subscribe to event notifications they are interested in. Data producers publish messages, usually about events that were observed, to the middleware. The middleware delivers those messages to consumers with matching subscriptions. In the architecture presented here, messages are produced by sensors and consumed by actuators, client applications, or value-added services. Value-added services themselves can produce new streams of data and, as such, also act as message producers.

Different approaches for matching and filtering messages have been proposed. The most widely adopted filtering method in the IoT context is topic-based filtering, where messages are published on a topic, usually a hierarchical string, and subscriptions are expressed using the same topics. When a new message enters the system, it is checked against ongoing subscriptions and is forwarded to all subscribed clients. Often, wildcards are supported to subscribe to a whole subtree of a parent topic. As such, the publish/subscribe pattern only enables the monitoring of sensors for new events, but in a pure publish/subscribe protocol, no direct messaging, such as for controlling actuators, is provided.

MQTT (Banks and Gupta, 2014) is emerging as the de facto standard protocol for IoT messaging. It was initially developed by IBM and standardized by OASIS in 2014, enabling royalty-free usage, which the European Interoperability Framework lists as one of the core principles of pan-European interoperability (IDABC, 2004). MQTT is a publish/subscribe protocol designed specifically for constrained devices and low-bandwidth, high-latency, and unreliable links, as often seen in cloud-centric IoT settings (Hamida et al., 2013; oneM2M, 2015). Various open-source implementations are available, notably the mosquitto broker* and client library, the Eclipse paho client library,† and the Apache ActiveMQ broker.‡ A slimmed-down variant MQTT-S (Hunkeler et al., 2008) was proposed as a port of MQTT to sensor nodes. MQTT is a pure publish/subscribe messaging protocol without direct messaging and uses topic-based filtering. It was shown to be well suited to IoT requirements and workloads (Happ et al., 2015). Because of its wide adoption, MQTT is considered a good candidate for IoT messaging as a practical example for a generic publish/subscribe system in the remainder of the chapter. The research prototype is therefore based around the mosquitto broker.§

The fundamental characteristics of a MoM, in general, and a publish/subscribe system, in particular include the decoupling of message producers and consumers (Eugster et al., 2003). These properties make the publish/subscribe pattern particularly useful for large-scale IoT deployments:

1. *Decoupling in time*: Message producers and consumers are decoupled in time; that is, they do not necessarily have to be connected to the publish/subscribe system at the same time.
2. *Decoupling in space*: Messages are not explicitly addressed to a specific consumer but to a symbolic address (channel or topic).
3. *Decoupling in thread*: Messaging is asynchronous, nonblocking.

The property of decoupling in space is helpful for constraint sensor data producers, because they do not need to know or take care of potential subscribers. This work already outlined two major limitations of the pattern, which both are a side effect of those decoupling properties, in particular the decoupling in space. The subscriber does not know if there are any publishers publishing on a given topic, how many of them there might be, or who the publisher of a given message might be. This hampers the discovery of sensor data producers, more specifically, the topics they will publish

* Mosquitto. An open source MQTT v3.1/v3.1.1 broker. 2016. http://mosquitto.org/
† Eclipse Foundation. Paho. 2016. https://eclipse.org/paho/
‡ Apache Software Foundation. ActiveMQ. 2016a. http://activemq.apache.org/
§ Mosquitto. An open source MQTT v3.1/v3.1.1 broker. 2016. http://mosquitto.org/

their data on. Common publish/subscribe systems, such as MQTT, do not offer discovery of publishers or topics. There are in principle three options by which the subscriber can obtain the information needed for issuing a subscription request: subscribers have to know topics to subscribe to in advance, have to negotiate suitable topics with data producers over another channel, or use some other nonstandard discovery approach using the underlying publish/subscribe system. Likewise, the publisher does not know if there might be any subscribers actually interested in messages on a certain topic. This lack of feedback from subscriber to publisher inhibits cross-layer optimization of network parameters. In the worst case, a sensor would send its data to a cloud-based messaging layer without any other component being interested in the data. A great deal of energy and traffic, and therefore cost, could be saved if the sensor is advised to stop sampling and enter a deeper sleep state to save energy. Those two challenges are tackled in the next sections.

10.3 DISCOVERY USING METADATA AND AGGREGATES

A crucial enabler of the IoT vision will be the development of well-defined schemes for sensor and actuator discovery, which minimizes, if not removes, the need for external human intervention. On the other hand, due to the diverse nature of devices in the upcoming IoT, and their capabilities, characteristics, and communication technologies, a discovery mechanism should enable, but must not rely on, the individual attributes sensors or actuators might have. This section presents a generic discovery approach based on the publish/subscribe architecture introduced. Devices advertise metadata as well as sensor measurement aggregates, and potential users or services can search using those attributes.

The discovery of sensing and actuating devices, as well as value-added services, is crucial mainly due to two reasons: Primarily, to allow retrieving data from sensors in publish/subscribe architectures, subscribers need to become aware of suitable topics to subscribe to. Second, a well-defined discovery scheme allows exposing cost estimates before each subscription, which can be used to encourage potential subscribers to use a sensor hardware subset that reduces the overall cost of sensing itself, as well as data communication. Further requirements are openness, scalability, and service reusability, which are achieved mainly through leveraging the existing publish/subscribe architecture. Additionally, several major challenges need to be considered:

1. *Dynamicity*: Sensors may join or leave the system at any time. In particular, this may be the case if sensors are mobile, so that they change their gateway frequently or lose connection altogether. Also, sensors can infrequently change attributes that were initially thought to be stable, such as position. When everyday objects will have sensors attached to them, this membership churn will have to be coped with by a suitable discovery mechanism.
2. *Context*: As highlighted before, a better understanding of the environment by fusion and analysis of collected sensor data is one central goal of the IoT. Without context, raw sensor data is not useful at all, since the context influences the reasoning about certain data streams. Providing context information thus plays a critical role in IoT discovery.
3. *Privacy and access control*: Sensor data providers have to be able to decide which information is publicly available and which data they want to keep private or restrict access to. This also includes the temporal and spatial accuracy or resolution of the data. The discovery approach should take into account those privacy concerns and enable the definition of constraints.

10.3.1 Sensor Advertisements with Metadata

This section presents one way of enabling discovery of sensor topics to subscribe to. The underlying problem is that potential publishers are not known in advance, neither to the publish/subscribe system nor to potential subscribers. In the publish/subscribe model, a common approach is for publishers to advertise their willingness to publish on a certain topic in the future using special advertisement messages.

For sensor discovery in IoT settings, a similar advertisement is used, where the sensor or a gateway device (by proxy) advertises on start-up, on updates to account for the dynamicity, and in a fixed interval not only its topic, but also additional metadata, such as position, owner, sampling interval, and cost. This information represents the relevant context for a specific sensor that can be used for sensor search and selection. In principle, those advertisements could be directly subscribed to by potential subscribers and appropriate sensors could be chosen by the subscriber and be subscribed to. The number of devices expected in a global IoT setting, however, prohibits this approach at such a large scale, as the number of advertisements would certainly overwhelm subscribers.

Instead, a registry of available sensors would be deployed, that can be queried to find appropriate sensors to subscribe to. In future systems, the registry service can additionally help ensure that privacy and access control restrictions are followed. For instance, the registry could return only a list of potential sensors the requesting party has access to or only topics that have data that can be used according to a permissive license. If data is transmitted unencrypted over the publish/subscribe network, this would not add significant security, since a subscription would still be possible when the right topic string is known. Future systems would need to additionally encrypt the data they transmit. The registry would then respond only with sensors that the client will be able to successfully receive data from. To prevent this registry from being overloaded with a large number of advertisement messages, suitable partitioning of the data in the database to different repository servers has to be applied. For instance, the topic tree could possibly be sharded between the registry nodes using an eventually consistent database model. This, however, is not the focus of this work.

Advertisement messages are part of some publish/subscribe protocols. Protocols widely used for the IoT do not often implement these advertisement messages (e.g., MQTT, Extensible Messaging and Presence Protocol [XMPP], or Advanced Message Queuing Protocol [AMQP]) (Happ et al., 2015). However, advertisements can be emulated using a dedicated topic or topic subtree, on which devices can announce their availability. The basic approach is shown in Figure 10.3: Sensors advertise themselves regularly on a special topic (1). At least one of the possibly distributed repository components in the cloud subscribes to each of the topics intended for advertising. On receiving an advertisement (2), the database is updated accordingly. If no advertisement is received for a predefined time-out value, the sensor is considered offline and will be deleted from the repository.

While different options exist, this work proposes letting potential subscribers query the sensor registry over the publish/subscribe system using a request/response pattern (3, 4). The registry responds using the publish/subscribe system (5, 6). The advantage of this approach is that it does not rely on external protocols and just uses the available publish/subscribe system for sensor search. In pure publish/subscribe systems, no request/response messaging is offered, so dedicated topics would have to be used to emulate a request/response pattern, which is a clear disadvantage, or the

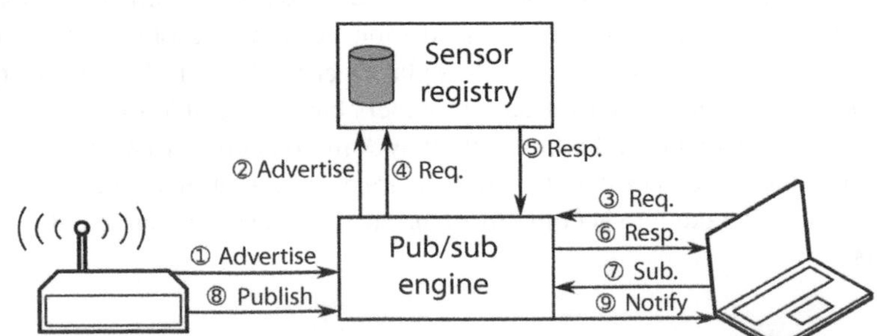

FIGURE 10.3 Discovery using advertisements. (© 2016 IEEE. Reprinted with permission, from Happ, D. and Wolisz, A. Limitations of the Pub/Sub pattern for Cloud based IoT and their Implications. CIoT 2016, Paris, IEEE.)

protocol would have to be extended with a request/response pattern. The subscriber can then proceed to subscribe to the sensor (7). On an event published by the data producer (8), the subscriber is notified (9).

The research prototype outlined in this chapter is based on MQTT, and the discovery approach is implemented as follows. Sensors and services are advertised on topics starting with the reserved string $ADV. The syntax is consistent with other reserved topics already in use; for instance, mosquitto MQTT brokers* use reserved topics starting with $SYS for broker statistics, such as number of publication messages received. After the $ADV prefix, the publisher adds the topic he is interested in providing data for. This way, clients interested in advertisement messages can subscribe to the wildcard topic $ADV/# to get all advertisement messages. Of course, an extended MQTT broker could also interpret the messages itself or only forward them to a predefined set of trusted registry servers.

```
{
  "MessageType": "Advertisement",
  "Name": "tkn/twist/128/temperature",
  "ID": "BB:8C:D1:F7:B7:92",
  "Type": "Sensor",
  "Device": "tkn/twist/128",
  "Gateway": "tkn/twist",
  "Manufacturer-Name": "TelosB Sensirion HT11 Temperature",
  "Datastream": {
    "Data-Semantic": {
      "unit": "°C"
    }
  },
  "Controlable-Parameters:": {
    "Frequency": {
      "Symbol": "f",
      "Min": "0.25",
      "Max": "86400"
    }
  },
  "Cost-Function": "c=256/f",
  "Room": "FT425"
}
```

Listing 1: Example of sensor advertisement

This general scheme is until now payload agnostic, so the payload of the message published to an advertisement topic does not affect the behavior of the brokerage. For sensor data, publishers would usually announce sensor metadata, such as a sensor name, location, owner, cost, and sampling interval. In the system outlined here, for example, the advertisements are given in JSON. An example is given in Listing 1. The only mandatory part of the message is the Name attribute, which describes the topic the sensor will publish on, and an ID, which is a unique identifier, such as a MAC address. If one device with one MAC address has multiple sensors, those must be advertised with different unique IDs. Additionally, sensors will usually provide a type, which can be "Sensor" or "Actuator," the device the sensor is part of and the gateway the sensor uses to connect to the platform. Also, the sensor can expose some semantics, such as the actual sensor hardware (Manufacturer-Name) and unit of provided data points. The sensor can additionally provide controllable parameters that can be adapted by the subscriber. Changing those parameters will have an associated cost to prevent users

* Mosquitto. An open source MQTT v3.1/v3.1.1 broker. 2016. http://mosquitto.org/

from always requesting the highest-quality service, in this example sampling frequency, which is given as the number of seconds between readings. Simple cost functions can be provided in the advertisement. Other attributes can be specified by the user at will, such as the "Room" attribute in this example.

10.3.2 Using Sensor Aggregates as Describing Metadata

So far, advertisements and metadata included in the registry cover only sensor characteristics that rarely change, such as name. There might also be a need for sensor search based on the actual sensor values the device senses. For instance, it may be necessary to monitor in more detail appliances that show an average temperature over the last few minutes above a certain threshold to prevent overheating.

An approach where sensor data is constantly sent to the cloud and stored and processed there will not be feasible when the number of connected devices increases as expected. The registry servers that are used for searches are therefore not able to maintain a list of up-to-date sensor data at all times. Hence, this work proposes that the gateway nodes hold the sensor data for future reference and only send aggregates to the cloud-based registry servers. Aggregates may be the average, maximum, or minimum value, or any other user-defined function that condenses several values of a time series into one value. Those update messages can then also be used for signaling that sensors are still alive.

Another part of the problem, though, is that the full time series of sensor values will not fit on constraint gateway devices, since their storage is usually very limited. In principle, there are three main strategies that can be used to limit the amount of data stored on the gateway. The simplest is to discard the oldest data in a first in, first out (FIFO) manner. Up until a certain time in the past, every sensor value is preserved; that is, the data is in the highest possible temporal resolution. The downside of this approach is that with every new sensor value, the oldest reading is discarded and inevitably lost. Another approach is so-called *culling*, where the temporal resolution of the data is decreased with an increasing amount of data points. While data is not lost, its resolution may be reduced to a point where no valuable features can be extracted by analysis. A combination of the two approaches is a multiresolution data store (Zhou et al., 2004). That is, data is stored in multiple ring-buffers with different resolutions. When data in one of the high-resolution ring-buffers is overwritten by new data, an aggregation function is used to combine data into an aggregate and added to a lower-resolution ring-buffer. In an IoT setting, it may be sufficient to only store the most recent data points in a high resolution and decrease the resolution by applying aggregation functions on older values while still maintaining the most prominent features.

Since a multiresolution data store already uses aggregation functions (minimum, maximum, average, and count) for combining older values, it is straightforward to use those aggregates for updating the metadata in the registry while at the same time providing a keep-alive mechanism. The result of that concept can also be seen as a form of multiresolution data store itself: the cloud-based registry always maintains a low temporal resolution representation of the current status of every connected sensor. This approach is shown in Figure 10.4. Multiple fog-based tiers could be introduced between gateways and the cloud to maintain higher-resolution representation and send aggregates to higher tiers.

In the prototypical implementation, the whisper multiresolution database* is used on the gateway side to store a time series locally. Whisper is the database used within graphite, an open-source enterprise-scale monitoring system. It is conceptually similar to RRDtool by Tobias Oetiker.† However, RRDtool only supports fixed intervals for data, while whisper also supports irregular

* Graphite Project. Whisper. 2016. https://github.com/graphite-project/whisper

† Oetiker, T. RRDtool: Round robin database tool. 2016. http://www.rrdtool.org

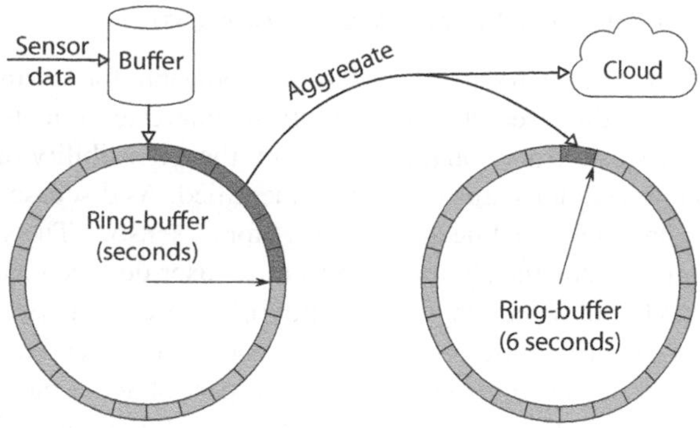

FIGURE 10.4 Multiresolution database to store and forward aggregates.

writes to the database. Since sensor data is expected to be potentially sampled at irregular intervals, whisper is used in the prototype. Whisper is written in python and writes data to fixed-size files on disk. It is therefore suitable to deploy on gateway devices with limited storage. Upon each aggregation, the corresponding aggregates are sent to the cloud-based registry in JSON format using MQTT. The prototype uses resolutions for the ring-buffers of 1 s, 15 min, 30 min, 1 hour, 1 day, and 1 week. The messages sent to the cloud have the form of an update to the metadata that is already present and is shown in Listing 2. Metadata updates include the aggregates as a data point with information about the time stamp, value, aggregation function, and time over which the aggregation function aggregated the values.


```
{
  "MessageType": "Update",
  "Name": "tkn/twist/128/temperature",
  "ID": "BB:8C:D1:F7:B7:92",
  "Datastream": {
    "Data-Points": [
      {
        "timestamp": "2016-06-01T09:45:00.000Z",
        "value": "21",
        "aggregation-function": "mean",
        "aggregation-time": "900"
      },
      {
        "timestamp": "2016-06-01T09:30:00.000Z",
        "value": "20",
        "aggregation-function": "mean",
        "aggregation-time": "1800"
      },
      {
        "timestamp": "2016-06-01T09:00:00.000Z",
        "value": "19",
        "aggregation-function": "mean",
        "aggregation-time": "3600"
      }
    ]
  }
}
```


Listing 2: Metadata update message to add sensor data aggregates to the registry

10.3.3 Sensor Search Based on Metadata and Aggregates

This work now focuses on the registry itself, which is responsible for storing the sensor metadata and aggregates and for providing search functionality to interested clients and services using a suitable query language. As a nonfunctional requirement, the extensibility of the data expressed in advertisements, as well as metadata status updates, is identified. As discussed earlier, the prototype should enable the sensor provider to define arbitrary tags for its sensors. The sensor search should be agnostic to the scheme of the data and should support those user-defined custom attributes.

To enable sensor search and selection over the publish/subscribe network, clients need to be able to query the registry directly. Request/response messaging is needed to enable clients requesting such metadata from the sensor registry, which is not part of some publish/subscribe systems, such as MQTT. One option to add this functionality to MQTT is to extend the existing protocol altering PINGREQ/PINGRESP messages, which are defined for implementing keep-alive messages between the broker and connected clients to detect offline clients. They usually do not have a payload, and therefore a length field is set to zero, but can easily be changed to include a payload with an appropriate length field. This solution, however, is not compatible with standard clients widely used today. Although in this chapter the approach is implemented as a proof of concept, the discovery scheme works for generic publish/subscribe systems, so that the prototype uses a pure publish/subscribe system without a dedicated request/response mechanism. As an example, the prototype therefore uses the standard MQTT protocol without any modifications. It again uses a reserved topic the sensor registry is subscribed to, on which the request can be published by clients, $REQ/registry. Requests have to include a response topic in the payload to which the client has previously issued a subscription to and the registry can respond on. A straightforward solution is that clients subscribe to a dedicated response topic on start-up. The message payload is a query for sensors using an existing query language.

In our prototype, sensor devices provide sensor advertisements as JSON and need to search characteristics for exact matches and ranges ($\geq$, $\leq$, etc.), for instance, to specify thresholds. The prototype uses Elasticsearch,* which is a NoSQL data store that natively stores schema-less JSON documents. Elasticsearch enables sharding of the data store among different servers, which is useful for scaling the platform. It allows querying the data store using a domain-specific language (DSL) over a RESTful interface and supports, among others, term, wildcard, regular expression, and fuzzy and range queries for searching the data store. An adapter proxies the requests between MQTT and the RESTful interface provided by Elasticsearch.

```
{
  "MessageType": "RegistryQuery",
  "Handle": 4135,
  "ResponseTopic": "/client/tkn/8861/response",
  "Filter": {
    "must": [
      {
        "match": {
          "Name": "temperature"
        }
      },
      {
        "match": {
          "Type": "Sensor"
        }
      },
      {
```

* Elastic. Elasticsearch. 2016. https://www.elastic.co/products/elasticsearch

```
          "range": {
            "Datastream.Data-Points.value": {
              "gt": 20
            }
          }
        }
      ]
    }
}
```

Listing 3: Example of a query from client to registry

An example of a query request that is sent to the MQTT end point with a payload that is partially proxied to Elasticsearch is given in Listing 3. The message has a handle and includes the response topic, so that the client can issue several requests to the registry at once and still distinguish between the queries. The query specifies a filter that is forwarded to Elasticsearch, which defines two exact matches and one range query on the data points.

10.4 EXPOSING APPLICATION REQUIREMENTS TO OPTIMIZE NETWORK PARAMETERS

Another major drawback of existing solutions is the rigid decoupling of sensors as measurement devices and value-added services or other subscribers as data sinks. In particular, using publish/subscribe systems, the publisher is usually not aware of active subscribers; that is, the publisher cannot adapt to a changing number of subscribers or changing requirements as is. As sensors in traditional sensor networks are usually battery powered, it is important that they operate in an energy-efficient manner. This is commonly achieved by reducing the sampling interval and entering a deep-sleep state for the time not needed for sampling or forwarding data over the wireless interface. Energy efficiency could thus be dramatically improved by exposing application-level requirements to the publisher, such as the desired sampling frequency, so that the sensors as well as the underlying network can be reconfigured accordingly to allow a longer sleeping period while still fulfilling user requirements. Here, a scheme where the requirements are embedded in publish/subscribe subscription messages is presented. Requirements are combined in the cloud layer by the query service, and if an update should be necessary, the service forwards a corresponding subscription to the publisher. The publisher can then adapt to those changing requirements by starting to sample sensor data or by revising the sampling interval or other networking parameters, such as the beacon interval. The proposed scheme with our own prototype using MQTT and IEEE 802.15.4 hardware is illustrated.

Considering the critical importance of sensor data in some use cases, for example, healthcare or medical applications, the proposed system needs to fulfill user and application requirements, but at the same time the corresponding network operator requirements. For instance, the user requirement for a specific latency between sensor measurement and data delivery may contradict the sensor owner's requirement of a long battery life. This work only considers user requirements and not explicitly the network operators' requirements. Also, user requirements are treated as a hint, and the system just offers best-effort networking service. At this point, it does not guarantee the fulfillment of any of the requirements specified.

10.4.1 Capturing Application Requirements

This section first defines the requirements the user should be able to specify. At this point in time, it only considers the subset of possible requirements that can be easily translated to parameters of underlying wireless sensor networks on one side, and publish/subscribe systems on the other side. Note that the user is not able to change network parameters exclusively, as parameters are often changed for the whole network. Instead, the user specifies abstract, high-level requirements for a certain sensor subset. The actual network parameter changes are very technology specific, and therefore are hidden from the user and performed by the IoT platform transparently. The main user requirements are the following:

1. *Frequency*: Specifies the minimum frequency of sensor measurements that are delivered to the user. Sensors that are not needed can be turned off or slowed down to not sample at all or only send heartbeats at a low frequency. The actual sensor frequency will not map directly to the frequency specified by one user, but will have to be adapted to fulfill the requirements of all users.
2. *Staleness*: Specifies the accepted staleness of data. This can be used by the system to overlap subscriptions for queries that are otherwise not overlapping, for example, that are not multiples of each other.
3. *Delivery guarantees*: Specifies semantic guarantees, such as at-most-once or at-least-once delivery. Raw sensor readings could, for instance, be subscribed to with low priority, that is, with at-most-once delivery. A virtual sensor that acts as a service to identify critical state changes, such as a temperature above a threshold, would be subscribed to with delivery guarantees, such as at-least-once delivery.
4. *Cost*: Specifies the willingness to accept a cost to fulfill the specified subscriptions with stated requirements. Application requirements have to include a cost to prevent users from always specifying the highest possible service level. While the cost should represent the actual monetary cost the provider has when offering sensor services, the sensor provider can specify an arbitrary mapping between requirements and an abstract cost metric.

10.4.2 Distributing Requirements over Publish/Subscribe Networks

Delivery guarantees are already given by most messaging systems, such as MQTT. The cost function is treated as another filter that determines if a subscription is accepted and fulfilled by the system or rejected. In our reference architecture, this can be seen as an infrastructure management module in the cloud that changes networking parameters, in this case semantic guarantees, of the system on the fly.

Traditionally, publish/subscribe systems have incorporated decoupling properties that do not expose subscriber information to publishers. This work proposes to soften those decoupling properties and provide publishers with hints about subscribers. In particular, subscription requests annotated with metadata about user requirements are relayed to publishers. That means that the semantic for the subscriber stays the same: it can issue a subscription with additional requirements. The publisher—in this case the gateway publishes for the actual sensor by proxy—has additional information at hand to adapt its network parameters according to users' requirements. This is an example of an infrastructure management module on the gateway that was mentioned in the discussion of the architecture.

The definition of user-level requirements to a cloud-based publish/subscribe system has an additional benefit: subscriptions can be aggregated in the cloud tier before relaying them to the device tier as a single query. This saves on computation, and therefore energy, on the devices on one side. On the other side, it also saves on upstream bandwidth to the cloud, which may especially be an important issue for gateways with poor connectivity, such as 3G modules.

As this work considers all requirements as hints rather than as hard limits, the focus lies on providing samples with a provided staleness at a certain frequency. The prototype achieves this by interpreting the topic string at the message broker and storing along with the subscription the additional attributes, in our case frequency and staleness. The querying service decides on each new subscribe or unsubscribe message that subscriptions overlap. The querying service is implemented as a module in the mosquitto broker. The service sorts subscriptions by their frequency.

The subscription with the smallest frequency always has to be fulfilled and is chosen to be forwarded to the gateway. The service then checks which other subscriptions can be met by this initial subscription, that is, if a multiple of the current frequency is sufficient to cover either the frequency or the staleness of the other subscriptions. The service marks those subscriptions as met by the initial subscription and starts the same scheme again for the next smallest subscription that is not yet met. It then forwards the chosen subscriptions as a batch to the corresponding gateway. On publications from the gateway, the broker delivers the message to only those subscribers that have not gotten an event notification message for the time specified by their frequency and staleness. That means subscribers have the impression that they exclusively set the frequency of the sensor, while potentially two or more subscriptions are served using a single subscription at the gateway level.

We now study how the metadata mentioned above can be incorporated into a subscription request. While the annotation can be done using other protocols, an exemplary implementation using MQTT is presented. MQTT subscription messages have a 2-bit topic length field; a variable-length topic string field; a 6-bit reserved field, which at the moment is only used for padding; and a 2-bit QoS field. There are three main alternatives to express the content metadata in MQTT. The first would be to use the existing reserved field. The second would be to extend the message format to include one or more new fields for the metadata. Another alternative would be to use the existing topic field and overload it with the metadata.

As the existing reserved field is currently not specified and only used for padding, it may be a good candidate for an extension, as brokers usually do not check the content of the field. Its length of 6 bits gives only 64 possible states, which is certainly not enough to express all requirements. Extending the protocol with a new field would allow the addition of arbitrary data. However, this would mean altering the protocol, making the new fork incompatible with other implementations. Since many developers would rather stick to readily available MQTT client libraries, this would be a clear disadvantage of the platform for developers.

Therefore, the topic field is used to introduce key/value pairs for user requirements. The prototype uses syntax analogous to the query string format that is used in HTTP GET requests as part of a uniform resource locator (URL) containing data that does not reflect the hierarchical path structure, but additional parameters. This approach enables users to specify the metadata described above, as well as further metadata, which might be added in the future. To a certain degree, this solution is backward compatible. That is, ordinary MQTT users can omit the metadata and still subscribe to a topic with default settings. The broker has to be changed to interpret the additional metadata information, if it is present, and set them to reasonable default values, if they are omitted. Publishers need to be changed to support subscription messages that are forwarded to them by the broker. Legacy MQTT clients could specify that they do not support subscription forwarding in the advertisement message.

```
[...]/light?frequency=5.0&staleness=0.5
[...]/light?frequency=15.0&staleness=1.5     →     [...]/light?frequency=5.0
[...]/light?frequency=28.0&staleness=5.0
```

Listing 4: Merging of overlapping subscriptions at the broker and the resulting subscription forwarded to the gateway

An example of three topic strings is given in Listing 4. The frequency and staleness are given as floating-point numbers. In this example, a multiple of the frequency of the first subscription is the frequency of the second subscription, so that the sensor only has to sample every 5 s to also fulfill the requirement of sampling every 15 s. The third subscription has a frequency requirement that is not a multiple of 5, but can still be fulfilled by the same low-level subscription because the staleness is sufficiently large. As the frequency is already dictated to be as low as 5 s by the first subscription, the third subscription can be fulfilled by forwarding the latest data point every 28 s, which will never be older than 5 s. Also note that additional delays in the sensor network and the publish/subscribe system are ignored here. This is due to the best-effort nature of the requirements that are treated only as hints.

10.4.3 Adaptation of Network Parameters Based on Requirements

As shown in the previous section, the publisher of sensor data will get feedback about subscribers. Two main parameters can be changed in common sensor networks to adapt for changing subscriber needs: (1) sampling can be switched on or off altogether, and (2) the sampling frequency can be changed to reflect the application requirements. The latter can also be used to adapt other lower-level protocol parameters, such as the beacon interval, robustness of modulation, or routing.

A command to enable or disable a particular sensor would be sent over a publish/subscribe protocol, such as MQTT, in the form of a forwarded subscription to the corresponding gateway. The gateway would then translate the command into a network and application-specific command that is sent to the corresponding devices. The actual implementation of the logic to switch the sensor to a low power state would be provided by the application that runs on the sensor node. That means that the sensor operator has to have a certain amount of control over the sensor node software to fully leverage the power-saving potential of the feedback the system enables. Off-the-shelf hardware may therefore not offer suitable interfaces to enable lower power operation adequately.

Information about the requested sampling frequency can likewise be used to adapt the time the sensor node is in a low power state between individual wake-ups for sensing and sending the data to the corresponding gateway. Our prototypical implementation runs on TelosB sensor nodes running TinyOS and using an IEEE 802.15.4 stack. It uses a message type field with a reserved message type for sampling rate adaptation. The sampling rate is given as a 64-bit unsigned integer in milliseconds. The device has to confirm command receipt with an acknowledgment. The device adjusts its timers to wake up according to the new sampling interval that was given by the gateway and enters a lower power state when not active.

Additionally, other lower-level network parameters could be varied in the future to adapt to application requirements. In IEEE 802.15.4, the beacon-enabled mode is frequently used. It defines superframes that are framed by beacons from the coordinator. The network coordinator can set the duration of active and inactive periods in each superframe by adjusting the beacon order (BO) and superframe order (SO). The active and inactive periods can be adapted to allow the devices to stay in a low power state a higher fraction of time with regard to the application requirements. A straightforward approach is presented in Neugebauer et al. (2005), where the BO is adapted based on sensor data frequency and an acceptable delay constant. Currently, this potential is not leveraged and neither BO nor SO is altered, but will be investigated in future research.

10.5 CONCLUSIONS

IoT has the potential to improve our everyday life by enabling real-time awareness and automatic adaptation to the physical environment around us. Sensing devices together with cloud-based services enable use cases such as smart metering, smart building, or smart factories. While offering fast networking, reliable distributed storage, flexibility, scalability, and easy-to-use programming interfaces, the existing cloud-based IoT architecture as of now cannot fully address the notable design challenges that result from the rigid coupling of components. This chapter presented an architecture for interconnecting sensors, actuators, and applications that additionally tackles the challenges of device and service discovery and network optimization for cloud-based IoT platforms. In the foreseeable future, the cloud computing paradigm will remain the main driver behind the success of IoT solutions. Using closer-to-the-edge distributed computing, such as fog computing, as an extension to the current cloud-based architecture will further strengthen the role of the cloud in the IoT context.

REFERENCES

Antonic, A., K. Roankovic, M. Marjanovic, K. Pripuic, and I. P. Zarko. 2014. A mobile crowdsensing ecosystem enabled by a cloud-based publish/subscribe middleware. In *2014 International Conference on Future Internet of Things and Cloud (FiCloud)*. Barcelona, Spain, 107–114.

Banks, A. and R. Gupta. 2014. MQTT version 3.1.1. OASIS Standard.

Bonomi, F., R. Milito, J. Zhu, and S. Addepalli. 2012. Fog computing and its role in the Internet of things. In *Proceedings of the First Edition of the MCC Workshop on Mobile Cloud Computing*, Helsinki, Finland, 13–16.

Bradley, J., J. Barbier, and D. Handler. 2013. Embracing the Internet of everything to capture your share of $14.4 trillion. White Paper. San Jose, CA: Cisco Systems.

Chappell, D. 2004. *Enterprise Service Bus*. Sebastopol, CA: O'Reilly Media.

Chui, M., M. Löffler, and R. Roberts. 2010. The Internet of things. *McKinsey Quarterly*.

Curry, E. 2005. Message-oriented middleware. In *Middleware for Communications*, edited by Q. H. Mahmoud, 1–28. Hoboken, NJ: Wiley.

Eugster, P. T., P. A. Felber, R. Guerraoui, and A.-M. Kermarrec. 2003. The many faces of publish/subscribe. *ACM Computing Surveys (CSUR)* 35 (2): 114–131.

Gubbi, J., R. Buyya, S. Marusic, and M. Palaniswami. 2013. Internet of things (IoT): A vision, architectural elements, and future directions. *Future Generation Computer Systems* 7 (29): 1645–1660.

Hamida, S. T.-B., E. B. Hamida, B. Ahmed, and A. Abu-Dayya. 2013. Towards efficient and secure in-home wearable insomnia monitoring and diagnosis system. In *13th International Conference on Bioinformatics and Bioengineering (BIBE)*, Chania, Greece, 1–6.

Happ, D., N. Karowski, H. Thomas, V. Menzel, and A. Wolisz. 2015. Meeting IoT platform requirements with open pub/sub solutions. *Presented at First International Conference on Cloudification of the Internet of Things (CIoT'15)*, Paris, France.

Hassan, Q. F. 2016. An outlook into novel concepts of high performance computing. In *Innovative Research and Applications in Next-Generation High Performance Computing*, edited by Q. F. Hassan, xxiii–lii. Information Science Reference. Hershey, PA: IGI Global.

Hassan, Q. F., A. M. Riad, and A. E. Hassan. 2012. Understanding cloud computing. In *Software Reuse in the Emerging Cloud Computing Era*, edited by H. Yang and X. Liu, 204–227. Hershey, PA: IGI Global.

Hunkeler, U., H. L. Truong, and A. Stanford-Clark. 2008. MQTT-S—A publish/subscribe protocol for wireless sensor networks. In *3rd International Conference on Communication Systems Software and Middleware and Workshops (COMSWARE '08)*, Bangalore, India, 791–798.

IDABC (Interoperable Delivery of European eGovernment Services). 2004. European Interoperability Framework for pan-European eGovernment services.

Krafzig, D., K. Banke, and D. Slama. 2005. *Enterprise SOA: Service-Oriented Architecture Best Practices*. Upper Saddle River, NJ: Prentice Hall Professional.

Mell, P., and T. Grance. 2011. *The NIST Definition of Cloud Computing*. Gaithersburg, MD: Computer Security Division, Information Technology Laboratory, National Institute of Standards and Technology.

Menzel, T., N. Karowski, D. Happ, V. Handziski, and A. Wolisz. 2014. Social sensor cloud: An architecture meeting cloud-centric IoT platform requirements. *Presented at Ninth KuVS NGSDP Expert Talk on Next Generation Service Delivery Platforms*, Berlin, Germany.

Neugebauer, M., J. Plonnigs, and K. Kabitzsch. 2005. A new beacon order adaptation algorithm for IEEE 802.15.4 networks. In *Proceeedings of the Second European Workshop on Wireless Sensor Networks*, 302–311.

oneM2M. 2015. MQTT protocol binding: Version: TS-0010-V1.0.1.

Papazoglou, M. P. 2003. Service-oriented computing: Concepts, characteristics and directions. In *Proceedings of the Fourth International Conference on Web Information Systems Engineering (WISE 2003)*, 3–12.

Riad, A. M., A. E. Hassan, and Q. F. Hassan. 2010. Design of SOA-based grid computing with enterprise service bus. *AISS* 2 (1): 71–82.

Rowe, A., et al. 2011. Sensor Andrew: Large-scale campus-wide sensing and actuation. *IBM Journal of Research and Development* 55 (1.2): 6:1–6:14.

Singh, D., G. Tripathi, and A. J. Jara. 2014. A survey of Internet-of-things: Future vision, architecture, challenges and services. In 2014 *IEEE World Forum on Internet of Things (WF-IoT)*, Seoul, Korea, 287–292.

Tschofenig, H., J. Arkko, D. McPherson, D. Thaler, and D. McPherson. 2015. Architectural considerations in smart object networking. RFC 7452. Fremont, CA: Internet Engineering Task Force.

Zhang, B., et al. 2015. The cloud is not enough: Saving IoT from the cloud. *Presented at Seventh USENIX Workshop on Hot Topics in Cloud Computing (HotCloud '15)*, Santa Clara, CA.

Zhang, M., T. Yu, and G. F. Zhai. 2011. Smart transport system based on "the Internet of things." *Applied Mechanics and Materials* 48: 1073–1076.

Zhou, X., S. Prasher, S. Sun, and K. Xu. 2004. Multiresolution spatial databases: Making web-based spatial applications faster. In *Proceedings of Advanced Web Technologies and Applications: 6th Asia-Pacific Web Conference (APWeb 2004)*, Hangzhou, China, 36–47.

11 The Emergence of Edge-Centric Distributed IoT Analytics Platforms

Muhammad Habib ur Rehman, Prem Prakash Jayaraman, and Charith Perera

CONTENTS

11.1 Introduction 213
11.2 Role of Analytics in IoT Systems 215
11.2.1 Descriptive Analytics 216
11.2.2 Predictive Analytics 216
11.2.3 Prescriptive Analytics 216
11.2.4 Preventive Analytics 216
11.3 Toward Device-Centric IoT Systems 217
11.3.1 Device-Centric Immobile IoT Systems 217
11.3.2 Device-Centric Mobile IoT Systems 218
11.4 A Speculated Multilayer Application Architecture 218
11.4.1 Tier 1: Data Stream Layer 219
11.4.2 Tier 2: Data Acquisition and Adaptation Layer 220
11.4.3 Tier 3: Data Preprocessing, Fusion, and Data Management Layer 220
11.4.4 Tier 4: Data Analytics and Knowledge Integration Layer 220
11.4.5 Tier 5: Security and Privacy-Preserving Data Sharing Layer 220
11.4.6 Tier 6: Actuation and Application Layer 221
11.4.7 Tier 7: System Management 221
11.5 Journey Toward Device-Centric Multilayer Architecture 221
11.5.1 MOSDEN Architecture 221
11.5.1.1 Achievements of MOSDEN in Connection with Speculated Architecture 222
11.5.2 CARDAP Architecture 223
11.5.2.1 Achievements of CARDAP in Connection with Speculated Architecture 223
11.5.3 UniMiner Architecture 224
11.5.3.1 Achievements of UniMiner in Connection with Speculated Architecture 225
11.6 Conclusion 227
References 227

11.1 INTRODUCTION

The phenomenal growth in IoT devices and systems has opened many new research avenues (Gubbi et al., 2013). This massive growth is leading toward gigantic data production and requires sophisticated systems to perform analytical operation and uncover useful insights from underlying data (Mukherjee et al., 2014; Hassan, 2016a). The topological settings of existing systems are based on three levels, namely, IoT devices, edge servers, and cloud data centers (Figure 11.1) (Satyanarayanan et al., 2015). At the first level, the IoT devices perform data collection operations by monitoring their surroundings using onboard sensors. In addition, the devices perform data filtration and actuation

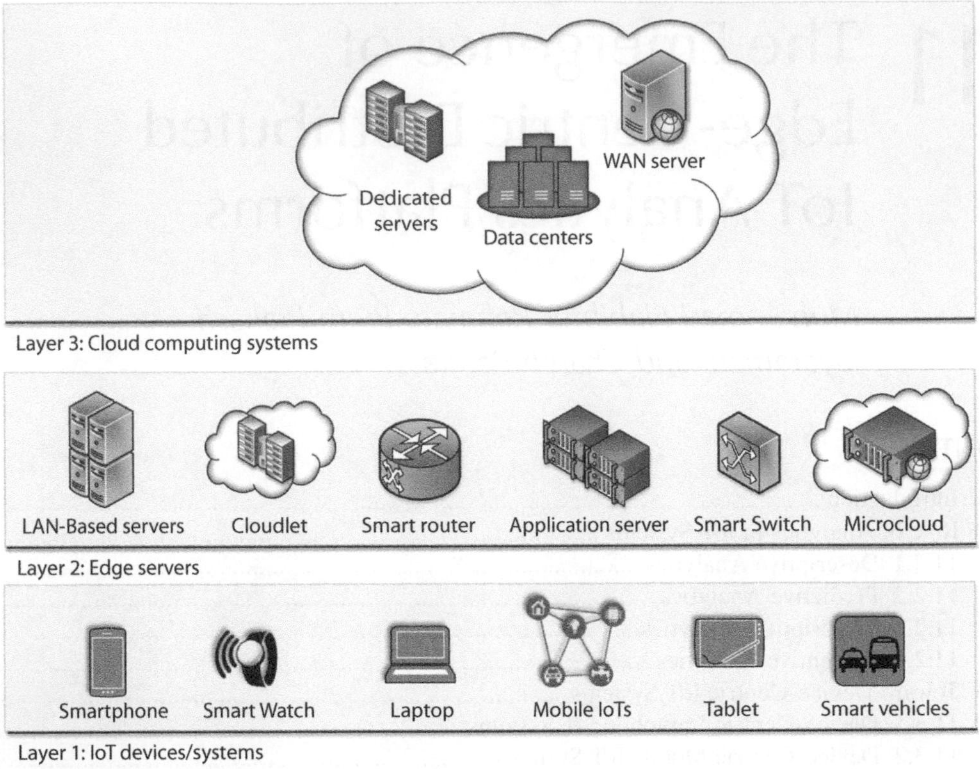

FIGURE 11.1 Topological setting of IoT systems.

operations to respond in the external environments. At the second level, the nearby edge servers collect data streams from connected IoT devices and perform local data processing to transfer reduced data streams in cloud environments. At the third level, the cloud data centers provide unbounded cloud resources for IoT devices and cloud-based IoT applications. The IoT applications span over all three levels, but control of application execution always remains at the cloud level (Satyanarayanan et al., 2015). This approach increases the dependency over Internet connections and enforces the devices to transfer data streams to cloud data centers prior to any analytics operations.

A device-centric distributed analytics system is presented in this chapter. The cloud-first approach increases the data communication cost and network data movement in cloud data centers (Lea and Blackstock, 2014; Hassan, 2016b). There exist many application areas where local analytics in IoT devices are beneficial and given priority over global analytics in cloud data centers. The device-centric IoT systems have multiple benefits (Rehman et al., 2014a). The IoT devices perform local analytics that reduce dependency on edge servers and cloud data centers. Also, the device-specific analytics operations are performed with minimum latency compared with cloud-based data analytics. Moreover, the integration of local knowledge patterns to form global patterns requires fewer cloud resources than processing raw data streams in cloud environments (Sherchan et al., 2012; Rehman and Batool, 2015).

Before moving further, first let us have a look at the operational view of IoT systems (Figure 11.2). The IoT systems work at four levels (Bonomi et al., 2012; 2014): (1) physical, (2) communication, (3) middleware, and (4) application. The physical layer is based on three level topological settings of IoT devices, edge servers, and cloud data centers. A plethora of devices and systems are involved at this stage in order to provide sensing, processing, and storage resources for IoT applications. The communication layer enables multiple communication interfaces and protocols for device-to-device and device-to-cloud communication. These communication interfaces include Wi-Fi, ZigBee, Z-Wave, 6LoWPAN, cellular, near-field communication (NFC), LoRaWAN, and Ethernet, to name

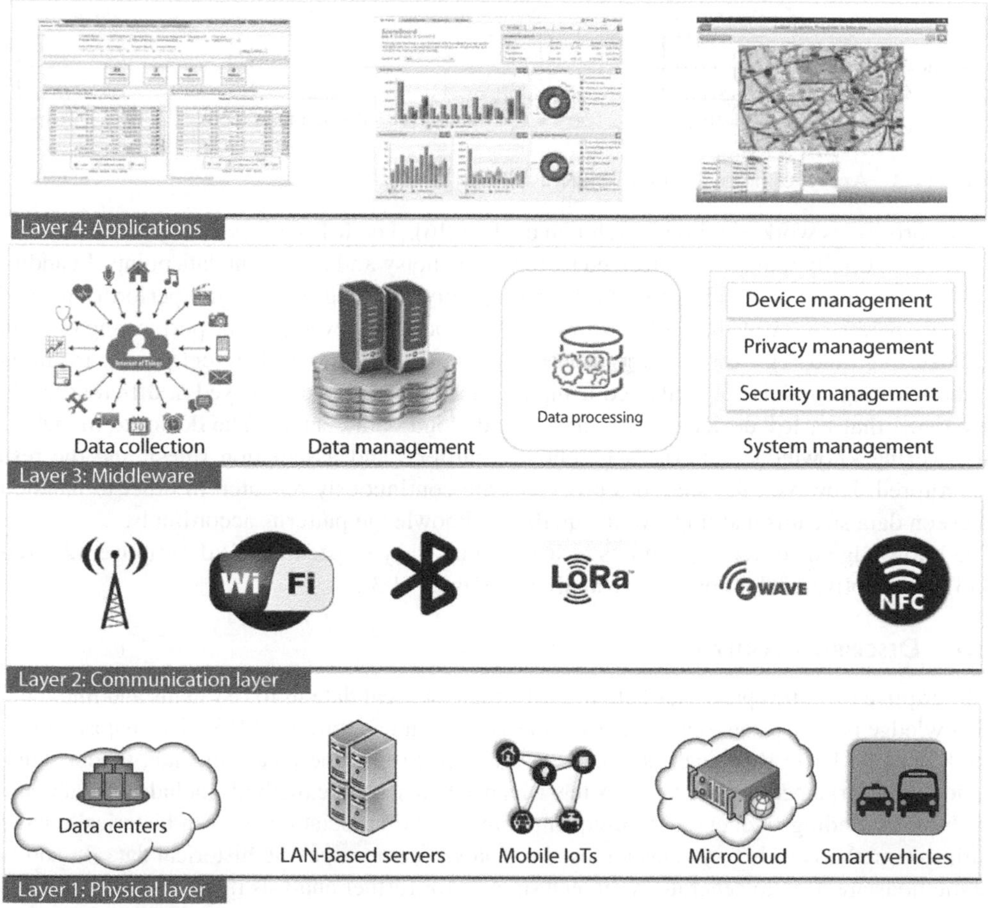

FIGURE 11.2 Operational view of an IoT system.

a few. The middleware layer enables the sensing, data transfer, data management, data storage, and data processing operations and uses device management, privacy, and security policies to perform end-to-end system management. Finally, the application layer provides the functionality to deploy different kinds of IoT and big data applications at both the IoT device end and the cloud data centers.

This chapter aims to present device-centric distributed IoT analytics systems and highlights three major contributions: Mobile Sensor Data Processing Engine (MOSDEN) (Perera et al., 2014), Context-Aware Real-time Data Analytics Platform (CARDAP) (Jayaraman et al., 2014), and UniMiner (Rehman et al., 2014a; ur Rehman et al., 2016). The rest of the chapter is organized as follows. Section 11.2 defines the role of analytics in IoT systems and discusses variants of analytics methods and cloud-based analytics systems for IoTs. Section 11.3 presents device-centric mobile and immobile IoT systems. Section 11.4 presents the speculated multitier architecture for device-centric distributed analytics systems. Section 11.5 presents an overview of MOSDEN, CARDAP, and UniMiner. Finally, Section 11.6 concludes the chapter.

11.2 ROLE OF ANALYTICS IN IoT SYSTEMS

Primarily, IoT devices collect a massive amount of continuously streaming data using onboard sensors (Gubbi et al., 2013). The analytics processes in IoT systems enable us to convert these raw data streams into actionable knowledge patterns (Satyanarayanan et al., 2015). Conventionally, the

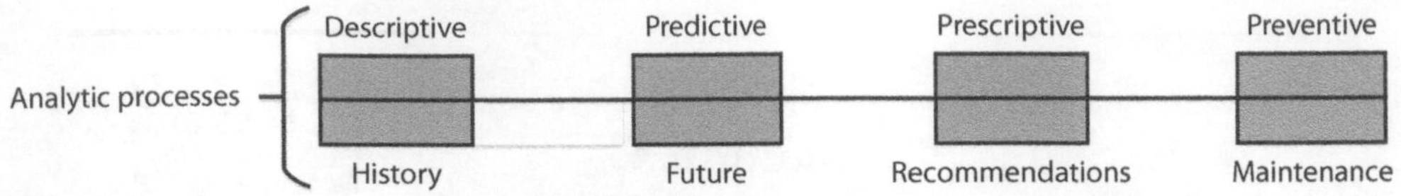

FIGURE 11.3 Analytics processes in IoT systems.

analytics processes work as follows (Rehman et al., 2016). The IoT devices and systems collect raw data streams initially that are transformed by removing noisy and irrelevant data points. In addition, useful features are extracted in order to develop quality learning models and perform knowledge discovery operations. A plethora of learning models could be developed for performing classification, clustering, and regression and finding association rules among different points. In the next step, the learning models are evaluated using test data streams and deployed at different levels in IoT systems, that is, IoT devices, edge servers, and cloud data centers. The deployed models help in discovering knowledge patterns (e.g., classes, clusters, and association rules), and the results are monitored. However, the analytics processes are continuously repeated in order to handle the unforeseen data streams and improve the quality of knowledge patterns accordingly.

The knowledge discovery operations in IoT systems vary to perform (1) descriptive, (2) predictive, (3) prescriptive, and (4) preventive analytics (Figure 11.3).

11.2.1 Descriptive Analytics

The descriptive analytics processes help in analyzing historical data in IoT systems and find the hidden knowledge patterns from sensor data streams (Delen and Demirkan, 2013). The majority of these methods include basic statistical methods that are used to find the mean, median, mode, standard deviation, and variance, to name a few. A few advanced prescriptive methods include the data mining algorithms for finding frequent and infrequent itemsets and association rules in historical data. The prescriptive methods in IoT systems are used to analyze device-specific historical data. In addition, these methods are used to generate event data streams for further analysis in cloud environments.

11.2.2 Predictive Analytics

The predictive analytics processes enable us to learn the characteristics of historical data and predict the future behavior of unforeseen data (Waller and Fawcett, 2013). Predictive modeling is the essence of predictive analytics, whereby different machine learning and data mining algorithms are developed to learn the behavior of historical data. These predictive models are further used to recognize, detect, find, and predict the behavior of newly incoming data streams. A benefit of predictive modeling in IoT systems is finding the future behaviors of devices. Primarily, machine learning and data mining methods for classification, clustering, and regression analysis are used for predictive modeling.

11.2.3 Prescriptive Analytics

The prescriptive analytics processes not only enable predictive modeling but also help in devising future courses of actions (Basu, 2013). The prescriptive models work by first developing the learning models from historical data, and based on the predictions, these models suggest further alternate actions in order to find the best possible solutions. Although the existing literature still lacks prescriptive modeling methods for IoT systems, these methods can optimize business operations, such as scheduling inventories and improving supply chain management systems.

11.2.4 Preventive Analytics

The preventive analytics processes monitor the performance of IoT devices (Wilkerson and Gupta, 2016). Preventive models are best suited for machine analytics applications, whereby these models

TABLE 11.1
Commonly Used Cloud-Based IoT Analytics Systems

No.	Platform	Web Address
1	Autodesk Fusion Connect	http://autodeskfusionconnect.com/
2	AWS IoT	https://aws.amazon.com/iot/getting-started/
3	Cisco	https://developer.cisco.com/site/iot/
4	Dell Statistica	https://software.dell.com/products/statistica/
5	GE Predix	https://www.predix.io/
6	Google Cloud IoT	https://cloud.google.com/solutions/iot/
7	IBM Watson	https://www.ibm.com/internet-of-things/
8	IBM Bluemix	https://console.ng.bluemix.net/
9	Intel	https://shopiotmarketplace.com/iot/index.html#/home
10	Kaa	http://www.kaaproject.org/
11	Microsoft Azure	https://azure.microsoft.com/en-us/
12	Pentaho	http://www.pentaho.com/internet-of-things-analytics
13	RTI	http://www.rti.com/
14	Saleforce	https://www.salesforce.com/iot-cloud/
15	SiteWhere	http://www.sitewhere.org/
16	Splunk	https://www.splunk.com/en_us/download-5.html
17	Tellient	http://tellient.com/index.html
18	ThingSpeak	https://thingspeak.com/
19	ThingWorX	https://www.thingworx.com/
20	VitriaIoT	http://www.vitria.com/iot-analytics

foresee the performance degradation issues well before the failure of machines. Preventive modeling improves preventive maintenance and is useful for conditions whereby the failure of certain IoT devices may degrade the overall performance of IoT systems. The preventive models are useful for manufacturing, assembly lines, surveillance and security, and mission-critical IoT systems.

Most of the existing IoT systems execute analytics processes using cloud-based computing, networking, and storage resources (Jayaraman et al., 2017). This approach lowers the computational and battery power consumption for data processing in IoT devices. However, it upsurges the data communication cost in terms of bandwidth utilization, energy consumption during data transfer, and network data movement in cloud data centers (Khan et al. 2015; 2014). In addition, the accumulation and aggregation of continuously streaming data in cloud data centers increase the cost of data processing for cloud services utilization, as well as the programming efforts to handle and process the raw streaming data. Table 11.1 enlists the commonly adopted cloud-based systems for IoT analytics; however, all these systems enable the cloud-centric approach.

11.3 TOWARD DEVICE-CENTRIC IoT SYSTEMS

Device-centric IoT systems are designed to delegate control of application execution at the device end by using IoT devices as the primary platform for performing data analytics. However, device-centric IoT systems are deployed as either mobile systems or immobile systems to address different kinds of challenges.

11.3.1 Device-Centric Immobile IoT Systems

The immobile IoT systems facilitate the static deployment of IoT devices and systems in virtually bounded communication areas, such as a local area network (LAN), personal area network (PAN), or body area network (BAN), or are connected to the Internet through a wide area network (WAN) (Fan et al., 2014). The IoT devices in immobile systems primarily use Ethernet connections for data communication, and hence do not need to counter the issue of persistent Internet connections. In addition, immobile devices are usually powered using direct current; therefore, the issue of limited battery power does not arise in these systems. The high availability of Internet connections and electrical power increases the utility of immobile systems, but a limited sensing range of the IoT device may induce noise or incompleteness in data streams. Therefore, immobile IoT systems need to deploy more sensing nodes in order to achieve maximum coverage in the sensing areas.

A few common applications of immobile IoT systems are the smart home network, patient monitoring systems in intensive healthcare units, smart parking, security and surveillance systems, and air quality monitoring systems. For example, in a smart home network, electrical appliances such as the television, refrigerator, and microwave oven are connected through a local edge server (i.e., IoT hub) in the home. The appliances can continuously sense, processes, and analyze the data streams using onboard computational elements, such as graphics processing units (GPUs) and secure digital (SD) memory cards. In the case of significant changes in the knowledge patterns or detection of specific events, the devices transfer the data streams to edge servers or cloud data centers. Similarly, the IoT devices can sense and process the biomarkers for patients and alert medics whenever a critical situation occurs. In the case of smart parking, the IoT devices can periodically monitor parking lots using infrared sensors and cameras and perform local analytics to detect empty spaces. Likewise, cameras and environmental monitoring devices are deployed on roadsides, in busy shopping and commute areas, and in other designated spaces to collect and process the continuously streaming data.

11.3.2 Device-Centric Mobile IoT Systems

Unlike immobile systems, the mobility in IoT devices involves several issues (Mavromoustakis et al., 2016). The size of the devices needs to be small in order to move easily. The devices need to enable multiple communication interfaces, such as Wi-Fi, Bluetooth, and GSM, for data communication in different networks. The device-centric mobile IoT systems are bounded to efficiently utilize limited onboard computational and battery power resources for maximum data processing, consuming minimum battery power. Although wide area coverage is a benefit of mobile IoT devices, the orientations and positions of devices significantly impact the quality of collected data streams, which indirectly affects the quality of knowledge patterns. Considering the mobility, limited resources, orientations, and positions of mobile IoT devices on the one end, and the issues of high bandwidth utilization and increased in-network data movement in data centers on the other, the device-centric approach helps to minimize the complexities in IoT systems.

The device-centric mobile IoT systems could be deployed in a wide range of applications areas. Mobile IoTs can help in the development of personal data analytics systems that are used for lifelogging, quantified self, personal data mining, personalized services, and so forth. In addition, mobile IoTs facilitate the uncovering of collective intelligence from the personal data of multiple users and apply it for mobile crowd-sensing, opportunistic sensing, and large-scale sensing applications. Device-centric IoT systems also enable us to preserve the privacy and security of devices and users by delegating complete control of application execution and data sharing at the device end. This approach helps to reduce the dependency on edge servers and cloud data centers, and hence minimize the risks relevant to the privacy and security of personal data.

11.4 A SPECULATED MULTILAYER APPLICATION ARCHITECTURE

This section presents the speculated application architecture of device-centric distributed data analytics applications in IoT systems (Figure 11.4). The architecture is based on seven tiers of

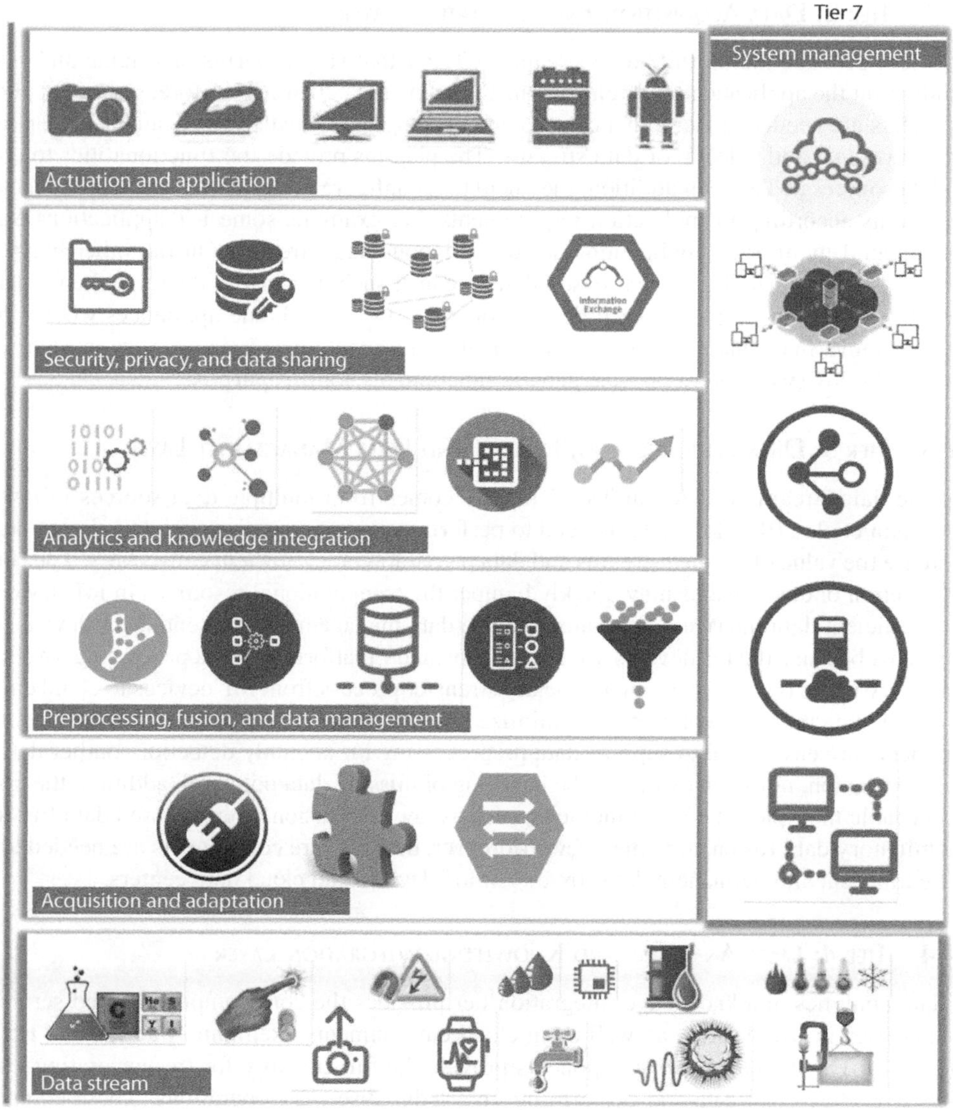

FIGURE 11.4 Speculated multitier application architecture.

components in order to convert raw data streams into useful and device-specific knowledge patterns. Moreover, the application components ensure secure and privacy-preserving data sharing among IoT devices and edge servers and cloud data centers. Also, the application architecture integrates several system management components in order to switch application execution among the three-layer architecture, which was presented earlier in Figure 11.1.

11.4.1 Tier 1: Data Stream Layer

IoT devices produce data streams from a plethora of sensory and nonsensory data sources, which include off-board and onboard sensors and online Internet-based web and social media data sources (Rehman et al., 2015). These data streams emerge in multiple forms, such as structured, unstructured, and semistructured data points having different data types, such as numerical, textual, multimedia, and unstructured signals. The IoT systems need to handle massive heterogeneity at this layer to enable the provision of maximum data at the upper tiers.

11.4.2 Tier 2: Data Acquisition and Adaptation Layer

IoT devices produce multiformat data streams at Tier 1 that vary in terms of volume and velocity, depending on the application requirements (ur Rehman et al., 2016). Therefore, generic application components are needed at Tier 2 that can provide the plug-ins for data sources and also help in handling the volume and velocity of data streams. The plug-ins provide the functionalities to connect with data sources at Tier 1. In addition, they help to reconfigure the data rates and size of incoming data streams according to application requirements. For example, some IoT applications, such as environmental monitoring applications, need continuous data streams. On the other hand, some applications need periodic or event-based data streams, such as alerting doctors when a patient's blood pressure increases from a specific level or triggering some home appliances when a person enters the home. In essence, the data acquisition and data adaptation strategies vary in each system; therefore, the IoT system must provide generic functionality for this purpose.

11.4.3 Tier 3: Data Preprocessing, Fusion, and Data Management Layer

Since the data stream collected at Tier 2 mainly comes from multiple data sources in raw form (ur Rehman et al., 2016), IoT systems need to perform data preprocessing and data fusion methods to increase the value of the data streams and data preparation for knowledge discovery. The continuous collection of data streams may quickly hamper the computational resources in IoT devices and require efficient data management strategies. The data management is essential for device-centric IoT systems because the IoT devices are used as a primary platform for data processing. In addition, the mobility of the devices may impact the data transfer process from IoT devices to cloud data centers; hence, data management helps to minimize missing data. A plethora of libraries and software components are envisioned to support data preprocessing for anomaly detection, outlier detection, feature extraction, noise removal, and the handling of missing data points. In addition, the libraries should enable multiple data fusion methods, such as raw data fusion, preprocessed data fusion, and discriminatory data fusion, to name a few. Moreover, the software components are needed to manage the transient and permanent data streams in IoT devices and cloud data centers.

11.4.4 Tier 4: Data Analytics and Knowledge Integration Layer

The data analytics and knowledge integration tier provides the core components and services for data analytics in IoT devices, as well as in cloud environments (Rehman et al., 2014a; Haghighi et al., 2013). The data analytics components provide the functionality for (re-)generating learning models from historical data and performing knowledge discovery operations. These components enable supervised, unsupervised, and semisupervised learning models for classification, clustering, association rule mining, and regression analysis. In addition, these components provide statistical data analysis methods for descriptive analytics. Previous studies show that analytics components significantly vary in computational power and resource consumption in IoT devices; therefore, the computational complexities of these components must be considered before designing device-centric IoT systems.

11.4.5 Tier 5: Security and Privacy-Preserving Data Sharing Layer

The openness of IoT devices, especially when operating in a mobile environment, increases the security and privacy concerns (Jayaraman et al., 2014). In addition, the analytics components at the device end produce more sensitive information after processing raw data streams. The security components at Tier 5 enable us to securely transfer the data streams within IoT devices and between IoT devices and cloud data centers (Daghighi et al., 2017; 2015). Privacy preservation is also essential due to the sensitivity of personal and device-centric data. A minor vulnerability of this data stream could easily lead toward compromises and disastrous situations. The privacy preservation components at this tier are useful in preserving privacy by enabling data hiding and anonymization algorithms.

11.4.6 Tier 6: Actuation and Application Layer

The acquired knowledge patterns at the device end are used for multiple purposes. The IoT devices themselves perform further actions using actuators, for example, switching home appliances on or off when people enter in smart homes. Similarly, the acquired knowledge patterns are required by big data applications whereby the data streams acquired from multiple IoT devices are processed and used for large-scale knowledge discovery, for example, big data applications in the smart city can use the knowledge patterns from citizens' mobile devices to monitor noise pollution in the cities. In essence, this tier enables the actuators and IoT applications and needs to be able to handle a massive amount of heterogeneity in terms of IoT devices and applications.

11.4.7 Tier 7: System Management

The system management tier provides the components that run in parallel with all the abovementioned tiers, resulting in the smooth execution of IoT applications (ur Rehman et al., 2016). This tier contains multiple application components for context management, resource management, peer-to-peer communication, device–cloud communication, and performing parallel data processing. The heterogeneity in IoT systems is massive in terms of devices and data processing platforms, that is, IoT devices, edge servers, and cloud data centers. The context management components provide the functionality of context collection, context processing, and inferring the right situations for data processing in heterogeneous IoT systems. The IoT devices usually operate in resource-constrained environments, such as bounded CPU and memory and limited battery power. The resource monitoring components in IoT devices enable us to monitor and profile onboard resources, and this information could be further utilized for finding the right platform for data processing. The communication components at this tier enable peer-to-peer communication in device-to-device and group settings. These components also provide functionality to enable communication between IoT devices and cloud data centers. The parallel data processing components are useful for scheduling data streams, performing computation and data offloading operations, synchronizing the data and knowledge patterns across IoT systems, and ensuring the complete data transfer by considering the energy efficiency in IoT devices.

11.5 JOURNEY TOWARD DEVICE-CENTRIC MULTILAYER ARCHITECTURE

The abovementioned multitier architecture involves massive heterogeneity at all tiers. In addition, the architecture supports a huge stack of operations. To this end, three variants of the proposed architecture are presented in the following sections, which leads toward the development of speculated architecture.

11.5.1 MOSDEN Architecture

MOSDEN works as middleware for resource-limited IoT devices (Perera et al., 2014). Primarily, MOSDEN performs data collection and data processing by lowering the programming efforts for application developers. MOSDEN is based on sensing as a service model and works as client-side tool in any Android device, including smartphones and other IoT devices (Perera et al., 2014). The strength of MOSDEN is its ability to reduce programming efforts, whereby users do not need to program the devices for data collection and data processing. MOSDEN is mainly useful for crowd-sensing and opportunistic sensing applications and implies both push-based

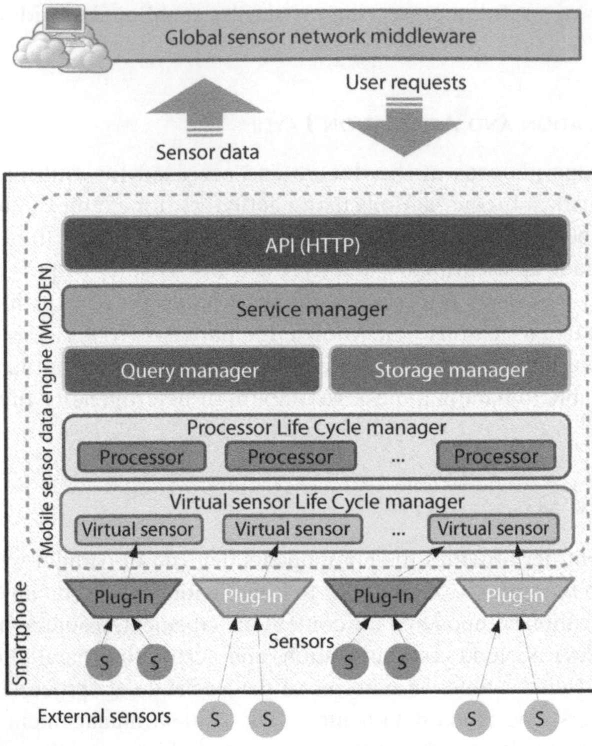

FIGURE 11.5 MOSDEN architecture.

and pull-based data streaming models. The essence of MOSDEN is the enablement of plug-in architecture so that developers can reuse, repurpose, and develop the plug-ins for interaction with hardware sensors. MOSDEN architecture is shown in Figure 11.5 and thoroughly presented in (Perera et al., 2014).

The plug-in layer in MOSDEN enables us to customize interfaces with different onboard and off-board sensors in IoT systems. The virtual sensor life cycle manager facilitates the creation of virtual sensors, that is, the software abstraction of each physical sensor, whereas the processor life cycle manager enables us to process the collected data streams and perform the data mining operations. MOSDEN also provides query manager and storage manager in order to manage the processed data streams. The service manager facilitates the connection with external device and cloud data centers. MOSDEN was tested in server and client modes and promising results were found in terms of energy consumption, efficient resource utilization, and scalability. It was observed that MOSDEN is suitable for data collection and processing in large-scale opportunistic sensing applications.

11.5.1.1 Achievements of MOSDEN in Connection with Speculated Architecture

The primary focus of MOSDEN design was on distributed processing, scalability, community-based development, and usability.

11.5.1.1.1 Distributed Processing

MOSDEN runs on a multitude of resource-limited IoT devices and supports distributed data processing in peer-to-peer settings (Perera et al., 2014). Since local data processing in IoT devices results in data reduction and reduced data communication, MOSDEN provides local analytics in IoT devices. In addition, it enables collaborative data processing in peer-to-peer networks. Using MOSDEN, multiple IoT devices can perform collaborative data processing without depending on remote and cloud-centric control for data processing.

11.5.1.1.2 Scalability

The plug-in architecture of MOSDEN results in the scaling up of the sensing and data processing operations. MOSDEN provides virtual sensor components that are used to program any type of data sources, whether they are physical or virtual data sources or onboard or off-board data sources. The MOSDEN plug-ins are easily installable and configurable at runtime. It was also observed during experiments that the MOSDEN plug-in uses about 25 KB of memory for an individual sensor; hence, a large number of sensors could be programmed and used even in resource-limited IoT devices. The strength of MOSDEN is its ability to remove unused plug-ins at runtime and also download new plug-ins from .apk files and Google Play. This approach not only enables a plug-and-play approach for scalability but also reduces the size of MOSDEN applications.

11.5.1.1.3 Community-Based Development

Hypothetically, MOSDEN design supports hundreds of thousands of data sources; therefore, MOSDEN architecture was designed to be generic and as simple as possible in order to engage the developers' community and ensure large-scale community-based development. To this end, MOSDEN application and samples source code is provided for developers. The developers just need to integrate the new data sources according to given guidelines. All the plug-ins are available at Google Play.

11.5.2 CARDAP Architecture

CARDAP enables mobile distributed data analytics (Khan et al., 2015). It works in a mobile cloud computing environment, where mobile devices perform local analytics for data reduction and cloud services further process the reduced data. CARDAP facilitates on-demand querying in local storage, separating sensing from analytics. Moreover, the issues pertinent to MOSDEN are dealt with by enabling smart processing in the mobile devices (Khan et al., 2015).

CARDAP architecture is based on the following five key components: (1) data stream capture, (2) analytics, (3) open mobile miner, (4) data sink, and (5) storage and query. The data stream capture component is based on virtual sensors and plug-ins. The virtual sensors capture data from nonphysical data sources, which include Internet-based and machine-resident data sources. On the other hand, the plug-ins provide the interface to acquire data from external data sources. CARDAP supports multiple onboard and off-board sensors and interacts with the IoT-enabled sensing components, for example, Raspberry Pi. The analytics component is based on the activity recognition component developed using StreamAR. The open mobile miner component provides lightweight data mining algorithms for CARDAP. The data sink component enables the uploading of data to external links, for example, cloud services, and finally, the storage and query component enables local data storage that can be accessed afterwards using the RESTful application program interface (API) over the Hypertext Transfer Protocol (HTTP). In addition, the authors proposed some cost models to evaluate the system performance. The cost models include the data transmission cost model (for data reduction) and the energy usage cost model (for energy consumption).

11.5.2.1 Achievements of CARDAP in Connection with Speculated Architecture

The CARDAP architecture facilitates distributed data processing and enables multiple data processing strategies. In addition, it enables data reduction at the device end.

11.5.2.1.1 Distributed Data Processing

Multiple devices can perform distributed data processing using peer-to-peer and device–cloud communication models. CARDAP clients are installed in mobile devices; however, nearer mobile devices or cloud servers are used as CARDAP servers. The data processing tasks are distributed among all participating devices and servers in client–server models. Primarily, client devices perform lightweight data processing using onboard computational resources and perform local data storage and query management for crowd-sensing applications (Figure 11.6).

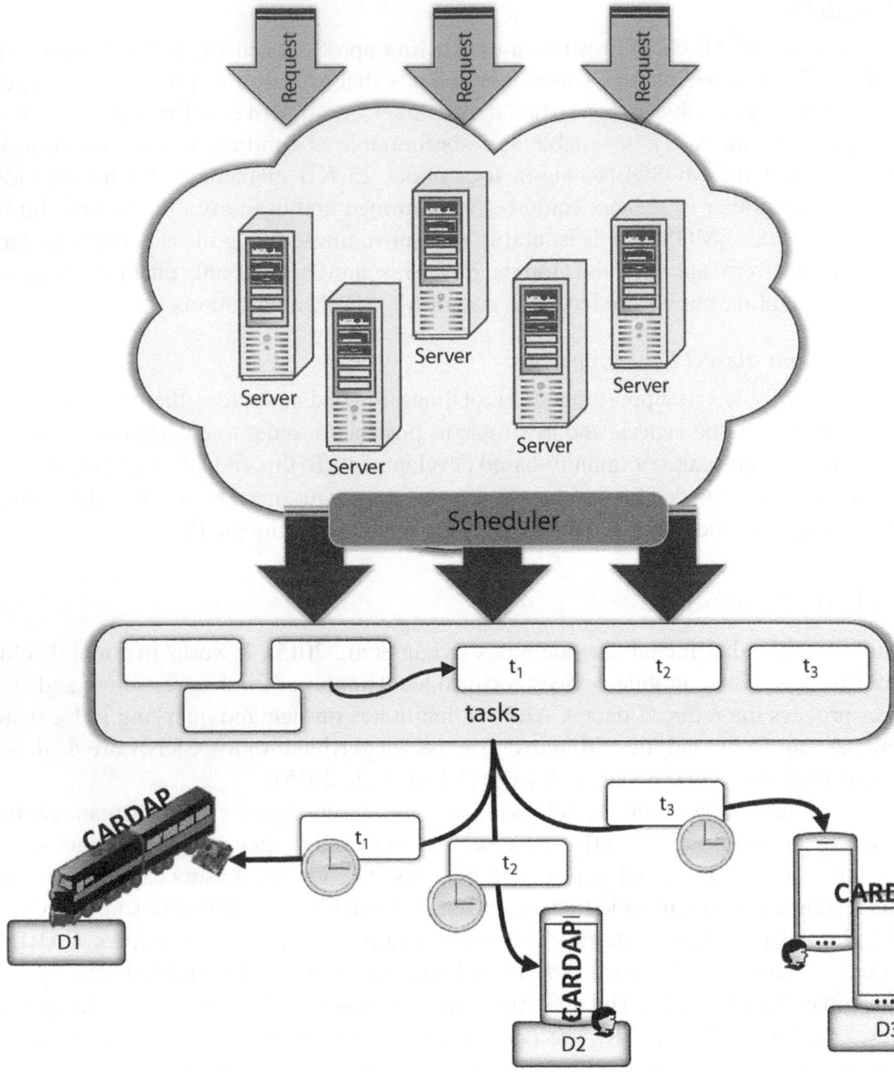

FIGURE 11.6 CARDAP architecture. (From Jayaraman, P.P., et al., presented at *Software: Practice and Experience*, 28–31, 2016, Prague, Czech Republic.)

11.5.2.1.2 Multiple Data Processing Strategies

CARDAP provides three data processing strategies. The first strategy works with a naïve approach, where all collected data is uploaded in the cloud in raw form. The second strategy is based on local analytics, where the mobile device performs local analytics and stores results locally for further queries. The third strategy is based on the combination of local analytics, data reduction, and opportunistic sensing, where the mobile device performs local analytics, stores results, and sends data when there is a significant change.

11.5.3 UniMiner Architecture

UniMiner is a three-tier architecture for distributed data stream mining in mobile edge cloud computing systems (Rehman et al., 2014a). The strength of UniMiner is its ability to support device-centric application execution in IoT systems. UniMiner is developed using a component-based application development approach in order to ease the programmability and extendibility of the proposed architecture. The UniMiner architecture works at three layers: (1) local analytics layer, (2) collaborative analytics layer, and (3) cloud-enabled analytics layer. Figure 11.7 presents components of UniMiner applications.

The local analytics layer mainly enables application components in the personal ecosystem of IoT devices. In addition, the control of application execution remains at the device end, and hence reduces dependability on a persistent Internet connection for device–cloud communication. The local analytics layer provides five modules for performing analytics operations using IoT devices. The data acquisition and adaptation module provides the functionalities to collect the data streams from various sensory and nonsensory data sources. In addition, this module enables us to control the volume and velocity of data streams in order to efficiently utilize onboard computational resources. The knowledge discovery module facilitates the performance of data preprocessing, data fusion, and data mining operations over continuously incoming streamlining data. The knowledge management module provides application components for knowledge integration and summarization. The visualization and actuation modules provide application components for on-screen data visualization, as well as data sharing in external environments, such as actuators in IoT systems, big data systems, and cloud data centers, to name a few. The system management module facilitates seamless application execution among mobile devices, mobile edge servers, and cloud data centers. The system management module provides components for context collection, resource monitoring, user profiling, and the offloading of computational tasks in edge servers and cloud data centers (Shuja et al., 2016).

The collaborative analytics layer provides components to perform device discovery operations to find adequate computational resources in peer mobile devices on the same LAN. In addition, it provides components for peer ad hoc network formation, data stream offloading, knowledge discovery, knowledge synchronization, and garbage collection operations in mobile edge servers. Finally, the cloud-enabled analytics layer of UniMiner provide cloud services to discover other services, data stream offloading, data preprocessing, data mining, knowledge management, and garbage collection for efficient storage resource utilization in cloud data centers. UniMiner's performance was tested using a real-world activity use case for activity detection. Further details about experimental evaluation are presented in (ur Rehman et al., 2016) for interested readers.

11.5.3.1 Achievements of UniMiner in Connection with Speculated Architecture

11.5.3.1.1 Distributed Data Processing

UniMiner provides a platform for distributed data processing among mobile devices and cloud computing systems. The architecture primarily utilizes onboard computational resources in mobile devices. However, it distributes the data processing tasks among other mobile devices in the locality, as well as remote cloud computing servers. Although efficient data processing in distributed settings is a challenging task, this approach helps in achieving maximum processing of streaming data in IoT environments.

11.5.3.1.2 Data Reduction

UniMiner ensures maximum data processing in mobile devices in order to minimize the data communication efforts in mobile cloud settings (Rehman et al., 2016; 2014b). In addition, the device-first strategy results in data reduction using mobile devices. The experimental evaluation revealed promising results, whereby UniMiner was able to reduce about 91% of the data stream in a single mobile device and about 98% of the data stream in multidevice settings.

11.5.3.1.3 Load Balancing

UniMiner enables load balancing strategies for efficient resource utilization in mobile devices and opportunistically utilizes computational resources from other mobile devices and cloud data centers. The load balancing strategies work on the basis of contextual information, the resource consumption of computational tasks, the availability of other devices for data processing in locality, and

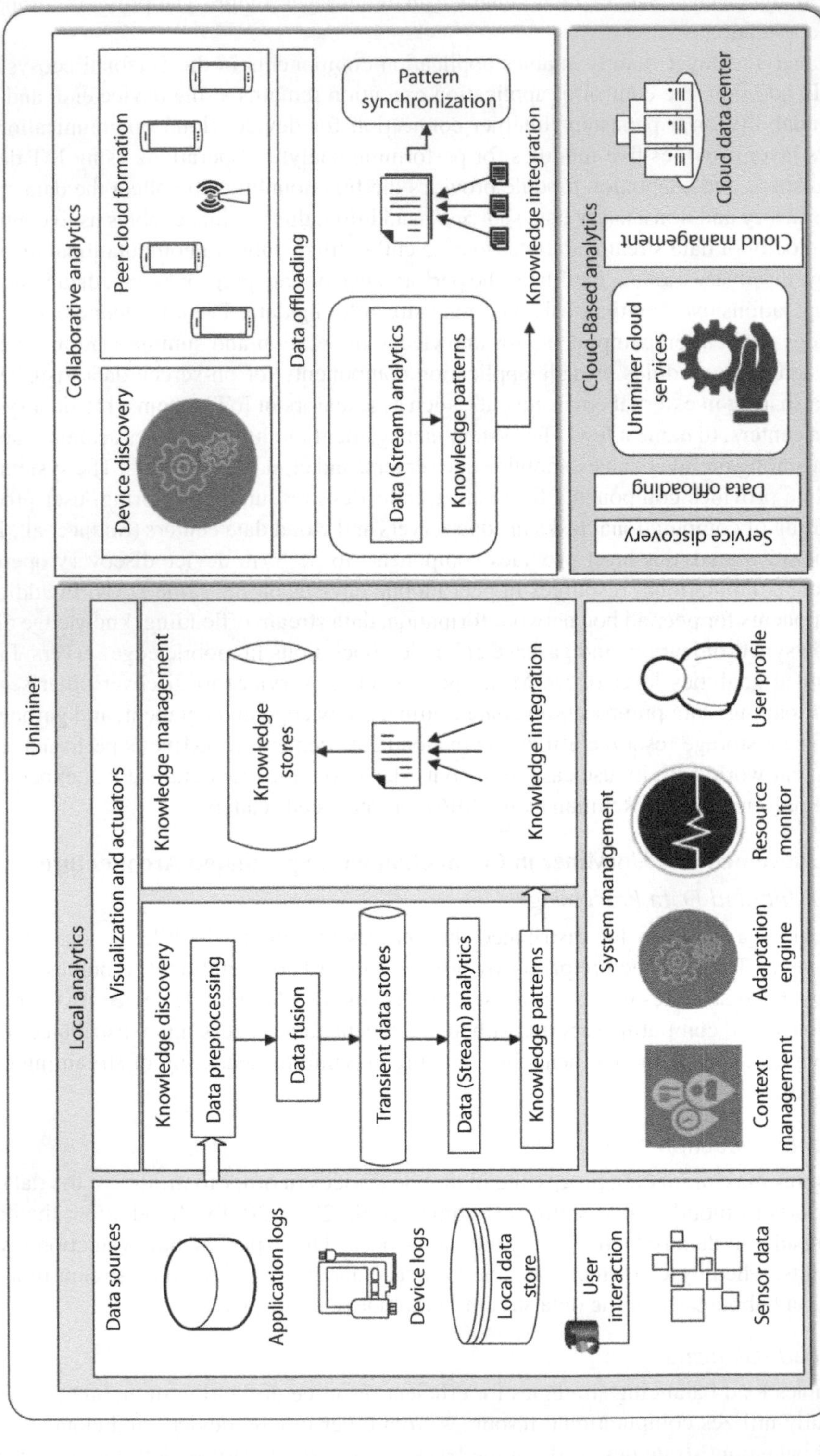

FIGURE 11.7 UniMiner architecture.

the availability of Internet connections. The main objective for load balancing strategies is to create a balance among energy efficiency, data reduction, and bandwidth utilization cost.

11.5.3.3.4 Device-Centric Data Analytics

UniMiner is the first study that enabled device-centric data analytics in a three-layer architecture. The architecture was designed to ensure maximum computational and storage resources for mobile devices in mobile cloud settings. In addition, the device-centric approach helps to delegate complete user control at the device end, whereby all decisions about data processing and switching among different layers are made by mobile devices.

11.6 CONCLUSION

The device-centric distributed IoT analytics systems utilize IoT devices and edge servers as primary application execution platforms. This approach not only reduces latency in real-time analytics applications but also lowers in-network data movement and cloud service utilization costs. In this chapter, we discussed the emergence of edge-centric distributed IoT analytics platforms. The speculated architecture defines the key components and opens new horizons for further exploration by future researchers. Although MOSDEN, CARDAP, and UniMiner represent the earlier investigations, much research effort is needed to fully realize the potential in this important research area. It is worth mentioning that device-centric distributed IoT analytics will lead toward the development of a new cloud-less and server-less ecosystem. Further, it will lead toward personalized, adaptive, and real-time analytics services for personal and community usage. Considering the commercialization and research-related opportunities, it is perceived that device-centric distributed IoT analytics systems will be a real game changer in both industry and academia.

REFERENCES

Basu, A. 2013. Five pillars of prescriptive analytics success. *Analytics Magazine*, 8–12.

Bonomi, F., et al. 2014. Fog computing: A platform for Internet of things and analytics. In *Big Data and Internet of Things: A Roadmap for Smart Environments*, 169–186. Berlin: Springer International Publishing.

Bonomi F., R. Milito, J. Zhu, and S. Addepalli. 2012. Fog computing and its role in the internet of things. In *Proceedings of the first edition of the MCC workshop on Mobile cloud computing*, Helsinki, Finland, August 2012, 13–16.

Daghighi, B., M. L. M. Kiah, S. Iqbal, M. H. Rehman, and K. Martin. 2017. Host mobility key management in dynamic secure group communication. *Wireless Networks*: 1–19.

Daghighi, B., M. L. M. Kiah, S. Shamshirband, and M. H. Rehman. 2015. Toward secure group communication in wireless mobile environments: Issues, solutions, and challenges. *Journal of Network and Computer Applications* 50: 1–14.

Delen, D. and H. Demirkan. 2013. Data, information and analytics as services. *Decision Support Systems* 55 (1): 359–363.

Fan, Y. J., Y. H. Yin, L. Da Xu, Y. Zeng, and F. Wu. 2014. IoT-based smart rehabilitation system. *IEEE Transactions on Industrial Informatics* 10 (2): 1568–1577.

Gubbi, J., et al. 2013. Internet of things (IoT): A vision, architectural elements, and future directions. *Future Generation Computer Systems* 29 (7): 1645–1660.

Haghighi, P. D., S. Krishnaswamy, A. Zaslavsky, M. M. Gaber, A. Sinha, and B. Gillick. 2013. Open mobile miner: A toolkit for building situation-aware data mining applications. *Journal of Organizational Computing and Electronic Commerce* 23 (3): 224–248.

Hassan, Q. F. 2016a. *Innovative Research and Applications in Next-Generation High Performance Computing*. Hershey, PA: IGI Global.

Hassan, Q. F. 2016b. An outlook into novel concepts of high performance computing. In *Innovative Research and Applications in Next-Generation High Performance Computing*, xxiii–lii. Hershey, PA: IGI Global.

Jayaraman, P. P., J. B. Gomes, H. L. Nguyen, Z. S. Abdallah, S. Krishnaswamy, and A. Zaslavsky. 2014. Cardap: A scalable energy-efficient context aware distributed mobile data analytics platform for the fog. In *East European Conference on Advances in Databases and Information Systems*, Springer, Prague, Czech Republic, 7 September 2014, 192–206.

Jayaraman, P. P., C. Perera, D. Georgakopoulos, S. Dustdar, D. Thakker, and R. Ranjan. 2017. Analytics-as-a-service in a multi-cloud environment through semantically-enabled hierarchical data processing. *Software: Practice and Experience*, 47 (8): 1139–1156.

Khan, A. U. R., M. Othman, A. N. Khan, S. Akhtar, and S. A. Madani. 2015. MobiByte: An application development model for mobile cloud computing. *Journal of Grid Computing* 13 (4): 605–628.

Khan, A. U.R., M. Othman, S. A. Madani, and S. U. Khan. 2014. A survey of mobile cloud computing application models. *IEEE Communications Surveys & Tutorials* 16 (1): 393–413.

Lea R. and M. Blackstock. 2014. City hub: A cloud-based iot platform for smart cities. In *2014 IEEE 6th International Conference on Cloud Computing Technology and Science (CloudCom)*, Singapore, 15 December 2014, 799–804.

Mavromoustakis, C. X., G. Mastorakis, and J. M. Batalla. 2016. *Internet of Things (IoT) in 5G Mobile Technologies*. Berlin: Springer.

Mukherjee, A., H. S. Paul, S. Dey, and A. Banerjee. 2014. Angels for distributed analytics in iot. In *2014 IEEE World Forum on Internet of Things (WF-IoT)*, Seoul, South Korea, 6–8 March, 565–570.

Perera, C., P. P. Jayaraman, A. Zaslavsky, D. Georgakopoulos, and P. Christen. 2014. Mosden: An internet of things middleware for resource constrained mobile devices. In 2014 *47th Hawaii International Conference on System Sciences (HICSS)*, Waikoloa, HI, 6 January 2014, 1053–1062.

Perera, C., A. Zaslavsky, P. Christen, and D. Georgakopoulos. 2014. Sensing as a service model for smart cities supported by Internet of things. *Transactions on Emerging Telecommunications Technologies* 25 (1): 81–93.

Rehman, M. H., A. Batool, and A. R. Khan. 2016. Big data analytics in mobile and cloud computing environments. In *Innovative Research and Applications in Next-Generation High Performance Computing*, 349–367. Hershey, PA: IGI Global.

Rehman, M. H., C. S. Liew, and T. Y. Wah. 2014a. UniMiner: Towards a unified framework for data mining. In *Fourth World Congress on Information and Communication Technologies (WICT)*, Bandar Hilir, Malaysia, December 2014, 134–139.

Rehman, M. H., C. S. Liew, and T. Y. Wah. 2014b. Frequent pattern mining in mobile devices: A feasibility study. *Presented at the 2014 International Conference on Information Technology and Multimedia (ICIMU)*, Putrajaya, Malaysia.

Rehman, M. H., C. S. Liew, T. Y. Wah, J. Shuja, and B. Daghighi. 2015. Mining personal data using smartphones and wearable devices: A survey. *Sensors* 15 (2): 4430–4469.

Rehman, M. H., V. Chang, A. Batool, and Y. W. Teh. 2016. Big data reduction framework for value creation in sustainable enterprises. *International Journal of Information Management* 36 (6): 917–928.

Rehman, M. H. and A. Batool. 2015. The concept of pattern based data sharing in big data environments. *International Journal of Database Theory and Application* 8(4): 11–18.

Satyanarayanan, M., et al. 2015. Edge analytics in the Internet of things. *IEEE Pervasive Computing* 14 (2): 24–31.

Sherchan W., P. P. Jayaraman, S. Krishnaswamy, A. Zaslavsky, S. Loke, and A. Sinha. 2012. Using on-the-move mining for mobile crowdsensing. In *2012 IEEE 13th International Conference on Mobile Data Management (MDM)*, Bengaluru, Karnataka, India, 23 July 2012, 115–124.

Shuja, J., A. Gani, M. H. Rehman, E. Ahmed, S. A. Madani, M. K. Khan, and K. Ko. 2016. Towards native code offloading based MCC frameworks for multimedia applications: A survey. *Journal of Network and Computer Applications* 75: 335–354.

ur Rehman M. H., C. Sun, T. Y. Wah, A. Iqbal, P. P. Jayaraman. 2016. Opportunistic computation offloading in mobile edge cloud computing environments. In *2016 17th IEEE International Conference on Mobile Data Management (MDM)*, Porto, Portugal, 13 June 2016, vol. 1, 208–213.

Waller, M. A. and S. E. Fawcett. 2013. Data science, predictive analytics, and big data: A revolution that will transform supply chain design and management. *Journal of Business Logistics* 34 (2): 77–84.

Wilkerson, G. and A. Gupta. 2016. Sports injuries and prevention analytics: Conceptual framework & research opportunities. *Presented at AMCIS 2016 Proceedings*, San Diego, Chili.

Part IV

IoT in Critical Application Domains

12 The Internet of Things in Electric Distribution Networks

Control Architecture, Communication Infrastructure, and Smart Functionalities

Qiang Yang, Ali Ehsan, Le Jiang, Hailin Zhao, and Ming Cheng

CONTENTS

12.1 Introduction 231
12.2 Current Control and Communication Provision in DNOs 233
12.3 AuRA-NMS-Based Electric IoT Architecture 236
12.3.1 Conceptual Architecture 236
12.3.2 IoT Framework for Distributed Control in AuRA-NMS 237
12.3.2.1 Unification of System Information and Standards 238
12.3.2.2 Distributed Intelligence and Function Integration 238
12.3.2.3 Open ICT Paradigm Living with Legacy System 239
12.4 Communication Standards, Protocols, and Requirements of Electric IoT 239
12.4.1 Communication Standards, Protocols, and Technologies 239
12.4.2 Communication Infrastructure Requirements 242
12.4.2.1 Timely Data Delivery and Differentiation 243
12.4.2.2 Data Availability, Robustness, and Redundancy 243
12.4.2.3 Flexibility, Scalability, and Interoperability 243
12.4.2.4 Communication Security 244
12.5 Case Studies 244
12.5.1 Case Study: 33 kV Meshed Networks 245
12.5.2 Case Study: 11 kV Radial Networks 247
12.6 Conclusions 250
Acknowledgments 251
References 251

12.1 INTRODUCTION

In recent decades, the advances in distributed renewable generation technology and environmental considerations have significantly reshaped the structure of power generation, transmission, and distribution. There is an increasing penetration of various forms of renewable energy sources (e.g., wind turbines, combined heat and power [CHP], and solar energy) in current medium-voltage power distribution grids. The UK government aims to provide 15% of national electricity supplies based on renewable energy by 2020, implying about 21 GW of generation from current medium-voltage distribution grids (e.g., 33 and 11 kV). Distributed generation is an approach that adopts small-scale technologies to produce electricity close to the end users of power. It can be defined as a variety of electrical power sources and technologies with limited capacity that can be directly connected to the distribution network and consumed by the end users. Distributed generators (DGs) may come from renewable sources like small-scale wind turbines, photovoltaic (PV) panels, micro-hydro systems, fuel cells, and biomass. Conventional DGs may include micro gas turbines, diesel engines, sterling engines, and internal combustion reciprocating engines. In many cases, DGs can provide lower-cost electricity and higher power reliability and security, with fewer environmental consequences, than traditional power generators. The DGs can generate the power and supply the customers locally without a long-distance power transmission and distribution process, which can effectively reduce the peak demand and minimize the network congestion from the centralized power utilities, as well as yield additional revenue (Lopes et al., 2007). However, it is known that a massive DG integration in medium- and low-voltage levels can introduce tremendous challenges, mainly due to the intermittent generation of renewable sources and limited available network monitoring and control functionalities. As a result, the distribution grid is no longer a passive system, but an active system interconnecting power generators and loads with bidirectional power flows and complex operational phenomena, for example, voltage rise effect, increased fault level, protection degradation, and altered transient stability (Lopes et al., 2007; Maurhoff, 2000). Most of the current distribution networks are managed via a centralized control at a control center of a distribution network operator (DNO) relying on supervisory control and data acquisition (SCADA) systems designed for the purpose of simple network operations. Due to the large geographical scale of distribution grids and the increasing population of DGs, such overall central control becomes inefficient, and even not practical. Therefore, a novel active network management (ANM) solution is required to maximize the DG connection capacity with acceptable cost penalties.

The power industry can greatly benefit from the IoT technology, which has the potential to reorient smart grids from different aspects. The IoT can be adopted to realize the desired benefits of the smart grid technology, such as energy conservation and cost reduction. This indicates that consumers, manufacturers, and utilities have to find novel ways to efficiently manage and control the system components. Over the past few years, a collection of research work, for example, "IntelliGrid"* and "SmartGrids"†, has been made to facilitate the smart power networks. The autonomous regional active network management system (AuRA-NMS n.d.)‡ is exploited as a cost-effective ANM solution by implementing distributed and intelligent active network control to enhance energy security and quality of supply. The key idea behind this IoT-based management solution is to devolve the management authority from a DNO control center to networked regional controllers deployed in the field to carry out management tasks in either an autonomous or cooperative fashion. In parallel, current advances of IoT create new opportunities for more direct integration of physical and computer-based systems, resulting in improved efficiency and economic benefits in system management. This motivates the exploitation of IoT-based approaches in the context of the AuRA-NMS project to carry out the scalable and intelligent management of energy networks.

This chapter exploits the IoT application in electric power distribution networks and focuses on the aspects of control architecture, communication infrastructure, and smart functionalities. The remainder of this chapter is organized as follows: Section 12.1 presents the background of utilizing IoT in electric power utility. Section 12.2 overviews the current communication provision and management approaches in DNOs. Section 12.3 explains the IoT-based management scheme for distribution networks. Section 12.4 discusses the related technologies, standards, and protocols. Section 12.5 carries out two case studies for meshed and radial distribution networks. Finally, Section 12.6 provides some conclusive remarks.

* IntelliGrid, Smart Power for the 21st century. Available at http://smartgrid.epri.com/IntelliGrid.aspx.

† SmartGrids, Vision and strategy for Europe's electricity networks of the future. Available at http://www.smartgrids.eu.

‡ Autonomous regional active network management system (AuRA-NMS). Available at http://gow.epsrc.ac.uk/NGBOViewGrant.aspx?GrantRef=EP/E003583/1.

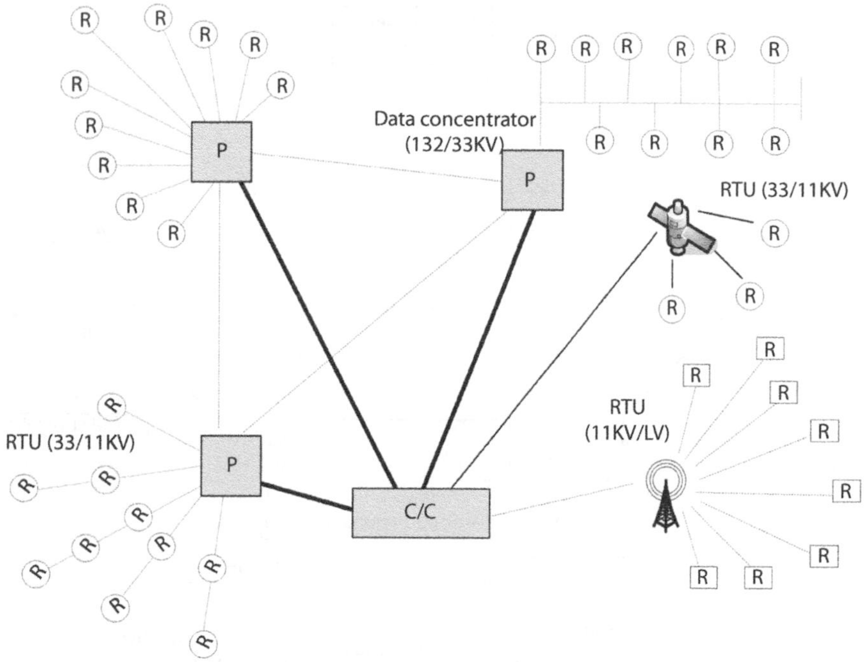

FIGURE 12.1 Current DNO's SCADA communication provision. C/C, DNO's control center; P, data concentration point; R, RTU; LV, low voltage.

12.2 CURRENT CONTROL AND COMMUNICATION PROVISION IN DNOS

In most current distribution networks, the system monitoring and control is mostly implemented by the use of a SCADA system in DNOs. The SCADA system generally consists of a master terminal (at a DNO's control center) and a large number of remote telemetry units (RTUs) located at geographically dispersed sites. Figures 12.1 and 12.2 illustrate the communication infrastructure provision and the SCADA system of current DNOs, respectively. These RTUs are able to collect network measurements from field sensors (analogue and digital) and deliver commands to control devices. All RTUs are connected together to the DNO's control center via heterogeneous communication channels with a diverse physical medium (e.g., leased digital fibers, private pilot cables, public switched telephone network [PSTN] lines, and radio, satellite, and mobile cellular networks) and limited link capacities, from a few hundred bits per second to a few thousand bits per second. These characteristics mean frequent communication media conversions, with an undesirable impact on their end-to-end availability and average bit error rate (BER) (Hauser et al., 2005). Furthermore, SCADA protocols are mostly proprietary and designed specifically with error detection and message retry mechanisms to guarantee data delivery under most circumstances. With respect to the communication topology, RTUs at lower-voltage sites (33/11 kV) are organized in a multidropped structure and connected to the data concentrators at higher-voltage sites (132/33 kV), where the channels are semimeshed with better robustness and reliability (e.g., triangulation and duplicated routes). At a lower-voltage level, DNOs may also directly connect remote RTUs with the control center in a star topology through point-to-point communication channels by using the communication technology that can cover a large geographical area (e.g., satellite or wireless mobile network).

It can be observed that SCADA communication systems in most DNOs are generally designed and implemented with a centralized architecture. In such an architecture, all RTUs deployed in the power distribution network sites send their up-to-date information (analogue and digital) through the

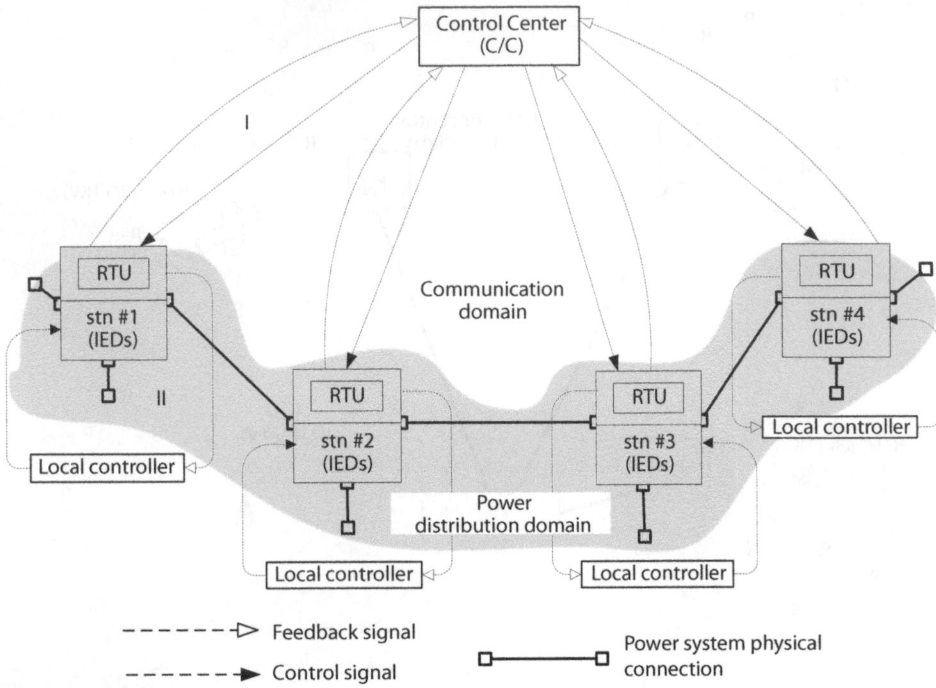

FIGURE 12.2 SCADA system and physical power distribution network.

underlying communication infrastructure to the control center periodically (e.g., every 10–20 s) to capture the network operation state. In the control center, these data are analyzed based on the system software programs (e.g., fast simulation platform) to detect any operating conditions (e.g., undervoltage or faults) that may require a corrective action (e.g., operating an on-load tap changer [OLTC]). However, such a long distance for the communication channel and limited bandwidth can result in substantial large data delivery latency in both directions, and hence cannot effectively support the advanced control and management functionalities with stringent real-time requirements (Hauser et al., 2005; Roberts, 2004).

As a response to the observation of communication inadequacy for a centralized control paradigm, DNOs have also adopted local control schemes, where management functions are carried out in network subsections, such as covering one or more substations. The controllers in the local control scheme merely monitor and control the devices in their own areas without the consideration of impacts on peer subareas during their operation. In addition, the control actions are mostly dictated by predefined hard-coded logic designed for some very specific operation conditions. Thus, the control functionalities are restricted to a small scope and are inflexible, and hardly lead to a globally efficient and economical operation.

Figures 12.3 and 12.4 present the communication infrastructure of the existing SCADA system in 33 and 11 kV power distribution networks, respectively. RTUs act as relays to collect network operational states from sensors (analogue and/or binary) and route control signals to actuating devices. In medium- and low-voltage distribution networks, the data acquisition is often carried out through a polling mechanism in a noncontinuous fashion, for instance, every 10–20 s at higher-voltage sites (e.g., 33 kV) and hours or even days at lower-voltage sites (e.g., 11 kV). Meanwhile, these RTUs may also deliver event-driven data, for example, field alarms. At the DNO control center, the collected operational state data are analyzed to detect anomalous conditions (e.g., undervoltage) that may require corrective actions (e.g., operating an OLTC), and the control center will send relevant control signals to remote network elements. In this way, DNOs use periodic polling and event-driven

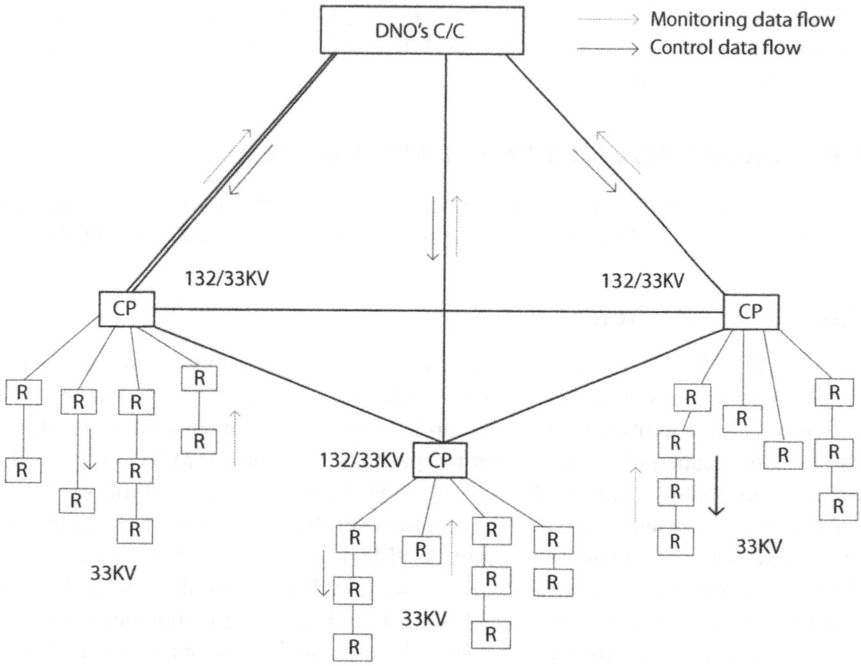

FIGURE 12.3 Communication infrastructure of existing SCADA system in current UK power distribution networks: 33 kV mesh network. C/C, DNO's control center; CP, concentration point; R, RTU.

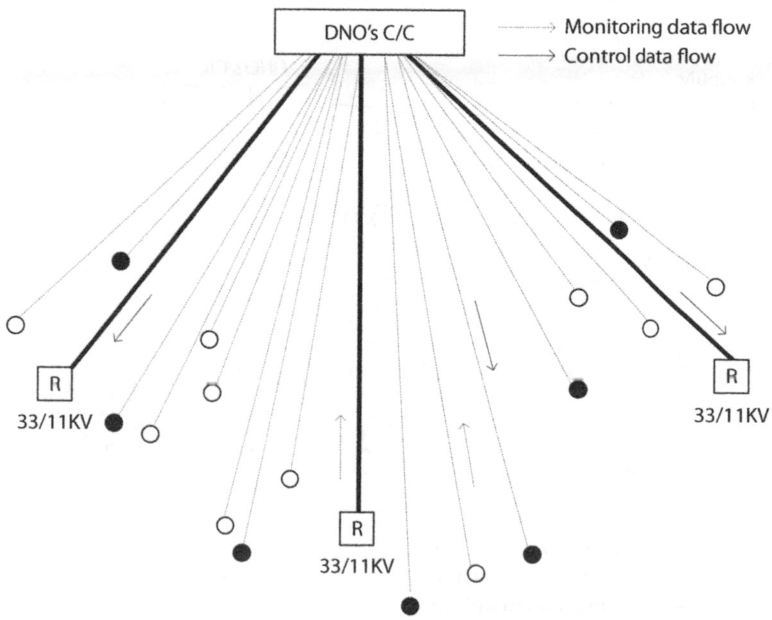

FIGURE 12.4 Communication infrastructure of existing SCADA system in current UK power distribution networks: 11 kV radial network. C/C, DNO's control center; R, RTU.

messages to ascertain the state of network operation. Currently, most SCADA systems still lack sensing and control capabilities at low-voltage levels (e.g., 11 kV), which makes their service largely dependent on manual operation.

12.3 AURA-NMS-BASED ELECTRIC IOT ARCHITECTURE

The AuRA-NMS management scheme takes advantage of the IoT paradigm to carry out the network management and control functionalities in an autonomous and cooperative fashion.

12.3.1 Conceptual Architecture

In recent years, AuRA-NMS* was investigated as a cost-effective solution through implementing distributed and intelligent ANM. This was implemented through an approach that devolves current centralized control functionalities from the control center to a set of regional controllers that are connected to communication channels across the power distribution grid to carry out distributed decision making and control tasks. In this solution, DNO's control center could still administrate these regional controllers when necessary to adapt the overall operation to meet certain objectives.

Figure 12.5 schematically illustrates the AuRA-NMS-based network management principle by using the 33 kV power distribution network as an example. It is shown that the 33 kV distribution network is partitioned into a set of regions (I, II, and III) and deployed with underlying communication infrastructure. Hardware controllers are installed in individual control regions, and they are able to access their regional devices to conduct monitoring and control through communication channels. Between regional controllers, communication channels (dashed lines) are established to enable their information exchange for collaborative decision making and operations in a larger scope, covering three regions. The communication within the region and among the controllers may operate over the existing or upgraded SCADA system, but elsewhere, new communication channels need to be provided among the peer regional controllers. Currently, a collection of wired (e.g., power line carrier and optical fiber) and wireless (e.g., private licensed or unlicensed radio, cellular and satellite) communication technologies are available to provide these channels, which need to be carefully

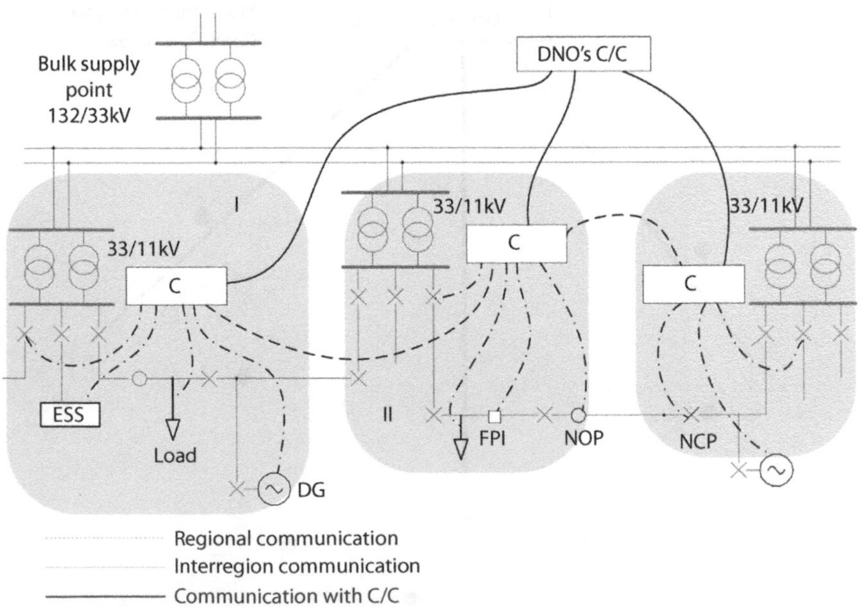

FIGURE 12.5 AuRA-NMS-based network management system in a 33 kV network. C, regional controller, C/C, DNO's control center; FPI, fault passage indicator; NOP, normally open point; NCP, network connection point.

selected and dimensioned based on cost–benefit analysis. In fact, the AuRA-NMS-based network management paradigm provides an IoT framework. The distributed networked controllers provide the computation and communication platform for the execution of a set of ANM functions, such as maintaining the voltage profile and managing faults. This approach effectively transforms the overall centralized control system into a number of autonomous and coordinated subsystems:

- *Autonomous local control*: The network monitoring and control actions are taken locally by the regional controllers in their own controlled regions. In the case that the scope of the anomalous condition does not exceed the regional boundary, the controller will carry out local control, for example, changing the transformer tap position or changing the generator load reference set point.
- *Coordinated wide area control*: In the case that the detected anomalous conditions could affect adjacent regions, the controllers in these involved regions that are only with partial view of their own regions could cooperate in finding suitable management solutions in a wider scope.

Such a distributed regional management approach aims to manage the active distribution grids in a timely and accurate manner, but not surprisingly, this imposes an obvious requirement of an efficient and reliable information and communication technology (ICT) system to support the local management, as well as the interactions among peer controllers. Current SCADA systems in the UK DNOs were mostly built several decades ago with a master–slave polling mechanism with limited channel bandwidth (from a few hundred to a few thousand bits per second) and proprietary communication protocols. In recent years, some ICT advances in the power industry have been reported that suggest a migration from current SCADA systems to an open IP-based architecture (e.g., (Mak and Holland, 2002; McClanahan, 2003; Hauser et al., 2005)), and the related issues of migration are also discussed, such as TCP not quite being suitable for system operation monitoring due to its unpredictable performance with nondeterministic latency, and a User Datagram Protocol (UDP)–based protocol is proposed for the monitoring system. This section proposes and studies the IP-based ICT systems and uses standard TCP/IP and UDP/IP transport protocols to carry different types of data traffic as appropriate. Previous studies highlighted the feasible communication architecture, requirements, and standards to facilitate AuRA-NMS, and pointed out the deficiency of current SCADA in supporting such a control mechanism (Yang et al., 2009a,b).

In such an IoT-based regional management solution, a variety of network control algorithms can be developed and expected to run over the networked regional controllers to conduct a range of power network control tasks, voltage control, automated restoration, and power flow management. The mechanism of local management and information exchange between controllers is determined by the designed control algorithms. While the control actions of the local controllers are not dictated by the DNO's control center via long-distance communication, the latter can also administrate the local control and wider area coordination when necessary to adapt the overall power network operation to meet a desired operational goal.

In summary, the regional control management system supports local autonomous network management in individual regions, as well as collaborative decision making when necessary through distributed intelligence across the networked controllers. In comparison with the conventional centralized control via the SCADA system, such a method aims to provide DNOs with a cost-effective tool to manage their networks in a more timely and accurate manner with significantly improved flexibility, robustness, and scalability to meet the operational challenges due to the massive penetration of DGs.

12.3.2 IoT Framework for Distributed Control in AuRA-NMS

This section presents the primary considerations of distributed control in AuRA-NMS and highlights three key aspects of designing an IoT framework for the smart control of electric medium-voltage power distribution grids.

12.3.2.1 Unification of System Information and Standards

A unification in information representation and standards across the power distribution network is essential to enable communication interoperability. The International Electrotechnical Commission (IEC) has made tremendous efforts to address this issue. IEC 61970-301 (IEC, 2003) and its extension, IEC 61968-11 (IEC, 2010), provide a standard for describing the power network elements and their interrelationships at an electrical level. They are collectively known as the Common Information Model (CIM), which aims to improve the interoperability between energy management systems (EMSs) from different vendors. In parallel, IEC 61850 (IEC, 2005) has been widely accepted for electric substation automation. It introduces the substation Ethernet to provide a fast communication platform and models substation equipment and functions as abstract objects. This can significantly improve the interoperability among connected intelligent electronic devices (IEDs). The recent effort attempts to expand the scope of IEC 61850 for wide area communication, for example, substation to substation and substation to control center (Brunner, 2008a). It can be envisaged that IEC 61850 can be used as a standard for network monitoring and make the information available to the control center or any other management system. The main research effort is to harmonize these standards, particularly CIM and IEC 61850 (e.g., (EVERIS-CIM, 2009)). Such efforts enable the monitoring, control, or protection applications developed on different platforms to represent information in a consistent format and lead to seamless communication throughout the various voltage levels across the overall distribution grid.

12.3.2.2 Distributed Intelligence and Function Integration

In AuRA-NMS, a variety of control functionalities can be integrated, and in particular, three control functions are investigated: voltage control, automated supply restoration, and power flow management (Davidson and McArthur, 2007). These control algorithms running at the networked regional controllers obtain power network measurements in their regions periodically and carry out control actions on demand. In the case of high penetration of renewable DGs, the network voltage profile needs to be efficiently maintained within the regulated limits under various operational conditions (e.g., load variation and intermittent output of DGs) by a set of control measures (e.g., OLTC control, DG power factor control, DG real power curtailment control, or a combination of these). On the other hand, fault management is of paramount importance to improve the reliability and quality of the power supply of an active power distribution network. The power supply restoration function aims to restore power supply to as many customers as possible as quickly as possible after a fault while meeting certain operational criteria (e.g., within feeder/switch ratings or minimum switching operations). Finally, the power flow management function aims to meet a set of operational constraints, for example, within thermal limits at a given level of distributed generation and contractual constraint, by operating switches, a charge/discharge energy storage system (ESS), a trip/trim DG, or a shed load. These control functions need to be efficiently operated over the hardware platform. In our suggested management solution, agent technology is adopted as a software platform to integrate and distribute these functionalities to achieve "plug-and-play" management (Davidson et al., 2008; Taylor et al., 2008), rather than mapping the agents to physical power network elements (e.g., DGs and bus bars). The JADE (JAVA Agent Development Framework, n.d.) platform is a well-known software framework for the development of intelligent distributed agents based on the Java programming language. The platform supports coordination between several agents, facilitates the communication between agents, and allows the services detection of the system. JADE has been considered an efficient platform and is widely adopted in the literature, as it provides various debugging tools, the mobility of code and content agents, and the possibility of parallel execution of the behavior of agents, as well as support for the definition of languages and ontologies. Figure 12.6 shows that three different functions (*a*, *b*, *c*) are encapsulated: Foundation for Intelligent Physical Agents (FIPA)–compliant (FIPA, 2002a) agents ($a_1 \sim a_n$), ($b_1 \sim b_n$), and ($c_1 \sim c_n$) over a Java Agent Development Framework platform, able to cooperate through exchanging Agent Communication Language (ACL) messages

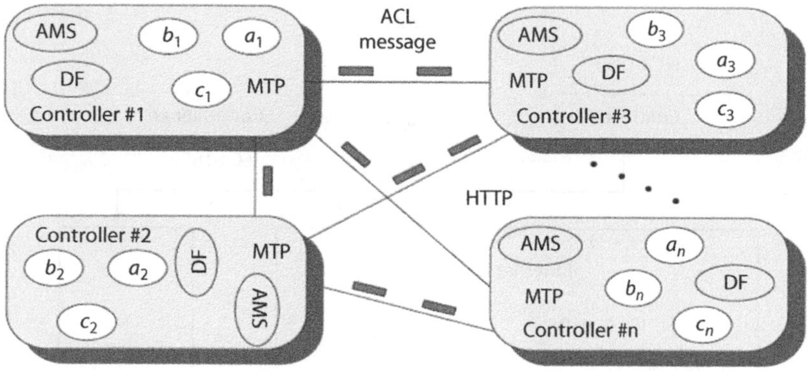

FIGURE 12.6 Function distribution and integration in a federated JADE platform.

via a Message Transport Protocol (MTP) (e.g., Hypertext Transfer Protocol [HTTP], TCP, or IP). The FIPA-compliant agent platform includes the Agent Management System (AMS), the Directory Facilitator (DF), and the Agent Communication Channel (ACC), which are automatically activated at the agent platform start-up. Such cooperation of multiple agents enhances the system robustness, as the platform could still operate when some intercontroller communication channels fail.

12.3.2.3 Open ICT Paradigm Living with Legacy System

It is vital to integrate the proposed IoT-based solution with the existing legacy management and communication system. As most of the DNOs operate their SCADA systems for a few decades with proprietary protocols and limited bandwidth, the existing infrastructures are not capable of supporting timely and rich data communication and the adoption of recent technology advances (e.g., Ethernet and TCP/IP) is required. The information exchange in an ANM system is expected to be based on a standard data model (e.g., IEC 61850). However, at present, IEC 61850–compliant devices and devices running proprietary protocols coexist in substations. Properly interfacing with the legacy system is crucial to provide system backward compatibility and transparent communication. One of the viable solutions is to integrate a protocol converter through a gateway into the substation SCADA system. Such conversion can be implemented through IEC 61850 and Object Linking and Embedding (OLE) for Process Control (i.e., OPC) technology by defining the mapping between IEC 61850 and the legacy protocols. The IEC 61850–complaint devices can be directly connected to the substation local area network (LAN). As a result, regional controllers that execute ANM control algorithms can communicate with all substation monitoring, protection, and control elements. Such a solution, as shown in Figure 12.7, provides DNOs a graceful roadmap toward ANM.

12.4 COMMUNICATION STANDARDS, PROTOCOLS, AND REQUIREMENTS OF ELECTRIC IoT

The issues of communication standards and protocols to facilitate IoT services in electric power distribution networks need to be discussed.

12.4.1 Communication Standards, Protocols, and Technologies

A unified data modeling approach across the distribution networks is essential to enable communication interoperability in NMS. To achieve this objective, many standardization efforts have been made by the IEC, notably IEC 61970, IEC 61968, and IEC 61850. In recent years, tremendous efforts have been made to address this issue, and two standards have been recommended: IEC 61970 (i.e., the CIM) and IEC 61850. CIM (IEC, 2010) is a semantic model, originally developed

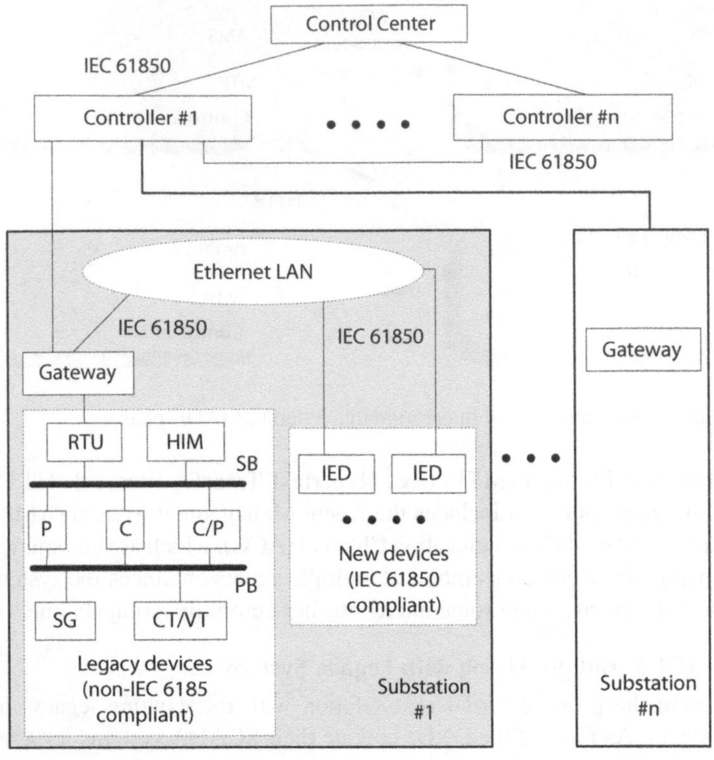

FIGURE 12.7 Interfacing with legacy devices in substations. HIM, human interfacing machine; P, data concentration point; C, regional controller; C/P, concentration point; SG, switchgear; CT/VT, current/voltage transformer.

to improve interoperability between EMSs from different vendors. CIM is object oriented and provides a consistent data definition and structure to describe different elements in power networks, and thus enables applications developed on platforms to represent and share information in a standard format. The IEC 61850 standard IEC (International Electrotechnical Commission), 2005 covers data model and communication protocols aiming to enhance interoperability and fast communication between substation IEDs from different vendors for substation automation. Like CIM, IEC 61850 is also object oriented and describes control devices and their data as abstract objects. Currently, efforts are being made to harmonize CIM and IEC 61850 models in power distribution networks (e.g., (Brunner, 2008a)), which will potentially enable seamless data communication between the substations and the control center. IEC 61850 specifies a set of generic abstract services that can be implemented by state-of-the-art communication protocols (e.g., Ethernet and TCP/IP). In addition, IEC 61850 abstract data models can be mapped to a number of application-level protocols: Manufacturing Message Specification (MMS), Generic Object Oriented Substation Event (GOOSE), and Sampled Measurement Values (SMV), which can run over TCP/IP systems or substation LANs using high-speed switched Ethernet, as shown in Figure 12.8. In regional network management systems, IEC 61850 is considered a standard to support the communication among all network devices across the entire power distribution network.

In the proposed IoT-based management architecture, the coordination of networked controllers can be implemented through multiagent system (MAS) techniques to obtain plug-and-play management. A more comprehensive discussion of using MAS technology in electric power systems can be found in (McArthur et al., 2007a) and (McArthur et al., 2007b). FIPA (2002a), as the de facto standard, is generally recommended for the multiagent technology, where a set of subsidiary FIPA standards cover different perspectives: agent language (FIPA-ACL (2002b)), message encoding (e.g., XML

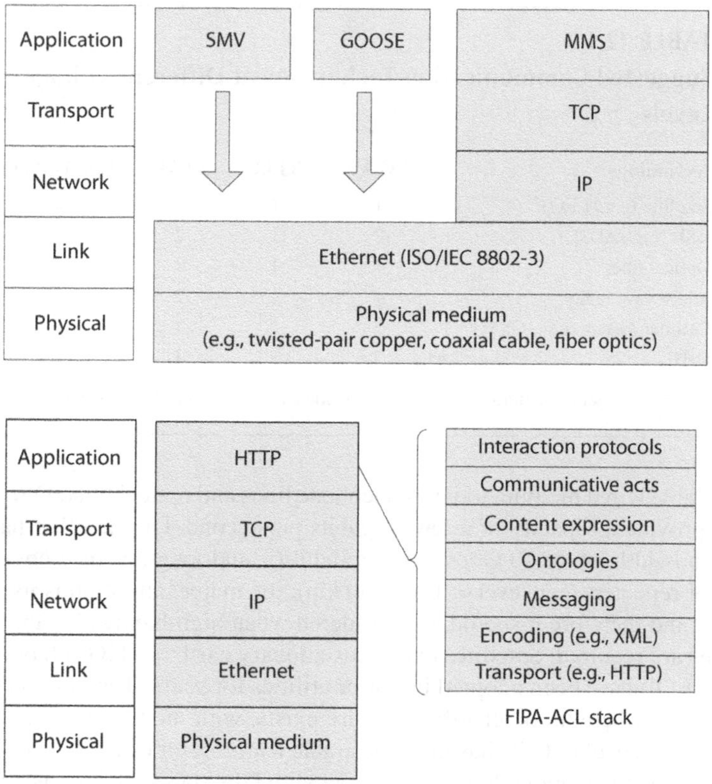

FIGURE 12.8 IEC 61850 standard and its service mapping, and FIPA standard and its relation to the TCP/IP suite.

(FIPA (Foundation for Intelligent Physical Agents), n.d.)), and message transport protocols (e.g., HTTP (FIPA, 2002c)), as shown in Figure 12.6. In regional control, the control function software agents located at a controller are expected to be able to communicate with other software agents across one or many other controllers in real time using HTTP, TCP, and IP transport protocols.

It should be noted that DNOs are often conservative when migrating from current proprietary protocols to TCP/IP for two major reasons (Birman et al., 2005): (1) TCP exhibits unpredictable performance with nondeterministic latency, and (2) open TCP/IP systems are more vulnerable to attacks. In recent years, many proposals (e.g., (McClanahan, 2003; Hauser et al., 2005)) aim to remove these concerns to utilize TCP/IP networking in power networks, and many current standard protocols, for example, Modbus and DNP3.0, contain extensions to be operated over TCP/IP and UDP/IP. Most of these efforts in the literature focus on utilizing TCP/IP technology to upgrade DNO's SCADA from a closed, proprietary system to an open, standardized one. In the suggested IoT management framework, TCP/IP can be adopted as an underlying technology to underpin the standard manner (e.g., IEC 61850 and FIPA) of data exchange and agent-based decision making among the distributed regional controllers.

Another key aspect of communication design in IoT-based management is to deploy suitable communication technologies. Previous studies (e.g., (Marihart, 2001; Egea-Lopez et al., 2005; Gungor and Lambert, 2006; Dhar and Tang, 1998)) provide a comprehensive study of communication technology for industrial applications. Among them, a number of technologies are particularly promising to be adopted in the regional ANM system, both wired medium (e.g., digital subscriber line [DSL] and optical fiber) and wireless medium (e.g., satellite, ultra-high-frequency [UHF] radio, microwave radio, and cellular system).

TABLE 12.1
Suggested Communication Technology at Different Voltage Levels

Technology	132 kV	33 kV	11 kV	Low Voltage
Satellite (e.g., LEO)	1	1	1	2
DSL (e.g., ADSL)	2	1	2	3
Optical fiber	1	1	2	3
Microwave radio	1	1	2	3
Cellular system (e.g., GSM)	2	2	1	1
UHF	3	3	1	2

Note: 1, generally suitable; 2, suitable in some cases; 3, generally not suitable.

The most available wired medium today is telephone lines and optical fiber. DSL technology over phone lines can provide a capacity of a few megabits per second. On the other hand, optical fiber supports high-bandwidth (up to 10 Gbps), high-reliability, and long-distance communication with a small number of repeaters (e.g., every 100–1000 km); the major limitation is its high renting and installation costs, and therefore it should be considered when high-bandwidth and stringent performance guarantees are required. Satellites (e.g., geostationary earth orbit [GEO] or a constellation of low earth orbit [LEO]) have been adopted in power utilities for years. They provide large coverage, including some areas where no other infrastructure exists, with moderate channel capacity (1200 bps to 1.2 Mbps). In particular, LEO has many desirable features, for example, significantly reduced round-trip propagation time (about 20 ms) compared with GEO (around 500 ms), better support for TCP/IP applications, and affordable cost. UHF radio (300–1 GHz) can provide point-to-point and point-to-multipoint communication with typical data rates of 9600 bps (full duplex) and 19.2 kbps (half duplex) at a range of 30–50 km. The reliability can be enhanced if the UHF band is licensed. The cellular systems (e.g., GSM) adopted by power utilities in the past decades to support system measurement and control can generally cover most of the utility assets, with a typical data rate of 9600 bps. Also, the current 3G/4G cellular network can provide much more bandwidth, several tens of megabits per second, and is a promising communication technology for power utilities. Finally, microwave radio (1–30 GHz) provides more capacity (a few to 155 Mb/s) with high reliability and is suitable for long-distance communication.

In summary, as the electric power distribution networks are often geographically large (including rural, urban, suburban, and some very remote sites), it will be more cost-effective to adopt a mixture of technologies. If the DNO sites are well within the network service provider's coverage, then the public communication network could be an option. Also, DNOs could build their private infrastructures to obtain full system control with increased flexibility, security, and reliability. When the sites are extremely remote and no other infrastructure is available, then satellite communication may be the only feasible solution. The candidate technologies at given voltage levels need to be carefully evaluated against various criteria, such as availability, bandwidth, reliability, security, and cost. In addition, the expected communication pattern also affects the selection (e.g., point-to-multipoint [regional sensing and control] or peer-to-peer [interregional coordination between controllers]). Table 12.1 shows the suggested voltage levels for the deployment for technologies of interest, based on (Roberts, 2004).

12.4.2 Communication Infrastructure Requirements

The philosophy of the regional network management system imposes many requirements on the underlying communication system. Some key requirements are summarized as follows.

12.4.2.1 Timely Data Delivery and Differentiation

In electric power networks, the control algorithms need to act in their designed timescales, on the order of milliseconds to minutes, to fulfill their functions. Certain control functions have stringent latency requirements; for instance, an undervoltage load-shedding control action needs to be operated within about 10 s (Gajic et al., 2005). However, the current communication infrastructure is mostly designed to support operations on timescales from a few seconds to a few hundred seconds (Roberts, 2004), and hence is inadequate to support such real-time and fast-acting functions. Also, some control functionalities may require slow actions; for example, OLTCs act in the timescale of tens of seconds, and transformer overheating may allow for several minutes before a control action is taken. As a result, the underlying communication system needs to be enhanced and properly managed to guarantee timely data delivery and service differentiation to meet diverse latency requirements.

12.4.2.2 Data Availability, Robustness, and Redundancy

The success of distributed control relies on high data availability (i.e., the data are accessible when needed at the right locations and in the expected formats (IEEE (Institute of Electrical and Electronics Engineers), 2005)). The advances in substation automation now support much faster data collection in substations and obtain more detailed operational information. However, these data cannot be made available to control center and peer substations due to communication inadequacy, which restricts the adoption of sophisticated control and protection functionalities (Tomsovic et al., 2005). Also, data often need to be available at multiple locations across a power distribution network ((IEEE (Institute of Electrical and Electronics Engineers), 2005). For instance, the status of a switchgear in one substation needs to be available at the control center, as well as at peer substations, to carry out potential collaborative protection and control actions. This implies that the communication system should be able to deliver data to multiple locations, for example, all devices in a communication subnetwork (i.e., broadcast) or a set of preselected devices (i.e., multicast). In addition, distributed control requires the underlying communication system, with a certain level of robustness and redundancy (e.g., backup channels and devices), to cope with potential communication failures (e.g., channel failure and device outage).

12.4.2.3 Flexibility, Scalability, and Interoperability

At present, the power distribution network is still growing, with an increasing number of measurement and control devices. Accordingly, the communication system should be flexible enough for easy expansion and reconfiguration to cope with such growth, for example, easily incorporating and managing communication with new added power network devices with minor configuration efforts (Goodman et al., 2004). The communication architecture and adopted technologies should be deployed in practice on a large scale without significant scalability hurdles. Also, interoperability across equipment and protocols from different vendors demands the adoption of unified communication standards or protocols (e.g., IEC 61850). Properly interfacing with the power elements' legacy is a major obstacle to provide system backward compatibility and a seamless communication across the entire network management system. One possible solution is to integrate a gateway functionality into the substation SCADA system to make it so that data from legacy power devices can be accessed as IEC 61850 data. Such functionality can be implemented through using IEC 61850 and OPC technology. This enables regional controllers that run control algorithms to communicate with legacy substation elements by defining the mapping between IEC 61850 and the legacy protocols used in the existing monitoring, protection, and control systems. Generic Substation Events (GSE) is a control model defined in IEC 61850 that provides a fast and reliable mechanism for transferring event data over entire substation networks, which are further subdivided into GOOSE and Generic Substation State Events (GSSE) (IEC (International Electrotechnical Commission), 2005). More specifically, GOOSE is a controlled model mechanism in which any format of data (status or value) is grouped into a dataset and transmitted within a time period of 4 ms, and a number of mechanisms are used to ensure a specified transmission speed and reliability; GSSE is an extension of the event transfer mechanism, and only status data can be exchanged through GSSE. It uses a status list (string of bits) rather than a dataset, as used in GOOSE. The installed devices in the IEC 61850–complaint substation can be directly connected to the substation LAN and communicate with each other using GOOSE and GSSE. Such communication backward compatibility can provide DNOs a graceful roadmap to their next-generation communication systems.

12.4.2.4 Communication Security

Data security is vital for the operation and control of power networks and has been widely studied (Ericsson, 2007; TCIP, n.d.; Energtics Incorporation, 2006). In particular, the research in (Ericsson, 2007) stated that data security in power utilities has a threefold objective, namely, confidentiality, availability, and integrity (CIA). Data confidentiality means allowing DNO operators to access power network elements while excluding all noncompliant parties. The distributed control of active distribution networks firmly relies on high data availability; that is, the data are accessible when needed at the right locations and in the expected formats (IEEE (Institute of Electrical and Electronics Engineers), 2005). The data often need to be available at multiple locations across a power distribution network; for example, the status of a switchgear in one substation needs to be available at the control center as well as at peer substations to carry out potential collaborative protection and control actions. This implies that the communication system should be able to deliver data to multiple locations, for example, all devices in a communication subnetwork (i.e., broadcast) or a set of preselected devices (i.e., multicast). On the other hand, the data are well protected in current DNO communication systems, as they are delivered through separate analogue circuits or private wires with proprietary protocols. However, data security becomes a critical issue when adopting an open communication paradigm, for example, TCP/IP, or sharing a communication system with other public network data traffic. The cyber-attackers can cause substantial damage by breaching the IoT infrastructure of a smart grid and manipulating the data transferred, thereby making the sensors make incorrect decisions. This can cause appliances to function in an undesired manner, resulting in widespread equipment damage, as well as considerable financial loss. Many forms of security threats are currently studied in the literature, for example, denial of service (DoS) attacks, eavesdropping attacks, and false information injection (Mo et al., 2012). An example of such an attack was the Stuxnet worm that corrupted the programmable logic controller circuit and hampered machinery operation in Iran, damaging a fifth of Iran's nuclear centrifuges.

12.5 CASE STUDIES

This section presents ICT system proposals for two example distribution grids with distinct structures (meshed and radial) and different voltage levels (33 and 11 kV), respectively. This study points out the limitations of existing communication provisions and evaluates the ICT proposals for a range of scenarios and configurations. It is important to evaluate the data delivery duration, which is defined as the total communication time to complete a measurement cycle (including polling events), a control command delivery, or a coordination process. DNOs require different traffic types to be delivered in different timescales to fulfill their functions. The general requirements are briefly described in (Gjermundrød et al., 2009) as "slow for post-event analysis, near real-time for monitoring and as close to real-time as possible for control or protection." The delivery time of data acquisition and control outside the substation is suggested to be within 1 s (IEEE (Institute of Electrical and Electronics Engineers), 2003). This chapter adopts this requirement in the assessment of the 33 kV network. At the lower-voltage level, the events in the 11 kV network are generally considered less critical, and hence the requirement is set at 10 s. For the same reason, the time requirement for coordination between peer controllers in the 33 and 11 kV networks is set at 10 and 100 s, respectively. The communication infrastructure simulation model is introduced for the performance assessment of the 33 and 11 kV networks (i.e., covering three grid substation areas and three primary substation

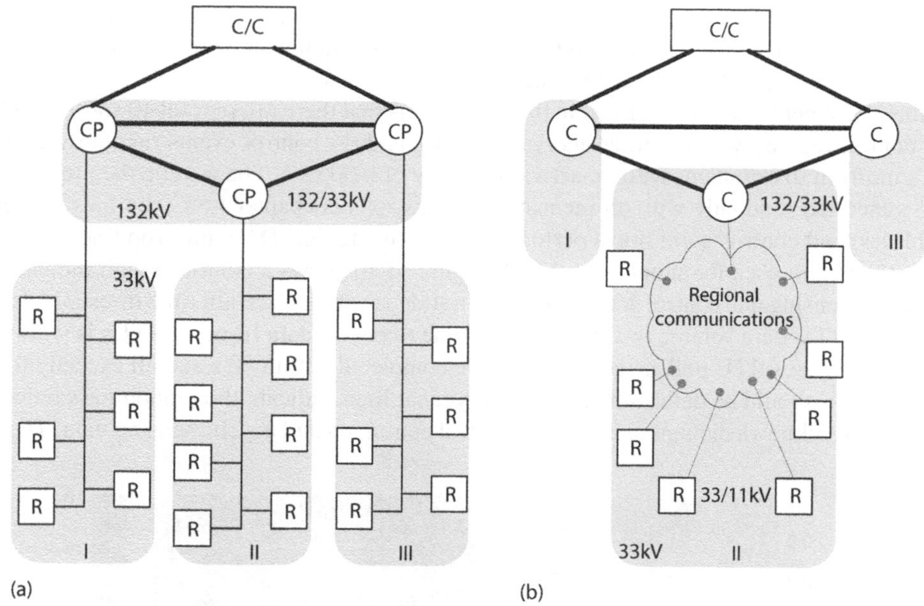

FIGURE 12.9 Thirty-three-kilovolt meshed network: (a) SCADA and (b) communication system for NMS. C/C, control center; R, 33/11 kV RTU; CP, data concentrate point; C, regional controller.

areas, respectively). All simulation experiments are carried out with the simulation time of 10,000 s (about 2.8 h) to capture the steady-state performance measurements (no significant difference in measurements was observed with a longer simulation time).

12.5.1 Case Study: 33 kV Meshed Networks

Figure 12.9a illustrates the structure of the SCADA system in the 33 kV meshed network covering three grid substation (132/33 kV) areas. At the 33 kV level, primary substation (33/11 kV) RTUs are organized in a multidrop structure via leased analogue phone lines or private wires, and RTU data are retrieved to data concentration points at grid substations by polling RTUs in sequence at fixed intervals, for example, 20 s. At the 132 kV level, grid substations are interconnected in a semimeshed topology by channels (e.g., microwave or leased digital circuits) with high reliability (through triangulation and duplicated links) and bandwidth, from 512 kb/s to a few megabytes per second.

However, the multidropped lines are often operated at a low data rate (e.g., 1200 b/s), as conventional SCADA data are limited, a few to tens of bytes. The restricted provision is also due to the line "retraining" mechanism: if excessive transmission errors are detected, the modem will interrupt the transmission and reevaluate the line, which could take significant time (e.g., a few seconds). Also, the multidropped lines have some other limitations: (1) a low degree of security, as every connected RTU may see every message, even those that are not destined for it; (2) a tedious process for line diagnosis and fault isolation, as the fault can be anywhere along the line; and (3) becoming bottlenecks, preventing immediate communication access when data traffic are heavy.

To conduct active control of the 33 kV network, the regional controllers situate at grid substations, with each controller covering one grid substation area (I, II, and III). The substation automation makes almost all the required measurements for ANM available at the 33/11 kV substation RTUs. However, both the low data rate and restricted structure prevent these data from being made available to the regional controllers in a timely manner. To solve this problem, Figure 12.9b shows that a new regional communication system (e.g., based on DSL technology) is adopted to replace the multidropped lines in region II to connect the controller and RTUs, which can be polled in the same manner as in SCADA. The SCADA channels provide the connectivity among controllers at grid substations. Thus, they may be used to carry the coordination data between peer controllers, if applicable; otherwise, additional channels need to be set up.

For the 33 kV network, it is assumed in the simulations that there are in total 10 substation RTUs in each control region, with a polling interval of 100 s, and the control events randomly occur following a uniform distribution, with an arrival intensity of 0.005 s^{-1}. The acyclic data in each RTU are also generated randomly with an intensity of 0.02 s^{-1}. This chapter first examines the sensing (dashed lines) and control (solid lines) performance if existing SCADA multidrop lines with a data rate of 1200 b/s are used for supporting ANM. Figure 12.10 shows a contour graph indicating the probability of sensing and control actions whose duration is within a certain time threshold. It shows that when the RTU data volume is 200 bytes, the time to obtain data from all RTUs is within 20 s, whereas only 70% of RTU polling activities can be completed within 70 s and all exceed 60 s. The sensing duration should be deterministic. The polygonal lines indicate the randomness induced by the presence of acknowledgment messages of control command and acyclic sensing data. The result

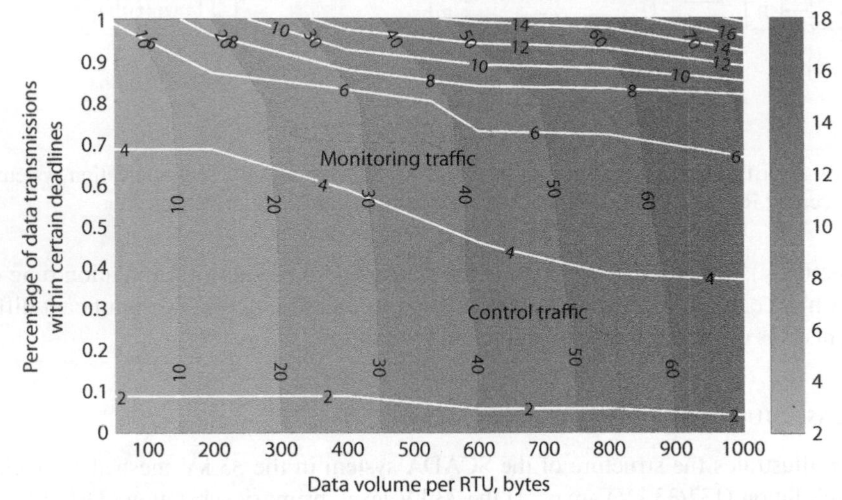

FIGURE 12.10 Sensing and control duration using multidrop lines.

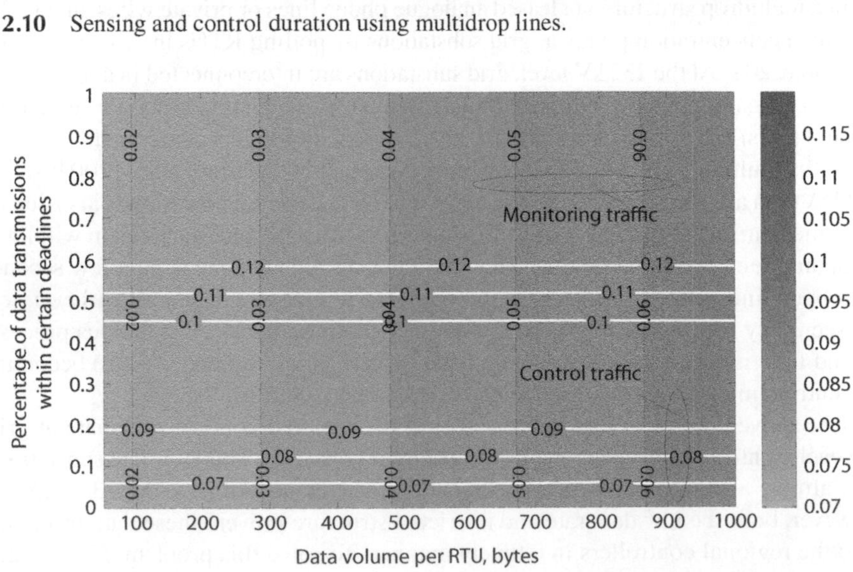

FIGURE 12.11 Sensing and control duration using ADSL system.

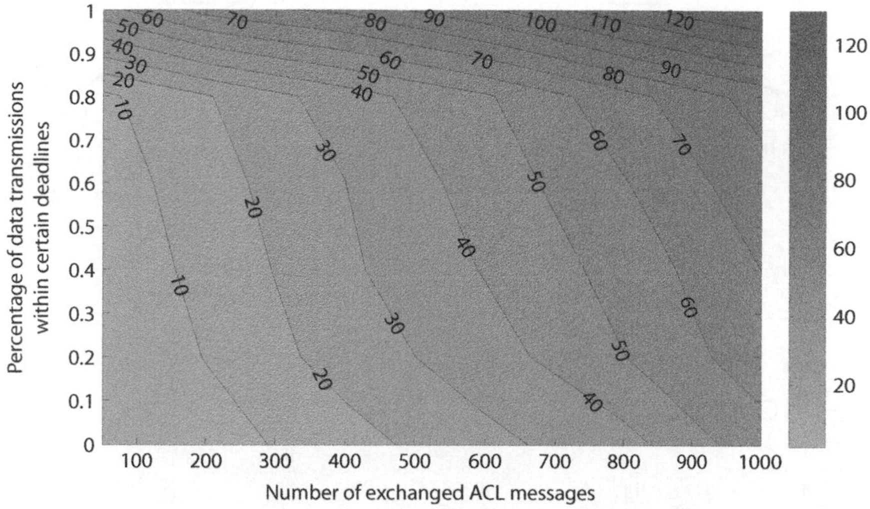

FIGURE 12.12 Coordination duration using existing SCADA provision (message length is 10 times 11 kV).

also shows that at a given RTU volume of 600 bytes, about 50% of the control command can be delivered within 4 s, and none exceed 14 s. The ANM functions require the substation data to be available to as many and accurate as possible in real time, and obviously the current low data rate of SCADA cannot meet the requirements. Figure 12.11 shows the performance if an asymmetric digital subscriber line (ADSL)-based system is adopted with a capacity of 1.544 Mb/s (T1 link) and a BER of 10 to 7. The high data rate efficiently accommodates the impacts of randomness and provides desirable performance; for example, the sensing duration is within 50 ms when 600 bytes is transmitted per RTU, and all control commands can be delivered within 130 ms.

In the 33 kV network, reliable communication connections are available in SCADA among 132/33 kV substations. The effectiveness of these channels to support coordination among controllers is evaluated. Assuming that two controllers are connected via a path comprising a microwave channel (1024 kb/s, with BER of 10^{-6}) and an optical fiber link (2048 kb/s, with BER of 10^{-15}) and the coordination events occur randomly following a uniform distribution with an intensity of $0.005\ s^{-1}$, Figure 12.12 shows the performance, with the number of exchanged messages increased from 100 to 1000. It is expected to use different numbers to indicate different degrees of complexity in coordination; that is, a more complex process could result in more message exchanges. It demonstrates that a coordination process with 800 messages can be completed within 10 s. Thus, these existing SCADA channels could be utilized if applicable.

12.5.2 Case Study: 11 kV Radial Networks

Figure 12.13a illustrates the SCADA provision in the 11 kV radial network, which comprises three primary substations (33/11 kV) areas. Topologically, the RTUs at primary sites and secondary units (11 kV or low voltage) are directly connected with the control center separately. For regional ANM of this 11 kV network, the control region is determined as each substation area and controllers are installed at 33/11 kV substations as the operation at the substation has the dominant influence on all its downstream feeders, as shown in Figure 12.13b.

There are two issues that need to solved. First, there are no existing connections between 11 kV network units and 33/11 kV substations and among substations. Therefore, additional communication infrastructure needs to be deployed among the sites. Another issue is that the number of measurements required by ANM algorithms to make a credible solution (e.g., voltage and current measurements on all feeders) can be significant in the 11 kV network due to its large number of

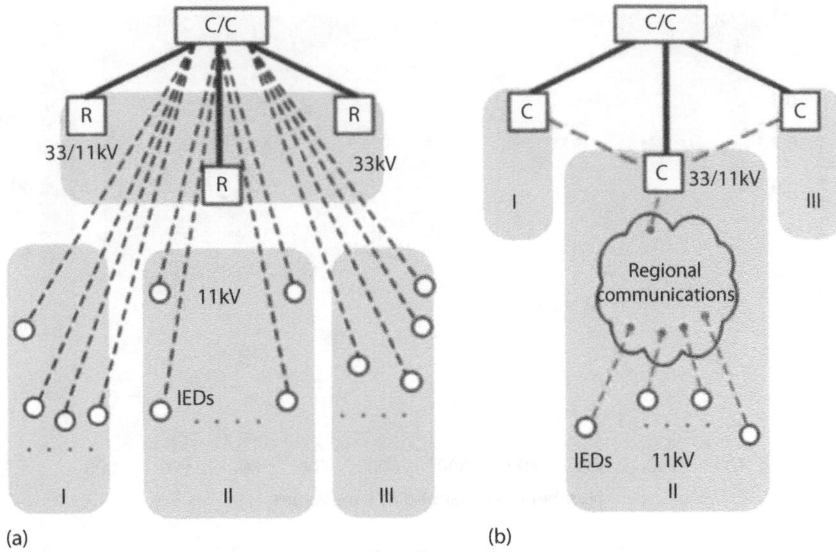

FIGURE 12.13 Eleven-kilovolt radial network example: (a) SCADA and (b) communication system for NMS. C/C, control center; R, 33/11 kV RTU; C, regional controller.

feeders and laterals. However, very few customer data are available due to the limited sensing and control capability. This implies that additional IEDs coupling with communication channels need to be installed on the 11 kV network, which is often not cost-effective in practice. One attractive solution is to adopt distribution state estimation (SE), which predicts the network operational state based on a set of pseudomeasurements and a small number of actual measurements obtained from some key locations. Take voltage control as an example, only a few measurements are needed; for example, tap position, DG voltage level, and power outputs. DNOs need to make a trade-off between adopting a sophisticated state estimator and deploying more sensing devices to make the solution cost-effective.

In regional communication, that is, between 11 kV network units and the regional controller, the measurement points could be potentially dispersive, and with a point-to-multipoint communication pattern, the wireless communication is particularly attractive to be adopted with flexible configuration and affordable costs. The key issue in designing a wireless system is to select a suitable medium access control (MAC) protocol to coordinate transmission among multiple devices sharing a common wireless medium with minimized collisions and access delay while maximizing throughput. The selection largely depends on the application, as different applications generate different traffic patterns. A variety of MAC protocols are available and can be simply categorized in two classes: fixed allocation and random access. The polling protocol is a widely used fixed-allocation approach that divides medium into a number of time slots, and the transmitter is assigned a fixed time slot for exclusive use of the channel in a round-robin manner. It ensures no data collision and reliable channel access, which is efficient for communication with devices with a stable and regular traffic pattern. Random access, for example, Aloha (Abramson, 1970) and carrier-sense multiple access (CSMA) (Lam, 1980), takes a contention-based approach, allowing devices to compete for transmission with a random back-off process with no guarantees of channel access and collision avoidance. In our study, the measurement data are regularly generated in 11 kV devices and need to be timely delivered to the controller. Therefore, the polling protocol is considered a suitable MAC protocol for wireless channel access.

Now, the performance assessment is carried out in the 11 kV case network deployed with a wireless communication infrastructure for regional management, and field measurements are collected using the polling protocol. It is assumed that the polling interval is 100 s, and the control events and the acyclic data in each region occur randomly with an intensity of 0.005 and 0.02 s^{-1}, respectively. The wireless channel is set with a propagation delay of 20 ms, capacity of 9600 b/s, and BER of

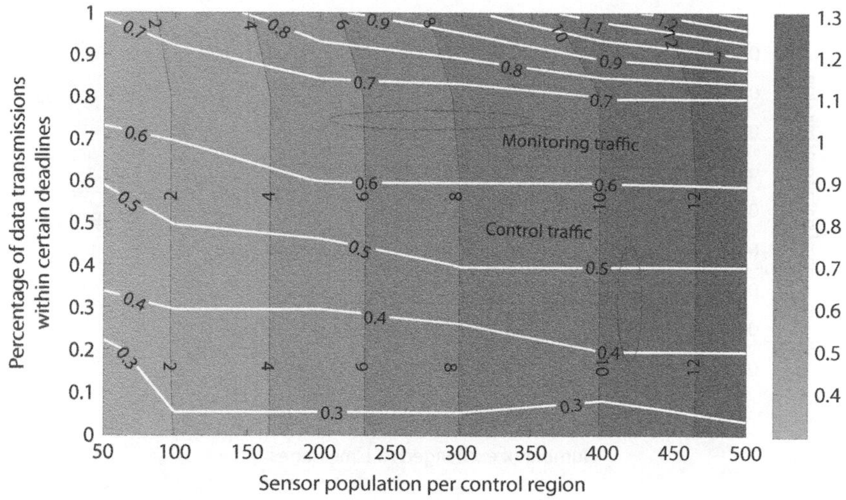

FIGURE 12.14 Sensing and control duration using wireless channel.

10^{-6}, which could represent a single-hop LEO channel or other wireless medium with similar characteristics. Figure 12.14 shows the sensing and control duration results, with the sensor population increased from 50 up to 500 per region. It suggests that all polling cycles exceed 10 s with 400 sensors, and the time can be reduced to 2 s if only 100 sensors are required. It implies that if the number of required measurement locations is reduced by adopting a well-designed state estimator, the key locations could be sampled at a fast rate with a small time for command delivery.

As wireless channels can exhibit various levels of BER due to environmental interference, Figure 12.15 shows the control performance against different BER values from 0 to 0.1, which falls in the range of (10^{-8}, 10^{-3}) in terms of BER, assuming a sensor population of 200 per region. It shows that the control performance degrades with increased error rates, which implies that a suitable wireless medium with acceptable BER needs to be selected to meet system requirements.

In the 11 kV network, it is considered that a hybrid system comprising a LEO network (1024 b/s, with BER of 10^{-5}) and terrestrial optical fiber (2048 kb/s) carries the coordination traffic between controllers with a propagation delay set to 200 ms to represent multiple LEO hops. The result (Figure 12.16) shows that exchanging 800 ACL messages can take up to 120 s, and only about half

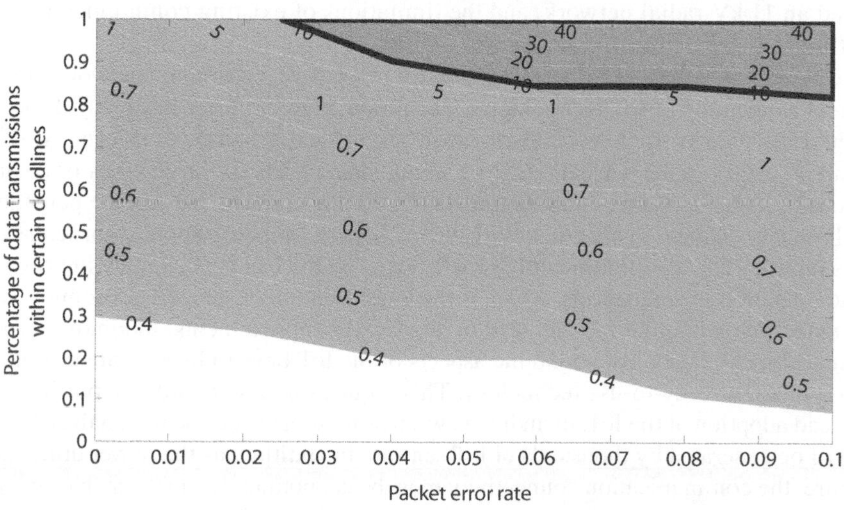

FIGURE 12.15 Control duration against different packet error rate values.

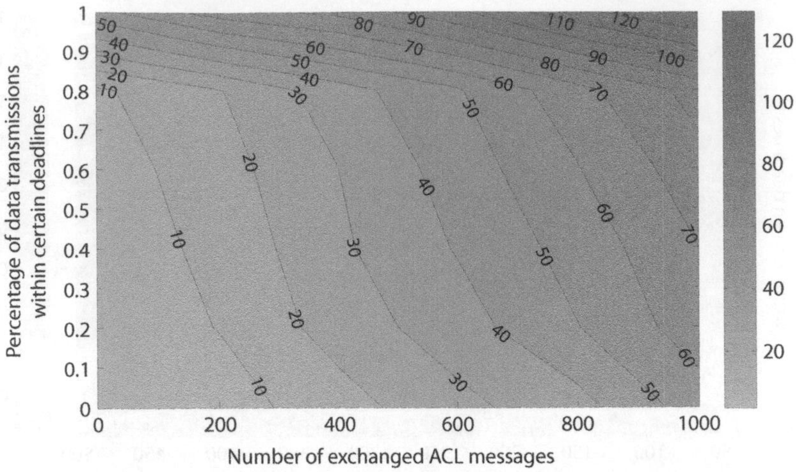

FIGURE 12.16 Coordination duration using hybrid system.

of all coordination events can be completed within 60 s. Such slow communication is partly due to the large propagation time of the LEO. Therefore, to achieve a better performance, the LEO channel may be replaced by a terrestrial system, if available, for critical sites.

12.6 CONCLUSIONS

This chapter discussed some key design issues of IoT-based electric distribution network management solutions. The key findings can be summarized as follows: the current SCADA system in DNOs can hardly meet the needs to facilitate ANM due to its constraints in both capacity and structure. Communication reinforcement is required to enhance the capability of timely network monitoring and control; the ANM system needs to adopt a unified data modeling standard to enable transparent communication across the distribution network where an interface with legacy network elements running proprietary protocols needs to be provided; a migration from DNO's proprietary communication via closed circuits to open TCP/IP-based systems is suggested, which is already becoming a trend for future energy networks. This chapter investigated communication system design and evaluation through two case networks, a 33 kV meshed network and an 11 kV radial network, and the limitations of existing communication provisions were identified.

It can be clearly seen that such distributed control of a power distribution network can be considered a typical application of IoT technologies. The power utility at large has many devices that can be considered IoT objects, such as reclosers, switches, capacitor banks, transformers, IEDs, smart sensors, and actuators in the substations. In general, smart grids for large cities or countries may have millions of home appliances and thousands of grid devices. The IoT has the potential to bring about revolutionary changes in our lives. With the IoT finding immense application across industries, the utilities industry can also significantly benefit since the IoT can considerably improve the development and operation of smart grids, which is the latest trend in developed economies. However, it should be noted that while the IoT can lead to large-scale improvements, like most emerging concepts, some technical, legal, and economic aspects of the IoT have to be exploited carefully before it becomes a mature, ready-to-use technology. This requires new standards supporting automation for widespread adoption of the IoT, including new software tools to efficiently analyze the myriad of data that will be generated by thousands of IoT sensors. In addition, as the power utility is a critical infrastructure, the communication connectivity must be economically viable, stable, and pervasive, and should comprise innovative routing algorithms for error-free data transfer.

ACKNOWLEDGMENTS

Part of this work was carried out when the author worked at Imperial College London under the AuRA-NMS partnership. This work is also supported in part by the Nature Science Foundation of Zhejiang Province (No. LZ15E070001). The authors also thank all the collaborators in the AuRA-NMS consortium.

REFERENCES

Abramson, N. 1970. The Aloha system—Another alternative for computer communications. In *Proceedings of the Fall Joint Computer Conference AFIPS*, Houston, TX, 37.

Birman, K. P., J. Chen, E. M. Hopkinson, R. J. Thomas, J. S. Thorp, R. V. Renesse, and W. Vogels. 2005. Overcoming communications challenges in software for monitoring and controlling power systems. *Proceedings of the IEEE* 9 (5).

Brunner, C. 2008a. IEC 61850 for power system communication. In *IEEE/PES T&D Conference and Exposition*, Chicago, IL, vol. 2, 1–6.

Brunner, C. 2008b. IEC 61850 for power system communication. In *Proceeding of IEEE/PES T&D Conference and Exposition*, Chicago, IL, Vol. 2, 1–6.

Davidson, E. M., and S. D. J. McArthur. 2007. Exploiting multi-agent system technology within an autonomous regional active network management system. *Presented at Proceedings of International Conference on Intelligent Systems Applications to Power Systems*, Toki Messe, Niigata, Japan.

Davidson, E., S. McArthur, C. Yuen, and M. Larsson. 2008. AuRA-NMS: Towards the delivery of smarter distribution networks through the application of multi-agent systems technology. *Presented at IEEE Power Engineering Society General Meeting*, Pittsburgh, PA, 20–24 July 2008.

Dhar, T. and T. Tang. 1998. Present and future communication technologies for distribution automation in rural Queensland, Australia. In *Proceedings of International Conference on Energy Management and Power Delivery, EMPD '98*, Singapore, 650–655.

Egea-Lopez, E. et al. 2005. Wireless communications deployment in industry: A review of issues, options and technologies. *Computers in Industry* 56 (1): 29–53.

Energtics Incorporation. 2006. Roadmap to secure control systems in the energy sector, *Prepared for U.S. Department of Energy and U.S. Department of Homeland Security*, January 2006. http://www.control-systemsroadmap.net/.

Ericsson, G. N. 2007. Toward a framework for managing information security for an electric power utility—CIGRE experiences. *IEEE Transactions on Power Delivery* 22 (3): 1461–1469.

EVERIS-CIM. 2009. EVERIS-CIM project: Harmonization of CIM and IEC 61850 models within the Framework DENISE (intelligent, secure and efficient distribution of electricity). developed for EVERIS, October 2007–March 2009.

FIPA (Foundation for Intelligent Physical Agents). 2002a. Agent management specification. http://www.fipa.org/specs/fipa00023/SC00023J.html.

FIPA (Foundation for Intelligent Physical Agents). 2002b. FIPA ACL message structure specification. http://www.fipa.org/specs/fipa00061/SC00061G.html.

FIPA (Foundation for Intelligent Physical Agents). 2002c. FIPA agent message transport protocol for HTTP specification. http://www.fipa.org/specs/fipa00084/SC00084F.html.

FIPA (Foundation for Intelligent Physical Agents). n.d. FIPA ACL message representation in XML specification, Specification No. SC00071. //www.fipa.org/specs/fipa00071.

Gajic, Z., D. Karlsson, Ch. Andrieu, P. Carlsson, N. R. Ullah, and S. Okuboye. 2005. Deliverable 1.5: Intelligent load shedding, Technical Report CRISP—Critical Infrastructures for Sustainable Power, 24 August 2005.

Gjermundrød, K. H., D. E. Bakken, C. H. Hauser, and A. Bose. 2009. Grid-stat: A flexible QoS-managed data dissemination framework for the power grid. *IEEE Transactions on Power Delivery* 24 (1): 136–143.

Goodman, F. et al. 2004. Technical and system requirements for advanced distribution automation. Technical Report, Electric Power Research Institute, Palo Alto, CA, June 2004.

Gungor, V. and F. Lambert. 2006. A survey on communication networks for electric system automation. *International Journal of Computer and Telecommunications Networking* 50 (7): 877–897.

Hauser, C. H., D. E. Bakken, and A. Bose. 2005. A failure to communicate: Next generation communication requirements, technologies, and architecture for the electric power grid. *IEEE Power and Energy Magazine* 3 (2): 47–55.

IEC (International Electrotechnical Commission). 2003. Energy management system application program interface (EMS-API)—Part 301: CIM base. IEC 61970, IEC, Geneva, November 2003.

IEC (International Electrotechnical Commission). 2005. Communications networks and systems in substations. IEC 61850, IEC, Geneva.

IEC (International Electrotechnical Commission). 2010. Application integration at electric utilities—System interfaces for distribution management—Part 11: Common Information Model (CIM) extension for distribution. IEC 61968-11, ed. 1.0, IEC, Geneva.

IEEE (Institute of Electrical and Electronics Engineers). 2003. Substation integrated protection, control and data acquisition communication requirements, IEEE Technical Report P152522003, IEEE, Piscataway, NJ.

IEEE (Institute of Electrical and Electronics Engineers). 2005. Standard communication delivery time performance requirements for electric power substation automation. IEEE 1646-2004, IEEE Power Engineering Society, Piscataway, NJ, February 2005.

JADE. n.d. JAVA Agent Development Framework. http://jade.tilab.com/.

Lam, S. S. 1980. A carrier sense multiple access protocol for local networks. *Computer Networks* 4: 21–32.

Lopes, J. A. P. et al. 2007. Integrating distributed generation into electric power systems: A review of drivers, challenges and opportunities. *Electric Power Systems Research* 77 (9): 1189–1203.

Mak, K.-H. and B. L. Holland. 2002. Migrating electrical power network SCADA systems to TCP/IP and Ethernet networking. *Power Engineering Journal* 16 (6): 305–311.

Marihart, D. 2001. Communication technology guidelines for EMS/SCADA systems. *IEEE Transactions on Power Delivery* 16 (2): 181–188.

Maurhoff, B. 2000. Dispersed generation—Reduce power costs and improve service reliability. *Presented at IEEE Rural Electric Power Conference*, Louisville, KY, 7–9 May 2000.

McArthur, S. D. J. et al. 2007. Multi-agent systems for power engineering applications—Part I: Concepts, approaches and technical challenges. *IEEE Transactions on Power Systems* 22 (4): 1743–1752.

McArthur, S. D. J. et al. 2007. Multi-agent systems for power engineering applications—Part II: Technologies, standards, and tools for building multi-agent systems. *IEEE Transactions on Power Systems* 22 (4): 1753–1759.

McClanahan, R. H. 2003. SCADA and IP: Is network convergence really here? *IEEE Industry Applications Magazine* 9 (2): 29–36.

Mo, Y. et al. 2012. Cyber–physical security of a smart grid infrastructure. *Proceedings of the IEEE* 100 (1): 195–209.

Roberts, D. 2004. Network Management Systems for Active Distribution Networks: A Feasibility Study. Technical Report URN 04/1361, DTI, Boulder, CO.

Taylor, P. et al. 2008. Integrating voltage control and power flow management in AURA-NMS. In *CIRED Smart Grids*, Frankfurt, Germany.

TCIP. n.d. TCIP: Trustworthy cyber infrastructure for the power grid. https://energy.gov/oe/downloads/tcip-trustworthy-cyberinfrastructure-power-grid.

Tomsovic, K., D. E. Bakken, V. Venkatasubramanian, and A. Bose. 2005. Designing the next generation of real-time control, communication, and computations for large power systems. *IEEE Special Issue on Energy Infrastructure Systems* 93 (5): 965–979.

Yang, Q., J. A. Barria, and C. A. Hernandez Aramburo, 2009a. A communication system for regional control of power distribution networks. *Presented at Proceedings of the 9th IEEE INDIN Conference*, Cardiff, Wales, June 24–26.

Yang, Q., J. A. Barria, and C. A. Hernandez Aramburo. 2009b. Communication infrastructures to facilitate regional voltage control of active radial distribution network. *Presented at Proceedings of the 52nd IEEE Midwest Symposium on Circuits and Systems*, Cancun, Mexico, August 2–5.

13 Satellite-Based Internet of Things Infrastructure for Management of Large-Scale Electric Distribution Networks

Qiang Yang and Dejian Meng

CONTENTS

13.1 Introduction ... 253
13.2 Distributed Control Approach for Smart Distribution Grid ... 257
13.3 LEO Network Characteristics and Modeling ... 258
13.3.1 LEO Constellation Characteristics ... 258
13.3.2 LEO Network Model ... 259
13.4 Communication Performance Assessment ... 264
13.4.1 Data Traffic Modeling ... 264
13.4.2 Numerical Results ... 264
13.4.2.1 Normal Operational Condition ... 265
13.4.2.2 Anomalous and Emergent Operational Condition ... 267
13.5 Conclusions ... 269
Acknowledgments ... 269
References ... 270

13.1 INTRODUCTION

In recent years, the enormous efforts in the pursuit of low-carbon energy provision and advances in distribution energy resources (DER) technology have led to a great boom in utilizing various forms of small-scale renewable distributed generators (DGs) (e.g., wind turbines, photovoltaics [PVs], and combined heat and power [CHP]) at the power distribution level (Harrison and Wallace, 2005). This makes the current supervisory control and data acquisition (SCADA)–based management paradigm with centralized control authority at the distribution network operator's (DNO) control center no longer efficient and practical and calls for novel active network management (ANM) mechanisms. IoT will deliver a smarter grid to enable more information and connectivity throughout the infrastructure and to homes. Through the IoT, consumers, manufacturers, and utility providers will uncover new ways to manage devices and ultimately conserve resources and reduce cost by using smart meters, home gateways, smart plugs, and connected appliances.

In many realistic IoT application scenarios, sensors and actuators are distributed over a very wide area. In certain cases, they are located in remote areas where they are not served by terrestrial access networks and, as a consequence, the use of satellite communication systems becomes of paramount importance for the Internet of Remote Things (IoRT). In recent years, in addition to the existing efforts (e.g., IntelliGrid* and SmartGrids†) on the exploitation of future "intelligent" or "smart" energy networks, autonomous regional active network management system (AURA-NMS)‡ aims to exploit a cost-effective ANM solution by implementing a distributed and intelligent control approach through a regional network management system to enhance energy security and quality of supply in UK medium-voltage (33 and 11 kV) distribution networks. The key idea behind this approach is to devolve the management authority from the centralized utility control center to a set of loosely coupled regional controllers deployed across the distribution network to carry out management tasks in either an autonomous or cooperative fashion, with controllers governing individual predefined control regions. The underlying communication infrastructure plays a paramount underpinning role in the realization of such an operation. In addition, IoT and machine-to-machine (M2M) applications have their own very unique features, such as diverse service requirements; group-based communications; low or no mobility; time-controlled, time-tolerant, small data transmission; secure connection; monitoring; priority alarm messages; low energy consumption; and low cost (Stankovic, 2014; De Sanctis et al., 2012; Lien et al., 2011). Figure 13.1 illustrates the heterogeneous communication infrastructure provision for current power utilities with a diverse physical medium (e.g., leased digital fibers, private pilot cables, public switched telephone network [PSTN] lines, and radio, satellite, and mobile cellular networks) with capacities from a few hundred bits per second to a few megabits per second.

Previous studies (Yang et al., 2009; Yang and Barria, 2009) presented the conceptual communication architecture in AURA-NMS and clearly identified the requirements and engineering challenges. It is concluded that current SCADA systems can barely meet these requirements due to limitations in many aspects, for example, constrained data rate, inflexible device organization (e.g., multidrop or daisy-chained lines), communication unavailability in the majority of low-voltage sites and poor scalability of technologies, and poor interoperability due to standard and protocol inconsistency. In addition, it is pointed out that the future "power utility intranet" needs to be independent from the public Internet and very likely based on the Internet protocol (IP) standard due to technology maturity, enhanced interoperability, and a low-cost and easy migration path, as suggested by many researchers (e.g., (Birman et al., 2005; Hopkinson et al., 2009)). These findings motivate our further exploration of new communication infrastructures with reliable, flexible, and future-proof technologies to enable timely and accurate autonomous as well as collaborative network management.

It is well known that current business operations that extend to geographically remote environments depend on satellites to provide the critical communication means to conduct remote facility monitoring and real-time asset management at unmanned sites and offshore platforms. A new breed of innovative IoT applications will emerge from the connectivity of intelligent devices. Expected to encompass billions of devices around the world, the potential scale of the IoT demands ubiquitous network coverage between satellite operators and carrier integrated services, even in remote locations. A number of unique characteristics make the satellite network an attractive technology in supporting the management of future active energy networks:

- The power distribution networks are generally geographically large (including rural, urban, suburban, and some very remote sites) and cover customers from a few million to a few thousand, or even less, with different degrees of communication availability. Satellites can reach and provide data service over all these areas regardless of the geographical conditions.
- The current utility underlying communication infrastructures consists of most available technologies, and hence the end-to-end path is with diverse communication media and physical characteristics (e.g., leased digital fiber, private wire, telephone lines, and satellite

* IntelliGrid, Smart power for the 21st century. Available at http://smartgrid.epri.com/IntelliGrid.aspx.

† SmartGrids, Vision and strategy for Europe's electricity networks of the future. Available at http://www.smartgrids.eu/.

‡ AuRA-NMS, Autonomous regional active network management system. Available at http://gow.epsrc.ac.uk/NGBOViewGrant.aspx?GrantRef=EP/E003583/1.

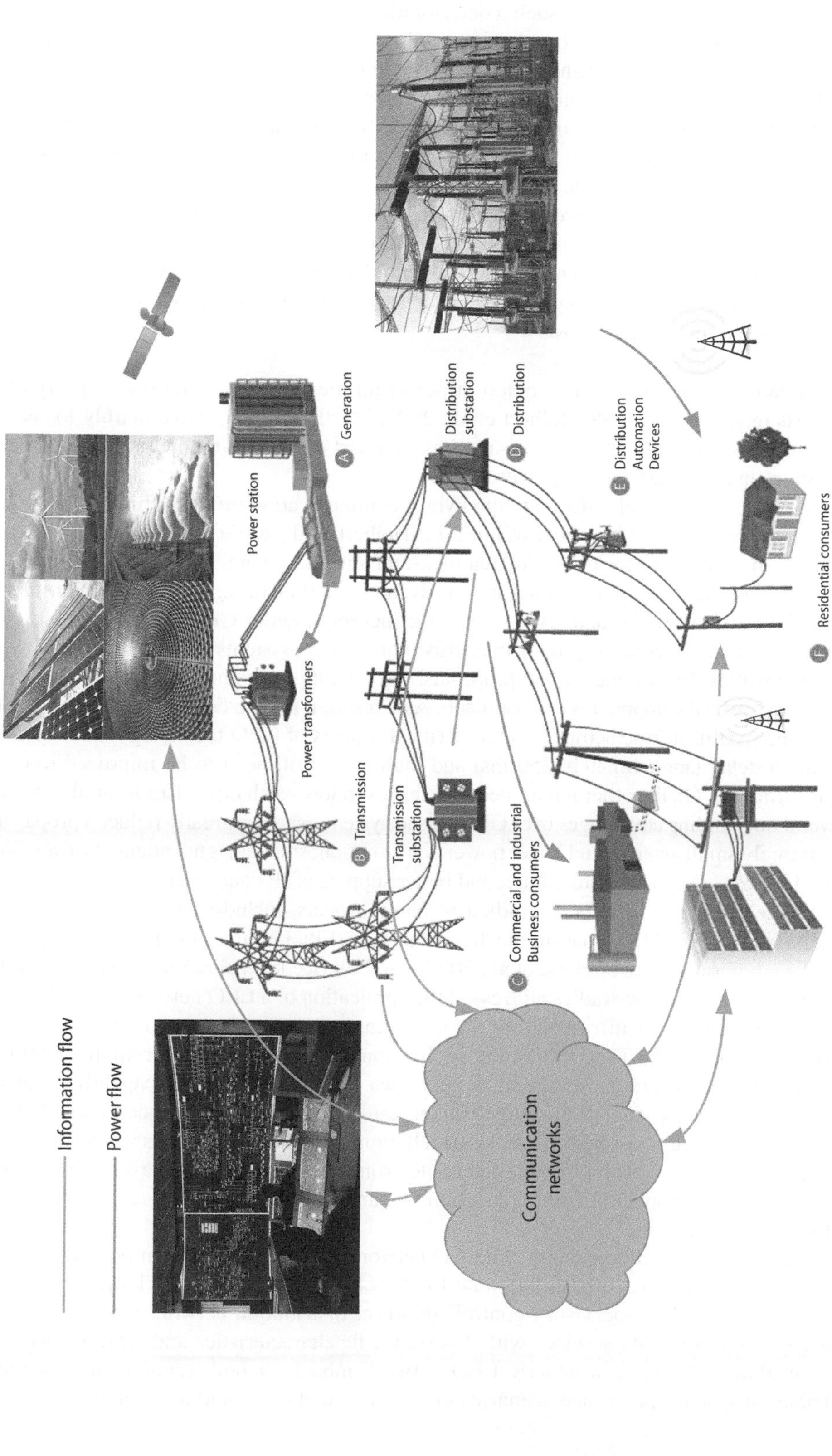

FIGURE 13.1 Communication infrastructure with heterogeneous technologies for smart power grid in an IoT paradigm.

and mobile radio). Frequent media conversion degrades the communication availability and reliability. The adoption of a satellite system in the underlying communication infrastructure can effectively remove such a deteriorating impact.
- The defined control regions across the power distribution network may be very heterogeneous in terms of the number and type of network elements, for example, DGs, substation transformers, and feeders, which results in data traffic over the underlying communication system with diverse demands and characteristics. Satellite networks have great flexibility and efficiency in bandwidth utilization to meet the requirements with bandwidth-on-demand capability and dynamic channel allocation.
- Current power distribution networks continue to grow, with an increasing number of monitoring and control devices. Satellite networks are flexible enough for easy expansion and reconfiguration to incorporate and manage communication with new added power network devices by simply installing a system interface at the device premises with an affordable installation and maintenance cost.

Satellite-based systems have been adopted in power utilities for years, and the majority of the research efforts (e.g., (Murty, 1998; Holbert et al., 2005; Marihart, 2001)) have mainly focused on the use of geostationary earth orbit (GEO) satellites as a part of the communication infrastructure to provide communication services in power utilities.

In (Murty, 1998), a terrestrial–GEO satellite hybrid communication infrastructure is proposed to support dedicated data services for power utilities. In (Holbert et al., 2005), the authors discussed the use of a satellite-based system for the wide area measurement, command, and control of power systems and suggested its application for improving the dynamic thermal rating of overhead transmission circuits. In (Marihart, 2001), the authors reviewed the incorporation of GEO satellites into existing monitoring, protection, and control systems in energy management systems (EMSs) and SCADA systems of power utilities. In aforementioned proposals, the satellite component is expected to operate in areas where terrestrial components cannot reach. As a result, it plays a complementary rather than major role in the overall infrastructure, as many critical aspects of GEO technology, notably excessive propagation delay (approximately 250 ms) and high costs, still need to be improved to ensure desirable performance. On the other hand, recent advances in low earth orbit (LEO) satellite systems provide several outstanding advantages over GEO-based systems, such as greatly reduced propagation distance, extremely small antenna and lower power consumption, smaller signal attenuation and lower cost, minimal impacts due to satellite failure, and better support of IP data services.

Although LEO networks have been studied in many aspects, including resource management (e.g., (Usaha and Barria, 2007)), routing mechanisms (e.g., (Ekici et al., 2001)), transport protocol performance evaluation (e.g., (Marchese et al., 2004)), and service level agreement (SLA) guarantee (e.g., (Ercetin et al., 2002)), few studies addressed the application of a LEO network as the key component of the communication infrastructure for managing critical large systems, for example, an energy network. In (Dhar and Tang, 1998), the application of LEO satellites for remote meter reading and rural distribution automation based on their two-way communication capability was proposed without providing the performance assessment results and analysis. In (Vaccaro and Villacci, 2005), the IP data service of a specific LEO constellation—Globalstar—was assessed through an experimental test bed with a prototype intelligent electronic device (IED). However, current investigations have focused on specific LEO constellations, and few studies have been carried out for generic LEO network analysis.

This chapter exploits the effectiveness of a LEO network as one of the key components of underlying communication infrastructure to facilitate IoT functionalities (i.e., network operation condition monitoring and remote supervisory control) in power distribution network management. The performance of a set of IP data services with diverse traffic characteristics and patterns over LEO networks is evaluated based on a generic LEO network model for both regular and emergency power distribution system operational scenarios under a range of delay and loss conditions.

The remainder of the chapter is organized as follows: Section 13.2 briefly introduces the distributed control solution for smart distribution network management. Section 13.3 overviews the LEO satellite network, discusses its relevant characteristics that may affect its performance, and presents the analytical model of a LEO network and the delay analysis for the network. Section 13.4 sets up the network and traffic scenarios for simulation experiments and provides numerical results and analysis. Finally, some conclusions are made in Section 13.5.

13.2 DISTRIBUTED CONTROL APPROACH FOR SMART DISTRIBUTION GRID

Figure 13.2 illustrates the current centralized control approach and AURA-NMS approach, respectively. The centralized control relies on DNO's SCADA for network monitoring and control tasks (Figure 13.2, i and ii). For the AURA-NMS approach, hardware controllers are equipped in individual predefined control regions (I, II, and III) that can access their local sensing and actuating devices via the communication connections among them, and the network monitoring and control actions are taken locally (Figure13.2, iii and iv). These networked regional controllers form the computation and communication platform for the execution of a set of ANM functions to conduct a range of network management tasks, for example, voltage control (Xu et al., 2009) to maintain the voltage profile within the regulated limits under conditions of load variation or intermittent output of DGs, power flow management (Dolan et al., 2009) to operate the network to within thermal limits at a given level of DG generation and contractual constraint, and automatic fault management to restore power supply to as many customers as possible as quickly as possible after a fault occurs.

In power distribution networks, substation remote telemetry units (RTUs) directly interface with the physical power network and collect these measurements from field sensors (analogue and binary) and forward control signals to actuating devices. The snapshots of network operation over time can be obtained through polling RTUs in a noncontinuous fashion with a reasonable polling rate. In addition, urgent field events or notification detected by the RTU, for example, alarms, are expected be delivered immediately to the controllers as unsolicited data service in a random manner. At the controller, collected network state information is analyzed to detect any operating conditions (e.g., voltage excursion) that may require a corrective action (e.g., operating an on-load tap changer), and

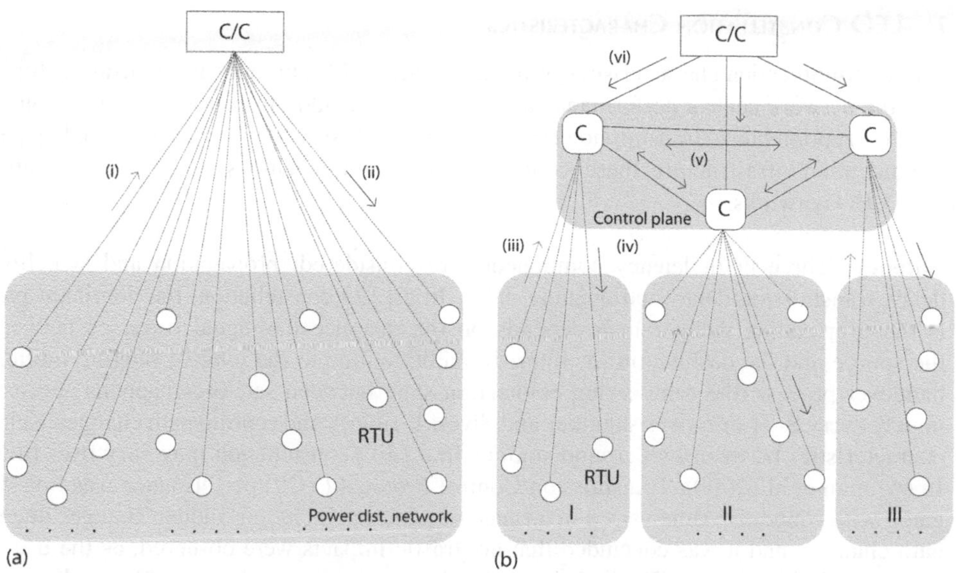

FIGURE 13.2 Network control approach: (a) centralized control via SCADA and (b) AURA-NMS approach. C/C, control center; C, regional controller.

one or a combination of multiple different control signals could be issued to control the actuating devices via the RTU.

The AURA-NMS approach exhibits both an autonomous and collaborative nature, which can be interpreted as follows: when the scope of the detected anomalous condition does not exceed the boundary of the region, the controller will carry out local control autonomously without consulting other controllers. However, autonomous control can be insufficient and wide area control may be needed in the case where the detected anomalous conditions could affect adjacent regions or potential control conflicts are detected; the controllers in these involved regions, which have only a partial view of the problem, will need to cooperate to derive suitable solutions through coordination (Figure 13.2, v). AURA-NMS suggests combining the multiagent theory into the management framework by encapsulating and distributing the management functions over different hardware controllers to accomplish complex tasks (Davidson and McArthur, 2007). The coordination is realized through agent message exchange among peer controllers. It is worth noting that although the controllers are expected to control the distribution network, they could still be governed by the DNO's control center (operator) to ensure that the overall operation meets certain utility objectives (Figure 13.2, vi).

13.3 LEO NETWORK CHARACTERISTICS AND MODELING

In general, a LEO network consists of a number of satellites organized in a constellation in orbits of 500–2000 km above the earth's surface. Currently, two types of LEO networks, namely, "little LEO" and "big LEO," are launched to support a wide range of communication services. The former, such as Orbcomm, Starnet, and Leosat, offer short-range narrowband nonvoice communication, for example, paging, vehicle tracking, messaging, and environmental monitoring. The latter, including Iridium, Globalstar, and Odyssey, focus on delivering real-time voice to personal handsets and low-bit-rate data service. Some proposals, for example, the Teledesic system, aim to provide a global "Internet in the sky" to offer Internet access for high-quality broadband service (tens of megabits per second). The Teledesic-like LEO network is considered particularly attractive for the management of future energy networks. A more detailed description of these LEO service providers can be found in (Lloyd's Satellite Constellations, n.d.).

13.3.1 LEO Constellation Characteristics

The major communication characteristics of the underlying IoT communication infrastructure that can affect the network control performance are latency, bandwidth, packet loss due to congestion, and packet corruption due to transmission errors. As the LEO constellation becomes the key part of the communication infrastructure, these parameters can exhibit dynamics over time due to intrinsic features of LEO networks.

- **Latency:** Three main latency components are considered: propagation and switching delay, transmission delay, and queuing delay. In a LEO constellation, the dominant part is the propagation delay, which depends on the distance the signal traverses between the source and the destination satellites. In addition, due to the relative satellite motion, handover process (the transferring connection state between satellites happens approximately every 8–11 min (Abrishamkar and Siveski, 1996)), and routing path changes, delay characteristics between two ground stations in a LEO constellation may vary over time. In (Allman et al., 2000), Transmission Control Protocol (TCP) performance was investigated with round-trip time varied in a range of patterns, including sudden changes due to path changes, and it was concluded that no drastic impacts were observed, as the use of a retransmission timeout (RTO) timer with a large minimum value can effectively avoid packet retransmission events. Therefore, we neglect the delay variation and assume that the propagation delay is constant during the overall communication process.

- **Channel bandwidth asymmetry:** Many designed LEO networks demonstrate asymmetric bandwidth on the forward and backward paths with distinct data rates due to economic factors; for example, satellite terminals are able to receive data at tens of megabits per second, but the data sending rate is limited to several hundred kilobits per second or even lower. This may degrade the performance of TCP, which depends on the channel characteristics in both directions, particularly when the channel congestion is present.
- **Transmission errors:** The occasional high bit error rate (BER) over satellite channels can cause packet corruption and their eventually being discarded, as they cannot be recovered at the end system. TCP deems this as an indication of communication congestion, and in turn undergoes a number of unnecessary switches to the slow-start phase with degraded throughput. In general, the satellite link exhibits a BER of as low as 10^{-8} on average and 10^{-4} in the worst case. Advanced modulation and coding techniques are often adopted to obtain even lower BER to deliver "fiber-like" service most of the time.
- **Communication congestion:** Intersatellite connections in most LEO networks use high-frequency radio or optical links that have sufficient bandwidth, for example, 25 Mb/s in Iridium and 155 Mb/s in Teledesic. Therefore, the bandwidth bottleneck resides at the links between the ground stations and their serving satellites, known as user data links (UDLs), due to restricted spectrum. Therefore, heavy congestion is generally not present in intersatellite links (ISLs). In addition, many advanced medium access control (MAC) protocols are available, taking advantage of both random access and reservation protocols with greatly improved scalability and flexibility, which exhibits excellent throughput and channel access characteristics (Peyravi, 1999).

In summary, it is envisioned that the LEO satellite constellation will be one of the key components of communication architecture to support IoT applications over a large geographical area. The future LEO satellite networks can be characterized as having a low BER, a high channel bandwidth with asymmetry, low but varied propagation delays (compared with GEO-based systems), and minimal channel access delay and congestion within the constellation. The following analysis and performance assessment are based on these assumptions.

13.3.2 LEO Network Model

Consider a polar-orbiting LEO satellite constellation consisting of N evenly separated (angular distance of $180°/N$) polar orbits (planes $p_1, p_2, \ldots, p_N$) that are across from each other only over the pole areas, and each orbit has M evenly separated satellites (angular distance of $360°/M$). In this chapter, it is assumed that each satellite $s_{i,j}$ (the jth satellite on the ith plane) in the constellation has four bidirectional ISLs with its four immediately neighboring satellites: two ISLs connecting to immediate up and down satellites in the same plane (intraplane ISLs) and two ISLs connecting to two neighboring satellites in the immediate left and right adjacent planes (interplane ISLs). As shown in Figure 13.3, these ISLs form a mesh network with a "seam" between the first plane (p_1) and last plane (p_N), where satellites in planes along side seams rotate in opposite directions.

On the same plane, all satellites move in the same circular direction. The intraplane ISLs are maintained at all times, and their lengths (and hence the propagation delay) are fixed all times (Ekici et al., 2001).

$$L_{ISL} = \sqrt{2} \cdot (R+h) \cdot \sqrt{1-\cos\left(\frac{360°}{M}\right)} \tag{13.1}$$

where R and h are the earth radius (6378 km) and LEO orbital altitude, respectively.

Between planes, interplane ISLs cannot always be maintained and are generally considered as unavailable under two certain circumstances: (1) when an interplane ISL is between counterrotating satellites across the seam, even though they may be very "geographically" close, and (2) when two

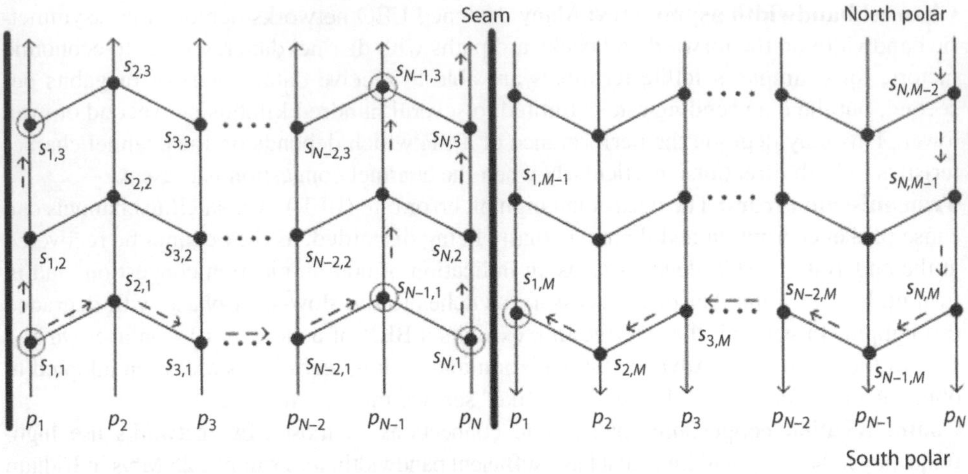

FIGURE 13.3 Polar LEO satellite constellation with four ISL connections.

satellites in adjacent planes move into the polar regions. The length of interplane ILSs (and hence the propagation delay) varies over time with the satellite movement and is related to the latitude (X) at which the interplane ISL resides (Ekici et al., 2001)

$$\overline{L}_{ISL} = \sqrt{2}\cdot(R+h)\cdot\sqrt{1-\cos\left(\frac{360^\circ}{2\cdot N}\right)}\cdot\cos X \tag{13.2}$$

It can be seen that $\overline{L}_{ISL}$ is longest when satellites are over the equator ($X=0$) and shortest when they are near the polar regions ($X \approx 90$).

The topological pattern in LEO satellite constellation changes over time with the satellite movement. Therefore, efficient routing mechanisms are generally required to select suitable end-to-end paths for connections and also maintain connectivity throughout the communication process. If the source and destination are not within the same satellite footprint, this causes packets to take multiple hops from the source satellite to the destination satellite within the constellation. Given the source–destination satellite pair ($s_{is,js}$, $s_{id,jd}$), the propagation distance in terms of number of hops, L_p, can be expressed as follows:

If $s_{is,js}$ and $s_{id,jd}$ are not separated by any constellation seams (i.e., satellites move in the same direction), for example, ($s_{1,1}$, $s_{1,3}$), ($s_{1,1}$, $s_{N-1,1}$), and ($s_{1,1}$, $s_{N-1,3}$) shown in Figure 13.3.

$$L_p = \begin{cases} \min\left\{\left|j_d - j_s\right|, M - \left|j_d - j_s\right|\right\}, i_s = i_d \\ \min\left\{\left|i_d - i_s\right|, N - \left|i_d - i_s\right|\right\}, j_s = j_d \\ \min\left\{\left|j_d - j_s\right|, M - \left|j_d - j_s\right|\right\} + \min\left\{\left|i_d - i_s\right|, N - \left|i_d - i_s\right|\right\}, i_s \neq i_d, j_s \neq j_d \end{cases} \tag{13.3}$$

If $s_{is,js}$ and $s_{id,jd}$ are separated by one of the two constellation seams (i.e., satellites move in the opposite direction), for example, ($s_{N,1}$, $s_{1,M}$) shown in Figure 13.3, the packets have to travel over one of the polar regions to reach the destination.

$$L_p = \min\left\{\left|j_d - j_s\right|, M - \left|j_d - j_s\right|\right\} + \min\left\{\left|i_d - i_s\right|, N - \left|i_d - i_s\right|\right\}, i_s \neq i_d, j_s \neq j \tag{13.4}$$

Compared with GEO systems, the lower altitude of LEO satellite constellation implies a significantly smaller footprint per satellite, which means more satellites are needed for global coverage. According to (Beste, 1978), the approximate number of satellites, $N{\cdot}M$, to obtain continuous global

TABLE 13.1
Number of LEO Satellites for Global Coverage

h(km)	*N*	*M*	*N·M*
600	7	15	105
800	7	11	77
1000	6	10	60
1200	6	8	48
1400	5	8	40
1600	5	7	35
1800	4	8	32
2000	4	7	28

coverage between any latitude X and the pole can be obtained by Equation 13.5. The number of satellites needed as a function of orbital altitude h (600–2000 km) is presented in Table 13.1 (assuming $\theta = 10°$).

$$N \cdot M \cong \frac{4 \cdot \cos X}{1 - \cos\psi}$$

$$1.3 \cdot N < M \cdot \cos X < 2.2 \cdot N \qquad (13.5)$$

$$\psi = \arccos\left(\frac{R}{R + h} \cdot \cos\theta\right) - \theta$$

where ψ and θ are the earth-centered half-cone angle of coverage for each satellite and the minimum elevation angle, respectively.

D	Packet delay on end-to-end path
D_T	Terrestrial network packet delay
$d_{T,t}$	Terrestrial network packet transmission delay
$d_{T,p}$	Terrestrial network packet propagation delay
$d_{T,s}$	Terrestrial network packet switching and processing delay
$d_{T,q}$	Terrestrial network packet queuing delay
D_S	LEO network packet delay
$D_{S,t}$	LEO network packet transmission delay
$D_{S,p}$	LEO network packet propagation delay
$D_{S,s}$	Terrestrial network packet switching and processing delay
$d_{S,up}$	LEO network uplink delay
$d_{S,down}$	LEO network downlink delay
$d_{S,ISL}$	LEO network ISL delay
$d_{s,q}$	LEO network packet queuing delay
p_s	Packet size (in bits)
S_{cap}	LEO satellite channel capacity
S_{vel}	Signal propagation velocity
b	Network intermediate node buffer size (in units of packets)
S_{BER}	Satellite channel BER
$P_{L,e}$	Packet loss rate due to transmission errors
$P_{L,c}$	Packet loss due to network congestion
h	LEO orbital altitude

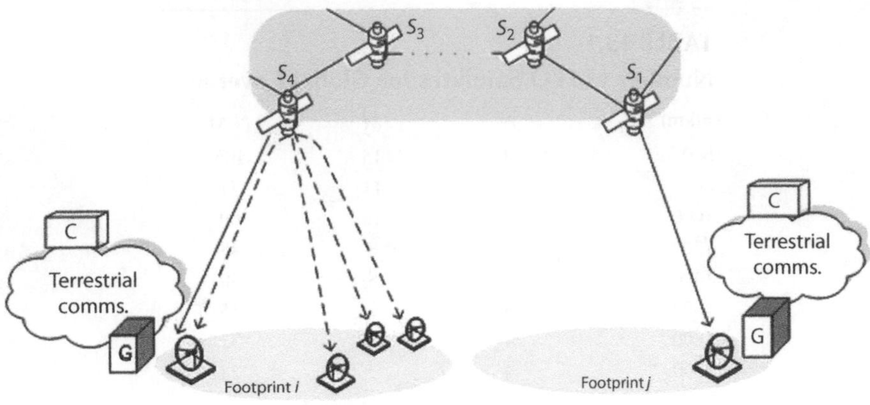

FIGURE 13.4 End-to-end communication via LEO satellite constellation.

Here, an analytic communication model is presented to characterize the packet delay and loss behaviors on the end-to-end path (including both terrestrial and space segments) between two communication terminals. This model does not incorporate the delay variation due to topology and routing changes or other LEO dynamic characteristics.

For the sake of clarity, the following notations are used:

Figure 13.4 illustrates the communication infrastructure, including LEO constellation, UDLs, gateways (acting as access points), and terrestrial segment. The packet delay on the end-to-end path comprises the delay that occurs in both terrestrial and LEO segments:

$$D = D_T + D_S \tag{13.6}$$

The delay in terrestrial segment can be expressed as

$$D_T = d_{T,t} + d_{T,p} + d_{T,s} + d_{T,q} \tag{13.7}$$

In this study, we assume that high-speed and reliable terrestrial links (e.g., asymmetric digital subscriber line [ADSL] or optical fibers) are adopted to connect with the field equipment in power distribution networks. Therefore, significant queuing delay ($d_{T,q}$) may occur on UDLs. Other delay components ($d_{T,t}$, $d_{T,p}$, and $d_{T,s}$) can be one or several magnitude less, and therefore can be neglected.

$$D_S = d_{S,t} + d_{S,p} + d_{S,s} + d_{S,q} \tag{13.8}$$

The transmission delay ($d_{S,t}$) is determined by the satellite channel capacity and packet size. The propagation delay ($d_{S,p}$) further consists of delays on up ($d_{S,up}$) and down ($d_{S,down}$) UDLs and ISL delay ($d_{S,ISL}$). In addition, as packets may pass through several satellites within the constellation, they will suffer additional queuing ($d_{S,q}$), switching, and processing ($d_{S,s}$) delays at each hop. Therefore, D_S can be further expressed as

$$D_S = \frac{p_s}{S_{cap}} + \frac{2 \cdot h}{S_{vel}} + \frac{L_{ISL}}{S_{vel}} + d_{S,s} + d_{S,q} \tag{13.9}$$

In LEO networks, packet transmission delay is very small and the buffer delay in the constellation can be disregarded due to minimal intersatellite congestion. Therefore, we consider switching time and signal propagation time to be the two major factors contributing to the end-to-end delay. The former is proportional to the number of satellites that the packet traverses on the path, which depends on the satellite constellation and adopted routing method. The latter depends on the orbital altitude and the number of ISL links.

Table 13.1 presents the number of satellites required, providing that global coverage decreases with increasing satellite altitude. Hence, a higher altitude reduces the switching delay component but leads to increased propagation delay. In (Gavish and Kalvenes, 1998), the authors discussed the impacts of LEO orbital altitude on communication performance and the altitude selection considerations from comprehensive perspectives. In our study, we are particularly interested in finding out the upper and lower delay bounds of LEO networks and further assessing the data communication services between these delay bounds for supporting AURA-NMS operation.

The best scenario occurs when source and destination terminals are within the same satellite footprint (i.e., no ISL links), and therefore it is composed of only the up- and downlink propagation time and a single satellite switching delay. The worst case happens when the terminals are on adjacent planes but separated by one of the two seams, across which there are no ISLs, so the packet has to traverse over one of the poles to reach the destination. The maximum propagation length (number of hops) and involved number of satellites are given in Equations 13.10 and 13.11.

$$L_{\max} = (N-1)\cdot \bar{L}_{ISL} + \left\lfloor \frac{M}{2} \right\rfloor \cdot L_{ISL} \tag{13.10}$$

$$S_{\max} = N + \left\lfloor \frac{M}{2} \right\rfloor \tag{13.11}$$

Based on the presented analytical model, Table 13.2 shows the calculated minimum and maximum propagation and switching delay in LEO networks with different altitudes (600–2000 km). Therefore, our following simulation experiments for assessing data delivery performance are based on the configuration that the propagation and switching delay in LEO network is within the range of 10–200 ms. Following the analysis of the end-to-end packet delay, another key aspect affecting the performance is packet loss. Two major types of loss are considered, namely, packet congestion loss due to buffer overflow in communication elements ($P_{L,c}$) and corruption loss ($P_{L,e}$) due to channel transmission errors. In general, satellite channels suffer both random and burst errors. The burst errors are mainly due to convolution encoding schemes and are not considered in this chapter. It is assumed that bit errors are independent and identically distributed (iid) as in the literature (e.g., (Zhu et al., 2006)) and the packet error rate can be related to the channel BER (S_{BER}) and the packet size (p_s) by the following approximation formula:

$$p_{L,e} = 1-(1-S_{BER})^{P_s} \tag{13.12}$$

TABLE 13.2
Minimum and Maximum LEO Propagation and Switching Delays, ms (assuming switching delay is 3 ms per satellite)

	Best Scenario			Worst Scenario		
h	$d_{S,p}$	$d_{S,s}$	$d_{S,p}+d_{S,s}$	$d_{S,p}$	$d_{S,s}$	$d_{S,p}+d_{S,s}$
600	4.0	3.0	7.0	133.8	42.0	175.8
800	5.3	3.0	8.3	136.6	36.0	172.6
1000	6.7	3.0	9.7	146.3	33.0	179.3
1200	8.0	3.0	11.0	150.7	30.0	180.7
1400	9.3	3.0	12.3	152.8	27.0	179.8
1600	10.7	3.0	13.7	145.6	24.0	169.6
1800	12.0	3.0	15.0	158.1	24.0	182.1
2000	13.3	3.0	16.3	150.2	21.0	171.2

13.4 COMMUNICATION PERFORMANCE ASSESSMENT

13.4.1 Data Traffic Modeling

To project suitable communication provision to make AURA-NMS function properly, the different types of data traffic to fulfill the IoT functionalities, along with their characteristics, need to be understood. Four major traffic types are considered:

1. Network monitoring: Network measurements (packet size of 32 bytes) collected by RTUs are periodically acquired by their designated regional controllers with a predefined polling rate.
2. Network alarms: Urgent field events or notifications (packet size of 60 bytes) should be reliably delivered to controllers immediately once generated.
3. Control commands: Control signals (packet size of 60 bytes) issued by regional controllers are delivered to the actuating devices via RTU to take certain actions, where a single control solution may consist of more than one control packet (a random number between 1 and 10 is used).
4. Coordination traffic: The controllers coordinating with peers in finding control solutions exchange a certain number of ACL messages (packet size of 1000 bytes).

These traffic types are expected to be delivered in different timescales to fulfill their functions. The general requirements are briefly summarized in (Gjermundrød et al., 2009) as "slow for post-event analysis, near real-time for monitoring and as close to real-time as possible for control or protection." This implies that in current best-effort communication infrastructure without inherent service prioritization or bandwidth reservation, selecting suitable transport protocols for different traffic types becomes paramount. It is concluded in (Hopkinson et al., 2009) that the performance of TCP used to deliver critical real-time information is unsatisfactory due to its congestion adaptation behavior, but a modified User Datagram Protocol (UDP) with reliability enhancement proves to be suitable. This is also confirmed by (Birman et al., 2005), as it introduced a UDP-based protocol (Astrolabe) for network monitoring.

For the data services delivered through the LEO-based IoT infrastructure, the power network operational condition monitoring data is carried by the UDP, while alarms and control commands are carried by the TCP with delivery acknowledgments and retransmission upon packet loss. In addition, the software agents operating among regional controllers use Hypertext Transfer Protocol (HTTP)/TCP as the underlying message transport protocol (FIPA agent message transport protocol for HTTP specification, 2002) for cooperative decision making. The primary performance metric is defined as data delivery duration, which is interpreted as the total data transmission time to complete a measurement collection cycle, a control command delivery, or a coordination process. Extensive simulation experiments based on ns2 (Network simulator, n.d.) are carried out to assess performance against a range of LEO propagation and switching delay and channel packet error rate conditions on network monitoring, alarms, and control and coordination traffic, respectively. All simulations are conducted with a simulation time of 10,000 s (about 2.8 h) to capture steady-state performance measurements, and the results are presented as contour graphs, with the numbers on contour lines representing 95% of the data delivery duration values corresponding to a given pair of LEO delay and packet error rate.

13.4.2 Numerical Results

This section presents the obtained simulation results to validate the communication performance of LEO satellites for the management of an electric distribution network under normal and anomalous conditions, respectively.

13.4.2.1 Normal Operational Condition

For all simulations, it is considered that the bandwidth of the LEO channels is asymmetric for carrying monitoring and control traffic with forward channel (controller to RTU) capacity of 2048 Kb/s and backward channel (RTU to controller) capacity of 16 or 64 Kb/s. The periodicity and volume of RTU data delivery to the controller are assumed to be very 10 s and 1000 bytes, respectively. The field alarms and control events are generated randomly following uniform distribution with an arrival intensity, λ, of 0.02 and 0.005, respectively, which is equivalent to the occurrence of alarm and control events every 50 and 200 s on average. For measurement, alarming, and control data packets, the LEO channel BER of (10^{-8}, 10^{-4}) results in a packet error rate in the range of (0, 0.06). For network monitoring traffic, this study emphasizes its timeliness and considers that the absolute data delivery guarantee is not essential when the sampling rate is high, for example, every several seconds. With respect to alarms and control commands, they require reliable delivery and can tolerate reasonable delay, for example, on-load tap changers act in the timescale of tens of seconds, and thermal phenomena (e.g., transformer overheating) may allow several minutes before a control action is taken.

Figure 13.5 shows the data delivery duration results for data retrieval from a single RTU against the LEO propagation and switching delay (from 10 ms up to 200 ms) and the packet error rate of (0, 0.06) for two different backward channel capacities, 16 Kb/s (dashed lines) and 64 Kb/s (solid lines), respectively. Here, the numbers shown on each contour line represent 95% of the data delivery duration values of all the data acquisition events; for example, for the 64 Kb/s channel, 95% of the RTU data acquisition can be completed within about 0.3 s when the LEO delay is 50 ms. The results demonstrate that the network monitoring duration increases along with the delay increases in the LEO network, and the impact of channel errors is minimal. They also show that a 64 Kb/s channel can provide much more desirable performance, for example, 1000-byte data can be retrieved in around 0.44 s in the worst condition.

Figure 13.6 illustrates the data delivery duration for delivering a control solution to the RTU, where it may consist of one or more control packets. Keeping the same previous channel asymmetry configuration, the result demonstrates that control performance degrades when delay and packet error rate increase. Two different backward channel capacities (16 and 64 Kb/s) have a performance that is very similar to that when using the same forward channel capacity, and the latter performs slightly better due to its higher bandwidth—hence a quicker delivery of acknowledgments

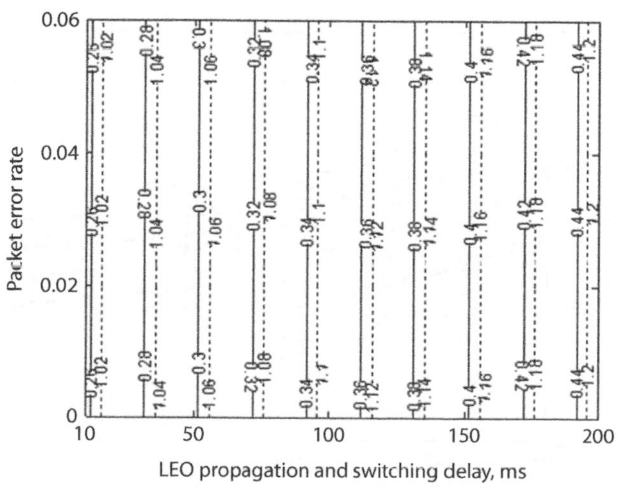

FIGURE 13.5 RTU data acquisition duration (s), satellite backward channel capacity: 16 Kb/s (dashed lines) and 64 Kb/s (solid lines).

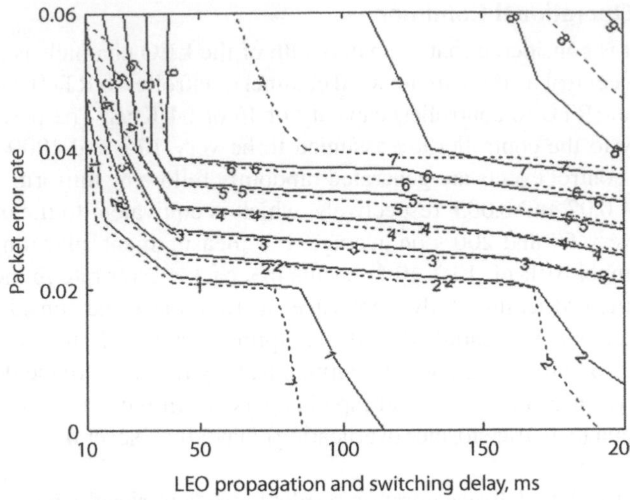

FIGURE 13.6 Control command delivery duration ($\lambda = 0.005\ s^{-1}$), satellite backward channel capacity: 16 Kb/s (dashed lines) and 64 Kb/s (solid lines).

of control packets. Both large delay and packet errors can deteriorate the LEO network throughputs. Any erroneous packet is treated as an indication of link congestion and consequently causes the TCP to decrease the data rate to avoid further congestion, and the increase of transmission rate after error detection needs to take several round-trips. It shows that under the condition of a small error rate (e.g., 0.02) and delay (e.g., 50 ms), the data delivery duration of a control solution (consisting of up to 10 control packets) can be within 1 s. Even in the worst scenario (delay of 200 ms and error rate of 0.06), the result is no more than 10 s. Therefore, such performance is considered to be able to meet the communication delivery requirements to support management functionalities.

Figure 13.7 illustrates the duration results of the field alarm delivery for the backward capacity of 16 Kb/s (dashed lines) and 64 Kb/s (solid lines), respectively. Each alarm for notification of field events is assumed to be a single packet. It shows that the 64 Kb/s channel outperforms the 16 Kb/s channel at all times, and when the satellite channel has a low error rate (<0.03), it makes the alarm

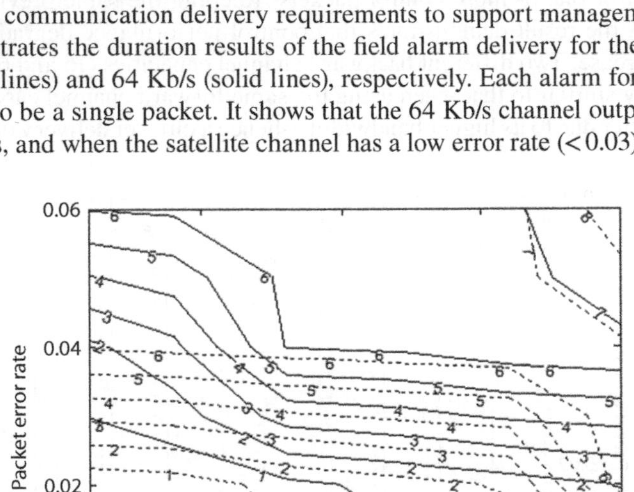

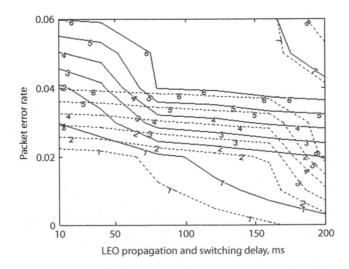

FIGURE 13.7 RTU alarm message delivery duration ($\lambda = 0.02\ s^{-1}$), satellite backward channel capacity: 16 Kb/s (dashed lines) and 64 Kb/s (solid lines).

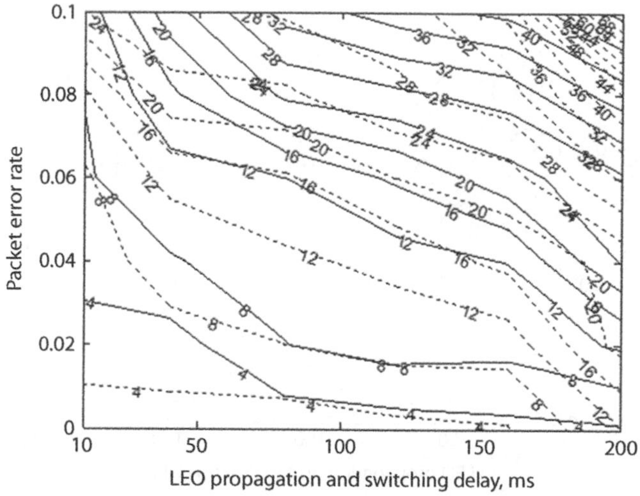

FIGURE 13.8 Coordination data delivery duration (200 messages and $\lambda=0.005\ s^{-1}$), satellite backward channel capacity: 16 Kb/s (dashed lines) and 64 Kb/s (solid lines).

delivery within 1 s, even under the worst delay condition (200 ms). Even under the circumstances of large delay and error rate, the delivery time is still within 10 s.

The coordination among peer controllers, which involves a process of agent message exchanges, is also examined. The satellite channel BER (10^{-8}, 10^{-4}) results in a packet error rate in the range of (0, 0.10) for the agent message. Here we consider that the channel is symmetric and two channel capacities, 1024 and 2048 Kb/s (E1), are examined for exchanging 200 messages between any pair of peer regional controllers, with the coordination event following uniform distribution with an arrival intensity of 0.005 (i.e., on average, one event every 200 s). Figure 13.8 shows the data delivery duration to complete a process of message exchange against a packet error rate from 0 up to 0.1 and a delay from 10 to 200 ms. Again, the performance degrades when the delay and packet error rate increase. Significant performance improvement can be observed by using the E1 channel when the LEO network delay and channel error rate are low. However, the benefit of using the E1 link becomes minimal when the delay and error rate are high, as the channel bandwidth is wasted with increased data transmission time.

13.4.2.2 Anomalous and Emergent Operational Condition

The LEO network performance for underpinning the power distribution network management under anomalous and emergent conditions (the occurrence of field alarms and control actions is frequent) is studied. It is considered that network monitoring is still with the periodicity of 10 s, while the alarms and control events are generated with intensities of 0.2 and 0.05 (equivalent to the occurrence of every 5 and 20 s, respectively). In this evaluation, no significant difference between the emergence condition and regular condition for RTU data acquisition duration can be observed, and hence the result is not presented.

Figure 13.9 illustrates the results of performance comparison between the emergent condition (solid lines) and regular condition (dashed lines) for control solution delivery with a forward and backward channel capacity of 2048 and 64 Kb/s, respectively. It shows that it takes longer to deliver the control commands under the emergent condition due to excessive traffic, and the results tell us that if the error rate is below 0.03, the time to deliver a control solution will be within 2 s. Figure 13.10 illustrates the delivery performance of field alarms when the arrival intensity is 0.2 compared with the regular condition ($\lambda=0.02\ s^{-1}$), with a channel capacity of 64 Kb/s. It shows that

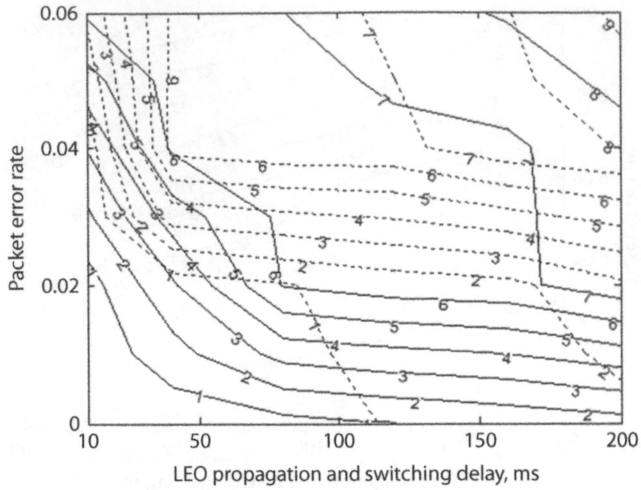

FIGURE 13.9 Control command delivery duration, with arrival intensity: λ=0.005 s⁻¹ (dashed lines) and λ=0.05 s⁻¹ (solid lines)

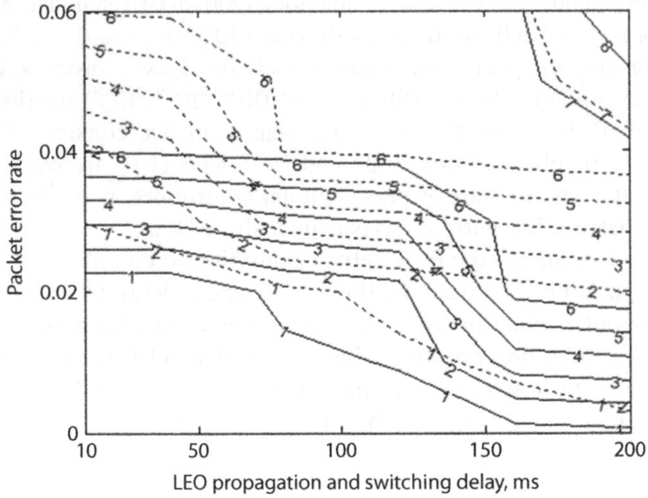

FIGURE 13.10 RTU alarm delivery duration, with alarm intensity λ=0.02 s⁻¹ (dashed lines) and λ=0.2 s⁻¹ (solid lines).

if the packet error rate of the satellite channel is sufficiently low (<0.2), then the alarm packets can still be delivered within about 1 s. Also, in the worst scenario with delay (200 ms) and error rate (0.06), the delivery time of alarms is within 10 s. The results indicate that the performance can still meet the delivery requirements under the emergent operating conditions. Finally, keeping the coordination event arrival intensity as $\lambda=0.005\ s^{-1}$, we increase the number of exchanged ACL messages from 200 to 500 to represent a more sophisticated coordination process across multiple control regions. Figure 13.11 shows that in the worst scenario, the data delivery duration for the coordination process is within 90 s (i.e., 1.5 min).

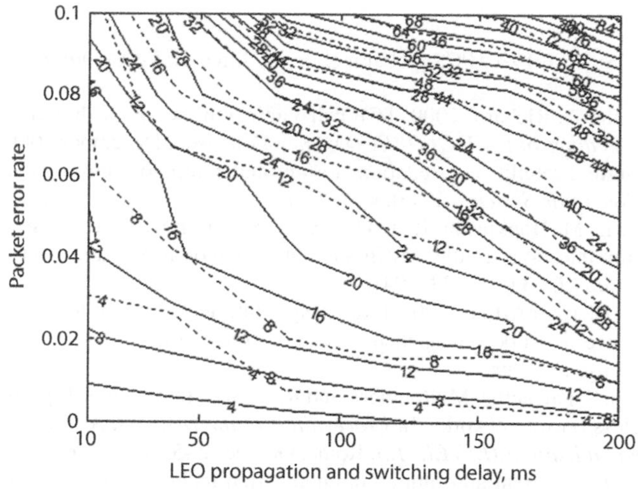

FIGURE 13.11 Coordination data delivery duration: 200 messages (dashed lines) and 500 messages (solid lines).

13.5 CONCLUSIONS

This chapter has presented an IoT-based infrastructure, including the satellite technology and its application in the management of smart electric distribution systems. The effectiveness of using standard IP-based services over a "best-effort" communication infrastructure with a LEO network as the key part for supporting a novel ANM solution of power distribution networks with DGs was investigated. This chapter highlighted the major characteristics of a LEO satellite network and discussed its potential impacts on delivering data traffic in the context of AURA-NMS. Rather than evaluating a specific LEO satellite constellation, a generic approach to characterize the delay behavior of LEO constellations based on an analytical model was presented. Through extensive simulation experiments for regular and emergent network traffic scenarios against a wide range of delay and loss conditions, the key observations are encouraging on the use of a LEO satellite-based system as the key component of the underlying communication infrastructure for managing the next-generation smart energy networks. Through the application of the IoT-based technologies, various intelligent services can be created and the development of most aspects of the smart grid can be further enhanced. It can be envisioned that the IoT will continue to drive up market demand for the integration of satellites into the communications mix and support various management tasks for industries, for example, civil engineering, transportation, and other applications for smart city. Carrier integration providers need to partner with a satellite operator that is able to provide the necessary technology integration support, as well as innovative hardware and flexible satellite infrastructure that are customizable to their users' needs.

ACKNOWLEDGMENTS

Part of this work was carried out when the author worked at Imperial College London under the AURA-NMS partnership. This work is also supported in part by the Nature Science Foundation of Zhejiang Province (No. LZ15E070001). The authors also thank all the collaborators in the AURA-NMS consortium.

REFERENCES

Abrishamkar, F. and Z. Siveski. 1996. PCS global mobile satellites, *IEEE Communication Magazine* 34 (9): 132–136.

Allman, M., J. Griner, and A. Richard. 2000. TCP behaviour in networks with dynamic propagation delay. *Presented at Proceedings of IEEE GLOBECOM*, San Francisco, December 2000.

Beste, D. C., 1978. Design of satellite constellations for optimal continuous coverage, *IEEE Transactions on Aerospace and Electronic Systems*. 14 (3): 466–473.

Birman, K. P., J. Chen, E. M. Hopkinson, R. J. Thomas, J. S. Thorp, R. V. Renesse, and W. Vogels. 2005. Overcoming communications challenges in software for monitoring and controlling power systems, *Proceedings of the IEEE* 93 (5): 1028–1041.

Davidson, E. M. and S. D. J. McArthur. 2007. Exploiting multi-agent system technology within an autonomous regional active network management system. *Presented at Proceedings of Intelligent Systems Applications to Power Systems*.

De Sanctis, M., C. Stallo, S. Parracino, M. Ruggieri, and R. Prasad, Interoperability solutions between smartphones and wireless sensor networks, In *Proceedings of the IEEE 1st AESS European Conference on Satellite Telecommunications (ESTEL '12)*, Rome, October 2–5, 2012, 1–6.

Dhar, T. and T. Tang. 1998. Present and future communication technologies for distribution automation in rural Queensland, Australia, *In Proceedings of the International Conference on Energy Management and Power Delivery*, 650–655.

Dolan, M. J., E. M. Davidson, et al. 2009. Techniques for managing power flows in active distribution network within thermal constraints. *Presented at Proceedings of CIRED*, Prague, June 2009.

Ekici E., I. F. Akyildiz, and M. D. Bender. 2001. A distributed routing algorithm for datagram traffic in LEO satellite networks. *IEEE/ACM Transactions on Networking* 9 (2): 137–147.

Ercetin, O., S. Krishnamurthy, et al. 2002. Provision of guaranteed service in broadband LEO satellite networks, *Computer Networks* 39: 61–77.

FIPA. 2002. FIPA agent message transport protocol for HTTP specification. http://www.fipa.org/specs/fipa00084/SC00084F.html.

Gavish, B., and J. Kalvenes. 1998. The impact of satellite altitude on the performance of LEOS based communication systems, *Wireless Networks* 4: 199–213.

Gjermundrød, K. H., D. E. Bakken, C. H. Hauser, and A. Bose. 2009. GridStat: A flexible QoS-managed data dissemination framework for the power grid. *IEEE Transactions on Power Delivery* 24 (1): 136–143.

Harrison, G. P. and A. R. Wallace 2005. Wallace, Maximizing renewable energy integration within electrical networks. *Presented at World Renewable Energy Congress*, Aberdeen, Scotland, May 22–27, 2005.

Holbert, K. E., G. T. Heydt, and H. Ni. 2005. Use of satellite technologies for power system measurements, command and control, *Proceedings of the IEEE* 93 (5), 947–955.

Hopkinson, K., G. Roberts, X. Wang, and J. Thorp. 2009. Quality of service considerations in utility communication networks, *IEEE Transactions on Power Delivery* 24 (3): 1465–1474.

Lien, S. Y., K. C. Chen, and Y. Lin. 2011. Toward ubiquitous massive accesses in 3GPP machine-to-machine communications, *IEEE Communications Magazine* 49 (4): 66–74.

Lloyd's Satellite Constellations. n.d. http://personal.ee.surrey.ac.uk/Personal/L.Wood/constellations/index.html.

Marchese, M., M. Rossi, and G. Morabito. 2004. PETRA: Performance enhancing transport architecture for satellite communications, *IEEE Journal on Selected Areas in Communications* 22 (2): 320–332.

Marihart, D. J. 2001. Communication technology guidelines for EMS/SCADA systems, *IEEE Transactions on Power Delivery* 16 (2): 181–188.

Murty Y. S. N. 1998. Hybrid communication networks for power utilities. *Presented at Proceedings of the Power Quality Conference*, June 1998.

Network Simulator. n.d. http://www.isi.edu/nsnam/ns.

Peyravi, H. 1999. Medium access control protocols for space and satellite communications: A survey and assessment, *IEEE Communications Magazine* 37 (3): 62–71.

Stankovic, J. A. 2014. Research directions for the Internet of things, *IEEE Internet Things Journal* 1 (1): 3–9.

Usaha, W. and J. A. Barria. 2007. Reinforcement learning for resource allocation in LEO satellite networks, *IEEE Transactions on Systems, Man, and Cybernetics* 37 (3): 515–527.

Vaccaro, A. and D. Villacci. 2005. Performance analysis of low earth orbit satellites for power system communication, *Electric Power Systems Research* 73: 287–294.

Xu, T., P. C. Taylor, et al. 2009. Case-based reasoning for distributed voltage control. *Presented at Proceedings of CIRED*, Prague, June 2009.

Yang, Q. and J. Barria. 2009. ICT system for managing medium voltage distribution grids. *Presented at Proceedings of the 35th Annual Conference of the IEEE Industrial Electronics Society, Energy and Information*, Porto, Portugal, November 3–5, 2009.

Yang, Q., J. A. Barria, and C. Hernandez Aramburo. 2009. A communication system for regional control of power distribution networks, In *Proceedings of the 7th IEEE International Conference on Industrial Informatics (INDIN2009)*, Cardiff, UK, June 2009, 372–377.

Zhu, J., S. Roy, and J. Kim. 2006. Performance modelling of TCP enhancements in terrestrial-satellite hybrid networks, *IEEE/ACM Transactions on Networking (TON)* 14 (4), 753–766.

14 IoT-Enabled Smart Gas and Water Grids
From Communication Protocols to Data Analysis

Susanna Spinsante, Stefano Squartini, Paola Russo, Adelmo De Santis, Marco Severini, Marco Fagiani, Valentina Di Mattia, and Roberto Minerva

CONTENTS

14.1 Introduction 273
14.2 Background 275
14.2.1 Smart City Context 275
14.2.2 Role of IoT as a Smart City Enabler 276
14.2.3 Smart Grids 276
14.3 Enabling Technologies: Communications 277
14.3.1 Communication Technologies for Capillary Networks in Smart Grids 277
14.3.1.1 Wireless Metering Bus 278
14.3.1.2 Unlicensed Low-Power Wide Area Networking 280
14.3.1.3 IEEE 802.11ah 282
14.3.2 Open Issues 282
14.3.2.1 Propagation Models for Coverage Estimation and Network Planning in Smart Metering Scenarios: Electromagnetic Issues 282
14.3.2.2 Power Consumption and Management in Capillary Network Devices 289
14.4 Machine Learning for Smart Gas and Water Grids 291
14.4.1 Load Forecasting 292
14.4.2 Leakage Detection 294
14.5 Future Perspectives: Cellular IoT 296
14.6 Conclusion 297
Acknowledgment 298
References 298

14.1 INTRODUCTION

According to the Water Partnership Program's analyses sponsored by the World Bank Group (2015), for the time being, more than 50% of the world population lives in urban areas. This percentage is expected to nearly double by 2030. While urbanization and industrial development contribute to the growth of cities, they also bring competing demands for resources, including water and natural gas, and contribute to increased pollution, lowering the supply of resources and their quality. More information and innovative approaches are needed to allow cities to better manage natural resources and promote sustainable growth. The drivers for the adoption of smart grid technologies in the water and natural gas management sectors are compelling: worldwide demand for water is expected to soar by 40% from current levels, according to the 2030 Water Resources Group (2009), and losses from unmetered water amount to $14 billion in missed revenue opportunities each year, according to the World Bank Group's analyses World Bank Group (2012). However, despite the abovementioned motivations, and differently from what happens in the natural gas sector, in the field of water grids management and monitoring, the situation is far from being mature. The adoption of the IoT paradigm in the water management sector would bring a number of advantages. Once in place, thousands of smart water meters and sensors could inform municipal authorities about events such as leaks (Fagiani et al., 2015), or transmit data about user consumption and storm water overflows, to enable real-time management of the water plants and energy savings. These technologies are only recently gaining popularity and will probably undergo a large market growth if institutions in developed countries will recommend or enforce the massive use of smart meters and monitoring systems.

From the user's perspective, domestic smart meters could provide households with information for optimizing water usage, or issue alerts for possible health threats related to water quality. From a social perspective, new services empowered by IoT and big data in the water field could stimulate a process of common consciousness toward the conservation of natural resources, by increasing the awareness of end users and promoting the reduction of consumption and waste; by enabling consumption forecasts, to avoid critical peaks and detect abnormal conditions, such as losses; and by empowering the end user to monitor the service provided by the utility, through direct communication to report anomalies, or to stay informed on how the service is delivered in anomalous conditions, such as when natural disasters occur. From a business-oriented perspective, data-driven services will enable the utilities to optimize the cost of water delivery and the quality of service by predicting the consumption patterns, simplifying the interactions between the utility and end user, integrating asset management and billing systems, and providing new value-added services and offers, such as safety and security alarms for condominiums and industrial plants; new supplier's approaches to the customer, based on consumer habits; and new offer paradigms, tailored to customers' needs. In the domain of smart gas grids, similar issues and opportunities may be found: both water and gas grids operate on similar physical entities (in the form of fluids), and rely on similar sensing technologies. The context is different and even much more mature for smart energy grids, which have experienced a *smart* revolution over the several years.

Given the above discussion, this chapter has two main goals: to highlight the challenges related to the IoT-oriented design and deployment of smart water grids in smart cities, by discussing enabling technologies, related constraints, and limitations, and presenting possible solutions (Spinsante et al., 2014; Gabrielli et al., 2014), and to discuss and unveil the potentially disruptive impact that data availability could provide in the field of water and gas management and related consumption forecasting, when powered with appropriate analytics and prediction strategies (Fagiani et al., 2014).

First, the chapter introduces the smart city concept and the role of IoT as a smart city enabler. This way, applications related to smart gas and water metering will be properly contextualized, providing background concepts and definitions, and the two main pillars on which they are built will be presented, that is, communication technologies and data processing techniques. The former will be addressed in the second section, spanning different technological issues, from network architectures to communications protocols to radio coverage estimation tools. In fact, a smart water or gas grid is made by meters and collectors (or concentrators), which receive the data recorded by the meters, usually placed inside the buildings. A collector can be located outside, in cabinets along the roads, or elsewhere. In planning the location of meters and collectors, one needs a reliable tool that can predict in an easy and fast way the channel attenuation. In the literature, there are many tools: some are more accurate but also more complicated (e.g., ray tracing); others are less accurate but easier to apply (e.g., empirical methods). The specific features of smart metering grids, however, limit the possibility of using well-known prediction tools for signal attenuation, and new approaches have to be provided. Similarly, the specific constraints of the nodes belonging to a smart water or gas grid require a power-saving design of the communication protocols, and suitable models to analyze the power consumption of the device and address it through properly designed task-scheduling algorithms. The second pillar in smart water and gas grid design is related to data management and processing, in order to enable knowledge extraction from raw data gathered through the grids. Through suitably designed machine learning algorithms, the possibility of implementing smart resource management will be discussed. In fact, once the necessary technological substrate is set up, to collect data from households and transfer them to suitable processing platforms, the available information on water consumption could be merged to other data, coming from gas meters, to forecast resource demands and improve the prediction performance, based on heterogeneous data and advanced analytics tools.

The rest of this chapter is organized as follows. Section 14.2 provides background information about the smart city context and the role of IoT as a smart city enabler. Smart water and gas grids are also presented, as a fundamental infrastructure contributing to the establishment of smart cities. Section 14.3 addresses the communication technologies needed to support smart gas and water grids: the concept of CNs is presented, together with different wireless technologies currently available to implement them, in an IoT-oriented approach. The same section also discusses the problems of estimating the radio coverage in the peculiar scenarios of gas and water grids, and ensuring extended lifetime to the nodes in the grids, by a power-efficient design of the communication protocols. Section 14.4 deals with the role of machine learning algorithms in exploiting the vast amounts of data collected from IoT-enabled smart grids, and establishing new services, like load forecasting and leakage detection, that would not be possible in traditional infrastructures. Section 14.5 highlights new directions in communication technologies for IoT applied to distributed infrastructures, like the so-called cellular IoT. Finally, the last section summarizes the lessons learned for the reader, paving the way for new technological assets and research directions, with a perspective on future technological evolutions, such as the next-generation IoT infrastructures.

14.2 BACKGROUND

Smart gas and water grids enabled by IoT represent a fundamental asset in the establishment of smart cities, as they allow an efficient, safe, and sustainable management of scarce resources (like water and natural gas), based on the proper exploitation of the information extracted from the data generated by sensors distributed over the grids. This section introduces the background concept of smart city, and highlights the role of IoT as a driver for the creation of smart grids.

14.2.1 Smart City Context

According to the World Migration Report 2015 (International Organization for Migration, 2015), more than 54% of people across the globe were living in urban areas in 2014. The number of people living in cities will almost double to some 6.4 billion by 2050, turning much of the world into a global city. Cities grow basically in three ways, which can be difficult to distinguish: through migration (whether it is internal, from rural to urban areas, or international migration between countries), through the natural growth of the city's population, and due to the reclassification of nearby nonurban districts. The increase of urban areas and the reduction of rural ones are expected to continue during the next decades, and the need to avoid such enormous agglomerations of people becoming uncontrolled entities has to be urgently tackled. The smart city may be seen as an approach to avoid the aforementioned risk, with the final aim to achieve sustainable and livable cities (Ballesteros et al., 2015).

A single and universally accepted definition of the smart city is not yet available, despite the concept itself having been around for several years (Telecommunication Standardization Sector of ITU, 2015). It can be identified as the means for the creation of sustainable economic development and high quality of life for multiple actors and stakeholders, by addressing a complex variety of key areas, like mobility, resource management, environment, people, economy, and government. Information and communication technologies (ICT) in general, and wireless and mobile technologies in particular, play an essential role in the challenging scenario of smart cities: they act as the "glue" connecting services and physical infrastructures, and allow orchestration of all the different interactions among them.

One of the most important elements pertaining to the smart city context is the smart grid, intended as the approach to the responsible management and operation of the city's energy networks. The smart grid is traditionally referred to as the power sector, and it encompasses the potential storage capacity for both electrical and thermal energy within the network (Luo et al., 2015), the intelligent demand side management (Wang et al., 2010), and the integration of decentralized generation into the grid (Jrventausta et al., 2010), with the new role of *prosumers* (Cai et al., 2017) identified for those who were traditionally acting as consumers only. By leveraging the innovation brought by the development of the smart grid, the smart city aims to optimize energy consumption and reduce pollution; from this perspective, efficiency improvements in water and waste management are also addressed, through the use of ICT (Thompson et al., 2013; Hornsby et al., 2017).

14.2.2 Role of IoT as a Smart City Enabler

IoT is seen as a new dimension of ICT, in which communication is maintained at any time, in any place, by anything and anyone, providing any service in any network (Minerva et al., 2015). IoT technologies can help cities to manage their resources better, monitor smart grids, deploy services as needed, and respond to emergencies in a more timely and efficient manner.

IoT platforms may be designed to address different vertical sectors, ranging from e-government to business- and enterprise-oriented applications. Ganchev et al. (2014) suggest an IoT platform that could serve as a generic architectural foundation for the development of a smart city. The core element of the platform, named integrated information center, is operated by an IoT provider and supports a number of underlying services (including energy, water, and gas supply; intelligent transportation services; city fire protection and security; and cooperative medical services), by means of physical resources, such as cloud computing and an Internet infrastructure (Kaur and Maheshwari, 2016).

As a communication infrastructure designed to provide unified, simple, and possibly economical access to a huge variety of public services, IoT may be seen as a smart city enabler. Zanella et al. (2014) introduce the definition of *urban IoT* for the application of the IoT paradigm to the urban context, where the availability of a large amount of data collected through pervasive technologies may create new synergies among traditionally separated services, increasing transparency to the citizens, and promoting their active involvement and participation. Data collected from the urban IoT may be processed to gather information aimed at optimizing the city management, improve the citizens' quality of life, and reduce costs. The IoT vision may also help in overcoming the issues related to the noninteroperability among the different technologies currently in use in urban infrastructures, which is a necessary prerequisite in creating a truly *smart* city.

14.2.3 Smart Grids

So-called smart grids represent a big promise toward the development of smart cities. In fact, the integration of ICT into existing distributed physical infrastructures opens new opportunities for improved management and maintenance of the grids, enhanced quality of services provided to citizens, and increased awareness about resource consumption and saving, not only from an economical perspective but also from an environment-preserving view (Monnier, 2013).

Most of the technological advances already took place in the energy sector, where the concept of a smart power grid has become a reality in several countries worldwide (Gellings, 2015). Even alone, once deployed in a city, a smart energy infrastructure brings three major advantages: the modernization of the power system, by enabling self-healing design, remote monitoring, and control; an increase in the consumers' awareness with respect to more responsible energy usage; and a safe and reliable integration of distributed and renewable resources (Geisler, 2013).

Among the largest consumers of energy in a city, it is possible to mention water utilities, due to water pumping operations needed to ensure water supply to the most remote users connected to the pipes. Such a condition is further exacerbated when undetected leakages (due to failures or illegal withdrawals) take place, which reduce in-pipe pressure and decrease the efficiency of the water distribution process. ICT could help address these challenges through the development of smart water grids, by networking automated monitoring and control devices (Mutchek and Williams, 2014). Water losses and inefficient use stand out as promising areas for applications of smart water grids (Spinsante et al., 2013; Squartini et al., 2013).

Smart grids rely on the availability of smart metering devices, as the terminal nodes connected to the grid (either power, gas, or water ones). Smart meters are the source for the data necessary to gather knowledge about electricity, gas, and water consumption; from these data, evolved services and functionalities may be enabled, pertaining to so-called smart buildings. For example, energy consumption can be adapted dynamically using smart metering devices, to balance the power generation and distribution in the smart grid. For the interested reader, an overview of smart metering projects completed in Europe, together with their aims and objectives, is provided in (Ivic et al., 2015).

14.3 ENABLING TECHNOLOGIES: COMMUNICATIONS

The concept of the smart city relies on a number of technological components spanning from the physical (PHY) layer (sensors and actuators, communication infrastructures) to the application layer, where information and knowledge are generated from the vast amount of data collected, and applied to inject smartness into the city.

14.3.1 Communication Technologies for Capillary Networks in Smart Grids

The acceleration in IoT deployments is forcing the evolution of the enabling communication standards and technologies. For example, several efforts are being carried out to add new techniques improving network performance in cellular standards, in order to address the peculiar needs of traffic patterns generated by IoT devices. CNs will be a fundamental part of the IoT development, enabling local wireless sensor networks to connect to and efficiently use the backhaul connectivity capabilities of cellular networks, through gateways (Sachs et al., 2014). This way, devices equipped with short-range radios only may actually use the cellular network facilities and gain wide area connectivity (Novo et al., 2015).

CNs may be defined as local area networks (LANs) acting as an extension of the wide area links (supported by cellular or satellite infrastructures), to provide connectivity to a great number of typically battery-powered and small devices, such as sensors or meters. The data is then sent over a wired or wireless CN to a server or application that translates the captured events into meaningful information for the user. The information transfer takes place with minimal or no human intervention (Singh and Huang, 2011). The deployment of short-range networks supports machine-to-machine (M2M) communications (also known as machine-type communications [MTC]) among many devices without overloading the main data links, traditionally designed for human-to-human (H2H) communications, with different requirements. By using CNs, many communicating devices may be organized into smaller groups, thus making their management easier. In addition, the use of short-range links, compared with long range ones, enables the reduction of the transmitted power, thus improving energy efficiency and reducing interference. The integration of M2M communications into existing and widely deployed networks, such as Third Generation Partnership Project (3GPP) Long Term Evolution (LTE) or other ones, poses several challenges, mainly due to the fact that those networks have been designed for different terminals, and to satisfy the requirements of traditional voice and data traffic generated by human intervention. For example, among the issues to solve, it is possible to mention the need to efficiently support and maintain connectivity of thousands or even millions of M2M devices, minimizing network overhead and preventing link congestion; the uplink-centric traffic generated by M2M devices, in contrast to the downlink-centric traffic of H2H communications; the problem of uniquely identifying and addressing each single M2M device, and how to enable data transmission over 3GPP networks, to devices that do not belong to those networks. CNs are typically used to connect non-LTE nodes to the LTE network, by the use of additional devices, named gateways. A gateway is equipped with an LTE-compliant radio interface to connect to the base station (BS) (also known as evolved NodeB [eNodeB] in LTE), and with any other wireless technology to communicate with IoT devices. A gateway may also apply policies on the data transmitted from end devices (EDs), like aggregation (Matamoros and Anton-Haro, 2013; Shariatmadari et al., 2015), in order to generate traffic streams toward the legacy cellular network that are more suitable to be delivered over the existing infrastructure. A simple illustration of the role of gateways in CNs is given in Figure 14.1.

Different technologies are envisioned to implement the local connectivity supported by CNs in smart gas and water grids. In general, wireless communications are going to be essential for the last-mile connectivity of grid devices, as it is expected that only a small subset of applications will require such a high network availability to rely on fixed and wired links, such as power grid applications, plant monitoring, or some kind of medical applications. Reusing wireless cellular infrastructures or Wi-Fi networks for sure is a cost- and time-saving approach; however, as mentioned above, some modifications to legacy technologies are needed, to face the new requirements of M2M communications (Kahn and Viswanathan, 2015; Nielsen et al., 2015).

14.3.1.1 Wireless Metering Bus

The Wireless Metering Bus (WM-Bus) protocol (Open Metering System Group, 2002–2011) is an open standard for automatic meter reading (AMR) at sub-gigahertz frequencies, and it is the basis upon which advanced metering infrastructure (AMI) installations are being deployed. In AMI, meter readings are collected without any operator's intervention, and actuation is also potentially enabled, from the utility back to the user's supply valve. WM-Bus defines the communication between water, gas, and heat meters and the so-called data concentrators, which typically feature the role of gateways toward the cellular backhaul network.

Thanks to the limited-overhead protocol, transmission-only modes (which do not require an *idle* receive phase), and long-range sub-gigahertz transmission bands, WM-Bus transceivers are low-energy demanding. The first WM-Bus release (EN 13757-4:2005) prescribed the use of the 868 MHz industrial, scientific and medical (ISM) and 468 MHz bands; the later version (EN 13757-4:2011) added new transmission modes at 169 MHz, with lower data rates. The lower 169 MHz frequency band enables a longer transmission range thanks to the inherently lower path losses, while the reduced data rates enable higher sensitivity at the receiver; this permits trading off the transmission power at the transmitter, and the covered distance.

The advantages of the 169 MHz band with respect to the 868 MHz one are implicitly related to the *narrowband transmission* concept. With a signal bandwidth limited to 25 kHz or less, the *N* mode introduces a much higher link budget, and provides extended-range solutions compared with the ones allowed at 868 MHz. The narrowband option brings performance improvement without significant limitations, because the amount of data to be transmitted in a metering scenario is very small and sporadic. Up to six channels can be allocated for the data exchange between the meter and concentrator, spaced by 12.5 kHz. Such a frequency division multiplexing (FDM) capability is exposed to potential adjacent channel interference phenomena, but cancellation filters onboard the

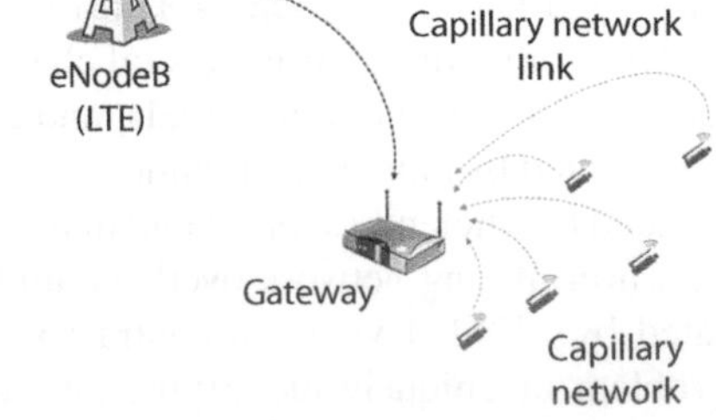

FIGURE 14.1 Role of a gateway in CNs.

WM-Bus transceivers reduce the interfering signal power level by 20 dB, compared with the power level of the channel central frequency.

Another clear advantage of the 169 MHz band is related to the reduced path loss (*PL*) experienced by the propagating radio signal. The path loss exponent *n* in the generalized Friis equation on propagation loss (Rappaport, 1996) varies according to the characteristics of the propagating environment. For $n=2$ and $n=3.5$ (free-space and urban area propagation, respectively), the comparison between the *PL* values at 169 and 868 MHz (at a parity of the antenna gains) confirms the better behavior of the radio transmission in the 169 MHz band, as shown in (Spinsante et al., 2014). Currently, the 169 MHz band is used only for remote control and smart metering, limiting the number of radio interferences. The main drawback related to the use of such a low frequency is the antenna size, which is sometimes too bulky for the form-factor requirements of the meters, which often leads to the use of nonperfectly resonant antennas.

Based on the specific application, there are combinations of communication modes for data concentrators and metering devices. These settings define the communication flow and the configuration of the radio channel. Table 14.1 lists the available communication modes.

The basic WM-Bus modes of interest are as follows:

T mode: Frequent transmission mode (several times per second or minute), 868 MHz, 100 kbps data rate from meter to gateway. In mode *T2*, the transmitter requires an acknowledgment (ACK), differently from *T1*.

S mode: Stationary mode (several transmissions per day), 868 MHz, 32.7 kbps data rate. In mode *S2*, the transmitter requires an ACK, differently from *S1*.

Further, in the 169 MHz band, the standard also foresees the following modes:

Nc mode: 169.431 MHz, 2.4 kbps data rate. *N2c* requires ACK; *N1c* does not.
Na mode: 169.40 MHz, 4.8 kbps data rate. *N2a* requires ACK; *N1a* does not.
Ng mode: 169.437 MHz, 38.4 kps data rate. Always requires ACK. The standard also foresees the following submodes:

- *N1a-f*: One-way transmission; the node transmits on a regular basis to a stationary receiving point. Single-hop repeaters are allowed.
- *N2a-f*: Two-way transmission; the node transmits like N1a-f, its receiver is enabled for a short period after the end of each transmission, and it gets locked on the received signal if a proper preamble and synchronization word are detected.

The WM-Bus link layer is compliant with EN 13757-4:2011.10. It provides data transfer between PHY and application layers, generates an outgoing cyclic redundancy check (CRC), and verifies CRCs for incoming messages. Further, the link layer provides WM-Bus addressing, acknowledges transfers for bidirectional communication modes, deals with WM-Bus frame formation, and verifies

TABLE 14.1
Operating Modes of Wireless M-Bus

S1	Unidirectional	In the stationary mode, the metering devices send their data several times a day. In this mode, the data collector may save power as the metering devices send a wake-up signal before transmitting their data.
S1-m	Unidirectional	Same as S1, but the data collector must not enter low-power mode.
S2	Bidirectional	Bidirectional version of S1.
T1	Unidirectional	In the frequent transmit mode, the metering devices periodically send their data to collectors in range. The interval is configurable in terms of several seconds or minutes.
T2	Bidirectional	Bidirectional version of T1. The data collector may request dedicated data from the metering devices.

incoming frames. Two frame formats are foreseen, named *A* and *B*, identified by a specific preamble and synch sequence. The standard specifies a number of predefined messages, used to manage operational conditions.

14.3.1.2 Unlicensed Low-Power Wide Area Networking

The rise of connected devices has placed an emphasis on low-power wireless communication that is able to cover wider areas than those typically associated with a personal area network (PAN) or LAN. These technologies are referred to as low-power wide area networking (LPWAN), intended for connecting low-cost (around 2 USD), low-power (at least 10 years of lifetime, if battery powered), and low-bandwidth devices. In addition, coverage (longer than 10 km) is one of the most critical performance metrics for LPWAN. These targets are addressed by using the sub-gigahertz radio bands and very low data rates to improve the sensitivity of receivers. Ultra-narrowband (UNB) radio signals are also used by some of the available solutions (Mikhaylov et al., 2016). The LPWAN landscape is quite dynamic, and primarily populated by industrial actors, that are competing to push their technologies into the global market. Among them, the most promising and affirmed proposals are those provided by *Sigfox* (2016), which operates both as technology and service provider for LPWAN; the Long Range (*LoRa*) alliance (2016), officially established at the Mobile World Congress 2015 in Barcelona (Spain); and the *Weightless* special interest group (SIG) (Weightless, 2016). In all the abovementioned cases, devices connected through LPWAN technologies belong to a network that is organized in a cellular-like fashion: the EDs are served by a central node acting as a BS. However, differently from cellular networks, in LPWANs most of the traffic takes place in the uplink, from the EDs to the BS (Mikhaylov et al., 2016).

The technology proposed by Sigfox addresses lowest-bandwidth applications with extremely tight energy budgets. It operates over the unlicensed sub-gigahertz frequency bands: 868 MHz in Europe and 900 MHz in United States. The communication solution provided by Sigfox does not require configuration, device pairing, or signaling operations; only very short messages are exchanged, thus originating very low energy consumption. This way, years of autonomy for battery-powered devices are ensured by the company.

Communications among devices in a Sigfox network are bidirectional, exploiting a UNB modulation. A pseudorandom frequency is chosen by a device when it has to send a message; no negotiation takes place with the BS, or signaling. The network is in charge of detecting the incoming messages, validating them, and avoiding duplication. Once these operations have been undertaken, the message is made available to third applications in the Sigfox cloud. This way, security features, like encryption or scrambling, may be applied according to specific custom requirements.

Application programming interfaces (APIs) are made available by Sigfox to fetch and use the data, like retrieving the list of devices associated with a specific type, retrieving the messages of a given device, and getting metrics about a device's traffic. The user may even subscribe to receive a Hypertext Transfer Protocol (HTTP) callback for every message received and processed by the Sigfox back end. Currently, the infrastructure is up and running in western Europe and San Francisco, with pilot programs in 20 countries, including South America and Asia.

LoRa technology is developed by Semtech, a chip manufacturer, and since it requires the use of the Semtech chip, it is not considered an open standard, but it has received consensus in the European market, with a number of current deployments (Rappaport, 2016).

LoRaWAN is designed to support low-cost, mobile, secure bidirectional communications for IoT and M2M applications. It is optimized for low power consumption, and to support large networks with millions of devices. The network architecture is typically deployed in a star-of-stars topology, in which gateways operate transparently to relay messages between EDs and a central network server in the back end. The LoRa network topology is schematically shown in Figure 14.2.

A standard Internet protocol (IP) connection supports communications between the gateways and the back-end server; EDs perform single-hop, bidirectional wireless communications to one or many gateways. Multicast is supported too, to enable over-the-air (OTA) software upgrade, or other mass distribution of messages to reduce the on-air communication time. In fact, LoRaWAN is

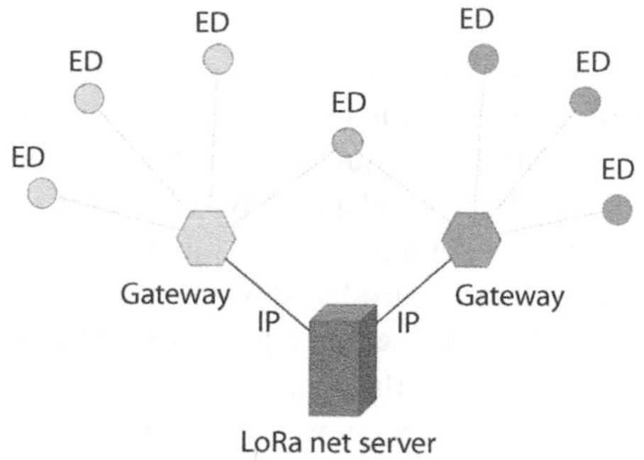

FIGURE 14.2 LoRa network topology.

intended for battery-operated devices, and can even run on energy harvesting technologies enabling the node mobility.

The communication between EDs and gateways is spread out on different frequency channels and data rates, from 0.3 to 50 kbps. The selection of the data rate is a trade-off between communication range and message duration. Thanks to the spread-spectrum technology, communications with different data rates do not interfere with each other, and create a set of virtual channels, increasing the capacity of the gateway. An adaptive data rate (ADR) scheme is applied for each device, to maximize both the EDs' battery lifetime and overall network capacity. Several layers of encryption allow us to address critical functions and communications. LoRaWAN provides several classes of EDs to address the different needs reflected in the wide range of applications: bidirectional EDs (Class A), bidirectional EDs with scheduled receive slots (Class B), and bidirectional EDs with maximal receive slots (Class C) (Mikhaylov et al., 2016; Petajajarvi et al., 2015).

The Weightless SIG specification consists of three different protocols. Weightless-W is designed to operate between 470 and 790 MHz of the TV white space spectrum, providing 1–10 Mbit/s throughput, subject to link budget and settings. Weightless-N is a UNB technology based on differential binary phase shift keying (BPSK). Weightless-P aims at LPWAN connectivity. The technology can be used over the broad range of license-exempt sub-gigahertz ISM bands, employing frequency and time division multiple access in 12.5 kHz narrowband channels. Frequency hopping is implemented, to ensure robustness to multipath and narrowband interference. EDs and BSs operate in a master–slave mode, where the BS is the master and the EDs are the slaves. This way, EDs do not need to locate themselves, nor do they need alternative channels to communicate with the white space database. BSs can readily locate themselves, or their location can be noted at installation, and they have backhaul connectivity, which allows them to interrogate white space databases. The databases will operate differently, accordingly to each country's specifications.

Weightless-P offers a bidirectional communications capability, with fully acknowledged two-way communications; it aims at comparable performance, network reliability, and security characteristics of 3GPP carrier-grade solutions, but with substantially lower costs, and less than 100 μW power consumption in the *idle* state, compared with more than 3 mW for the best cellular technologies. A maximum transmit power of 17 dBm allows for an integrated power amplifier. A flexible channel assignment is implemented, to allow frequency reuse in large-scale deployments; depending on the device link quality, an ADR from 200 bps to 100 kbps is possible, to optimize the radio resource usage, together with transmit power control (both downlink and uplink), to reduce interference and maximize network capacity. Lower data rates with channel coding provide a similar link budget as other LPWAN technologies, and up to a 2 km cell radius in urban environments. Time-synchronized BSs support efficient radio resource scheduling and usage.

Authentication to the network is available, as well as security of the transmitted messages, provided by Advanced Encryption Standard (AES) 128/256 encryption.

14.3.1.3 IEEE 802.11ah

The IEEE 802.11 standard, better known as Wi-Fi, was originally designed to provide broadband wireless Internet access for devices that generate rather heavy traffic. On the contrary, IoT encompasses a myriad of devices featuring MTC traffic, with quite different requirements. In 2010, the IEEE 802.11 Task Group started to address the problem of adapting Wi-Fi to MTC requirements, aiming to define a sub-gigahertz license-exempt amendment named 802.11ah, to support sensors and IoT applications (Khorov, 2015). Then, the Wi-Fi Alliance introduced Wi-Fi HaLow as the designation for certified Wi-Fi products incorporating IEEE 802.11ah technology (Wi-Fi Alliance, 2016).

Due to the scarcity of available bandwidth at sub-gigahertz frequencies, a new PHY layer was designed, based on the IEEE 802.11ac amendment for high-throughput WLANs (IEEE Standards Association, 2013). The medium access control (MAC) layer has been revisited too, to increase the system throughput. The 802.11ah standard has defined new compact frame formats to reduce the protocol overhead, thanks to the possibility of neglecting backward compatibility with other 802.11 systems, operating in totally different frequency ranges. Supported data rates go from 150 kbps to 347 Mbps, thus covering application requirements ranging from meter–concentrator communications to extended-range Wi-Fi. The PHY design is inherited from 802.11ac, scaling some of its features: for example, the same signal waveforms are used, but with 2–16 MHz bandwidth, maintaining the same number of subcarriers, with a tone spacing between adjacent subcarriers of 31.25 kHz. The resulting orthogonal frequency division multiplexing (OFDM) symbol period is 40 ms long, with a guard interval of 8 ms. In order to ensure a transmission range of 1 km at a minimum data rate of at least 100 kbps in outdoor IoT applications, the link budget has been increased by means of several design strategies, for example, using a sub-gigahertz frequency, which implies a reduced free-space path loss in signal propagation; a 10 times narrower channel bandwidth than for classic Wi-Fi systems; and robust coding schemes. Instead of increasing the transmission range, this enhancement can be used to lower the transmit power of a sensor device, and thus the node energy consumption, reducing costs too.

As the connected IoT world starts to take shape, a vast range of devices, objects, and systems will need communication support. Cellular networks will act as connectivity providers in this scenario: some things will connect directly to them; others will exploit the short-range radio technologies of the CNs reviewed above. Cellular networks will provide pervasive and global connectivity both outdoors and indoors by connecting CNs through special gateways, and additional functionalities, such as self-configuring connectivity management and automated gateway selection.

14.3.2 Open Issues

Despite the theory behind radio propagation being well known and established, and several tools for radio network planning having been created a long time, there are some specific issues that need to be addressed for radio coverage and planning in application scenarios related to smart cities, due to specific changes in operation conditions.

14.3.2.1 Propagation Models for Coverage Estimation and Network Planning in Smart Metering Scenarios: Electromagnetic Issues

The design of a smart metering network requires knowledge of the propagation channel and, in particular, of the electromagnetic signal path attenuation between the transmitter and the receiver. The path attenuation depends on the free-space loss and on the additional losses due to reflection, diffraction, and fading, that is, phenomena existing in the propagation channel. The path loss *PL* is defined as the ratio between the transmitted and the received power:

$$PL_{dB} = \frac{P_{RX}}{P_{TX}} = G_{RX,dB} + G_{TX,dB} + 20\log\left(\frac{\lambda}{4\pi r}\right) + A_S \tag{14.1}$$

where:

G_{RX}, G_{TX} are the transmitting (TX) and receiving (RX) antenna gains
r is the distance between the receiver and the transmitter
λ is the signal wavelength
A_S is a term accounting for additional attenuation

In telecommunication systems where the environment is large and complex, it is possible to accurately calculate path loss using software based on geometrical optics and theory of diffraction (Barclay, 2003), which allows us to approximate the electromagnetic propagation through rays. This technique is valid if the signal wavelength is small compared with object dimensions involved in the propagation (building dimension, distance between the source and near objects, etc.). The accuracy of the calculation, however, is counterbalanced by the demand for large computing resources and long simulation time. For this reason, the development and planning of cellular networks have led to the creation of different types of propagation models that would allow a fast and efficient evaluation of attenuation to be considered for the network design. In particular, in the past, when the computing capacity was reduced, several so-called empirical models based on the application of simple formulas were developed. The choice of the appropriate model depends on the type of coverage: micro- and macrocell (Saunders, 1999). The main difference between these two types of cell is the transmitting antenna position that, in a macrocell, is usually placed over the rooftop, and aims to cover areas of approximately 1 km radius if in an urban environment, and even more if in a suburban and/or rural environment. The propagation occurs mainly for diffraction over the rooftop. On the other hand, in microcells the transmitter is placed below the rooftop, although higher than the receiver; consequently, the type of transmission occurs essentially by reflection, scattering, and diffraction from vertical walls. The signal arrives at the receiver rather being attenuated, if it undergoes several reflections and refractions. For this reason, the type of propagation allows us to obtain only small cells (some hundred meters of radius).

The smart metering network has its specificities that make it different from the classical cell phone networks. For metering purposes, the transmitters (meters) are placed inside homes, or in the external facades of buildings; for this reason, the transmitter height can vary from the ground position to the last floors of buildings. The system that receives the information from the meters (the concentrator) can be placed at the street level (inside the operator service booths) or on higher supports, such as poles. Moreover, both the receivers and the transmitters are in fixed positions, and so it could be more useful to know the exact attenuations in each position of the meters than the average attenuation, typically output by empirical predictive models. For this type of network, the distinctions between micro- and macrocells are not so clear as in the case of the cellular phone network. In fact, the concentrator is considered the receiver, and it can be placed over the rooftop, and so it is higher than the transmitters (meters), or immersed in the environment in a higher or lower position, with respect to the meters, depending on their positions. In this section, we refer to smart metering systems that use sub-gigahertz frequencies for the transmission of data from meters to the concentrator, and to the case in which the meters are placed outside. In this frequency band, many predictive models can be applied. The following example of a smart metering scenario shows their application for smart network planning.

We consider two different situations in terms of frequency and concentrator position:

900 MHz and concentrator placed above rooftop
169 MHz and concentrator immersed in urban environment

For both frequencies, the ray-tracing algorithm, which accurately calculates the signal attenuation, is applicable. In fact, the dimensions of the buildings and the distance among them are large, compared with the wavelength considered.

The use of the 900 MHz frequency with a concentrator placed above the rooftop can be exploited for covering large areas of urban environment with a single concentrator that collects data from meters scattered throughout the interested area. Limitation to the number of meters is mainly due to the system capacity. In Figure 14.3, a scenario in which the concentrator is located on top of a building and the meters are placed at ground level is considered. This environment describes an area of Ancona, Italy, and the attenuation of the signal has been measured along the path reported in Figure 14.3a.

This situation is very similar to that of the cell phone network; therefore, you can use the models for mobile telephony, even if you are in a situation in which both antennas are located in fixed positions. The models typically used in mobile phones are the Okumura model (Okumura et al., 1968), which bases its prediction on a series of graphs; the Okumura–Hata (Hata, 1980) model, which approximates the Okumura curves with simple formulas; and the Cost 231 Walfisch–Ikegami

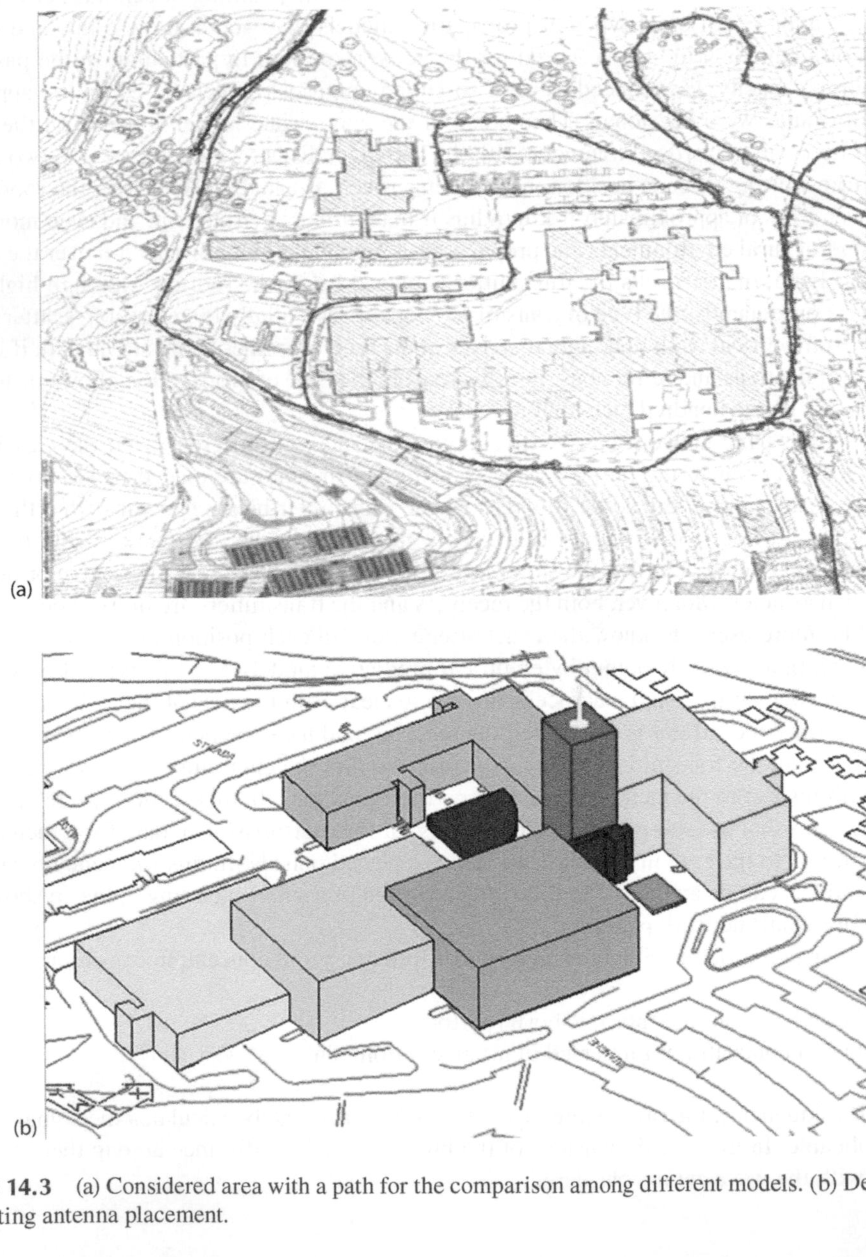

FIGURE 14.3 (a) Considered area with a path for the comparison among different models. (b) Details of the transmitting antenna placement.

(Commission of the European Communities and COST Telecommunications, 1999) model. These three models are applied to the scenario of Figure 14.3, together with a ray-tracing simulation, and the results are reported in Figure 14.4.

In particular, Figure 14.4a shows the comparison between the empirical models and the measurements made along the path evidenced in Figure 14.3a with regard to the received power versus the distance from the transmitter. Figure 14.4b, instead, shows the comparison between the same measurements and simulations performed with a software tool based on the ray-tracing technique. From Figure 14.4, it is possible to see that the empirical models perform better far from the antennas. Above 800 m, in fact, the deviation from the measurements is limited within a range of 10 dB. Below these distances, empirical models give very approximate prediction with much more deviation from the measurements. On the other hand, ray tracing gives good results in the vicinity of the antenna, which worsen when moving away from it. This seemingly strange behavior, for a deterministic model, is actually due to the limitations of computing power and memory that do not

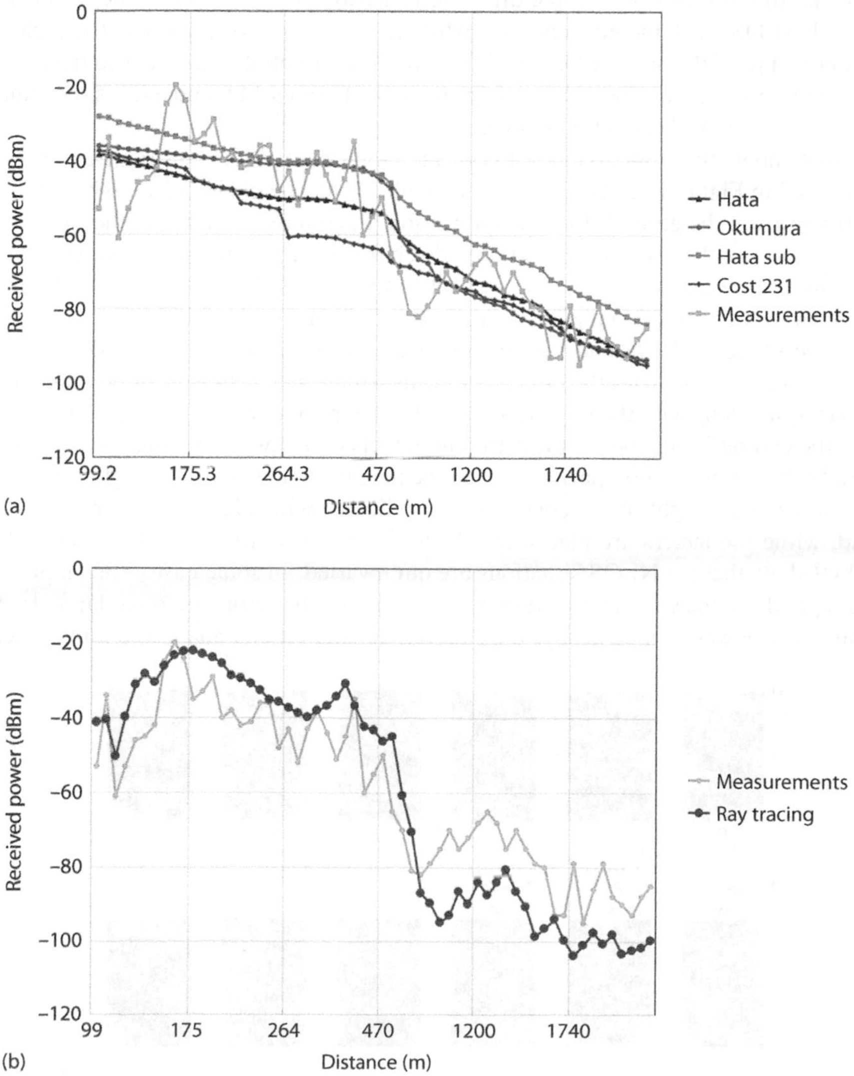

FIGURE 14.4 (a) Comparison between measurements and empirical models of the received power in the measurements campaign of Figure 14.3a. (b) Comparison between measurements and ray-tracing software (Armonica) of the received power in the measurement campaign of Figure 14.3a.

allow taking into account all the rays necessary for an accurate calculation in a region far from the antenna.

From this simple example, it can be concluded that in the case of microcellular coverage, but with the antenna of the concentrator placed above the rooftop, it is necessary to use deterministic predictive methods such as ray tracing. In the case of macrocellular coverage, the classical empirical models are still valid, when used over 1 km. When using the 169 MHz frequency and a concentrator immersed in an urban environment, the possibility of exploiting empirical models is limited. Among them, it is possible to mention the Okumura–Hata model, the dual-slope model developed for microcells (Barclay, 2003), and the CEPT-SE21 model (Electronic Communications Committee, 2004) developed for different contexts, but specific for the considered frequency. The Okumura–Hata model has actually been developed for situations in which the transmitting antenna is placed above the rooftop, but in order to analyze its usability out of its range, the comparison presented in the following considers it too.

Figures 14.5 through 14.7 show three different scenarios. In the first case, the meters are always in line of sight (LOS) with the concentrator, while in the other two situations, the meters are partially in LOS and partially not in LOS (NLOS) with the concentrator. In order to have a high number of propagation situations, the urban environment of Figures 14.6 and 14.7 is the same, but the locations of the meters and collector change.

Figure 14.8 shows the comparison between the empirical models and ray tracing for the scenario considered in Figure 14.5. The transmitter is placed at a height of 3.2 m, while the meters are placed at 0.5 m from the ground. In this comparison, the reference model is the ray-tracing result that (as also shown in the above example) provides fairly accurate attenuation values. As shown in the figure, empirical models, which take account of the building's presence along the path, which creates reflection and diffraction phenomena, provide quite different results from ray tracing, while the simple free-space model provides attenuations that differ by less than 10 dB. In this example, since the meters are visible with the concentrator, the free-space path loss is the main contribution to the propagation, along with the reflection from the ground and by vertical walls. Figures 14.9 and 14.10 show the comparison between the different models in the two situations shown in Figure 14.6 and 14.7. In both figures, the comparisons when the meters are lower (a) and higher (b) than the collector are shown. The height of the concentrator (indicated with TX in the figures) is 3.2 m above the ground, while the meters are placed at 0.1 and 7.5 m above the ground, respectively. Figures 14.8 and 14.9 show that the NLOS situations are quite varied. In some cases, the empirical models predict with good accuracy the attenuation, while in others, the errors are quite large. In particular, the Okumura model was applied always considering the highest antenna as the transmitter. For this

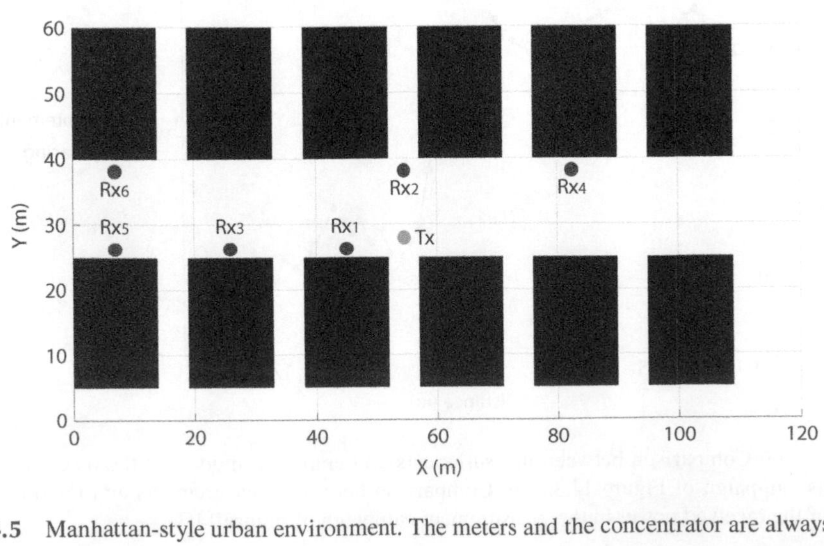

FIGURE 14.5 Manhattan-style urban environment. The meters and the concentrator are always visibile.

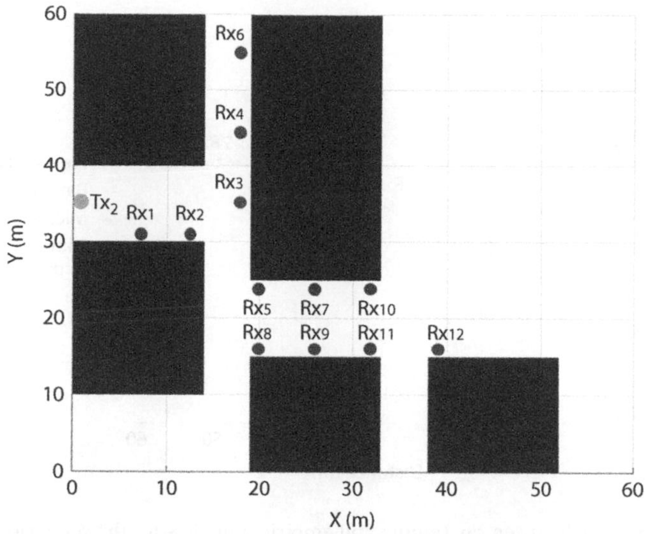

FIGURE 14.6 Urban scenario with LOS and NLOS situations. In this situation, only three meters are in LOS.

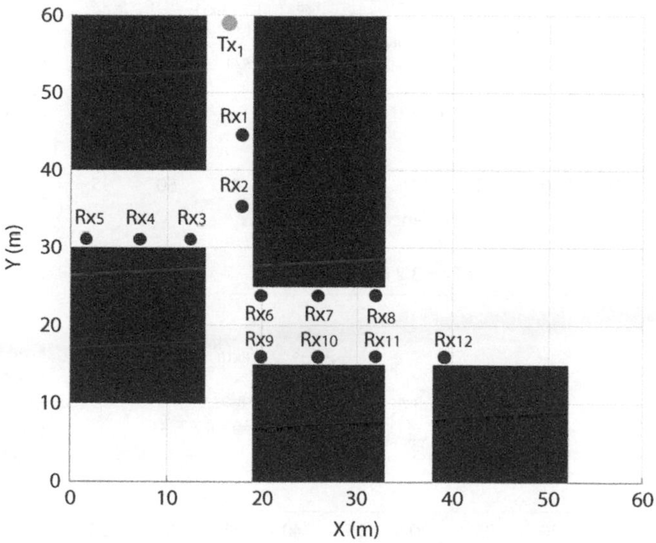

FIGURE 14.7 Same scenario as in Figure 14.6, but with the collector and meters placed at different positions. In this situation, only two meters are in LOS.

reason, when the meters are placed lower than the collector, it becomes the transmitter in the model, while for the opposite situation, the meters become the transmitters. By this trick, it is possible to obtain better results, compared with ray tracing, even if the errors can in some cases reach 20 dB. Concerning the CEPT model, the situation is very similar to that of Okumura. The "dual-slope" model provides the best results when the meters are placed high compared with the floor, while providing more errors when the meters are placed almost at the ground.

From the example reported above, it is clear how it is not possible to identify a model able to give acceptable attenuation values in all situations. The limitation of empirical models, to calculate the losses in typical situations of smart metering networks, makes them quite unreliable and would

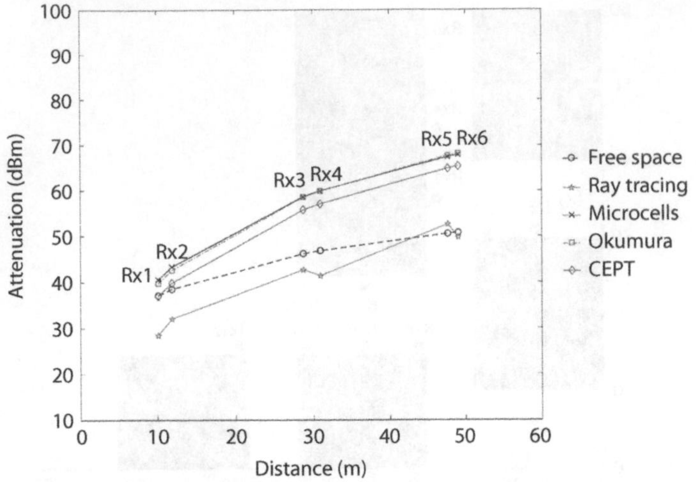

FIGURE 14.8 Comparison between ray tracing and empirical models for the scenario in Figure 14.5.

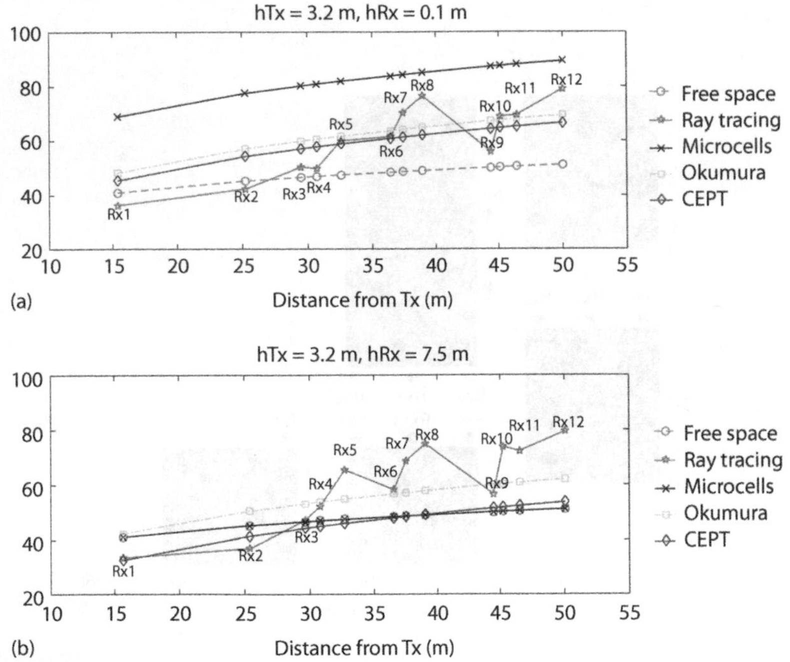

FIGURE 14.9 Comparison between ray-tracing and empirical models for transmitter TX1, placed at a height h_{tx}=3.2 m. The meters are placed at two different heights, h_{rx}=0.1 and 7.5 m.

force us to realize networks with systems that radiate more power than what really is necessary. This aspect, which turns out to not be critical for cell phone networks, which use efficient power control mechanisms, would be too expensive for simple systems such as smart meters, where communication takes place primarily in one direction (meter–collector), and where the maximum irradiated power is limited, due to power constraints of the node. For smart metering applications, a more precise knowledge of the attenuation would allow better planning of the location and number of meters to be included in the network. For this reason, the ray-tracing model is the best technique in terms of energy efficiency and network planning. But it has inherent limitations due to

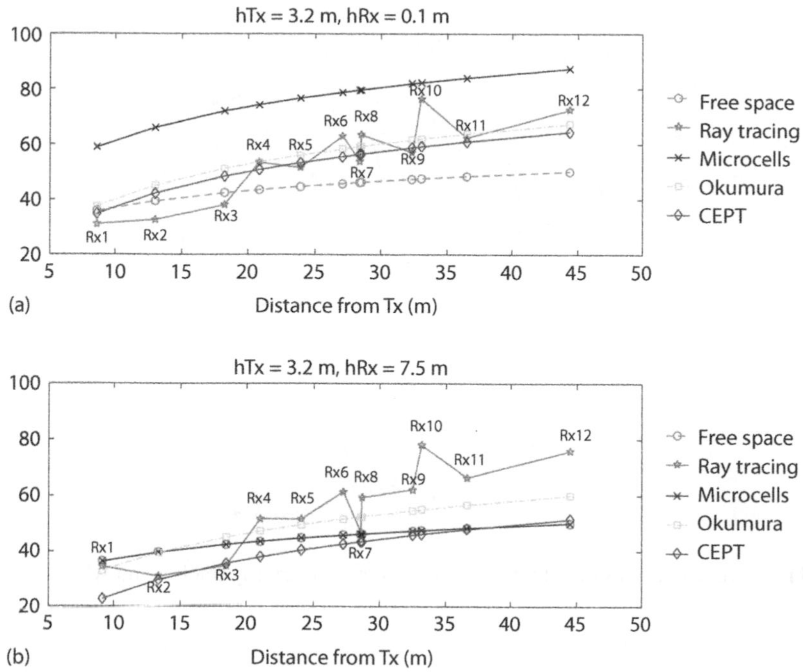

FIGURE 14.10 Comparison between ray-tracing and empirical models for transmitter TX2 (left), placed at a height h_{tx} = 3.2 m. The meters are placed at two different heights, h_{rx} = 0.1 and 7.5 m.

the required computational and memory resources. In this regard, in the past propagation models have been proposed that are based on the use of artificial neural networks (ANNs) trained with ray-tracing algorithms. This way, fast and reliable attenuation calculation in urban environments can be obtained. (Cerri et al., 2004; Cerri and Russo, 2006), the ANN approach has been proposed and applied in a real situation, to calculate the attenuation for a typical cellular phone network.

The most significant parameters, which determine the wave propagation in the environment, have been identified. They are based on the ray propagation algorithm typical of ray tracing. The parameters are essentially the angles of incidence of the dominant rays on the vertical walls and the distances covered by the same. The ANN makes use of these parameters as input and returns as output the signal attenuation at the considered point. The network was trained by calculating the attenuation with ray tracing in different urban situations, which enables the recovery of different types of propagation. The ANN was then applied to a realistic situation completely different from the ones used to train the ANN, and a strong ability to simulate the attenuation has emerged. Figures 14.11 and 14.12 show the scenario considered and the comparison between the ray-tracing and ANN simulations. The transmitter is placed 34 m above the ground, higher than the buildings that surround it. The attenuation is calculated along the path shown in the figure, at a height of 1.5 m above the ground. The comparison shows that the neural network accurately follows the attenuation calculated by the ray tracing, thus demonstrating its accuracy. The advantage, however, turns out to be the speed of execution and the reduced amount of memory required.

14.3.2.2 Power Consumption and Management in Capillary Network Devices

LPWANs can meet the communication requirements of a broad range of IoT applications, while at the same time ensuring low power consumption, to enable end points to remain connected for an extended period of time. Energy efficiency is a decisive metric to select the right technology for MTC, as EDs are typically scattered in wide areas and battery replacement may be an issue. However, energy-efficient design may be challenging, as it relates to system reliability, the rate of

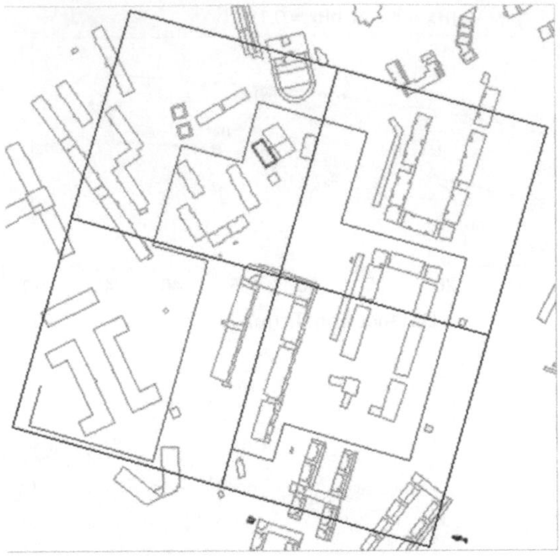

FIGURE 14.11 Urban area considered for the comparison between ray tracing and ANN.

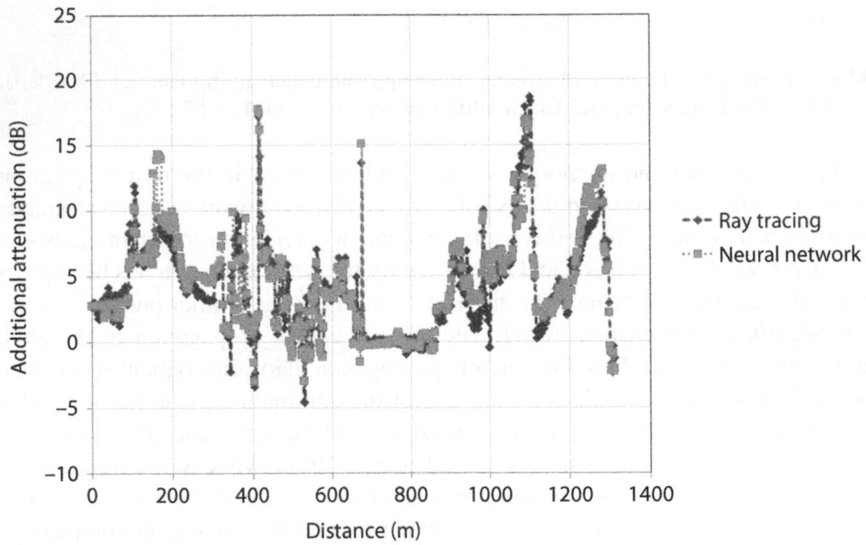

FIGURE 14.12 Comparison of the additional attenuation calculated with ray tracing and the neural network for the urban path reported in Figure 14.10.

data exchange, and the hardware design constraints of the radio chip (Ali et al., 2015). If the transmission link is not reliable enough, this may result in frequent retransmissions, implying longer active time for the devices, and greater consumption. The same effect originates from a long period of continuous activity of the radio front end, so that a reduction of the power expenditure may be obtained through a reliable communication protocol requiring a low duty cycle. In fact, IoT devices consume a high amount of power when they are in the *active* state, and a low amount of power when in the *sleep* (or *idle*) state.

MAC protocols have to rely on duty cycling to achieve energy efficiency, periodically putting the ED into a *sleep* state. However, this condition has to be balanced with the need to avoid delays in network transmissions, occurring when a node has to wait for another device to wake up and be ready to receive data. Duty-cycling coordination may be performed by gateways in hybrid and CNs, to minimize the joint cost of energy consumption and end-to-end delay (Li et al., 2015). Similarly, application-assisted power saving for devices may be supported, to keep them in the *sleep* mode as long as possible (Taneja, 2014).

The transmission success rate affects the power consumption profile of devices, as well as the latency in accessing a radio resource. From this perspective, protocols shall be designed with the aim of reducing latency for MTC, which is, among others, one of the main issues still preventing LTE from being widely used as a communication infrastructure for IoT (Koc et al., 2014). In typical IoT scenarios, a large number of devices need to be simultaneously served, in a wide covered area. This requires suitable channel access mechanisms, to share the common communication resource among them. Of the two basic approaches to shared channel access, that is, reservation based and contention based, the latter is the most adequate for MTC. Due to limited LPWAN channel bandwidth, reservation-based approaches cannot accommodate a huge number of EDs; further, they require synchronization among nodes, which gives rise to energy expenditure. Contention-based approaches, also known as random access methods, such as the well-known carrier-sense multiple access with collision avoidance (CSMA/CA) and Aloha-based protocols adopting direct-sequence spread-spectrum (DSSS) techniques at the PHY layer, do not require synchronization, and enable the nodes to sleep, in order to save energy. On the other hand, idle listening, collisions, and overhead may require additional power to the ED (Xiong et al., 2015).

The choice of a proper network topology also contributes to power saving at the EDs. The star topology with only a single hop is a quite common choice in LPWAN and CNs, as the direct communication between an access point (AP), or concentrator, and each ED in the coverage area helps minimize the transmission latency, and also avoiding unnecessary packets transmitted for routing or multihop communications.

Power management approaches, based, for example, on task scheduling to optimize operations execution with energy availability, are usually applied in devices that have to perform several functions, featuring some complexity. They are not usually found in low-cost devices like those used in IoT, whereas they may be implemented in devices exploiting an energy harvesting–based power supply, as discussed in (Rao et al., 2015; Severini et al., 2013; Li et al., 2014).

14.4 MACHINE LEARNING FOR SMART GAS AND WATER GRIDS

In recent years, machine learning has become of paramount importance in different fields, as a tool to gather knowledge and information from the vast amounts of raw data collected by sensors in IoT platforms. Machine learning algorithms allow us to automatically classify data, and discriminate the relevant situations and conditions from the ordinary ones. As an example, with respect to smart gas and water grids, the introduction of IoT smart meters is simplifying the automatic harvest of great amounts of consumption data, which can be exploited to provide innovative services like real-time billing, user profiling, and resource management, at different scales (looking at single users, buildings, or even districts). Therefore, the application of innovative algorithms on these data is a fast-growing interest. The prediction of future demands and the optimal exploitation of energy resources represent undoubtedly important issues to face from both the utility and consumer sides. On the one hand, service providers need to know in advance the demand trend in order to apply efficient management strategies, for example, to optimize the resource storage, as well as monitor the losses along distribution pipes. On the other hand, end users may need to promptly identify undetected leakages in the household network to contain, or altogether avoid, the waste of billed resources and damages to their property.

These issues can be modeled as regression and one-class classification problems that can be solved effectively by means of machine learning techniques. The regression is adopted in prediction problems, and it aims to identify a model in order to generate accurate estimates by using only the information from recent data, whereas the one-class classification, used in leakage detection problems, has the goal of characterizing the normal behavior of the data in order to discriminate any unknown anomalies. Moreover, with the recently introduced smart metering systems, which can collect and store the consumption data of an entire utility grid, the machine learning approaches are able to greatly improve their usefulness, since the same approach can address a single house, as well as a district or the entire grid.

Therefore, in order to better investigate the advantages of smart sensing points in the residential network of water and natural gas, an analysis of the state of the art concerning machine learning approaches for load forecasting and leakage detection is presented. For both fields, recent technological solutions appearing in the literature are presented and discussed.

14.4.1 Load Forecasting

Among load forecasting techniques, the support vector machine (SVM) has been widely exploited to deal with forecasting problems. Among its variations, the least-squares SVM (LS-SVM) has raised strong interest, and recently, Ji et al. (2014) and Zhu and Chen (2013) have proposed the ameliorate teaching-learning-based optimization (ATLBO) and the quantum particle swarm optimization based on phase encoding (PQPSO), for the selection of the optimum parameters in urban water consumption forecasting procedures.

Specifically, the ATLBO (Ji et al., 2014) has improved the standard teaching-learning-based optimization (TLBO). By selecting and keeping the best elite individuals (learners) for the next generation, ATLBO enhances the convergence ability of its predecessor. An adaptive teaching factor has also been introduced in the teaching phase, and additionally, the self-monitoring mechanism in the learner phase has been improved. The proposed enhancements have allowed us to gain a better regression precision than TLBO, particle swarm optimization (PSO), and grid search. The water dataset was collected in Shanghai (Qingcaosha water supply system, line 2) from June 1, 2012, to May 31, 2013, and heterogeneous information was also used, that is, the maximum and minimum temperatures, the precipitations, and the holiday information. The overall results for each technique have been evaluated in terms of mean relative error (MRE) and mean square error (MSE).

The novel method PQPSO (Zhu and Chen, 2013) has improved the quantum particle swarm optimization (QPSO) approach by expressing the state information as phases of a qubit, rather than as a linear combination of 0 and 1, and by introducing a method of adaptive adjustment of the inertia factor. The proposed method has demonstrated better prediction accuracy and computing speed than the approaches based on SVM and LS-SVM. The system performance has been evaluated on a database composed of water consumption and meteorological data from April 1 to July 10, 2010. The adopted evaluation criteria were the relative error (RE) and the MRE.

Recently, due to the promising improvements accomplished by the deep belief network (DBN) and recurrent neural network (RNN), new attention has been paid to neural network applications.

Concerning urban water prediction, a recent innovation has been proposed by Zhu and Xu (2012). Specifically, the authors have combined the QPSO algorithm with a radial basis function (RBF) neural network, in order to calculate the parameters (weights) of the net and to achieve a higher accuracy level of the prediction. The effectiveness of the approach has been confirmed by experimental results, as well as the higher convergence speed with respect to the standard RBF network. The urban water consumption and meteorological data from April 1 to July 10, 2010, have been adopted as a dataset. The evaluation criteria were the RE and mean square relative error (MSRE).

A neural network, with a logistic transfer function, has also been adopted by Azari et al. (2012), in order to predict the daily and monthly gas consumption. The approach has been tested by adopting heterogeneous data. Specifically, for the daily prediction of the meteorological parameters, the gas consumption data for the previous 5 days and the meteorological parameters forecasted for the prediction day have been adopted as input information, whereas only the monthly effective temperature for the previous and predicted months and the gas consumption for the previous month have been used for the monthly prediction. The tests have been performed with the gas consumption and meteorological data of Tehran from March 21, 2001, to August 8, 2005, and the only evaluation criterion adopted was the RE.

Shabri and Samsudin (2014) have exploited a neural network in order to compose a hybrid integrating empirical mode decomposition (EMD) to forecast monthly water demand series. In the EMD method, the original data are decomposed into a sum of intrinsic mode function (IMF) components with individual intrinsic timescale properties. Experimental results have confirmed the better performance of the EMD method combined with the neural network (EMDANN), over the use of the neural network only. Moreover, an autoregressive integrated moving average (ARIMA) model has also been adopted in the EMD, but both the EMD-ARIMA and ARIMA approaches have provided worse results than EMD-ANN. The Batu Pahat city water consumption from January 1995 to December 2011 was used to test the approaches, and the root mean square error (RMSE), mean absolute error (MAE), and coefficient of correlation (R) were adopted as evaluation criteria.

Another well-known method, adopted to address the prediction problem, is the regressive model theory and its variations. Recently, contributions in that field have been provided by Quevedo et al. (2014), Brown et al. (2015), and Akpinar and Yumusak (2013).

ARIMA and basic structural and exponential smoothing models have been used by Quevedo et al. (2014), in order to produce hourly water prediction. The results of the models have been compared by means of the explained variance (EV), MAE, MSE, and mean absolute percentage error (MAPE). The dataset has been composed of sampled values from 100 pressure-sensing points in the Barcelona water network.

As pointed out by Brown et al. (2015), many factors need to be taken into account in order to obtain a proper prediction. Therefore, in order to produce more accurate long-term predictions of natural gas, the authors have proposed a novel "detrending" algorithm. The detrending approach, based on linear regression, exploits temperature information, specifically the heating degree day, the change in heating degree day, and the cooling degree day. A comparison against the state-of-the-art detrending approach has proven the performance improvement of the proposed approach in the forecast of daily gas consumption. The adopted dataset, composed of a natural gas consumption series from a U.S.-based local distribution company and temperature data, spans over 15 years, and the RMSE and the weighted mean average percentage error (WMAPE) were adopted as evaluation criteria.

As with the previous contribution, in order to achieve a higher accuracy with an ARIMA model, Akpinar and Yumusak (2013) have proposed the removal of the "cycling component" from the data series. The better performance of the models without a cycling component has been proven in the experimental results, evaluated by means of the relative absolute error (RAE), the MAPE, the RMSE, and the standard percentage error (PE). The tests have been performed by using the daily gas consumption from 2009 to 2012 of Sakarya, Turkey.

Bakker et al. (2013) have presented an adaptive forecasting model to predict the short-term water consumption. Heterogeneous information, that is, static calendar data, has been adopted to improve the accuracy of the prediction. The model has been composed as a regression model in which the main contributions are updated at each new input data. These main contributions are the average forecasted demand for the next 48 h, the normal forecasted demand for the 15 min step, and the extra sprinkle forecasted demand for the 15 min step. The datasets of urban water demand are collected over six different areas in the period 2006–2011 with a sample rate of 15 min. The model performance has been evaluated by using the RE, MAPE, relative root mean square error (RRMSE), and determination coefficient (R^2).

The grey system theory has been widely adopted as well, in combination with other well-known techniques, as recently presented by Wan et al. (2014) and Wang et al. (2014a,b).

For instance, Wan et al. (2014) have proposed a combination of the grey model and the Markov chain to predict the annual natural gas demand. The results have been evaluated in terms of predicted value and difference between predicted and actual values.

In their first contribution, Wang et al. (2014a) adopted the grey theory to enhance the differential evolution (DE) algorithm, by producing a novel model called Step-DE-GM. Whereas in the latter contribution (Wang et al. 2014b), the grey model was combined with a back-propagation neural network in order to forecast the urban water consumption and deal with an insufficient amount of data, required for a robust optimization, or train, of the network. The used dataset includes historical annual data from 2001 to 2012, and is composed by the datasets used in both the contributions.

A nonlinear combination forecasting model, based on a generalized dynamic fuzzy neural network (GD-FNN), has been proposed by Chen et al. (2014) to predict daily gas demand. The fuzzy rule, which defines the FNN structure, is not predetermined, and it may change during the learning process. Moreover, the elliptic basis function has been adopted in order to allow more flexibility and a wider range of nonlinear transformation. The daily gas load of Hardin, China, from April 17 to June 29, 2008, was used to evaluate the performance, and the RE was selected as the evaluation criterion.

As seen so far, many contributions have presented approaches aiming to provide forecasts with high accuracy levels for urban demands of water and natural gas. Actually, none of the contributions target the consumption forecast of a single house; that is, none can be used with the data collected by domestic metering systems, and thus representing the demands of a single household. As reported by Fagiani et al. (2015), this is partially due to the lack of suitable databases, because either they are not available, as well as not publicly available, or they do not provide suitable data, due to low time resolution and/or too short of a time series.

Moreover, in literature very promising techniques and results are often presented, but in many aspects, the information supplied is inadequate to provide a comprehensive and objective comparison among the contributions. First, for various contributions, it is difficult to have a clear understanding of the achieved performance, and homogeneous evaluation criteria among the contributions are not provided. Second, most of the contributions have not adopted common databases, for the same problem of availability cited above.

For those reasons, Fagiani et al. (2015), in order to fulfill a comprehensive comparison among state-of-the-art techniques, have executed experiments on short-term predictions of water and natural gas by adopting two common datasets. The evaluated techniques encompass genetic programming (GP), support vector machine for regression (SVR), ANNs, echo-state networks (ESNs), DBNs, and extreme learning machine (ELM). The forecasting experiments have been conducted by adopting two publicly available datasets, representing the consumption of living (or domestic) and office building environments. Moreover, for the living environment, the performance has been evaluated by also considering heterogeneous information.

With regard to this, for domestic consumption the best predictions of natural gas consumption have been achieved with SVR, by including both water consumption and temperature information, whereas the ANN performs better for water prediction, without the need for additional information.

For the office building environment, the ANN achieved the best performance with both water and natural gas consumption.

14.4.2 Leakage Detection

In literature, many contributions have been produced to address the leakage and fault detection problem in the industrial environment. But, being mainly aimed at oil and natural gas pipelines, they are based on data collected at high sampling rates and depend on intrusive and/or manual (i.e., operated by person) detection techniques. Furthermore, the data could be also collected by multiple sensing points arranged along the pipeline. In the urban distribution network of the utilities, however, sampling rates are usually low, whereas sensing points cannot be arranged along each branch of the piping. In fact, they are usually placed at very specific points and, of course, at the end user's home. Because of these conditions, techniques aimed at the industrial environment are not suited to address the living one.

Therefore, only state-of-the-art contributions, suitable for water and natural gas in a living environment, are presented. Moreover, in order to provide a comprehensive view of all the suitable approaches, contributions based on district network data have been also reported.

Common computational intelligence techniques, such as ANN and support vector regression (SVR), have been exploited by Nasir et al. (2014). To achieve the estimated position and size of water leakages, an EPANET* (Rossman, 1993) simulation of a residential network has been performed, and the raw data, acquired by two pressure sensors, two differential pressure sensors, and two flow sensors, have been used to predict the leakage parameters. The performance has been evaluated in terms of MSE and squared correlation error coefficient (R^2). The proposed quasi-static analysis confirmed the good behavior of the SVM and its resilience to sensor measurement errors.

SVM has been also exploited by Salam et al. (2014), in order to analyze the pressure change pattern when a leakage occurs, and therefore identify the leakage position and size. A real network system of a district area has been reproduced with EPANET (Rossman, 1993), and all the pressure data collected at the junctions have been used. The overall performance has been reported in terms of RMSE.

In order to generate a set of rules to categorize the features data, as *leakage* or *leakage-free*, Gamboa-Medina et al. (2014) have used the well-known decision tree–based algorithm named C4.5, by J. R. Quinlan. Water pressure data, at high sample rate, have been collected from a controlled experimental laboratory circuit, and a set of four features have been extracted. The single features and their combinations have been evaluated in terms of receiver operating characteristic (ROC) and area under the curve (AUC).

Among the leakage detection techniques applied to large networks, monitoring of the minimum night flow (MNF) is widely used. Recently, Alkasseh et al. (2013) have exploited a multiple linear regression method to correlate the overall loss and the number of connections, the total length of pipe, and the weighted mean age of the pipe of the network. The difference between the actual MNF and the estimated one, which could allow us to establish the presence of a leakage, has been evaluated by using R and R^2.

Even fuzzy logic has been applied to both detection and localization of leakages in water networks by Sanz et al. (2012). The proposed fuzzy inductive reasoning (FIR) approach has been applied to data collected by two pressure sensors located in a district network, whereas the leakage data have been synthetically created by means of the EPANET (Rossman, 1993) software. The detection performance has been evaluated by pointing out the total amount of leakages detected in the various experiments.

The only contribution that reports an approach developed directly for domestic water systems has been presented by Oren and Stroh (2013). The authors have proposed a mathematical model, based on the definition of threshold values regulated by means of average domestic water usages. The approach exploits the data acquired in one sensing point, but neither exhaustive experiments nor evaluation criteria have been presented.

Even the novel change detection test (CDT), developed by Boracchi and Roveri (2014), has been developed adopting only the information collected in a single flow-sensing point in the Barcelona water distribution network. The approach allows the detection of structural changes in the time series, and tests with different types of manipulations have been carried out, that is, leakage, sensor degradation, source change, and stack-at. The evaluation criteria have been directly derived from the ones applied in novelty detection: false positive rate (FPR), false negative rate (FNR), and detection delay (DD).

As seen so far, in the recent literature on leakage detection aimed at residential and district networks, none of the contributions have taken into account the natural gas. Moreover, most of the contributions (Nasir et al., 2014; Gamboa-Medina et al., 2014; Alkasseh et al., 2013; Sanz et al., 2012), have developed approaches by assuming that the input data were composed of flow and/or pressure data collected in multiple sensing points. Therefore, these approaches are unsuitable for residential application, where only one sensing point is available. In addition, even the suitable

* Software that models the hydraulic and water quality behavior of water distribution piping systems.

ones (Oren and Stroh, 2013; Boracchi and Roveri, 2014) show a few shortcomings from the computational approach standpoint: in the former, a "real-time" identification of the leakages is not performed, whereas the latter approach seems to lack appropriate experimental validation.

An approach suitable for a water and natural gas residential network has been presented by Fagiani et al. (2016) and Global Mobile Suppliers Association (2016), where the authors have proposed statistical modeling, exploiting a Gaussian mixture model (GMM) and hidden Markov model (HMM). The approaches allow us to perform real-time monitoring of the network status, by verifying step-by-step the presence of a leakage. In the first contribution (Fagiani et al., 2015), only the flow information has been used, and a set of suitable features have been selected. The false detection rate (FDR), the true detection rate (TDR), the ROC, and the AUC, have been adopted as evaluation criteria. The experimental results have proven the suitability of the approach for both water and natural gas. The HMM has achieved the best performance by assuming input data at 1 min of the sample rate.

In the latter contribution (Fagiani et al., 2016), by adopting the simulation tool EPANET (Rossman, 1993), the introduction of pressure information has been evaluated. The experimental results have proven the validity of the combination of flow and pressure features, confirming that it is possible to discriminate a leakage with a very low error rate, even by using the GMM model. Specifically, the adoption of flow and pressure information can be used to apply leakage detection techniques to consumption data collected at very low sampling rates, in the order of 30 min, thus lowering the computational burden, as well.

14.5 FUTURE PERSPECTIVES: CELLULAR IoT

The fourth-generation (4G)/LTE technology is significantly contributing to closing the digital divide, delivering mobile broadband in the most efficient way, in both developing and developed economies globally, as reported by the Global Mobile Suppliers Association (2016). Together with the ever-rising performance achievements and successes of LTE and LTE-Advanced systems and device capabilities, it is equally important that other LTE user terminals are available to meet the needs of developing markets, where cost factors and flexibility are particularly important.

Existing cellular networks already offer very good area coverage in mature markets. However, many potential "connected objects" are located in vast remote areas, far away from the next cellular BS. When coverage is available, it is often weak, which requires the device transmitter to operate at high power, draining the battery. In addition, cellular networks are not optimized for applications that occasionally transmit small amounts of data. A battery life of several years, combined with an inexpensive device, cannot be realized on existing cellular standards, as they do not support the required power-saving mechanisms. Additionally, mobile devices working on GSM, 3G, and LTE are designed for a variety of services, including mobile voice, messaging, and high-speed data transmission. However, IoT applications just require low-speed but reliable data transfer, and an appropriate level of reliability (Minerva, 2014). Therefore, using cellular devices for IoT applications requiring low capacity means using devices that are too expensive for the application, also due to practical aspects, such as ease of installation, or risk of theft.

New PHY layer solutions, MAC procedures, and network architectures are needed to evolve the current LTE cellular systems to meet the demands of IoT services (Ratasuk et al., 2015). Within the 3GPP, several efforts have been undertaken, to include the necessary amendments in the upcoming LTE standards release (3GPP Release 13 (Third Generation Partnership Project (3GPP), 2015), which will support enhanced machine-type communications (eMTC), and new MAC and higher-layer procedures provided by extended discontinuous reception (DRX), together with the so-called narrowband IoT (NB-IoT) (Rico-Alvarino et al., 2016).

The design of the eMTC amendment has to account for several requirements: most of the existing LTE PHY layer procedures should be reused, but new features are needed, to reduce the cost and power consumption of user equipment (UE) and at the same time extend coverage. eMTC should be deployed with the existing infrastructure, simply by means of a software update at the eNodeB, so that eMTC UE may coexist with legacy LTE mobile stations. For cell search and initial access, eMTC UE uses the same signals and channels as a legacy LTE UE. The maximum channel bandwidth for eMTC is reduced to 1.08 MHz, corresponding to 6 LTE resource blocks (RBs); an eMTC UE performs narrowband operations to transmit and receive physical channels and signals. A predefined set of six contiguous RBs in which an eMTC UE can operate is established as a new frequency unit, named *narrowband*.

Compared with eMTC, NB-IoT will further decrease the bandwidth requirements to 180 kHz, also pushing a reduction of the device complexity, but decreasing the available peak data rate (around 50 kbps for uplink and 30 kbps for downlink). In fact, while eMTC targets higher-data-rate and possible mobility requirements (e.g., for applications based on the use of wearables), NB-IoT will support limited mobility procedures, and a very low data rate, thus keeping the possibility of reusing existing GSM or LTE spectrum. A classical scenario for adopting NB-IoT is smart metering, as well as smart monitoring of water and heat distribution plants.

NB-IoT is designed to support three different deployment scenarios, not necessarily within the same spectrum as LTE: guard band, in-band, and stand-alone. The stand-alone deployment mainly uses new bandwidth, whereas the guard-band deployment exploits the bandwidth reserved in the guard band of the existing LTE networks. Finally, in-band deployment makes use of the same RBs in the LTE carrier of the existing LTE network.

New physical channels are needed in NB-IoT, for synchronization, broadcast information, and random access, due to the featured bandwidth reduction, as well as new reference signals in downlink, for channel estimation, tracking, and demodulation.

14.6 CONCLUSION

IoT is steadily becoming a fundamental technological enabler for a number of different applications, services, and scenarios. Among them, the smart city has attracted the attention of the scientific and technical communities for several years, as it offers the possibility to tackle relevant issues, affecting citizens and users on a large scale, by innovative approaches at different levels.

This chapter provided an overview of the enabling technologies for IoT-oriented smart water and gas grids, by approaching both the communications and networking-related issues, and the opportunities opened by the application of analytics and machine learning techniques to the data provided by the grids.

Two main directions in the design of smart networks for ICT-enabled grids emerge: the former adopts the CN paradigm, according to which low-power wide area networks exploiting the subgigahertz frequencies are backhauled by cellular networks, thus providing long-range connectivity to devices that do not natively belong to the cellular domain. The latter envisions the most recent amendment approved by the 3GPP for the LTE standard, which paves the way to future NB-IoT, with suitable profiles for machine-type communications enabled within the legacy LTE network.

The selection of the network architecture that better fits the requirements of smart water and gas grids for metering applications relies on the propagation channel characterization, which needs to be revisited, with respect to well-established models used to predict the radio coverage in wireless cellular networks. From the scenarios analyzed and the results reported, it is evident that the knowledge of the static channel is a critical point for the planning of smart metering networks. In particular, the additional attenuation knowledge within urban areas can be obtained with uncertainties that can reach values higher than 20 dB when empirical models are used, if compared with more accurate deterministic models. Because of the peculiarity of the smart metering networks, it is not always possible to introduce compensation margins of these uncertainties in the channel characterization; therefore, it might be preferable to use a more accurate model, although more complex to implement. An alternative could be the use of the ANN technique in order to have a fast and accurate calculation of the attenuations. With regard to the uncertainties related to the dynamic variability of the channel, the smart metering network is not critical. Usually the type of information sent from the meters to the collector is very simple, and it is not transmitted continuously, but with a certain repetition that can be predetermined, or on demand by the collector. In the former case, the data will be lost, whereas in the latter case, if an acknowledgment of receipt is provided by the concentrator, the transmission will be repeated; if not, the data will be lost. In both cases, the type of data transmitted can be lost without any significant reduction of network performance.

Finally, a detailed state-of-the-art analysis of machine learning techniques for data processing in the context of smart water and natural gas grids has also been provided. In recent years, increasing interest has been registered on a worldwide scale from this perspective, as confirmed by the many commercial engineering solutions already available on the market and the many contributions that have appeared in the scientific literature so far. Load forecasting and automatic leakage detection problems, which undoubtedly represent the most relevant issues for advanced monitoring purposes, have been specifically addressed. Besides the review of the most performed machine learning techniques in this field, the authors also reported the most used databases for training and testing the employed data-driven algorithms, together with the adopted evaluation criteria and indexes. In doing this, a certain emphasis was given to highlight the actual miss of a benchmarked approach in the literature to comparatively evaluate the new proposed algorithmic solutions. Even though some interesting attempts have been recently made in this sense, such an issue will surely represent an asset for the scientific community in the next years.

ACKNOWLEDGMENT

The authors of this chapter acknowledge the support of the project "TWIST: Tecnologie WIreless eterogenee e Sostenibili nelle smarT cities del futuro"—Ricerca Scientifica di Ateneo 2014 (in Italian), promoted by the Dipartimento di Ingegneria dell'Informazione at Università Politecnica delle Marche (Ancona, Italy).

REFERENCES

Akpinar, M. and N. Yumusak. 2013. Forecasting household natural gas consumption with ARIMA model: A case study of removing cycle. In *2013 7th International Conference on Application of Information and Communication Technologies (AICT)*, Azerbaijan, Baku, October 2013, 1–6.

Ali, A., W. Hamouda, and M. Uysal. 2015. Next generation M2M cellular networks: Challenges and practical considerations. *IEEE Communications Magazine* 53 (9): 18–24.

Alkasseh, J. M. A., M. N. Adlan, I. Abustan, H. A. Aziz, and A. B. M. Hanif. 2013. Applying minimum night flow to estimate water loss using statistical modeling: A case study in Kinta Valley, Malaysia. *Water Resources Management* 27(5): 1439–1455.

Azari, A., M. Shariaty-Niassar, and M. Alborzi. 2012. Short-term and medium-term gas demand load forecasting by neural networks. *Iranian Journal of Chemistry and Chemical Engineering* 31(4): 77–84.

Bakker, M., J. H. G. Vreeburg, K. M. van Schagen, and L. C. Rietveld. 2013. A fully adaptive forecasting model for short-term drinking water demand. *Environmental Modelling & Software* 48(0): 141–151.

Ballesteros, L. G. M., O. Alvarez, and J. Markendahl. 2015. Quality of experience (QOE) in the smart cities context: An initial analysis. In *2015 IEEE First International Smart Cities Conference (ISC2)*, Guadalajara, Mexico, October 2015, 1–7.

Barclay, L., ed. 2003. *Propagation of Radiowaves*, 2nd ed., Stevenage, UK: IEE.

Boracchi, G. and M. Roveri. 2014. Exploiting self-similarity for change detection. In *2014 International Joint Conference on Neural Networks (IJCNN)*, Beijing, China, 3339–3346.

Brown, R. H., S. R. Vitullo, G. F. Corliss, M. Adya, P. E. Kaefer, and R. J. Povinelli. 2015. Detrending daily natural gas consumption series to improve short-term forecasts. In 2015 *IEEE Power Energy Society General Meeting*, July 2015, 1–5.

Cai, Y., T. Huang, E. Bompard, Y. Cao, and Y. Li. 2017. Self-sustainable community of electricity prosumers in the emerging distribution system. *IEEE Transactions on Smart Grid* 8 (5):2207–2216. doi:10.1109/TSG.2016.2518241.

Cerri, G. and P. Russo. 2006. Application of an automatic tool for the planning of a cellular network in a real town. *IEEE Transactions on Antennas and Propagation* 54: 2890–2901.

Cerri, G., M. Cinalli, F. Michetti, and P. Russo. 2004. Feed forward neural networks for path loss prediction in urban environment. *IEEE Transactions on Antennas and Propagation* 52: 3137–3139.

Chen, H., Z. Wang, and P. Yu. 2014. Study on combination forecasting of gas daily load based on the generalized dynamic fuzzy neural network. In *2014 33rd Chinese Control Conference (CCC)*, Nanjing, China, July 2014, 6235–6239.

Commission of the European Communities and COST Telecommunications. 1999. Digital mobile radio towards future generation systems. Cost 231 Final Report. ITU-T Focus Group on Smart Sustainable Cities FGSSC.

Electronic Communications Committee. 2004. Compatibility between existing and proposed SDRs and other radiocommunication applications in the 169.4–169.8 MHz frequency band. ECC Report 55. October 2004.

Fagiani, M., S. Squartini, L. Gabrielli, M. Pizzichini, and S. Spinsante. 2014. Computational intelligence in smart water and gas grids: An up-to-date overview. In *2014 International Joint Conference on Neural Networks (IJCNN)*, Beijing, China, July 2014, 921–926.

Fagiani, M., S. Squartini, L. Gabrielli, S. Spinsante, and F. Piazza. 2015. A review of datasets and load forecasting techniques for smart natural gas and water grids: Analysis and experiments. *Neurocomputing* 170: 448–465.

Fagiani, M., S. Squartini, M. Severini, and F. Piazza. 2015. A novelty detection approach to identify the occurrence of leakage in smart gas and water grids. In *2015 International Joint Conference on Neural Networks (IJCNN)*, Killarney, Ireland, July 2015, 1–8.

Fagiani, M., S. Squartini, R. Bonfigli, M. Severini, and F. Piazza. 2016. Exploiting temporal features and pressure data for automatic leakage detection in smart water grids. In *2016 IEEE World Congress on Computational Intelligence (WCCI)*, Vancouver, Canada, 5, July 2016, 295–302.

Gabrielli, L., M. Pizzichini, S. Spinsante, S. Squartini, and R. Gavazzi. 2014. Smart water grids for smart cities: A sustainable prototype demonstrator. In *2014 European Conference on Networks and Communications (EuCNC)*, Bologna, Italy, June 2014, 1–5.

Gamboa-Medina, M. M., L. F. Ribeiro Reis, and R. Capobianco Guido. 2014. Feature extraction in pressure signals for leak detection in water networks. *Procedia Engineering* 70 (0): 688–697.

Ganchev, I., Z. Ji, and M. O'Droma. 2014. A generic IoT architecture for smart cities. In *25th IET Irish Signals Systems Conference 2014 and 2014 China-Ireland International Conference on Information and Communications Technologies (ISSC 2014/CIICT 2014)*, Limerick, Ireland, June 2014, 196–199.

Geisler, K. 2013. The relationship between smart grids and smart cities. IEEE Smart Grid Newsletter Compendium, May 2013. http://www.egr.msu.edu/aesc310/resources/SmartCities/Smart%20Grids%20Smart%20Cities%20IEEE.pdf.

Gellings, C. W. 2015. A globe spanning super grid. *IEEE Spectrum* 52 (8): 48–54.

Global Mobile Suppliers Association. 2016. Status of the LTE ecosystem. June 2016. http://gsacom.com/paper/status-of-the-lte-ecosystem/.

Hata, M. 1980. Empirical formula for propagation loss in land mobile radio services. *IEEE Transactions on Vehicular Technology* 29: 317–325.

Hornsby, C., M. Ripa, C. Vassillo, and S. Ulgiati. 2017. A roadmap towards integrated assessment and participatory strategies in support of decision-making processes. The case of urban waste management. *Journal of Cleaner Production* 142 (Pt. 1): 157–172.

IEEE Standards Association. 2013. IEEE standard for information technology—Telecommunications and information exchange between systems—Local and metropolitan area networks—Specific requirements. Part 11: Wireless LAN medium access control (MAC) and physical layer (PHY) specifications. Amendment 4: Enhancements for very high throughput for operation in bands below 6 GHz. http://standards.ieee.org/getieee802/download/802.11ac-2013.pdf.

International Organization for Migration. 2015. World migration report 2015: Migrants and cities, new partnerships to manage mobility. http://publications.iom.int/system/files/wmr2015_en.pdf.

Ivic, N. S., O. Ur-Rehman, and C. Ruland. 2015. Evolution of smart metering systems. In *2015 23rd Telecommunications Forum Telfor (TELFOR)*, Belgrade, Serbia, November 2015, 635–638.

Ji, G., J. Wang, Y. Ge, and H. Liu. 2014. Urban water demand forecasting by LS-SVM with tuning based on elitist teaching-learning-based optimization. In *26th Chinese Control and Decision Conference (2014 CCDC)*, Changsha, China, May 2014, 3997–4002.

Jrventausta, Pertti, Sami Repo, Antti Rautiainen, and Jarmo Partanen. 2010. Smart grid power system control in distributed generation environment. *Annual Reviews in Control* 34 (2): 277–286.

Kahn, C. and H. Viswanathan. 2015. Connectionless access for mobile cellular networks. *IEEE Communications Magazine* 53 (9): 26–31.

Kaur, M. J. and P. Maheshwari. 2016. Building smart cities applications using IoT and cloud-based architectures. In *2016 International Conference on Industrial Informatics and Computer Systems (CIICS)*, United Arab Emirates, March 2016, 1–5.

Khorov, E., A. Krotov, and A. Lyakhov. 2015. Modelling machine type communication in IEEE 802.11ah networks. In *2015 IEEE International Conference on Communication Workshop (ICCW)*, London, United Kingdom, June 2015, 1149–1154.

Koc, T., S. C. Jha, R. Vannithamby, and M. Torlak. 2014. Device power saving and latency optimization in LTE-A networks through DRX configuration. *IEEE Transactions on Wireless Communications* 13 (5): 2614–2625.

Li, Y., K. K. Chai, Y. Chen, and J. Loo. 2015. Duty cycle control with joint optimisation of delay and energy efficiency for capillary machine-to-machine networks in 5G communication system. *Transactions on Emerging Telecommunications Technologies* 26 (1): 56–69.

Li, Y., Z. Jia, and X. Li. 2014. Task scheduling based on weather forecast in energy harvesting sensor systems. *IEEE Sensors Journal* 14 (11): 3763–3765.

LoRa Alliance. 2016. https://www.lora-alliance.org/.

Luo, X., J. Wang, M. Dooner, and J. Clarke. 2015. Overview of current development in electrical energy storage technologies and the application potential in power system operation. *Applied Energy* 137: 511–536.

Matamoros, J. and C. Anton-Haro. 2013. Traffic aggregation techniques for environmental monitoring in M2M capillary networks. In *2013 IEEE 77th Vehicular Technology Conference (VTC Spring)*, Dresden, Germany, June 2013, 1–5.

Mikhaylov, K., J. Petaejaejaervi, and T. Haenninen. 2016. Analysis of capacity and scalability of the LoRa low power wide area network technology. In *22th European Wireless Conference, European Wireless 2016*, Oulu, Finland, May 2016, 1–6.

Minerva, R. 2014. From Internet of things to the virtual continuum: An architectural view. In *2014 Euro Med Telco Conference (EMTC)*, Naples, Italy, November 2014, 1–6.

Minerva, R., A. Biru, and D. Rotondi. 2015. Towards a definition of the internet of things (IoT). IEEE Internet Initiative, May 2015. https://iot.ieee.org/images/files/pdf/IEEE_IoT_Towards_Definition_Internet_of_Things_Issue1_14MAY15.pdf.

Monnier, O. 2013. A smarter grid with the Internet of things. Texas Instruments White Paper. http://www.ti.com/lit/ml/slyb214/slyb214.pdf.

Mutchek, M. and E. Williams. 2014. Moving towards sustainable and resilient smart water grids. *Challenges* 5 (1): 123.

Nasir, M. T., M. Mysorewala, L. Cheded, B. Siddiqui, and M. Sabih. 2014. Measurement error sensitivity analysis for detecting and locating leak in pipeline using ANN and SVM. In *2014 11th International Multi-Conference on Systems, Signals Devices (SSD)*, Castelldefels-Barcelona, Spain, 2014, 1–4.

Nielsen, J. J., G. C. Madueo, N. K. Pratas, R. B. Srensen, C. Stefanovic, and P. Popovski. 2015. What can wireless cellular technologies do about the upcoming smart metering traffic? *IEEE Communications Magazine* 53 (9): 41–47.

Novo, O., N. Beijar, M. Ocak, J. Kjllman, M. Komu, and T. Kauppinen. 2015. Capillary networks—Bridging the cellular and IoT worlds. In *2015 IEEE 2nd World Forum on Internet of Things (WF-IoT)*, Milan, Italy, December 2015, 571–578.

Okumura, Y., E. Ohmori, T. Kawano, and K. Fukuda. 1968. Field strength and its variability in VHF and UHF land-mobile radio service. *Review of the Electrical Communication Laboratory* 16 (9–10): 825–873.

Open Metering System Group. 2002–2011. Communication systems for meters and remote reading of meters. Parts 1–5.

Oren, G. and N. Y. Stroh. 2013. Mathematical model for detection of leakage in domestic water supply systems by reading consumption from an analogue water meter. *International Journal of Environmental Science and Development* 4 (4): 386–389.

Petajajarvi, J., K. Mikhaylov, A. Roivainen, T. Hanninen, and M. Pettissalo. 2015. On the coverage of LPWANs: Range evaluation and channel attenuation model for LoRa technology. In *2015 14th International Conference on ITS Telecommunications (ITST)*, Copenhagen, Denmark, December 2015, 55–59.

Quevedo, J., J. Saludes, V. Puig, and J. Blanch. 2014. Short-term demand forecasting for real-time operational control of the Barcelona water transport network. In *2014 22nd Mediterranean Conference of Control and Automation (MED)*, Palermo, Italy, June 2014, 990–995.

Rao, V. S., R. V. Prasad, and I. G. M. M. Niemegeers. 2015. Optimal task scheduling policy in energy harvesting wireless sensor networks. In *2015 IEEE Wireless Communications and Networking Conference (WCNC)*, Istanbul, Turkey, March 2015, 1030–1035.

Rappaport, T. S. 1996. *Wireless Communication*. Upper Saddle River, NJ: Prentice Hall.

Ratasuk, R., N. Mangalvedhe, and A. Ghosh. 2015. Overview of LTE enhancements for cellular IoT. In *2015 IEEE 26th Annual International Symposium on Personal, Indoor, and Mobile Radio Communications (PIMRC)*, Hong Kong, August 2015, 2293–2297.

Rico-Alvarino, A., M. Vajapeyam, H. Xu, X. Wang, Y. Blankenship, J. Bergman, T. Tirronen, and E. Yavuz. 2016. An overview of 3GPP enhancements on machine to machine communications. *IEEE Communications Magazine* 54 (6): 14–21.

Rossman, L. A. 1993. *The EPANET Water Quality Model*. Vol. 2, Coulbeck, B. ed. Somerset, England: Research Studies Press Ltd. Software available at www.epa.gov/nrmrl/wswrd/dw/epanet.html.

Sachs, J., N. Beijar, P. Elmdahl, J. Melen, F. Militano, and P. Salmela. 2014. Capillary networks—A smart way to get things connected. *Ericsson Review*. https://www.ericsson.com/res/thecompany/docs/publications/ericsson_review/2014/er-capillary-networks.pdf.

Salam, E. U., M. Tola, M. Selintung, and F. Maricar. 2014. A leakage detection system on the water pipe network through support vector machine method. In *2014 Makassar International Conference on Electrical Engineering and Informatics (MICEEI)*, Makassar, Indonesia, November 2014, 161–165.

Sanz, G., R. Perez, and A. Escobet. 2012. Leakage localization in water networks using fuzzy logic. In *20th Mediterranean Conference on Control Automation (MED)*, Barcelona, Spain, 646–651.

Saunders, S. 1999. *Antennas and Propagation for Wireless Communication Systems*, 1st ed. Chichester, UK: Wiley.

Severini, M., S. Squartini, and F. Piazza. 2013. Energy-aware lazy scheduling algorithm for energy-harvesting sensor nodes. *Neural Computing and Applications* 23(7): 1899–1908.

Shabri, A. and R. Samsudin. 2014. A new approach for water demand forecasting based on empirical mode decomposition. In *2014 8th Malaysian Software Engineering Conference (MySEC)*, Langkawi, Malaysia, September 2014, 284–288.

Shariatmadari, H., P. Osti, S. Iraji, and R. Jntti. 2015. Data aggregation in capillary networks for machine-to-machine communications. In *2015 IEEE 26th Annual International Symposium on Personal, Indoor, and Mobile Radio Communications (PIMRC)*, Hong Kong, China, August 2015, 2277–2282.

Sigfox. 2016. http://www.sigfox.com/en/.

Singh, S. and K. L. Huang. 2011. A robust M2M gateway for effective integration of capillary and 3GPP networks. In *2011 Fifth IEEE International Conference on Advanced Telecommunication Systems and Networks (ANTS)*, Bengaluru, India, December 2011, 1–3.

Spinsante, S., M. Pizzichini, M. Mencarelli, S. Squartini, and E. Gambi. 2013. Evaluation of the wireless m-bus standard for future smart water grids. In *2013 9th International Wireless Communications and Mobile Computing Conference (IWCMC)*, Cagliari, Italy, July 2013, 1382–1387.

Spinsante, S., S. Squartini, L. Gabrielli, M. Pizzichini, E. Gambi, and F. Piazza. 2014. Wireless m-bus sensor networks for smart water grids: Analysis and results. *International Journal of Distributed Sensor Networks* 2014: 16. Article ID 579271.

Squartini, S., L. Gabrielli, M. Mencarelli, M. Pizzichini, S. Spinsante, and F. Piazza. 2013. Wireless M-bus sensor nodes in smart water grids: The energy issue. In *2013 Fourth International Conference on Intelligent Control and Information Processing (ICICIP)*, Beijing, China, June 2013, 614–619.

Taneja, M. 2014. A framework for power saving in IoT networks. In *2014 International Conference on Advances in Computing, Communications* and *Informatics (ICACCI)*, Greater Noida, India, September 2014, 369–375.

Telecommunication Standardization Sector of ITU. 2015. Smart sustainable cities: An analysis of definitions. ITU-T Focus Group on Smart Sustainable Cities FG-SSC, October 2015.

Third Generation Partnership Project (3GPP). 2015. Evolution of LTE—Release 13. February 2015. http://www.3gpp.org/release-13.

Thompson, F., A. H. Afolayan, and E. O. Ibidunmoye. 2013. Application of geographic information system to solid waste management. In *2013 Pan African International Conference on Information Science, Computing and Telecommunications (PACT)*, Lusaka, Zambia, July 2013, 206–211.

Wan, X., Q. Zhang, and G. Dai. 2014. Research on forecasting method of natural gas demand based on GM (1, 1) model and Markov chain. In *2014 IEEE 13th International Conference on Cognitive Informatics Cognitive Computing (ICCI*CC)*, London, United Kingdom, August 2014, 436–441.

Wang, P., J. Y. Huang, Y. Ding, P. Loh, and L. Goel. 2010. Demand side load management of smart grids using intelligent trading/metering/billing system. In *IEEE PES General Meeting*, Minneapolis, MN, July 2010, 1–6.

Wang, W., J. Jiang, and M. Fu. 2014a. An enhanced differential evolution based grey model for forecasting urban water consumption. In *2014 33rd Chinese Control Conference (CCC)*, Nanjing, China, July 2014, 7643–7648.

Wang, W., J. Jiang, and M. Fu. 2014b. A grey theory based back propagation neural network model for forecasting urban water consumption. In *2014 33rd Chinese Control Conference (CCC)*, Nanjing, China, July 2014, 7654–7659.

Water Resources Group. 2009. Charting our water future. Economic frameworks to inform decision-making. www.2030waterresourcesgroup.com.

Weightless. 2016. http://www.weightless.org/.

Wi-Fi Alliance. 2016. http://www.wi-fi.org/discover-wi-fi/wi-fi-halow.

World Bank Group. 2012. The 2012 World Bank annual report. http://siteresources.worldbank.org/EXTANNREP2012/Resources/8784408-1346247445238/AnnualReport2012_En.pdf.

World Bank Group. 2015. Water security for all: The next wave of tools—2013/14 annual report. http://documents.worldbank.org/curated/en/2015/01/23990364/water-security-all-next-wave-tools-201314-annual-report.

Xiong, X., K. Zheng, R. Xu, W. Xiang, and P. Chatzimisios. 2015. Low power wide area machine-to-machine networks: Key techniques and prototype. *IEEE Communications Magazine* 53 (9): 64–71.

Zanella, A., N. Bui, A. Castellani, L. Vangelista, and M. Zorzi. 2014. Internet of things for smart cities. *IEEE Internet of Things Journal* 1(1): 22–32.

Zhu, X. and B. Xu. 2012. Urban water consumption forecast based on QPSORBF neural network. In *Eighth International Conference on Computational Intelligence and Security*, Guangzhou, China, 2012, 233–236.

Zhu, X. and J. Chen. 2013. Urban water consumption forecast based on PQPSO-LSSVM. In *2013 Ninth International Conference on Natural Computation (ICNC)*, San Diego, CA, July 2013, 834–837.

15 The Internet of Things and e-Health
Remote Patients Monitoring

Assim Sagahyroon, Raafat Aburukba, and Fadi Aloul

CONTENTS

15.1 Introduction 303
15.2 e-Health System Monitoring Architecture 304
15.3 Medical Sensors in RPM 305
15.4 RPM: Application Scenarios 306
15.4.1 Clinical Applications: RPM in the Field 307
15.4.2 Industrial Platforms in Support of RPM 308
15.5 RPM Enabling Technologies 309
15.5.1 Wireless Body Area Network for RPM 309
15.5.2 Cloud Computing: An Enabling Technology for RPM 310
15.5.2.1 Cloud Deployment Models for Healthcare 312
15.5.2.2 Healthcare Framework as a Service 312
15.6 Security and Privacy in Remote Health Applications 313
15.7 Issues Facing RPM Penetration 315
15.8 Conclusions 317
References 317

15.1 INTRODUCTION

According to McKinsey Global Institute research (Bauer et al., 2016), IoT might have an impact on the global economy that could reach trillions of dollars by the year 2025, and one of the promising industries where IoT is expected to have a great impact is healthcare. By some predictions, it is expected that spending on IoT-based solutions related to healthcare might reach $1 trillion by the year 2025; the motivation here is to primarily provide everyone with personalized, accessible, and on-time healthcare services. Adding IoT features to medical devices improves the quality and effectiveness of services delivered to patients with chronic conditions, and to those who are in need of constant supervision (Kaa, IoT development platforms, 2016).

Telemedicine (Di Cerbo et al., 2015; Jonathan and Charles, 2015) is the reliable and effective remote delivery of different healthcare services over the telecommunications infrastructure. In its search for more efficient and cost-effective ways of doing business, the healthcare industry is lending strong support to this field, which comes with the promise of increased efficiency in delivering care to patients and of lowering costs. A main category of telemedicine is remote patient monitoring (RPM). RPM allows patients with chronic diseases to be monitored in their homes through the use of devices that collect data about blood sugar levels, blood pressure, or other vital signs. The data can be reviewed instantly by remote caregivers. Thereby physicians can interact remotely with their patients in real time, providing advice and care for the patients while they remain in their homes, or perhaps in a remote medical facility that lacks experts in certain medical specializations.

The IoT is an emerging paradigm that is well positioned to play an important role in a variety of healthcare applications. These applications might vary from simply assisting in preventing diseases, at one end of the spectrum, to where they can be used in managing chronic diseases, on the other end. This paradigm is supported by significant advances in sensor and connectivity technology that pave the way for various successful deployments in healthcare-related applications. RPM is an application field that can hugely benefit from IoT and its supporting technologies. Some of the benefits gained by the different stakeholders when RPM solutions are deployed include (Aegis Corporation, 2016)

Benefits for patients

- An improved quality of life and better health outcomes
- Real-time feedback and timely interventions
- Minimization of the chances of emergencies and readmissions by simply extending care to patients in their homes
- Less time spent in hospitals

Benefits for healthcare providers

- Extension of clinical environments to patients' homes after they are discharged
- Continuous collection of patients' health data regardless of their location
- Support for an increased level of accuracy for clinical monitoring readings, particularly readings that would otherwise be provided by the patients themselves
- In due time, an increased level of trust and reliance that physicians place on data
- Reduced costs from readmissions and reduced hospital stays

Benefits for insurance payers

- Better visibility of patient compliance practices
- More accountability from patients and care providers
- Reduced costs of care

In essence, this chapter discusses the role of RPM in the future of healthcare delivery. The key components that are required for the deployment of RPM-based solutions, as well as the current issues that need to be addressed to fully realize the benefits of this paradigm shift in delivering care, are discussed. The rest of the chapter is organized as follows: In Section 15.2, we describe the making of a web-based architecture within an IoT framework to facilitate RPM applications and discuss the underlying technologies. We follow that with a discussion of medical sensors in Section 15.3, and in Section 15.4, we continue to introduce the readers to some of the clinical trials in remote monitoring and how the industry is approaching this medical application from a commercial perspective. In Section 15.5, we review wireless body area networks (WBANs) and their use in RPM, and discuss the basis for using cloud computing as an enabling technology. Section 15.6 presents an IoT framework as a service within the cloud, followed by a discussion of security and privacy aspects in Section 15.7. In Sections 15.8, we summarize the possible hurdles that might slow down the introduction of RPM in the medical sector.

15.2 e-HEALTH SYSTEM MONITORING ARCHITECTURE

The use of the IoT in healthcare requires a web-based architecture to guarantee information delivery on demand and to change the traditional healthcare system into smart networked healthcare that depends on the connected devices and that is largely automated by applying policies, intelligence, and monitoring services. The architecture depicted in Figure 15.1 illustrates the essential layers of such a system.

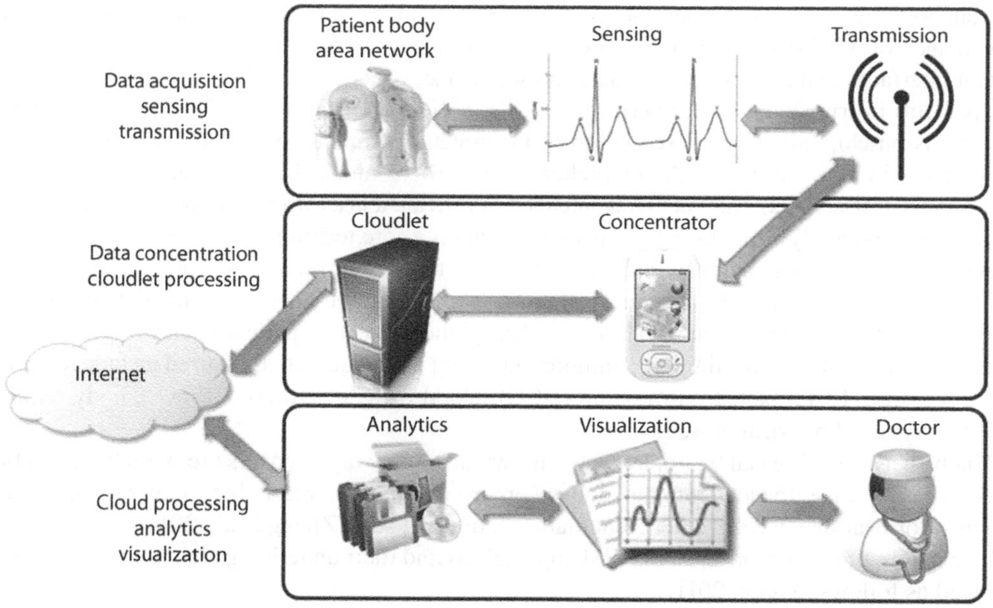

FIGURE 15.1 Components of an RPM application. (From Hassanalieragh, M., et al., presented at Proceedings of the IEEE International Conference on Service Computing, New York, 2015.)

At the top layer, medical sensors are the source of medical data by sensing and measuring vital signs, such as temperature, electrocardiography (ECG), and oxygen saturation (SpO_2). Collectively, these sensors (placed on the body or sometimes implanted) form a WBAN. A gateway or an aggregator acquires data from these sensors and connects to the network where the aggregate data is transmitted to a data center or the cloud, ideally in real time. Transmission of collected data over long distances is carried out using broadband communications networks or Wi-Fi. The sensor units communicate with the gateway using different low-power protocols, such as ZigBee and Bluetooth. The gateway provides the needed functionalities of device management and protocol conversion. The sensors in the data acquisition layer form an IoT-based architecture, as each individual sensor is uniquely identified and its data can be accessed through the Internet via this gateway (Hassanalieragh et al., 2015; Hu et al., 2013). The cloudlet unit is typically utilized to complement or support the storage and processing capabilities of the gateway or concentrator. At the next layer (cloud layer), the main tasks performed are storage, analytics, and visualization. During the analytics phase, sensor data is used, along with stored e-health medical records, to diagnose a number of health conditions and diseases. Furthermore, and to fully benefit from the IoT application in healthcare, attractive and easy-to-understand visualization techniques are a must. Visualization techniques convert collected data and analytics into a digestible and easy-to-grasp format. This in turn enables physicians and other decision makers to assess difficult concepts or identify new patterns, for example, by investigating graphs, images, and charts.

15.3 MEDICAL SENSORS IN RPM

Sensors are a key component of IoT. The ability of today's smart and wearable sensory devices to collect health-related data on their own without human intervention reduces the risk of error and improves the implementation of healthcare systems. Advances in sensors and embedded technologies are allowing for the collection, recording, and analysis of data in ways that was not possible before, and hence paving the way for better patient-centered care, as well as reduction in medical cost. From an application point of view, microelectromechanical sensors (MEMSs) offer several benefits that support the increasing penetration of MEMS technology into the medical applications area. Advantages of MEMSs include their low power consumption, smoother interaction between their silicon interfaces and patient body tissues, and small size, which enables less invasive (and therefore less painful) instrumentation. Furthermore, today's fabrication technology allows for the integration of systems-on-chip, allowing for sensors to be integrated with microcontrollers and radio-frequency transmitters in a single device. A comprehensive review of medical sensors technology is beyond the scope of this chapter. This section is intended to provide a brief overview and emphasize the fact that sensing and analysis of physiological signals remain a core technique that is still used in clinical screening and diagnosis, and therefore a cornerstone to the successful deployment of RPM systems.

The primary objective of unobtrusive sensing is to enable the continuous monitoring of physiological and biochemical parameters while, ideally, the subject or patient is carrying on with his daily normal life. Typically, the very common signs that are frequently measured include heart rate, blood pressure, ECG, ballistocardiogram (BCG), blood oxygen saturation (SpO_2), body temperature, posture, and physical movement.

Unobtrusive sensing can be implemented in two different ways: sensors are worn by the patient, or else sensors are embedded into the ambient environment or as smart objects interacting with the patients, for example, a chair, intelligent mattress, or toilet seat (Zheng et al., 2014).

The nature of the sensors used in RPM applications and their underlying technology can be summarized as follows (Thusu, 2011):

- Pressure sensors that can be used in sleep apnea machines, kidney dialysis equipment, oxygen concentrators, ventilators, infusion and insulin pumps, blood analyzers, respiratory monitoring and blood pressure monitoring equipment, and intelligent mattresses for monitoring bedbound patients.
- Temperature sensors that are used in remote temperature monitoring, sleep apnea machines, ventilators, kidney dialysis machines, blood analyzers, and digital thermometers.
- Applications for flow sensors include oxygen concentrators, sleep apnea machines, ventilators, and respiratory monitoring.
- Image sensors can be used in external observation applications where images can be taken in the remote patient location and then uploaded to physicians and other stakeholders using the network.
- Motion sensors, such as accelerometers, are used in the design of defibrillators and heart pacemakers, blood pressure monitors, patient movement monitoring, detection of falls, and other integrated health monitoring equipment.
- Biosensors can be used to monitor blood glucose and for cholesterol testing.
- Radio-frequency identification (RFID)–enabled, GPS, or Wi-Fi tracking devices for Alzheimer's or dementia patients.

The blend of sensor technology with wireless technology has led to the establishment of WBANs. The IEEE 802.15 Task Group (IEEE Standards, 2016) is responsible for developing this standard that is optimized for low-power devices and for serving different applications, including telemedicine and consumer electronics. As depicted in Figure 15.1, at the upper layer BANs provide the needed infrastructure that enables the efficient and cost-effective transmission of sensory data via the network. This in turn leads to the realization of practical and reliable RPM systems.

15.4 RPM: APPLICATION SCENARIOS

A recent report by Spyglass Consulting Group (Baum, 2015) indicated that due to various population health initiatives, adoption of RPM solutions will continue to grow where the primary goal is to support a large patient population with complex chronic conditions, including congestive heart failure (CHF), diabetes, chronic obstructive pulmonary disease (COPD), and hypertension. Health providers will continue to deploy RPM solutions facilitated by the emergence of IoT and other supporting technologies. In the following subsections, we discuss real-life clinical trials related to RPM, followed by a brief introduction to some of the commercial platforms available for use in RPM applications. Both subsections are meant to highlight recent strides made in the field, and demonstrate its applicability to the benefit of society at large.

15.4.1 Clinical Applications: RPM in the Field

In the case of elderly patients, it was observed that following a major surgery, functional recovery has always been problematic. This is attributed to the fact that elaborate surgical procedures and extended periods of hospitalization often lead to weakness and lack of mobility (Cook et al., 2013). To assess postsurgery recovery of its patients, the Mayo Clinic used accelerometers and pedometers to remotely monitor the activity of patients who had cardiac surgery performed on them. Collected data was transmitted wirelessly, aggregated, and projected onto a provider-viewable dashboard. Researchers concluded that wireless monitoring of mobility after major surgery creates an opportunity for early identification and intervention in individual patients and could serve as a tool to evaluate and improve the process of care and to affect postdischarge outcomes. In yet another interesting RPM clinical trial, researchers at the Universities of Padova, Montpellier, and Virginia and at Sansum Diabetes Research Institute (Kovatchev et al., 2013; Appelboom et al., 2014) have used patients with type I diabetes to experiment with and assess the feasibility of using a wearable artificial pancreas at home while using the patient's smartphone as a computational platform, and also to provide for a phone-based closed-loop control to vary the strength of injected insulin. The study monitored remotely the data from 20 subjects and concluded that a smartphone is capable of operating as an outpatient closed-loop control device, delivering performance that is comparable to that of a similar in-hospital setting, but using a laptop configuration instead.

Heart failure continues to be a major burden on the healthcare system. As the number of patients with heart failure increases, the cost of hospitalization alone is contributing significantly to the overall cost of this disease. RPM can be used to manage and optimize care delivery. In one study in the United Kingdom (Eurohealth, 2009), heart failure patients with a mean age of 71 years were telemonitored daily for a period of 6 months. While demonstrating no sharp difference in hospitalization (for any cause), the study concluded that there was a clear reduction in the number of emergency room visits, clinic reviews, and unplanned hospitalizations. This confirms that as a result of the remote monitoring and the daily collection of physiological data, physicians were able to intervene early and in a planned manner, and therefore minimize the risks of another heart failure attack. A randomized controlled trial was conducted in the National Taiwan University Hospital to study the effect of remote monitoring on patients diagnosed with COPD who had been discharged from the hospital (Ho et al., 2016). During the 2 months following discharge, a telemonitoring group of patients had to report their symptoms daily. The primary outcome measure was time to first readmission for COPD exacerbation. At the end of the trial period, it was concluded that the time to first readmission for COPD exacerbation was significantly increased in the group that was remotely monitored when compared with a regular group of discharged patients.

Due to recent advances in communication technology, new options are now available for following up on patients implanted with pacemakers and defibrillators. In a study that is somewhat different in nature (Varma et al., 2015; Freeman and Saxon, 2015), remote monitoring of patients wearing pacemakers and defibrillators has been carried out. The idea is to remotely monitor device functionality and gather patient clinical status using smart sensors. The study reported a clear reduction in mortality rate in those patients who were remotely monitored when compared with those who had only in-person follow-up. The study also demonstrated that there was a survival advantage for patients who frequently used remote monitoring compared with those who used it less frequently.

15.4.2 Industrial Platforms in Support of RPM

The global RPM devices market size was valued at $546.8 million in 2014 (http://www.grandviewresearch.com/). The rising demands of a growing geriatric population and their need for independent living and quality healthcare will further propel the need for these RPM platforms. In this section, we introduce some of the healthcare RPM solutions that make use of commercially available platforms.

For RPM, Vivify Health Inc. (Vivify Health Corp., 2016) offers an integrated package containing 4G tablets and wireless sensors where all devices are remotely managed and supported by its secured and Health Insurance Portability and Accountability Act (HIPAA)–compliant security protocols. The platform allows for data collection, video conferencing, and patient-guided care plans, plus a few additional features. Recently, the University of Pittsburgh Medical Center rolled out a large-scale RPM initiative that involves around 800 patients with CHF, each of whom has been provided with a free kit containing RPM technology from Vivify Health. In another RPM project, Children's Hospital of Alabama has deployed Vivify Health Solution to closely monitor infants with congenital heart disease at home.

Kaiser Permanente, an integrated managed care group based in California, partnered with IT solutions provider Cognizant to test a solution based on Azure IoT services that connects medical and health devices, such as blood pressure monitors, glucose readers, and wearable "bracelet" monitors, to smartphones. Functioning as gateway devices, the smartphones send data to the secure cloud for integration with an existing analytics and data visualization program that can run in a Kaiser Permanente data center. Clinicians can access this data via a central dashboard for a holistic, near-real-time view of a patient's health and activities (Microsoft Corp., 2016).

In other deployments, Honeywell Inc. LifeStream Manager RPM software has been used to help reduce avoidable readmissions, improve patient care, and better manage the telehealth program by integrating telemonitoring data into a single view (Honeywell Corp., 2016). With LifeStream Manager, care providers can view the health status of monitored patients and receive alerts when data falls outside of established parameters. Health Net Connect (Health Net Connect Inc., 2016) VideoDoc solution enables doctors to conduct a complete physical exam remotely using Dell Venue 11 Pro tablet's touchscreen. Vital information, such as temperature, blood pressure, blood glucose, and weight, is encrypted and securely transmitted to a HIPAA-compliant portal for later review by healthcare professionals, from any web-capable device. VideoDoc has proven to be of great benefit, especially in chronic disease management situations.

Alere Connect (Alera Inc., 2016) offers over-the-counter products, such as glucose meters, blood pressure monitors, weight scales, and pulse oximeters, that can use Bluetooth to communicate with an Alere HomeLink gateway that in turn connects to the company's Connected Health platform, with integrated products and services for health management. These various integrated tools connect patients to their provider, and enhance care by delivering timely information from patients to their physicians. The Alere HomeLink has a 7-inch touchscreen display that allows patients to submit responses to disease management questions, along with their test measurements. Recently, Anthem Blue Cross health provider selected Sentrian's Remote Intelligence Platform (Sentrian Corp., 2016) to reduce preventable hospitalization for its members. The Sentrian Remote Patience Intelligence platform is designed to prevent avoidable hospitalization by leveraging utilization of remote biosensors and machine learning techniques to remotely detect deterioration in patients' health before it becomes acute. The initial implementation is focused on patients with COPD and other concomitant conditions.

Almost 80% of patients suffering from atrial fibrillation (AF) are treated with warfarin. It is really critical that patients follow a planned therapy since suboptimal warfarin therapy management might require immediate medical intervention. Roche, a biotechnology company, in collaboration with Qualcomm Life (Qualcomm Corp., 2016), has deployed a home monitoring solution for AF patients. It uses the cloud-based 2net telehealth platform to deliver accurate and near-real-time data to Roche's CoaguCheck link portal. The goal is to make self-monitoring easy while providing clinicians with critical data to better manage patients and make informed interventions.

15.5 RPM ENABLING TECHNOLOGIES

Technology advances in fields such as sensor design and information and communication technology have resulted in a convergence of enabling technologies that can be utilized to effect new modalities of healthcare delivery. For example, today's network technologies can be used to allow patients to remotely access medical expertise from wherever they are, thereby enhancing the quality of care they can receive. Nevertheless, there are still challenges that need to be addressed to accelerate the deployment of technology-driven solutions to solve today's urgent health-related problems. In this section, we discuss WBANs and cloud as enabling technologies, and provide insight into their use within the context of RPM applications.

15.5.1 Wireless Body Area Network for RPM

Communication technologies in support of the networking infrastructure of an IoT-based RPM system can be broadly categorized into long-distance and short-distance technologies.

Long-distance infrastructure-oriented technologies include WLANS, cellular, Wi-Fi, and satellite-based networks. Short-range ubiquitous communication technologies in support of RPM and WBAN formation include Bluetooth, ZigBee, and Z-Wave. A WBAN consists of tiny smart and lightweight sensors located on the patient's body, integrated with the clothing (e-textiles), or sometimes implanted beneath the skin. The primary objective is to use wireless technologies to support patient monitoring in an unobtrusive, reliable, and cost-effective manner, thus providing personalized sustainable services to patients (Ragesh and Baskaran, 2011). WBANs designed to support healthcare-related applications such as RPM are still in the development stages but are considered to be a cornerstone to the successful deployment of healthcare-integrated services. Typically, a WBAN architecture consists of an intra-BAN communications tier in reference to communication between body sensors, and between body sensors and the concentrator or personal server (Figure 15.1). Inter-BAN communications involve the link between the concentrator and one or more access points. A beyond-BAN tier allows connectivity to the Internet using global system for mobile communications (GSM), General Packet Radio Service (GPRS), and other broadband technologies (Negra et al., 2016).

For medical applications, such as RPM, there are certain attributes and desirable requirements of WBANs, which include (de Schatz et al., 2012; Filipe et al., 2015)

- Miniature form factor, use of standards-based protocols, and patient-specific customization
- Low-power operation, energy-efficient design, and sound energy management policies
- Network quality of service that allows for the reliable transmission of medical data
- Fault tolerance in case a sensor node fails—a backup node should come to life in support of the ongoing activities
- Security and encryption of sensitive data related to personal health
- Mobility support by allowing WBAN users to move around without impacting performance

Different WBAN architectures have been proposed in the literature; the coverage of each is beyond the scope of this chapter. For a detailed review, readers are referred to (Filipe et al., 2015). The main protocols that are used in healthcare-related applications include

- **Bluetooth (Bluetooth standard, 2016):** A widely used short-range communication standard with a data rate of up to 1 Mbps, and a reasonable security level. A major advantage is its ability to allow a wide range of Bluetooth-enabled devices within the same vicinity (approximately 10 m range) to communicate with each other and without the need for a line-of-sight positioning. Most of today's commercial devices are Bluetooth enabled. This technology is currently in widespread use in hospitals, medical offices, assisted living facilities, and homes.

- **Bluetooth Low Energy (Bluetooth standard, 2016):** Provides ultra-low-power consumption, a data rate of up to 1 Mbps, and a range of 10 m. It consumes only 10% of the power consumed by Bluetooth, extending its battery life by sleeping and waking up when it needs to send data. These features, plus a few others, make it quite suitable for latency-critical WBAN applications like alarm generation and emergency response. It is a promising technology; however, it is not yet supported by many devices on the market.
- **ZigBee (IEEE 802.15.4) (IEEE standardization projects, 2016):** This standard builds on the established IEEE 802.15.4 standard for packet-based wireless transport. It has a data rate of up to 250 kbps and a coverage and range of 10–30 m. It was developed to provide low-power, wireless connectivity for a wide range of network applications concerned with monitoring and control. It uses Advanced Encryption Standard (AES) methods for encryption.
- **IEEE 802.15.4 (IEEE standardization projects, 2016):** The most widely adopted point-to-point communication standard for low-rate, long-battery-life wireless personal area networks. It has a 128-bit security support for authentication and guarantees the integrity and privacy of data. However, its low data rate (250 kbps) is a shortcoming when it comes to designing large-scale and real-time WBAN medical applications.
- **IEEE 802.15.6 (IEEE standardization projects, 2016):** In recent years, WBANs have moved to the forefront as a key technology in providing real-time health monitoring of patients and in managing many chronic diseases. The main task of the IEEE 802.15.6 group is to establish a communication standard optimized for low-power in-body or on-body nodes to serve a variety of medical and nonmedical applications (IEEE standards, 2016). The IEEE 802.15.6 standard defines three physical layers: the narrowband (NB), ultra-wideband (UWB), and human body communications (HBC) layers. Based on the type of application, and the network that is to be implemented, the PHY layer can be varied accordingly. This standard is a step forward (Negra et al., 2016) in wearable wireless sensor networks, as it is designed specifically for use with a wide range of data rates, less energy consumption, a low range, an ample number of nodes (256) per body area network, and different node priorities according to the application requirements.

In recent years, there have been ongoing efforts to natively integrate the Internet protocol (IP) into WBAN packets; therefore, the underlying network infrastructure will be transparent to the application. This allows native connectivity between the Internet and wireless sensor networks, enabling smart objects to participate in the IoT. Focusing efforts in this direction has resulted in the development of the IPv6 over Low-Power Personal Area Networks (6LoWPAN) specification. 6LoWPAN is an international open standard developed by the Internet Engineering Task Force (IETF) that enables building the wireless IoT using IEEE 802.15.4 and IP together in a simple, well-understood way. It enables the efficient use of Internet protocol version 6 (IPv6) over low-power, low-rate wireless networks on simple embedded devices through an adaption layer and optimization of related protocols (Tabish et al., 2013; Cao et al., 2010; Mainetti et al., 2011).

Various WBAN projects using different architectures and protocols related to healthcare have been reported in recent years (Filipe et al., 2015). Table 15.1 summarizes the main features of some of these projects, including the nature of the applications, operational environment, and type of protocol used.

15.5.2 Cloud Computing: An Enabling Technology for RPM

An IoT-based platform provides solutions based on the integration of networking technology with hardware and software. It facilitates communication between different entities, such as devices to devices and devices to individuals. Advances in information and communication technology have led to impressive innovation in three layers of technology, namely, the cloud, data and communication networks, and devices (World Economic Forum, 2016).

TABLE 15.1
Examples of WBAN Projects

WBAN Project	Operational Environment	Application	Standard
CodeBlue (Filipe et al., 2015)	30-node ad hoc sensor network	Medical care and disaster response	IEEE 802.15.4
LOBIN (Filipe et al., 2015)	Hospital environment	Monitoring of physiological parameters (ECG, heart rate, temperature, etc.)	IEEE 802.15.4
Body inertial sensing network (Filipe et al., 2015)	Hospital environment	Body movement monitoring: Provides data for 3 degrees of freedom of orientation in real time	Bluetooth
Unobstructive body area networks (Filipe et al., 2015)	Hospital and disaster events, residential monitoring, motion activities	Identify postures and movements with alarm issuance	IEEE 802.15.1
MEDISN (Filipe et al., 2015)	Dedicated wireless sensor network in hospital	Emergency detection	IEEE 802.15.4
6LoWPAN monitoring system (Mainetti et al., 2011)	Lab environment	Monitoring of ECG, temperature, and acceleration	6LoWPAN

In distributed RPM applications, sensory data collected by the various gateways eventually needs to be transferred to the cloud for long-term storage. The collected data can be used as part of the patient electronic medical records (EMRs). EMRs in healthcare refer to the storage of all healthcare data and information in electronic formats with the associated information processing and knowledge support tools necessary for managing the health enterprise system (Hannan, 1996).

Cloud computing is a paradigm that enables on-demand access to a pool of computational resources to consumers over the Internet (Mell and Grance, 2011). This can provide an advantage for the healthcare domain where computation resources can be provisioned and provided through virtualization. Moreover, any healthcare computation need can be provisioned on demand based on the computational resource requirement by healthcare providers or any running applications, such as data analytics and patient monitoring applications.

The cloud computing features that can deliver on the enhancements of EMRs and RPM include

- **Broad network access:** Healthcare providers and patients access cloud services on any client or end-point device from anywhere over a network, such as the Internet or an organization's private network. For instance, patients can access their health records through a browser using any device. Moreover, the integration of IoT devices that are connected to doctors, patients, or clinics can happen over the Internet. In the cloud, network-accessible capabilities go beyond applications. Cloud computing enables healthcare providers and consumers to access essential data center capability from any place and on any device. Cloud solutions provide access to data, computation, storage, and facilities, such as data backup and recovery.
- **Resource pooling:** Resources such as storage, processor, memory, and network bandwidth are pooled to serve multiple consumers, such as hospitals, administrators, patients, and doctors. Resource pooling enables IT resources to be dynamically assigned, released, and reassigned according to consumer demand. For example, a patient might have IoT devices that capture specific readings related to his or her health. Those devices require storage resources that are provisioned from the cloud and are allocated to the patient to store his or her health readings. The data stored can be utilized by other systems, such as a healthcare monitoring system. Computing resources can be provisioned based on the specific computing requirements for monitoring and

analytics systems. This enables cloud providers to achieve high levels of resource utilization and to flexibly provision and reclaim resources when they are not in use anymore.

15.5.2.1 Cloud Deployment Models for Healthcare

The U.S. National Institute of Standards and Technology (NIST) (Mell and Grance, 2011) presented four deployment cloud models: public, private, hybrid, and community clouds. The choice of the adequate cloud deployment model depends on the required security and privacy compliance against data being stored, processed, and disseminated.

- **Public cloud:** In this deployment model, healthcare providers, among other cloud users, may share common resources. Moreover, data storage and processing might travel across borders and could violate legal compliance.
- **Private cloud:** This is an infrastructure that is set up for the use of healthcare providers. When compared with a public cloud, a private cloud offers the healthcare domain a greater degree of privacy and control over the cloud infrastructure, applications, and data. Hence, it can ensure healthcare legal compliance of data being processed, stored, and disseminated.
- **Community cloud:** This is an infrastructure that is set up for sole use by a group of healthcare entities, such as all hospitals, clinics, medical labs, and pharmacies within a country. The community cloud provides the different entities with the needed infrastructure and services. The data processing, storage, and dissemination between different participants within the community cloud must adhere to the legal compliance rules and regulation.
- **Hybrid cloud:** In healthcare, the hybrid cloud is a possible deployment model with close attention to sensitive data transfer between clouds. Data that is not defined as sensitive by the regulatory entity could be transferred and processed by a public cloud provider. This can be done for different purposes, such as full resource consumption within the private cloud. In this case, instead of rejecting requests, resources can be freed by utilizing the public cloud resources.

15.5.2.2 Healthcare Framework as a Service

While healthcare has always incorporated the use of a broad range of medical devices, typically within critical or long-term care facilities, there are a growing number of devices that are becoming available to hospitals, doctors, patients, and administrators that can be readily attached to their point-of-service or mobile devices. Medical devices, in general, were not designed for interoperation with other medical devices or computation systems. This presents the need for advancement within the cyber-physical architecture. This includes the ability for patients to monitor their own condition and self-manage, or for providers, to monitor patients' conditions remotely, alerting them when abnormal observations are detected. These devices may be an integrated part of a telehealth service, allowing specialists and other providers to have access to biometrics, imagery, and interactive virtual visits across great distances or anywhere patient access to necessary medical expertise does not exist (Vermesan et al., 2015). Analytics in healthcare must be capable of collecting and correlating vast amounts of data from a number of sources to deliver insights that inform health practices and service delivery. Figure 15.2 depicts the layered architecture of cloud computing services within the context of healthcare. The figure shows the following layers:

- **Infrastructure as a service (IaaS):** The healthcare domain requires computational resources for healthcare monitoring and analytics. The cloud IaaS will provide the required computational resources. Moreover, the computational resources required for data input from doctors, nurses, receptionist, and so forth, can be eliminated by utilizing the cloud IaaS, where virtualized computational resources are provisioned and accessed over the Internet or the local network, depending on the cloud deployment model adopted.
- **Platform as a service (PaaS):** The platform provides the runtime environment needed to deploy healthcare as a service (HaaS).

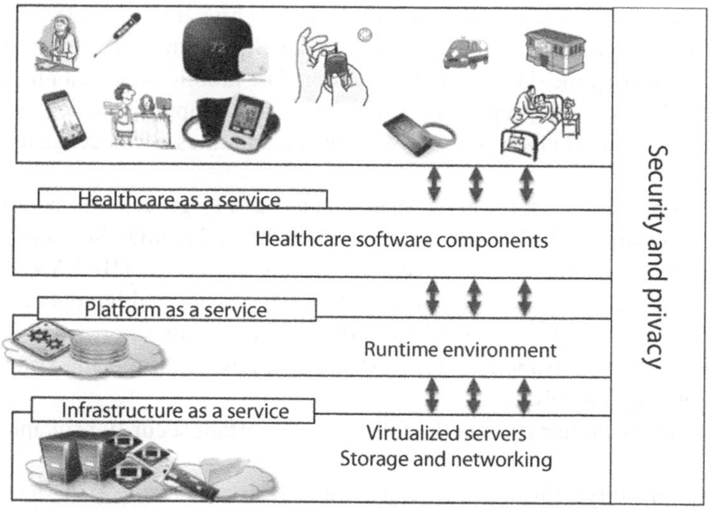

FIGURE 15.2 Healthcare as a service layer in cloud computing.

- **Healthcare as a service:** This layer provides the essential services that enable the integration of IoT devices, health monitoring, analysis, and coordination between devices and other systems, as well as any healthcare-related applications.
- **Security and privacy:** This is a cross-layer that ensures that the cloud services, infrastructure resources, data acquisition, processing, and dissemination adhere to the relevant policies and legal requirements. It also provides the mechanisms that guarantee confidentiality, integrity, and availability.

Healthcare data and services can be accessed using web services that allow a client application to request data and computations and a service provider to return responses. Different data formats, such as plaintext, Extensible Markup Language (XML), and JavaScript Object Notation (JSON), can be used to communicate client requests and service responses. The web services are accessible by users (healthcare providers and users) by specifying the URLs, and the users must have the required permission to execute the services and access the data. Web services enable those client applications to communicate by providing web service interfaces that use standard web protocols. Web services provide a standard means of interoperating between different software applications running on a variety of platforms. The web services are primarily based on a Simple Object Access Protocol (SOAP) and/or representational state transfer (REST) (Thomas et al., 2013).

15.6 SECURITY AND PRIVACY IN REMOTE HEALTH APPLICATIONS

In RPM applications, sensors continuously capture and stream data related to the monitored patients. Such data can include *biological* data (e.g., body temperature, blood pressure, and heartbeat rate), *environmental* data (e.g., GPS location), or *behavioral* data (e.g., walking speed rate, typing speed, and sound level). The data is either stored locally, for example, on the mobile phone or uploaded to remote servers for analysis and storage. The data can also be shared with other users, such as medical physicians and medical insurance companies. The large amount of collected data, in addition to the large number of varied users that can have access to the data, and the ability to store and analyze the data remotely raise security and privacy issues for the monitored users (Zubaydi et al., 2015; Plachkinova et al., 2015). Ideally, users should be able to control what data is being collected, how frequently it is collected, where it is stored or analyzed, who it is shared with, and what others can do with the data.

Note that mobile phones perform the work of a "data aggregator" by collecting data from sensors (e.g., speaker, accelerometer, and mic), or external Bluetooth-based connected sensors (e.g., oximeter), and transmitting the data to a remote server. The mobile phone can be replaced by other wireless electronic devices or aggregators, such as tablets or computers, which exist at the patient's site. The discussion below relates to any wireless device and does not necessarily focus on mobile phones.

Several regulations have been proposed many years back to govern and control the privacy of personal health information. The U.S. Department of Health and Human Services proposed HIPAA (Pieper, 2004) in 1996 to safeguard medical information. While the HIPAA regulation is widely enforced, the fast advances in technology and electronic health (e-health) applications over the past years have made some of the HIPAA guidelines out of date. Furthermore, many of the users have basic knowledge about e-health applications and technology, which limits their understanding of the privacy and security risks involved.

Any e-health application has to meet the following important security principles:

- **Data confidentiality:** Ensures that only authorized users can read the medical data. Authorized users can include the patient, physician, nurse, insurance company, hospital management, medical researchers, and family members. The use of advanced encryption protocols ensures that medical data stored or in transit over the network can only be read by authorized users. The encryption keys can be shared among the authorized users. The NIST (2016) recommends various effective encryption algorithms that use keys of 128+ bits to encrypt and decrypt the medical data. NIST also recommends various algorithms that allow authorized users to securely exchange the encryption keys. Authentication and authorization measures play an important role in controlling who has access to the medical data and what the users can do with it.
- **Data integrity:** Ensures that only authorized users can modify the medical data. In order to prevent the tampering of medical data, security measures have to be imposed to protect the medical data from being illegitimately modified. Access control lists (ACLs), hashing, and malware protection measures can be used to enforce the integrity of the collected medical data.
- **Data availability:** Ensures that the medical data is available to be accessed by the authorized users whenever needed. Health applications are time critical, and depending on the severity of the collected data, some medical data has to be shared instantly with specialized physicians or emergency respondents. Data availability can be handled by having active backup programs, additional servers, and disaster recovery plans (DRPs).

In order to meet the above three security principles, the following measures can be implemented:

- **Secure wireless networks:** Since mobiles, tablets, and computers typically transmit medical data over wireless networks, data security can be increased by enabling strong wireless encryption protocols, such as Wi-Fi Protected Access II (WPA2) (Wi-Fi Alliance, 2016). While the wireless encryption protocol, for example, WPA2, encrypts only the data transmitted between the wireless unit (i.e., mobile, tablet, or computer) and the wireless access point, it is recommended that a virtual private network (VPN) (Microsoft TechNet, 2001) be used between the wireless unit application and the end server, which ensures end-to-end encryption and covers both the wireless and wired network. It includes the use of *https* in browsers, which is convenient for users and free to use.
- **Physical data encryption:** Enabling data encryption on the wireless device and the server is important and can help secure the medical data in the case of a physical attack in which the wireless device or server is stolen. Physical security measures, such as locks, can be applied to the wireless device and servers.

- **Authentication:** Strong access control measures, such as random difficult passwords or personal identification numbers (PINs), two-factor authentication, and biometrics, are recommended to be enabled on the wireless device and server to limit access to authorized users. Passwords or PINs, which fall under the "what you know" authentication systems, can be long and include random uppercase alphabets, lowercase alphabets, symbols, and numbers to make them harder to guess. Two-factor authentication forces the user to use two items to authenticate into the system. That typically consists of a password or PIN, in addition to a token or smart card, which falls under the "what you have" authentication systems. Since many wireless devices (e.g., mobile phones and tablets) today have fingerprint scanners, they can also be enabled to verify the identity of the user. Other biometric options, which fall under the "what you are" authentication systems, can be voice recognition or eye scanners.
- **Malware protection:** Wireless devices and servers must be equipped with the latest antivirus, firewall, and intrusion detection systems (IDSs) that can block malware from attacking the systems and alert the user of any possible attacks. Malware includes a long list of potential attackers, such as viruses, worms, Trojans, and back doors. Malware attacks can be *active*, in which the malware will modify or damage the medical data, or *passive*, in which the medical data is read and forwarded to the attacker's machines without the knowledge of the user, hence attacking the user's privacy. Passive attacks are more difficult to detect given that the data is not modified. Malware can also lead to denial of service (DoS) attacks, which disable the wireless device or server, hence stopping legitimate users from accessing the systems.
- **Security awareness:** Given the significant growth of technology and mobile applications in the past few years, users must be continuously educated on how to securely use the electronic devices such as mobile phones and tablets. Topics can include what applications to download and from which trusted sources, what private information to share and with whom, what authentication systems to employ, and so forth.
- **Phone location tracking and remote data deletion:** Given the small size of mobile phones and tablets, they can easily be lost or stolen. Mobile phones today allow the authorized users to track the phone's location via GPS. They also allow authorized users to remotely connect to the mobile phone and wipe out any sensitive data. These options should be enabled, and the mobile phone must have a valid Internet connection via Wi-Fi or 3G/4G in order to remotely access the mobile phone.
- **Patch management:** Users should continuously update the wireless unit operating system and applications.

15.7 ISSUES FACING RPM PENETRATION

Despite strong prospects for the use of the IoT and RPM in the health sector, there are still challenges and a certain level of skepticism (Sundmaeker et al., 2010) that will determine and shape how clinical care could benefit from this paradigm shift in healthcare delivery.

Some of the issues and concerns facing RPM include

- Reluctance or hesitation by traditional healthcare providers to transition to this new mode of operation. We expect in due time, and because of the cost-effectiveness of RPM, that the majority of providers will eventually seek to make use of the benefits offered by deploying RPM solutions in their network. Similarly, the reluctance of patients, especially the older generation, to interface with and use RPM-related technology can be overcome through education and awareness.
- To fully benefit from any IoT-based technology, security and privacy issues need to be addressed. In such systems, data collection, mining, and provisioning are all performed

over the Internet, and therefore the opportunity exists for unauthorized entities to intrude. Patients' privacy must be guaranteed to eliminate the possibility of identification and tracking. The problem is exacerbated further by the fact that in IoT applications, most of the communications are wireless, thus making eavesdropping relatively simple; additionally, IoT devices used in RPM are characterized by low energy, low on-device memory, and low processing capabilities, which makes it hard to implement complex security schemes on them (Sagahyroon, 2017). In short, the flow of big data in the ever-increasing IoT applications, and the security issues that will continue to surface require intensive research in such areas as dynamic trust, security, and privacy management (Yin et al., 2016). New and innovative solutions are still required to provide an acceptable level of security regardless of the limitation in resources in terms of energy availability and limited computational capabilities.

- Another concern is the accumulation, handling, and reliable analysis of big data created by the voluminous flow of information from the patients' population. Streamlining the automation and analysis of this big data and defining a practical mechanism of alerting physicians and other stakeholders in a timely manner is essential. Managing this data also brings the challenges of mining this information and knowledge extraction to the forefront of research activities. Furthermore, even though the cost of storage is getting lower, we still need to reduce the cost of storing this data by developing intelligent algorithms that can help in removing redundant data.
- Regulators may view some of the available mobile monitoring applications on the same level as medical devices—hence the need, for example, for approval by entities such as the Food and Drug Administration (FDA) for it to be used in clinical settings. This in turn will lead to unwarranted delays. Stakeholders need to synchronize with regulating bodies and provide guidelines that would allow for the efficient deployment of monitoring applications.
- Another concern is the need for standardization, as well as device and data interoperability. A number of standardization efforts continue to take place. Major contributors include the IETF and the European Telecommunications Standards Institute (ETSI). As an example of the ongoing efforts, the Institute of Electrical and Electronics Engineers (IEEE) has an active project to develop a standard for an architectural framework for IoT adaptation. "This standard defines an architectural framework for the IoT, including descriptions of various IoT domains, definitions of IoT domain abstractions, and identification of commonalities between different IoT domains. The architectural framework for IoT provides a reference model that defines relationships among various IoT verticals (e.g., transportation, healthcare, etc.) and common architecture elements" (IEEE standardization projects, 2016). It is expected that this architectural framework will promote cross-domain interaction, aid system interoperability and functional compatibility, and enhance the growth of the IoT market. Medical data that is typically exchanged between different health providers should be formatted following standards such as the electronic health record (EHR) standards. Additionally, work continues on developing standards and practices that enable the integration of data from sensors across devices, users, and domains to allow for the creation of the types of applications and services that would maximize the benefits from the IoT paradigm, and eventually lead to a better quality of life.
- In health-related IoT deployments, the design of wearable sensors and power consumption continues to be an issue. The question of how to perfectly achieve unobtrusiveness and monitor patients is still an open problem (Yin et al., 2016) since comfort while monitoring is a primary objective. Work on exploring the use of multifunctional sensors designed using lighter material, such as fabric or carbon fiber, continues with promising results. The continuous need for energy to power sensors and other devices continues to be a problem. Rechargeable batteries require frequent recharging that might burden patients. Thus, minimizing energy consumption is a primary constraint. Currently, solutions such as energy harvesting and solar power are being investigated, while research efforts that

focus on developing communication protocols and sensors with low energy consumption are continuing.

- An increase in RPM deployments within an IoT framework will eventually lead to scalability issues (in terms of the number of monitored patients) that might arise at different levels, including naming and addressing or identification of devices. Also, because of the high level of interconnection among many entities, scalability issues related to networking and data communication will surface as well. Finally, the expected increase in the number of healthcare services and service execution options the need to handle the various heterogeneous resources will eventually lead to scalability issues in service provisioning and management (Miorandi et al., 2012).

15.8 CONCLUSIONS

There is a major trend in healthcare delivery characterized by an increase in the use of technology to better patients' lives and improve quality of care while reducing the associated costs. RPM applications that make use of the recent technological advances hold great promise and can be of great benefit to both patients and the health sector. Clinical trials have concluded that proper monitoring can elongate life by identifying high-risk patients, reduce hospital readmissions, and optimize the utilization of clinical resources. Despite the existence of some obstacles, and the need to address privacy and security risks, we expect this trend to continue, and for IoT-based solutions such as RPM to play a critical role in delivering cost savings and noticeable health outcome gains. The chapter emphasized the great potential of RPM in healthcare and discussed the technical aspects that are essential for its successful deployment. The emerging role of industry in RPM is presented, and the case for continuous research effort to mitigate the effects of the current hurdles is made by shedding light on the current technological and human limitations.

REFERENCES

Aegis Corporation. 2016. Aeris IoT services and healthcare focus on the patient. Retrieved October 12, 2016, from www.aeris.com/for-enterproses/healthcare-remote-patient-monitoring/.

Alera Inc. 2016. Products and Services. Retrieved October 12, 2016, from http://www.alere.com/en/home.html.

Appelboom, G., et al. 2014. Smart wearable body sensors for patient self-assessment and monitoring. *Archives of Public Health* 72: 28.

Bauer, H., M. Patel, and J. Veira. 2016. The Internet of things: Sizing up the opportunity, McKinsey & Company. Retrieved October 12, 2016, from http://www.mckinsey.com/industries/high-tech/our-insights/the-internet-of-things-sizing-up-the-opportunity.

Baum, S. 2015. Survey: Remote patient monitoring shifting from point solution to disease specific, patient engagement, November 16, 2015. Retrieved October 12, 2016, from http://medcitynews.com/2015/11/remote-patient-monitoring/.

Bluetooth standard. 2016. Retrieved October 12, 2016, from www.bluetooth.com.

Cao, H., V. Leung, C. Chow, and H. Chan. 2010. Enabling technologies for wireless body area networks: A survey and outlook. *IEEE Communications Magazine*, 84–93.

Cook, D. J., J. E. Thompson, S. K. Prinsen, J. A. Dearani, and C. Deschamps. 2013. Functional recovery in the elderly after major surgery: Assessment of mobility recovery using wireless technology. *Annals of Thoracic Surgery* 96: 1057–1061.

de Schatz, C., H. Medeiros, F. Schneider, and P. Abatti. 2012. Wireless medical sensor networks: Design requirements and enabling technologies. *Telemedicine and e-Health Journal* 18 (5): 394–399.

Di Cerbo, A., J. Morales-Medina, B. Palmieri, and T. Iannitti. 2015. Narrative review of telemedicine consultation in medical practice. *Patient Preference & Adherence* 9: 65–75.

Eurohealth. 2009. Vol. 15, No. 1, pub. London School of Economics, UK.

Filipe, L., F. Fdez-Riverola, N. Costa, and A. Pereira. 2015. Wireless body area network for healthcare applications: Protocol stack review. *International Journal of Distributed Systems* 2015 (1).

Freeman, J. V. and L. Saxon. 2015. Remote monitoring and outcomes in pacemaker and defibrillator patients. *Journal of the American College Cardiology* 65 (24): 2601–2610.

Hannan, T. J. 1996. Chapter 12: Electronic Medical Records. In *Health Informatics: An Overview*. Churchill Livingstone, Australia, 133–148.

Hassanalieragh, M., A. Page, T. Soyata, G. Sharma, M. Aktas, G. Mateos, B. Kantarci, and S. Andreescu. 2015. Health monitoring and management using Internet of things (IoT) sensing with cloud-based processing: Opportunities and challenges. *Presented at Proceedings of the IEEE International Conference on Service Computing*, New York.

Health Net Connect Inc. 2016. Remote Patient Monitoring. Retrieved October 12, 2016, from http://www.healthnetconnect.com/offerings/remote-patient-monitoring/

Ho, T. -W., et al. 2016. Effectiveness of telemonitoring in patients with chronic obstructive pulmonary disease in Taiwan—A randomized controlled trial. *Scientific Reports* 6, Article 23797.

Honeywell Corp. 2016. Life Care Solutions. Retrieved October 12, 2016, from https://www.honeywelllifecare.com/.

Hu, F., D. Xie, and S. Shen. 2013. On the application of the Internet of things in the field of medical and health care. *Presented at Proceedings of the IEEE International Conference on Cyber, Physical and Social Computing and IEEE Internet of Things*, Washington, DC.

IEEE standardization projects. 2016. Retrieved October 12, 2016, from http://standards.ieee.org/develop/project/2413.html.

IEEE standards. 2016. Retrieved October 12, 2016, from https://standards.ieee.org/findstds/standard/802.15.6-2012.html.

IEEE Standards. 2016. Retrieved October 12, 2016, from http://www.ieee802.org.

Jonathan, N. and D. Charles. 2015. Telemedicine spending by Medicare: A snapshot from 2012. *Telemedicine and e-Health* 21 (8): 686–693.

Kaa, IoT development platforms. 2016. Retrieved October 12, 2016, from http://www.kaaproject.org/healthcare/.

Kovatchev, B., et al. 2013. Feasibility of outpatient fully integrated closed-loop control. *Diabetes Care* 36: 1851–1858.

Mainetti, L., L. Patrono, and A. Vilei. 2011. Evolution of wireless sensor networks towards the Internet of things: A survey. *Presented at Proceedings of the International Conference on Software, Telecommunications and Computer Networks*, Split, Croatia.

Mainetti, L., L. Patrono, and A. Vilei. 2011. Evolution of wireless sensor networks towards the Internet of things: A survey. *Presented at Proceedings of the International Conference on Software, Telecommunications and Computer Networks*, Dubai, UAE.

Mell, P. and T. Grance. 2011. *NIST definition of cloud computing*. Gaithersburg, MD: National Institute of Standards and Technology.

Microsoft Corp. 2016. Customer Stories; Kaiser Permanente. Retrieved October 12, 2016, from https://www.microsoft.com/en/servercloud/customer-stories/kaiser-permanente.aspx.

Microsoft TechNet. 2001. Virtual private networking: An overview, September, 4, 2001. Retrieved October 12, 2016, from https://technet.microsoft.com/en-us/library/bb742566.aspx.

Miorandi, D., S. Sicari, F. De Pellegrini, and I. Chlamtac. 2012. Internet of things: Vision, applications, and research challenges. *Ad Hoc Networks* 10: 1497–1516.

Negra, R., I. Jemili, and A. Belghith. 2016. Wireless body area network: Applications and technologies. *Proceedings of the Second International Conference on Recent Advances on Machine-to-Machine Communications*, Madrid, Spain.

NIST (National Institute of Standards and Technology). 2016. Healthcare: Standards and testing. Retrieved October 12, 2016, from http://healthcare.nist.gov/.

Pieper, B. 2004. An overview of the HIPAA security rule. *Journal of the American Optometric Association* 75 (11): 654–657.

Plachkinova, M., S. Andres, and S. Chatterjee. 2015. A taxonomy of mHealth apps: Security and privacy concerns. *Proceedings of the Hawaii International Conference on System Sciences*, Hawaii, 3187–3196.

Qualcomm Corp. 2016. Qualcomm Life. Retrieved October 12, 2016, from https://www.qualcomm.com/news/releases/2015/01/29/roche-and-qualcomm-collaborate-innovate-remote-patient-monitoring.

Ragesh, G.K. and K. Baskaran. 2011. A survey on futuristic health care systems: WBANs. *Presented at International Conference on Communication Technology and System Design*, India.

Remote patient monitoring devices market analysis: A grand view research report. Retrieved October 12, 2016, from http://www.grandviewresearch.com/.

Sagahyroon, A. 2017. Remote patients monitoring: Challenges. *Presented at Proceedings of the 7th IEEE Annual Computing and Communications Conference*, Nevada.

Sentrian Corp. 2016. Remote patients intelligence, Retrieved October 12, 2016, from http://sentrian.com/anthem-blue-cross-picks-sentrians-remote-intelligence-platform/.

Sundmaeker, H., P. Guillemin, P. Friess, and S. Woelffl. 2010. Vision and challenges for realizing the Internet of things, CERP-IoT, European Commission, Brussels.

Tabish, R., A. Ben Mnaouer, F. Touati, and A. Ghaleb. 2013. A comparative analysis of BLE and 6LOWPAN for U-Healthcare applications. *Presented at Proceedings of the IEEE GCC Conference*, Doha, Qatar.

Thomas, E., P. Richardo, and M. Zaigham. 2013. *Cloud Computing: Concepts, Technology, & Architecture*. Upper Saddle River, NJ: Prentice Hall.

Thusu, R. 2011. Medical sensors facilitate health monitoring, *Sensors Online*, April 2011.

Varma, N., et al. 2015. Relationship between level of adherence to automatic wireless remote monitoring and survival in pacemaker and defibrillator patients. *Journal of the American College Cardiology* 65 (24).

Vermesan, O., P. Friess, P. Guillemin, R. Giaffreda, H. Grindvoll, M. Eisenhauer, M. Serrano, K. Moessner, M. Spirito, L.-C. Blystad, and E. Z. Tragos. 2015. Internet of things beyond the hype: Research, innovation and deployment. In *Building the Hyperconnected Society*, edited by O. Vermesan and P. Friess. Niels Jernes Vej, Denmark: River Publishers.

Vivify Health Corp. 2016. Vivify Pathways. Retrieved October 12, 2016, from http://www.vivifyhealth.com.

Wi-Fi Alliance. 2016. Discover Wi-Fi security. Retrieved October 12, 2016, from http://www.wi-fi.org/discover-wi-fi/security.

World Economic Forum. 2016. The global information technology report 2012: Living in a hyperconnected world. Retrieved October 12, 2016, from http://akgul.bilkent.edu.tr/webindex/Global_IT_Report_2012.pdf.

Yin, Y., Y. Zeng, X. Chen, and Y. Fan. 2016. The Internet of things in healthcare: An overview. *Journal of Industrial Information Integration* 1: 3–13.

Zheng, Y. -L., et al. 2014. Unobtrusive sensing and wearable devices for health informatics. *IEEE Transactions on Biomedical Engineering* 61 (5): 1538–1554.

Zubaydi, F., A. Saleh, F. Aloul, and A. Sagahyroon. 2015. Security of mobile health (mHealth) systems. *Presented at 15th IEEE International Conference on Bioinformatics and Bioengineering (BIBE)*, Belgrade, Serbia, November 2015.

16 Security Considerations for IoT Support of e-Health Applications

Daniel Minoli, Kazem Sohraby, Benedict Occhiogrosso, and Jake Kouns

CONTENTS

16.1 Introduction 321
16.2 Iot Security Challenges in e/m-Health Applications 324
16.3 Iot Regulatory (And Security) Requirements In e/m-Health Applications 326
16.4 Iot Architectures for E/M-Health Security 329
16.5 (Layer-Oriented) Iotsec Mechanisms 336
16.5.1 Overall (All Layers) 337
16.5.2 Lower Layers (Fog Networking Layer) 337
16.5.2.1 Body Area Networks 337
16.5.2.2 Traditional Personal Area Networks 337
16.5.2.3 Hybrid (Home–Public) Hotspot Networks 338
16.5.2.4 Low-Power Wide Area 338
16.5.2.5 Cellular Solutions (4G/5G) 338
16.5.2.6 Other Solutions 339
16.5.2.7 Security Considerations 339
16.5.3 Hardware Level 339
16.5.3.1 Trusted Execution Environment 339
16.5.3.2 Intel TXT 342
16.5.3.3 Other Approaches 344
16.6 Near-Term Trends Related to E/M-Health Security 344
16.7 Conclusion 345
References 345

16.1 INTRODUCTION

e-Health benefits from the IoT; in fact, there is evidence that mobile and/or remote healthcare monitoring made possible by IoT-based systems often can improve the clinical outcome of patients, while at the same time reduce costs and optimize healthcare personnel productivity. Proponents contemplate an "end-state" environment where mobile health (m-health) monitoring systems reduce the interval between the onset of a medical condition in an outpatient situation and the diagnosis of the underlying issue. These applications utilize one or more (IoT-based) sensors worn by the patient that enable the collection of a number of patient's parameters to be transmitted in real time to a monitoring system for analysis and diagnosis. Sensors of interest in healthcare include biosensors (e.g., temperature and blood pressure), as well as other sensors (e.g., position, motion, velocity, acceleration, video, acoustic, and radiation sensors). Currently, almost 50% of all patients in U.S. hospitals are not monitored with continuous telemetry; by allowing for continuous monitoring of patients, IoT devices (specifically those allowed under the medical body area network [MBAN] rubric) can help doctors respond more quickly in emergency situations. The addition of mobility to the telemetry also improves the overall hospital experience; in-home patient care can also be enhanced (Buckiewicz, 2016). Pharmaceutical companies and clinical research organizations (CROs) can also reduce clinical trials costs with enhanced data collection mechanisms enabled by IoT-based e/m-health wearable systems. The global over-65 population is expected to rise up to 1 billion by 2020; it follows that devices used in the management of age-related illnesses will experience significant penetration in the next few years. Security is an absolute requirement for all these applications.

Figure 16.1 depicts the taxonomy of IoT technology elements in the e-health arena (and also the taxonomy of security concerns). e-Health is also referred to as telehealth, telecare, and m-health (this term is particularly relevant when the focus is on mobile users and mobile communications); the term e/m-health is used below. In-hospital IoT applications include, but are not limited, to smart pills, smart beds, radio-frequency identification (RFID)-based medication management, RFID-based asset tracking, and RFID-based asset transportation. Outside the hospital (after the person leaves the hospital), routine medical care activities for a patient, for example, patient data recording, raw data analysis, and data storage, are increasingly being automated, typically utilizing cloud-based systems. This trend is expected to accelerate in the near future with the use of IoT-based technology. With an aging population worldwide, the increased onset of chronic diseases, such as diabetes, heart conditions, and high blood pressure, is driving the demand for medical devices that facilitate advanced mobile monitoring. IoT-powered devices make possible real-time patient monitoring (e.g., vital signs, blood pressure, and medication delivery), and enable medical personnel to quickly respond to the patient's transient medical situation. The deployment of advanced IoT-enabled (connected) medical devices (also in conjunction with the broader use of smartphones) and the use of sophisticated software analytics facilitate expedited patient testing, increased accuracy, mobility/portability, and ease of device usage. The IoT penetration is just one component of the overall drive toward healthcare automation; the other components include big data analytics, cloud services, and artificial intelligence (AI) (e.g., make note of IBM's Watson platform for drug discovery, oncology, and clinical trials tracking).

In 2014, medical expenditures represented 17.5% of the U.S. gross domestic product (GDP) (about $3.0 trillion, or at $9523 per capita), and that share of GDP is expected to rise to 19.6% of GDP by 2024 (CMS Program Statistics, 2015). This is a large opportunity space for automation. At the same time, medical errors are the third highest cause of death in the United States, responsible for an estimated 250,000 patient deaths a year (this being about 10% of the annual total) (Sternberg, 2016); while there are many structural causes for this predicament, improved monitoring as afforded by the IoT can help healthcare providers work to reduce these problematic numbers. The worldwide e/m-health IoT market is expected to reach over $400 billion by 2022 according to market research firms, with a compound annual growth rate (CAGR) of almost 30% against the 2014 base of about $20 billion (Staff, 2016). With half of that market being in the United States, this represents a $500 per capita *annual* investment for each person in the country. According to Ericsson's 2016 Mobility Report, the number of *mobile* IoT connections will overtake phone subscriptions by 2018 (with a CAGR of 23% from 2015 to 2021, at which time there will be 16 billion network-connected mobile IoT devices, in addition to an even larger population of nonmobile IoT devices) (Ericsson Mobility Report, 2016). Thus, the expectation is that, as an allocated average, there will be half a dozen IoT devices per person in the world by the end of this decade, some of which will certainly be medically related devices. Therefore, this topic is of interest to a broad set of stakeholders, which include the healthcare industry, the technology developers, the service providers, and the end users and patients.

In general terms, the wearable device comprises the following three segments: remote patient monitoring, home healthcare, and sports and fitness. The remote patient monitoring segment is additionally taxonomized as follows:

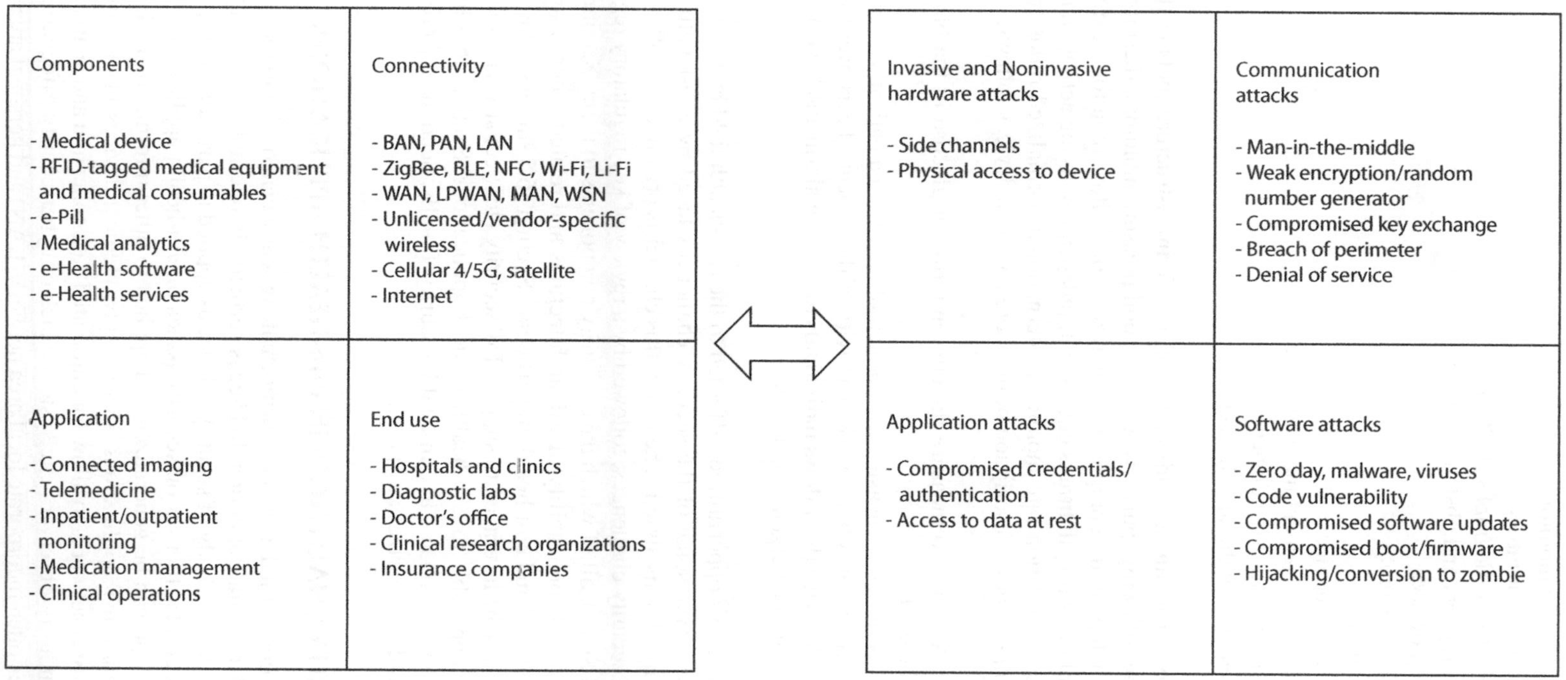

FIGURE 16.1 Taxonomy of e/m-health IoT elements and IoTSec issues.

- Wearable vital signs monitors
  - Wearable heart rate monitors
  - Wearable activity monitors
  - Wearable electrocardiographs (ECGs)
- Wearable fetal monitors and obstetric devices
- Neuromonitoring devices
  - Electroencephalographs
  - Electromyographs
- Therapeutic wearable device
  - Wearable pain management medical devices
  - Glucose/insulin monitoring devices
  - Wearable respiratory therapy devices

Healthcare monitors that can take advantage of an IoT infrastructure include the above-cited devices, as well as the following: pulse oximeters, blood pressure monitors, thermometers, weighing scales, glucose meters, body composition analyzers, peak flow monitors, cardiovascular fitness and activity monitors, strength fitness equipment, independent living activity hubs, medication monitors, basic ECGs, respiration rate monitors, international normalized ratios (blood coagulation), and insulin pumps. A brief description of the more common devices follows:

- *Glucose meter*: A device to measure the concentration of glucose in the blood (e.g., for patients with diabetes)
- *Pulse oximeter*: A device to measure the amount of oxygen in a patient's blood
- *ECG*: A device that records the electrical activity of the patient's heart over time
- *Alerting device*: A device that allows individuals to issue an alarm and have a conversation with a caretaker in an emergency situation

In this chapter, we assess opportunities afforded to the e/m-health field by the evolving IoT, and at the same time, we survey some of the security challenges faced by e/m-health applications; a number of solutions are also discussed. The rest of this chapter is organized as follows. Section 16.2 assesses some of the security challenges, followed by a review of key regulatory regimens for e/m-health (in the United States), all of which have security components to them (Section 16.3). Section 16.4 reviews the need and applicability of IoT architectures and related security architectures, as they apply to IoT in general and e/m-health in particular. Section 16.5 surveys some available security mechanisms that can be leveraged to support IoT security (IoTSec) at various architecture layers, and are directly applicable to e/m-health vertical applications. Section 16.6 highlights some noteworthy near-term trends related to e/m-health security. Finally, Section 16.7 provides a conclusion to this chapter and summarizes key points.

16.2 IoT SECURITY CHALLENGES IN e/m-HEALTH APPLICATIONS

There is a growing recognition that cybersecurity challenges associated with IoT systems are critical and need careful and immediate attention. IoTSec is particularly relevant to e/m-health applications. The term *IoTSec* refers to the body of security science as applied in particular and specifically to the IoT ecosystem. This chapter focuses on security issues in e/m-health applications; however, many of the presented concepts can be generalized and applied to other information and communication technology (ICT) and/or IoT applications. Increasingly, medical devices in the hospital and elsewhere are connected over networks to allow for automated data collection and analytics and improve patient care. Some industry observers cite systemic lack of cybersecurity safeguards in healthcare. (IoT) medical devices often incorporate local area network (LAN) and personal area network (PAN) networking and other software that make them, in fact, vulnerable to cybersecurity threats. A recent study published by the IoT Security Foundation (IoTSF) (iotsecurityfoundation.org) found that less than 10% of all IoT products on the market are designed with adequate security. Cybersecurity in an e/m-health environment is more important than, perhaps, in some other applications not only for the obvious reasons, but also because of the large body of regulations, tied to various penalties or reprimands if violated. Refer back to Figure 16.1.

There has been a large number of health record breaches in recent years. According to statistics collected by the firm Risk Based Security (RBS), there were 281 breaches of medical institutions just in 2015, ultimately compromising 9.9 million user medical records (1642 breaches of medical institutions in this decade alone, exposing more than 58 million customer records). Across all vertical industries, others quote even larger numbers, with an average of 13 breaches a day and 10 million records compromised each day (Staff, 2015). Furthermore, a recent (2016) advisory by the U.S. Department of Homeland Security identified in excess of 1400 vulnerabilities that could be exploited in third-party software (open-source code) used in a connected hospital supply cabinet that stores and dispenses medical products (George, 2016).

In the context of this discussion, there are some specific challenges that have to be addressed in the process of endeavoring to provide robust cybersecurity to IoT-based e/m-health solutions. Some of these challenges do not *a priori* appear to impact or relate to security, but, in fact, they do. Some of these challenges include, but are not limited to, the following (Al-Fuqaha et al., 2015; Granjal et al., 2015; Lai et al., 2015):

- *Attack surface*: A large attack surface exists in the IoT environment; it encompasses the network, the software, and the physical end points. Specifically, every communication link (and network element) and end point—from the IoT device to the aggregation and/or analytics server—is part of the network attack surface. All software code in the IoT ecosystems has, in principle, exploitable vulnerabilities—again, from the IoT device nodes to the aggregation and/or analytics server. At the physical level, if a device is not protected (e.g., in its normal mode, such as a camera on a pole, or in a disturbed mode if, e.g., a device is lost or stolen), an attacker can access the device via physical attacks, and from there the rest of the IoT/IT ecosystem. A medical IoT-based system has the same large ecosystem of possible cybersecurity infraction.
- *Low-complexity devices*: A relatively simple chipset may be used in the sensor or end-point device. This limits the amount of onboard computing that can be undertaken, including computing power needed for encryption, firewalling, and deep packet analysis. Medical devices used outside of a medical institution generally have the same issue.
- *Limited onboard power (short battery life)*: Mobile devices typically have small batteries that provide only limited electrical power. Power conservation algorithms may apply where the device limits either the time, duration, or extent of active computational functions. This could be limiting the amount of computing that can be undertaken, not only for intrinsic tasks, but also for security algorithms, including encryption, virus or malware scans, and deep packet analysis. Medical devices used outside of a medical institution, especially mobile devices, generally have the same issue.
- *Uncontrolled environments*: Devices may be in an open environment and physically tampered with, stolen, or lost (hard statistics collected by RBS show that in many industries—including healthcare—up to 30% of all compromised records are due to lost systems or media (www.riskbasedsecurity.com)). A misplaced medical device worn by a patient could, in principle, contain extensive private health information.
- *Mobility management mechanism (MMM)*: The mobility of devices (including roaming on open networks, when outside of one's home) requires an MMM that not only consumes valuable computing resources but also may place the device on some "foreign" network of unknown security status. Some MMM algorithms (e.g., Mobile IP or Mobile IPv6 [Internet Protocol version 6]) can be computationally complex, draining scarce onboard resources, such that trade-offs between computing and battery power have to be undertaken in regard to mobility and/or security). Medical devices that support vital functions

will require robust connectivity; thus, a significant amount of device resources may be allocated to this function.

- *Continuous operation*: IoT devices almost invariably are always-connected/always-on; therefore, they are in principle more susceptible to cybersecurity attacks (periodic reauthentication may be needed). One can assume that medical devices require such continuous operation.
- *Focused vulnerabilities*: Many IoT configurations make use of gateways to connect the devices to the larger network; this IoT gateway represents a concentrated point of attack. Gateways act as an edge device; typical functionality includes data ingestion, aggregation, cashing, and local storage. In some cases, gateways also process or summarize data and may generate alerts on behalf of the downstream devices. Similarly, there are other points of concentration that represent attractive target points of attack, such as the data repositories. IoT-based medical systems are expected to make use of this aggregation methodology at some critical gateway points. In addition, denial of service (DoS) impacting some critical access points (or portals) could be problematic in a remote patient monitoring application.
- *Vendor or user lack of concern*: Lack of emphasis on IoTSec by both the vendors and the end users, as noted earlier. Additionally, in general, regulatory guidance for compliance evolves much more slowly than the morphing wave of cybersecurity threats. While the healthcare industry is ahead of other industries in this context, additional work remains to be done to completely cement security as a fundamental requirement across all segments of the industry.

There are other issues that have a secondary impact on security. These include, but are not limited to, the following:

- *Lack of agreed-upon end-to-end standards, heterogeneity of environments*: IoT systems being deployed in the short term tend to be vendor specific; wide-ranging, comprehensive standards have not been developed, have matured, or have been implemented. This often limits the usage of off-the-shelf security solutions. Many medical IoT systems are currently vendor specific.
- *Lack of overall system architecture (covering all aspects of the ecosystem, such as connectivity, routing, and analytics)*: IoT systems being deployed in the short term tend not to follow an accepted overall (layered) architecture. A layered architecture would enable concept or function simplicity or standardization and the ability to integrate systems (including security systems) from various vendors, some of whom may be specialized in a given function (such as security), all of this resulting into a fragmented functional environment. Many medical IoT systems suffer from the same closed architecture issues.
- *Large population of users*: There will be a large number of devices in the system, requiring scalable solutions. Some researchers are of the opinion that existing approaches to security based on smaller end-point populations may not scale to the size of the IoT ecosystem—for example, key management is more challenging as the user community becomes large, or an authentication mechanism based on RADIUS or DIAMETER may become bogged down as the number of users accessing the server or servers becomes large. With an aging population worldwide, the number of people with remote monitoring devices is expected to increase over time.

16.3 IoT REGULATORY (AND SECURITY) REQUIREMENTS IN e/m-HEALTH APPLICATIONS

In the United States, there are extensive federal laws as well as individual state laws regarding medical records collection, retention, and access, although these may be somewhat older in nature and pre-IoT. Table 16.1 provides a partial list of applicable federal regulation. As it can be seen,

TABLE 16.1
U.S. Federal Laws Relating to Medical Records Collection, Retention, and Access (Partial List)

Regulation	Brief Explanation
42 CFR (Code of Federal Regulations) Part 2 (Electronic Code of Federal Regulations, n.d.)	1987-enacted regulations related to the confidentiality of records of patients undergoing drug and alcohol abuse treatment and prevention. It limits the use and disclosure of these patient records and identifying information.
Federal Information Security Management Act (FISMA) (Computer Security Resource Center, n.d.)	This 2002 act requires federal agencies (and entities that are in possession of federal information, such as contractors supporting the agency) to provide security protections for all information that the agency has collected or stored.
General provisions applicable to Part 164—Security, Breach Notification, and Privacy Rules (Part 164, Subpart A) (Government Publishing Office, n.d.)	45 CFR § 164.102 (2007) Entities that creates or receives protected health information (PHI) must comply with the requirements in § 164.105.8 covering privacy and security requirements. 45 CFR § 164.102 applies to health plans, healthcare clearinghouses, and healthcare providers who transmit any health information in electronic form in connection with a transaction covered by the HIPAA rules.
Genetic Information Nondiscrimination Act (GINA) of 2008 (Equal Employment Opportunity Commission, n.d.)	Intrinsically, this act protects a person's genetic data from being used by health plans, employers, and others in a discriminatory manner. Confidentiality is an element of this act.
Health Insurance Portability and Accountability Act of 1996 (HIPAA) (Department of Health and Human Services, n.d.)	This act has five section (titles). Title II, which contains relevant security regulations, is composed of seven subtitles aimed at combating fraud, waste, and abuse. Subtitle F (HIPAA administrative simplification provisions) relates to regulations controlling the use and disclosure of IIHI. It drives the adoption of standard transaction formats, code sets, and unique identifiers. Importantly, the provisions establish guidelines for the security and privacy of IIHI. Other titles (I, III, IV, and V) deal with (among other topics) portability, continuation of coverage, and tax deductions.
Medicaid (Title XIX of the Social Security Act) (Social Security Administration, n.d.)	Title XIX of the Social Security Act, enacted in 1965, establishes regulations for Medicaid. It contains several provisions related to the acquisition, use, and disclosure of enrollees' information. Specifically, patients have the right to privacy with regard to medical treatment and clinical records.
Patient Protection and Affordable Care Act (Department of Health and Human Services, n.d.)	This 2010 legislation requires that states establish American health benefit exchanges through which insurers will offer "qualified health plans." It states that, among other requirements, stakeholder entities must comply with the electronic health records requirements, as described in the Social Security Act (administrative provisions).
Proposed HIPAA regulations implementing HITECH (Department of Health and Human Services, n.d.)	HIPAA has a provision (known as the HIPAA Privacy Rule and the HIPAA Security Rule) related to standards aimed at protecting the privacy and security of IIHI. In 2010, the Department of Health and Human Services (HHS) sought to implement the guidelines for privacy and security required by HITECH. The Privacy Rule protects all IIHI/PHI held or transmitted by a covered entity or its business associate, in any form or media (electronic, paper, or verbal). Encryption is not required under the Privacy Rule or the Security Rule, but it is one of a number of identified means to protect PHI.
Breach Notification Rule table (Part 164, Subpart D) (Federal Register, n.d.)	An elaboration of HIPAA § 164.400 that defines a breach as the access, acquisition, use, or disclosure of PHI not consistent with the Privacy Rule. This rule requires healthcare entities, following discovery of a breach of PHI, to notify each individual whose information has been "accessed, acquired, used, or disclosed as a result of such breach."

(Continued)

TABLE 16.1 (CONTINUED)
U.S. Federal Laws Relating to Medical Records Collection, Retention, and Access (Partial List)

Regulation	Brief Explanation
Common Rule (Department of Health and Human Services, n.d.)	The Common Rule is a set of a federal regulations related to research involving humans. A form of the rule was published in 1979, with some elaboration in 1991. Subpart A of the Common Rule is generally for all research involving human subjects, while Subparts B–D include additional protections applicable to special populations (e.g., pregnant women). In 2011, the HHS and the FDA sought to modernize and strengthen this set of rules.
Enforcement Rule table (Part 160, Subparts C–E) (Department of Health and Human Services, n.d.)	An elaboration of HIPAA § 160.300 related to the regulation that covered entities must keep records and submit compliance reports.
Family Educational Rights and Privacy Act (FERPA) (Family Educational Rights and Privacy Act (FERPA), n.d.)	FERPA aims at protecting the privacy of information related to student education records; it must be followed by education institutions that receive federal funds (e.g., public elementary and secondary schools, private and public colleges, and universities).
Health Information Technology for Economic and Clinical Health (HITECH) Act (Department of Health and Human Services, n.d.)	This act (part of the American Recovery and Reinvestment Act of 2009 [ARRA]) directed the healthcare industry to adopt health information technology (HIT). In the context of ARRA, the HITECH Act defined a framework for facilitating HIT adoption and use; it also endeavored to motivate healthcare providers and hospitals (via financial incentives) to adopt certified electronic health record technology (CEHRT).
Privacy Rule table (Part 164, Subpart E) (Department of Health and Human Services, n.d.)	An elaboration of HIPAA § 164.500 adding a provision where the Privacy Rule applies to business associates with respect to the PHI of a covered entity.
Security Rule table (Part 164, Subpart C) (Department of Health and Human Services, n.d.)	An elaboration of HIPAA § 164.302 inserting reference to business associates in the definitions of administrative safeguards and physical safeguards.

there are many regulations related to protecting individually identifiable health information (IIHI) (also referred to in some regulations as "protected health information" [PHI]). IoT-based medical systems, whether used in the collection, transmission, analysis, or storage of this data, must comply with these regulations. Specifically, a panoply of cybersecurity requirements applies to IT systems in general, and e/m-health systems in particular, even more so to IoT-based systems. These requirements include confidentiality (e.g., full data life cycle encryption); data integrity, including digital signatures; trust; service; and system availability, authorization, and authentication.

Consistent with the medical security predicament, the U.S. Food and Drug Administration (FDA) issued a draft cybersecurity guidance in early 2016 aimed at addressing the security of IoT e/m-health devices that contain firmware, software, or programmable logic (U.S. Department of Health and Human Services, Food and Drug Administration, Center for Devices and Radiological Health, Office of the Center Director, Center for Biologics Evaluation and Research, 2016). The goal of this FDA guidance is to encourage medical device manufacturers to implement cybersecurity measures to assess and remediate vulnerabilities in their devices. Until recently, cybersecurity defenses as they relate to medical devices were generally assumed to be addressed by the end-user institution. The FDA guidance is shifting the responsibility to the developers of the medical devices, aiming for them to provide intrinsic security mechanisms. Device manufacturers are urged to rethink their product development approaches and incorporate automated vulnerability testing into their processes; they are also encouraged to run realistic penetration tests to assess how impervious their products are to attacks. This resilience is needed not only for the initial release of the product but also for ongoing software updates. New vulnerabilities may be (unwittingly) introduced during any software or firmware upgrades in the form of "open doors" that allow malware, Trojans, and other intrusive, invasive, or hostile software to infect the IoT medical devices. Notice, however, that guidance is not a *de jure* regulation; thus, there are no penalties for the manufacturers for failure to implement the identified best practices. This FDA document is, however, a first step in addressing IoTSec in the medical arena.

16.4 IOT ARCHITECTURES FOR E/M-HEALTH SECURITY

IoTSec is a major consideration in IoT system design in general and certainly for e/m-health applications, and the security architecture must support at all times a system state composed of secure components, secure communications, and secure asset access control to any and all assets in the IoT ecosystem under consideration. An architecture is a logical reference that provides consistency in the context of the development of system, solution, and application architectures. Architectures are usually defined for the "user plane" (the protocol stack that defines the data flow), the "control plane" (the protocol stack that defines the session establishment), and the "management plane" (including security as well as other management functions). An architecture framework makes available consistent definitions for the system under consideration, its decompositions and design patterns, and a defined vocabulary of terms to describe the specification of implementations, thus enabling design and implementation options to be compared. Figure 16.2 depicts graphically the IoT environment in terms of convergence (or lack thereof) regarding standards and architectures. Critical development initiatives required in order for the IoT to fully take off include the following: (1) the need for interoperability standards—these standards are facilitated when there is a reference architecture; (2) the need to understand that there may be networking differences between regular applications and IoT applications (e.g., due to population sizes, limited nodal complexity, limited power, and mobility); and (3) the need to acknowledge that reliable, ecosystem-wide security is of paramount importance. Currently, many approaches, solutions, and technologies are application (use case) specific; solutions apply to vertical silos. The industry goal is to facilitate horizontally applicable solutions (standards, architectures, and security mechanism) that are not parochial to specific use cases.

A number of IoT architectures have emerged in the recent past (e.g., Arrowhead Framework, ETSI High Level Architecture for M2M, Internet of Things Architecture [IoT-A], ISO/IEC WD 30141 Internet of Things Reference Architecture [IoT RA], and Reference Architecture Model for Industrie 4.0 [RAMI 4.0]), but there is not yet an industry-wide acceptance of any one particular framework; furthermore, none are specific to e/m-health. The lack of agreed-upon architectures and standards up to the present has not only frustrated the broad IoT deployment per se, but also impeded the full integration of security mechanisms in the IoT space. This predicament can have a negative effect in the deployment of IoT systems in the e/m-health field, because the lack of an agreed-upon architecture imposes limitations: security in general and security architectures in particular depend on the availability of an underlying overall ICT architecture, to build required capabilities upon a common, well-defined baseline. Obviously, the goal is to protect IoT devices and related assets from malicious attacks. Ultimately, architectures simplify the discussion of the system's building blocks and how these interrelate to each other, systematizing these functional blocks and fostering standardization. Architectures, frameworks, and standards enable seamless, even plug-and-play connectivity and operation. As a major vertical application of IoT, e/m-health systems will need to use an agreed-upon architecture; as noted in Chapter 17 written by these same authors, architectures facilitate standardization, which in turn fosters commoditization, lower run-the-engine (RTE) costs, and accelerated rollout of a service or function. The IoT ecosystem benefits from defining a usable architecture, and e/m-health applications in turn benefit from a well-defined IoT environment.

Some of the proposed IoT architectures do include security considerations, but most include security as a homogenous vertical stack. In reality, security mechanisms supporting confidentiality, integrity, and availability are needed at each of the architecture layers, and as a bare minimum, encrypted tunnels, encryption of data at rest, and key management are critical parts of IoT/IoTSec desiderata, if not absolute imperatives. As noted, capabilities such as security and management are often included in published architectures as vertical stacks that cut across multiple layers; in the Open Systems IoT

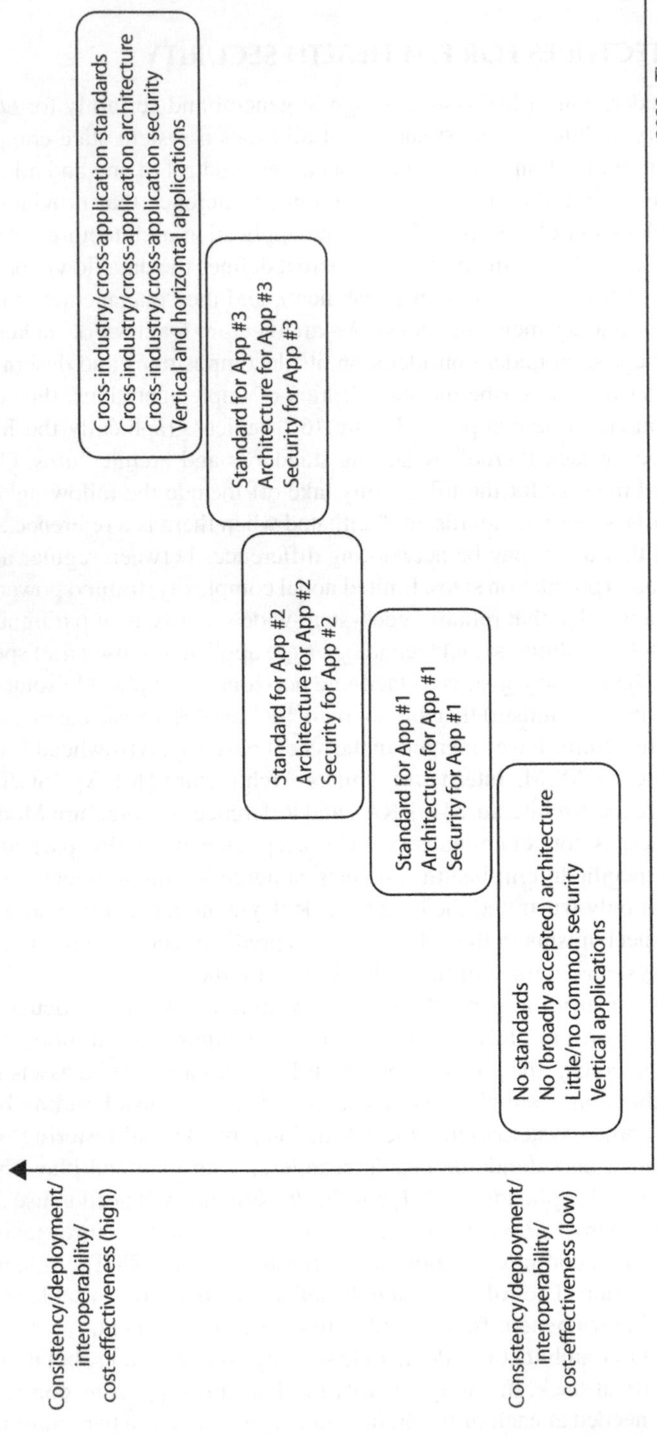

FIGURE 16.2 Evolution of IoT standards, architectures, and security over time.

Reference Model (OSiRM)—to be discussed later—the capabilities are included at each layer. In this model, the mechanisms are specific to and optimized for each layer; although the mechanisms may have the same functional purpose (e.g., authentication), these in-layers mechanisms address the functionality of intrinsic security with the layer in question (e.g., authenticate the node, authenticate the access point, authenticate the IP subnetwork from which the data is coming, and authenticate the core network or extranet from which the data is arriving at the analytics engine). In addition, some layers may have some additional mechanisms if deemed appropriate (e.g., mechanisms beyond the ones discussed by us below). There are advantages and disadvantages in either model. In the former, the security is kind-of tacked on at the side, as a separable product or functionality, at a discrete (future) point in time, say by a specialized vendor; also, there is minimal redundancy of functions replicated at each layer. This is generally the approach that has been used in the ICT industry to date; unfortunately, this model has weaknesses, such as the chronic, daily IT breaches attest (with billions of business records compromised each year, as noted earlier). The latter (baking in the capabilities at each layer) is more complex, requires that each layer vendor implement the features, and may require more processing power, but intuitively one can see that it is a tighter and more reliable model. It also provides "security in depth" by implementing redundant safety checks at several points in the system, architecture, or model. Security is always a balancing act.

In 2014, the National Institute of Standards and Technology (NIST) published a Framework for Improving Critical Infrastructure Cybersecurity (NIST (National Institute of Standards and Technology), 2014) that offers some guiding principles for the administration of cybersecurity that are equally well applicable to the IoT in general and e/m-health in particular. The framework operates to *identify* risks, *protect* critical assets, *detect* cybersecurity infractions, *respond* to such infractions, and *recover* from a possible breach or intrusion. However, an IoT-specific set of architectural constructs is advantageous, both in general and for e/m-health in particular. Figure 16.3 extends this model to include some added granularity; Table 16.2 provides some more detail. All IoT systems should take these requisite tasks under considerations, especially those dealing with mission-critical infrastructure and e/m-health.

To enhance the IoTSec capabilities, the authors recently introduced a seven-layer IoT architecture model, which we refer to as the OSiRM, that highlights the importance of security (Minoli and Sohraby, 2017). In practical terms, layer-specific mechanisms are needed. See Tables 16.3 and 16.4.

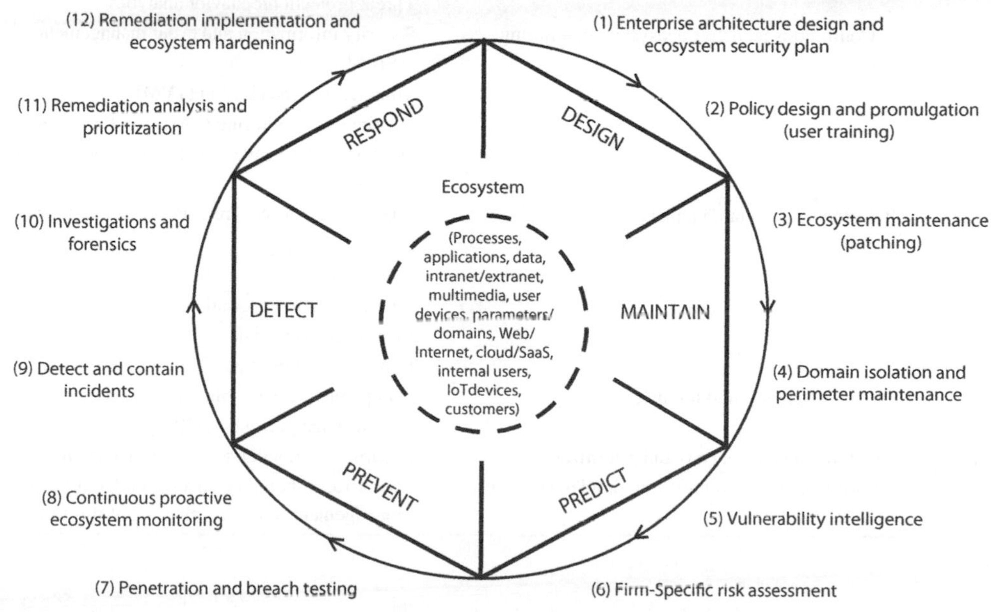

FIGURE 16.3 Extended security framework applicable to IoT-based e/m-health.

TABLE 16.2
Recommended Security-Related Tasks for IoT in General and e/m-Health in Particular

Category	Main Task	Subtask (for entire ecosystem)
1. Design	1. Enterprise architecture design and ecosystem security plan	Enterprise architecture framework tools
		"Security plan" tool
		Asset inventory
	2. Policy design and promulgation (user training)	Policy development/maintenance tool
		Encryption of data at rest, e-mail
		Policy/legal (e.g., Payment Card Industry Data Security Standard [PCI DSS]) compliance
2. Maintain	3. Ecosystem maintenance (patching)	Assess system issues (vulnerability management)
		Antivirus/malware tools
		Certificate management
	4. Domain isolation and perimeter maintenance	Unified threat management (UTM) firewall, Intrusion Detection System [IDS], Intrusion Prevention Systems [IPS], Virtual Private Network [VPN], antimalware gateway/web contents
		Web proxies
		Wireless and bring your own device (BYOD) management
		Voice over Internet protocol (VoIP), multimedia, IoT management
3. Predict	5. Vulnerability intelligence	Study environment at large
		Security vulnerability patch management
		Network access control (NAC)
	6. Firm-specific risk assessment	Vulnerability database tool
		Risk assessment management
4. Prevent	7. Penetration and breach testing	Network penetration testing
		Phishing testing/training
		Threat protection/behavior analytics
	8. Continuous proactive ecosystem monitoring	Security information and event management (SIEM)
		Vulnerability management (VM)
		Continuous monitoring (CM), NIST 800-137 (FISMA), Open Web Application Security Project (OWASP)
5. Detect	9. Detect and contain incidents	IDS/IPS (can be covered under NAC or UTM)
		Cloud computing
		Web filter
		Web application firewall
		Identity management
		Fraud prevention
	10. Investigations and forensics	Computer forensics solution
		Data leakage prevention (DLP) solution
6. Respond	11. Remediation analysis and prioritization	Security information and event management
	12. Remediation and mitigation implementation	The process should be covered under vulnerability management (no. 8 in column 2 of this table)

TABLE 16.3
OSiRM Architecture

Layer	Description	e/m-Health Application	Basic In-Layer Security Mechanism
Layer 7 (applications layer)	This layer encompasses a vast array of horizontal and/or vertical applications or "application domains" (as described in terms use cases). As is the case for Layer 1, effectively the list of applications is "unlimited" in scope. Applications include e/m-health, smart cities, smart building, smart grid, intelligent transport, surveillance, sensing (including crowd sensing), intelligent production, and logistics, to name just a few.	The e/m-health application has specific requirements, e.g., regulatory requirements, security requirements, and reliability and availability requirements. However, to the extent that it can share functionality or solutions used in other vertical applications (if appropriate and/or enhanced for this specific application), it helps control costs and accelerate the broad deployment of the technology.	e-Health user case-level authorization and authentication; encryption and key management; trust and identity management (specific mechanisms for this segment of the ecosystem).
Layer 6 (data analytics and storage layer)	This layer encompasses the data analytics and storage functions.	This layer describes functions and possible standardization of medical-related analytics engines. These are specific to the medical discipline supported by the application, e.g., a glucose analysis tool, an oximeter analysis tool, and ECG analysis tool.	Storage and analytics applications' (enterprise-based and/or cloud-based) authorization and authentication; encryption and key management; trust and identity management (specific mechanisms for this segment of the ecosystem).
Layer 5 (data centralization layer)	This layer supports the data centralization function. This corresponds to the traditional core networking functions of modern networks. It includes institutionally owned (core) networks; industry-specific extranets; public, private, or hybrid cloud-oriented connectivity; and Internet tunnels. These networks are generally composed of carrier-provided connectivity services and infrastructure and entail wireline and/or wireless links.	This layer describes the functionality of a core network used to provide, say, city-wide or campus-wide medical/health services in the context of an IoT-based application. Typically, it would not be a network dedicated just to the medical services (otherwise, it would be quite expensive) but would be shared with other applications. Naturally, security is critical.	Core network authorization and authentication; encryption and key management; trust and identity management (specific mechanisms for this segment of the ecosystem).

(*Continued*)

TABLE 16.3 (CONTINUED)
OSiRM Architecture

Layer	Description	e/m-Health Application	Basic In-Layer Security Mechanism
Layer 4 (data aggregation layer)	This layer supports the data aggregation function. This function may entail some kind of data summarization or protocol conversion (e.g., mapping from a thin, low-complexity protocol used by the IoT clients in consideration of low-power predicaments, to a more standard networking protocol), as well as the edge networking capabilities. The data aggregation function is typically handled in a gateway device. Edge networking represents the outer tier of a traditional network infrastructure, the access tier, employing well-known networking protocols.	There may be data summarization points in a network that is used to support IoT medical applications and/or shared with other applications.	Data aggregation (network) authorization and authentication; encryption and key management; trust and identity management (specific mechanisms for this segment of the ecosystem).
Layer 3 (fog networking layer)	This layer supports fog networking, that is, the localized (site- or neighborhood-specific) network that is the first hop of the IoT client ("device cloud") connectivity. Typically, fog networking is optimized to the IoT clients' operating environment and may use specialized protocols. It could be a wired link (e.g., on a factory LAN, say in a robotics application) or wireless (on a wireless LAN, also optionally including infrared links, e.g., Li-Fi).	This layer supports the initial communication link used by the medical devices, as listed in Figure 16.1, e.g., ZigBee, BLE, Wi-Fi, and cellular links.	Fog network/edge network authorization and authentication; encryption and key management; trust and identity management (specific mechanisms for this segment of the ecosystem).

(*Continued*)

TABLE 16.3 (CONTINUED)
OSiRM Architecture

Layer	Description	e/m-Health Application	Basic In-Layer Security Mechanism
Layer 2 (data acquisition layer)	This layer encompasses the data acquisition capabilities. It is composed of sensors (appropriate to the "thing" and the higher-layer application), embedded devices, embedded electronic, sensor hubs, and so on. Layers 1 and 2 could be seen as being in symbiosis in the IoT world in the sense that things "married" with sensors become the IoT clients or end points. The collected information might be data parameters, voice, video, multimedia, localization data, and so on.	This layer encompasses local concentration devices (base stations) that collect the local signals and data in preparation for additional packaging for upstream transmission. It could be a home-based device that aggregates all the IoT signals in a home, also including signals from the medical devices, or an in-hospital in-room device that aggregates the various monitors worn by a patient.	Aggregation link authorization and authentication; encryption and key management; trust and identity management (specific mechanisms for this segment of the ecosystem).
Layer 1 (things layer)	This layer is composed of the universe of things that are subject to the automation offered by the IoT. Clearly, this is a large domain, including, e.g., people (with wearables, e/m-health medical monitoring devices, etc.), smartphones, appliances (e.g., refrigerators, washing machines, and air conditioners), homes and buildings (including heating, ventilating, and air-conditioning and lighting systems), surveillance cameras, vehicles (cars, trucks, planes, and construction machinery), and utility grid elements. Effectively, this list is unlimited in scope.	This layer encompasses medical devices for both inpatient and outpatient applications, as listed in Figure 16.1, e.g., glucose meter, pulse oximeter, ECG monitor, and event alerting device.	Device-level authorization and authentication; encryption and key management; trust and identity management (specific mechanisms for this segment of the ecosystem).

TABLE 16.4
OSiRM-Assisted Transition from "As Is" IoTSec (e/m-health) Environment to a Target "To Be" Environment

Layers	Device	Typical Status Quo (As Is)	Target (To Be)/OSiRM Assisted
Lower layers	Sensors (sensor to base station communication)	• No encryption • Weak encryption • Weak protocols • No passwords • Weak passwords • Weak OS • Weak applications	• Strong encryption • Robust protocols • Use of Transport Layer Security • Device health checks • Stronger OS • Stronger applications • Strong ID • Strong user interface • Memory isolation • Firmware over the air (FOTA) • Hardware root of trust (RoT) • TEE
	Base station/gateway	• All the communication issues listed above • Hacked device keys • Side-channel attacks • Weak network element OS/memory leakage	• Hardware RoT • Secure Boot • TEE • Trusted firmware • Secure clocks and counters • Anti-rollback mechanisms • Secure key storage • Strong encryption/cryptography
Upper layers	Data servers/cloud (communication with base station)	• No encryption • Weak encryption • Weak protocols • Weak OS • Weak application	• Strong encryption • Strong protocols • Strong OS
	Key server (public key infrastructure [PKI])	• Weak encryption • Weak protocols • Weak OS • Disclosure OK keys	• Secure key provisioning • Key rotation

The OSiRM is a general framework that can be used for all IoT applications, and that is indeed its value and the goal of an architectural framework, but in fact, the OSiRM can be directly applied to the e/m-health environment without loss of generality—many functions in an e/m-health environment are similar (even identical) to functions in other vertical or horizontal IoT applications, and indeed, this is desirable because common functionality and solutions facilitate the deployment of the technology in the specific domain in question (e/m-health) and reduce costs.

16.5 (LAYER-ORIENTED) IoTSEC MECHANISMS

Fundamentally, IoTSec requires the ability to (1) identify IoT devices and their administrative entities (e.g., a gateway), (2) protect the information flow between those devices and their administrative entities, and (3) prevent device hijacking. The use of an IoTSec architecture will assist technology developers and user (medical) institutions comply with the regulations cited above. In particular, it was noted earlier that scalable solutions are needed: a layered "building block" approach intrinsic in the OSiRM allows designers to utilize methods that can scale from low-cost microcontrollers to high-performance platforms.

16.5.1 Overall (All Layers)

OSiRM includes three security-related mechanism realms that effectively exist independently at each layer, as needed:

- Authorization and authentication
- Encryption and key management
- Trust and identity management

Thus, the following mechanisms are intrinsically supported in a layer-oriented manner by the model: tamper resistance by specifying physical protection of devices; user identification by confirmation of the entities involved in a transaction; ensured services with protection against DoS; system-wide secure communication via strong encryption; system-wide management of secure content via data integrity mechanisms; system-wide secure network access, avoiding man-in-the-middle attacks; and strong perimeter (also known as boundary) security for all ecosystem domains. Additionally, in this OSiRM model there will be optimized differences for a given security function at different layers, as well as specializations that may occur with the type of thing and/or type of application.

16.5.2 Lower Layers (Fog Networking Layer)

A number of connectivity solutions exist for the fog area, along with some technology-specific security mechanisms (particularly for transmission confidentiality). These solutions include the ones described in the subsections that follow.

16.5.2.1 Body Area Networks

MBANs are more a concept than a technology per se. MBANs are low-power wideband networks that support the interconnection of multiple body-worn medical sensors to a control device also on a person's body or a nearby "controller" or "hub." In the United States, the Federal Communications Commission (FCC) recently allocated specific spectrum frequencies for MBANs targeted to both inpatient (in-hospital) and outpatient environments. The inpatient application is expected to experience more rapid deployment; given that they support the mobility of patients, they can be highly advantageous in situations where patients have to be transported to various parts of a hospital for various tests or procedures. MBANs interconnect inexpensive disposable body-worn sensors that allow active monitoring of a patient's health, including blood glucose, blood pressure, and electrocardiogram readings. The FCC has been planning to set aside 40 MHz of dedicated spectrum in the 2360–2400 MHz band specifically for wireless medical devices; these devices operate under a "license-by-rule" basis, bypassing the need to apply for individual transmitter licenses.

16.5.2.2 Traditional Personal Area Networks

PANs include ZigBee, Bluetooth Low Energy (BLE), and near-field communication (NFC) technologies.

- ZigBee enables the deployment of low-power, low-cost wireless monitoring and control end points based on the IEEE 802.15.4 standard (complemented with the personal, home, and hospital care profile); this protocol and supportive infrastructure was designed for local connectivity with the goals of simplicity and efficient use of power (where devices operate on commonly available batteries and do not require battery replacement for years). IEEE 802.15.4 defines a reliable radio physical (PHY) layer and medium access control

(MAC) layer; ZigBee per se defines the network, security, and application framework that uses the underlying IEEE PHY/MAC protocol.

- BLE is a low-power version of Bluetooth (IEEE 802.15.1); as such, it is capable of operating for up to a year from a small button battery—the data rate and radio range are lower than those of Bluetooth itself, but the low power and long battery life make it suitable for short-range monitoring in e/m-health applications.
- NFC supports contactless communication between devices (e.g., a card or a smartphone) where a user waves the device over an NFC-based reader to transfer information without the devices making contact.
- Other systems have also emerged or are emerging—for example, IEEE 802.15.6-2012, Body Area Network; IEEE 802.15.4j, Low-Rate Wireless Personal Area Networks (LR-WPANs), Amendment 4: Alternative Physical Layer Extension to Support Medical Body Area Network (MBAN) Services Operating in the 2360–2400 MHz Band; ISO/IEEE 11073, Personal Health Data (PHD) Standards; and ETSI TR 101 557.

16.5.2.3 Hybrid (Home–Public) Hotspot Networks

These networks typically support connectivity to various cloud-based servers when used in an IoT context. A landline-based network usually (but not always) exists to provide backhaul connectivity. These networks also include the newly standardized low-power Wi-Fi known as Wi-Fi HaLow™—these solutions afford cost-effective, energy-efficient build-out, ideal for e-health and mobility. They generally (but not always) support only relatively low bandwidth. In particular, Wi-Fi HaLow is based on IEEE 802.11ah technology; it operates in the 900 MHz band and affords longer-range, lower power connectivity to Wi-Fi-enabled devices. This technology supports power-efficient applications in the smart home, the connected car, digital healthcare, and other applications; clearly, it is positioned for the low power connectivity needed for wearables and similar use cases. Wi-Fi HaLow's range is approximately twice that of traditional Wi-Fi, while also providing a more robust connectivity, for example, the ability to more easily penetrate structures, walls, or other barriers. The goal is to support multivendor interoperability, strong security, and easy setup (including intrinsic support of IP and IoT). Devices that support Wi-Fi HaLow will also operate in the traditional 2.4 and 5 GHz bands.

16.5.2.4 Low-Power Wide Area

These are generally short-range (up to a few miles) technologies, with relatively low bandwidth, utilizing unlicensed radio spectrum. They are typically vendor-specific wireless systems. Also, they are subject to channel interference. Typically, there is no intrinsic security included, and confidentiality must, in general, be implemented by a device-level encryption or tunneling mechanism. Unlicensed non-Third Generation Partnership Project (3GPP) low-power wide area (LPWA) IoT wireless technologies include the following (Moyer, 2015):

- Platanus (<1 km; 500 kbps)
- OnRamp (4 km; up to 8 kbps)
- Weighless-N (up to 5 km; up to 100 kbps)
- Telensa (up to 8 km; low)
- NWave (10 km; up to 100 bps)
- Amber Wireless (up to 20 km; up to 500 kbps)
- LoRa (15–45 km suburban, 3–8 km urban; up to 50 kbps)
- SIGFOX (50 km suburban; 100 bps)

16.5.2.5 Cellular Solutions (4G/5G)

These solutions, as applied to IoT, offer a large footprint or range (especially considering roaming capabilities) and high throughput, but they have a relatively high-end-point or subscription cost; they make use of licensed spectrum. New cellular technologies, specifically 5G, are expected to become available toward the end of the decade; 5G cellular has a 10–15 km coverage, uses licensed spectrum, and supports throughputs in the gigabits per second range (the battery life for IoT applications is targeted at ~10 years). Cellular systems already support a number of IoT-specific (proposed) services, including (Ericsson, 2016)

- LTE-M Rel-13 (~10 km, licensed spectrum, 1 Mbps, battery ~10 years)
- Narrowband NB-LTE Rel-13 (~15 km, licensed spectrum, 0.1 Mbps, battery ~10 years)
- Narrowband EC-GSM Rel-13 (~15 km, licensed spectrum, 0.01 Mbps, battery ~10 years)
- Narrowband IoT (NB-IoT)

  3GPP recently adopted the NB-IoT LPWA system that defines a new radio access PHY for cellular IoT. NB-IoT aims for improved indoor coverage, a large number of low-throughput devices, low latency, very low device cost, and low device power consumption. NB-IoT is expected to replace the previous NB-LTE and NB-CIoT proposals.

16.5.2.6 Other Solutions

Satellite services offer global-reach IoT capabilities, but have a high-end-point or subscription cost. Wireline IoT solutions can offer high bandwidth but suffer from limited mobility and a relatively high connectivity cost.

16.5.2.7 Security Considerations

In the context of security, some of the protocols listed above (notably ZigBee, BLE, and Wi-Fi HaLow) offer MAC layer encryption in support of first-hop confidentiality, while others do not, and thus the developer or technology provider must provide encryption tools. Even when providing first-hop confidentiality, end-to-end confidentiality must be ensured. As implied in the OSiRM model discussed earlier (and illustrated in Figure 16.4), strong security measures for authorization and authentication, encryption and key management, and trust and identity management must be implemented at each layer of the model and end to end. Table 16.4 (partially inspired by (Wallace, 2016)) depicts an OSiRM-assisted transition from an "as is" IoTSec (e/m-health) environment to a target "to be" environment.

16.5.3 Hardware Level

Some security capabilities are best handled at the hardware level. A number of such capabilities have emerged in recent years. The Trusted Execution Environment (TEE) is of particular importance to (IoT) applications that deal with sensitive user data, including e-health-related data (such as user real estate locations, user real estate contents [e.g., jewelry], and medical claims). In addition, Intel Trusted Execution Technology (TXT) is a system hardware technology that deals with verifying and/or maintaining a trusted operating system (OS). Both of these technologies are discussed next, since they have relevance to the processing environment that may be utilized by e-health companies.

16.5.3.1 Trusted Execution Environment

The TEE is a secure ecosystem of the IoT device processor or of the application system (trusted applications [TAs]), offering the capability of isolated execution of authorized security software. In doing so, TEE provides end-to-end security by enforcing protected execution of authenticated code; it also ensures confidentiality, authenticity, privacy, system integrity, and data access rights (GlobalPlatform, 2015). In general, IoT devices are single-purpose devices and may have a specifically chosen OS or application script. However, a user may also utilize his or her smartphone as a device and/or to access IoT-related data repositories. In this instance, the situation may occur where a number of other, often poorly secured, applications have been downloaded, which can cause cross-domain contamination. Mechanisms are thus needed to allow trusted access to device,

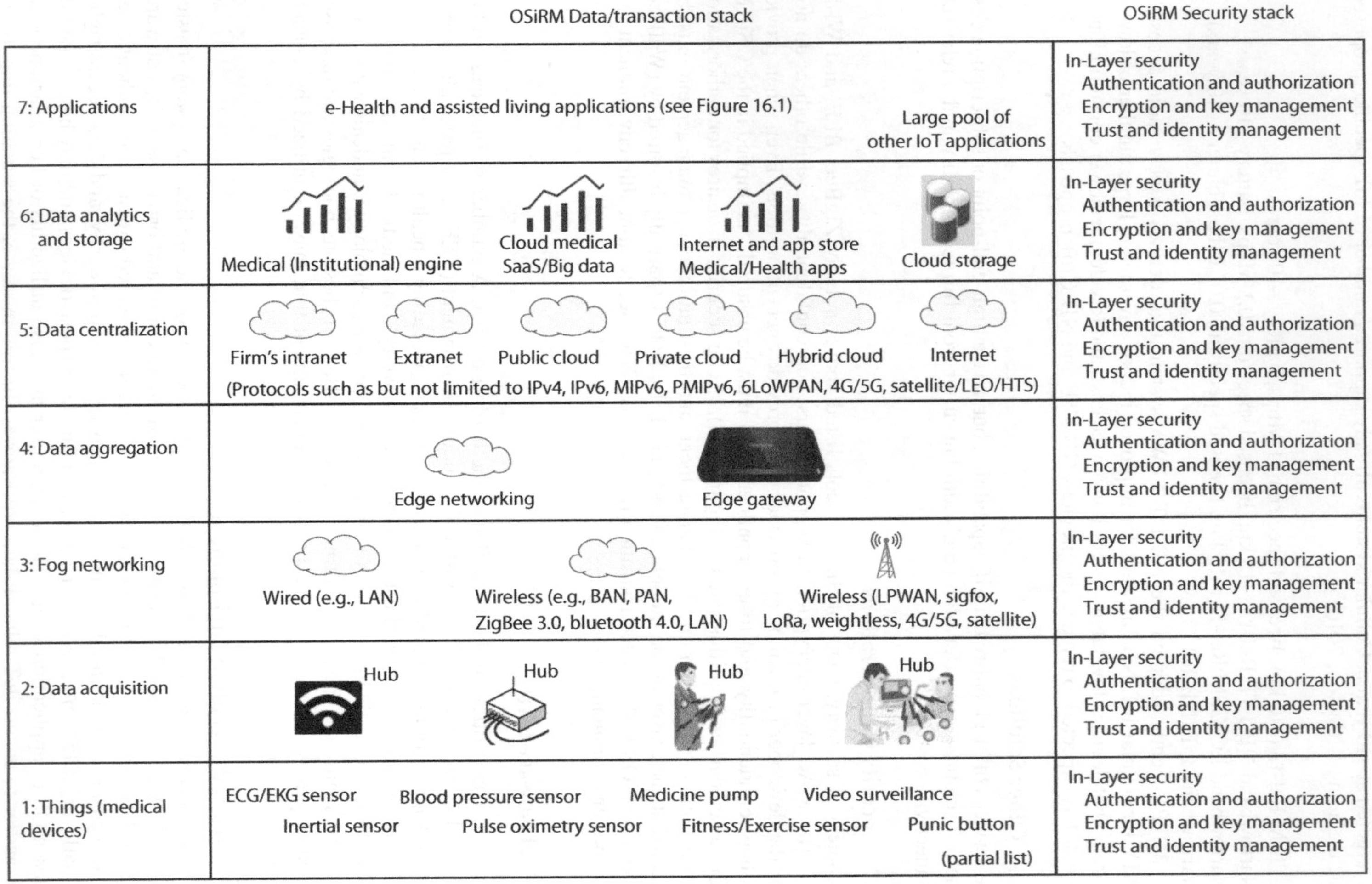

FIGURE 16.4 OSiRM, as applied to e/m-health environments.

system, or data resources. Service providers and device manufacturers (e.g., OEMs) have the challenge of protecting applications at various concentric domains, such as attacks against (1) the device's OS, (2) a device-resident application, (3) a device's (and/or user's) credentials (e.g., authenticating the correct user to the correct service), (4) transmission integrity and privacy, and (5) the data-at-rest content. The TEE isolates secure applications and screens them from malware and viruses that might be injected (or downloaded) inadvertently. TEE-based approaches may be leveraged to address these concerns.

Naturally, a key desideratum for TEE is to minimize the device or application processing overhead while at the same time providing controlled access to a large amount of processor memory. In particular, TEE ensures that sensitive data, such as insurance-related transactions, is processed, stored, and protected in a trusted isolated environment. Developers have defined the main OS on an IoT device or an application server as the "Rich OS" (e.g., Windows and Android). The TEE is a lean OS-like environment that resides aside the Rich OS (this is also known as the Rich Execution Environment [REE]). It is designed to extend the level of protection against attacks that may have been generated in the Rich OS (e.g., from malware), for example, by taking control of access rights. TEE assumes that nothing coming from the Rich OS is trustworthy. TEE domiciles sensitive applications that are best isolated from the Rich OS, and it maintains all cybersecurity credentials and data manipulation in the lean TEE rather than in a larger Rich OS. In this paradigm, sensitive functions are meticulously defined and assigned to the TEE in the form of TAs (while also integrating physical tamper-resistant mechanisms).

The TEE concept was at first defined in 2007 by the Open Mobile Terminal Platform (OMTP) forum. TEE standardization is critical to avoid application, hardware, and industry fragmentation; standardization enables simplified implementation, improves interoperability, and reduces costs. GlobalPlatform is an organization of more than 130 member firms developing standards. In support of an effort to bring some industry standardization to the TEE environment, GlobalPlatform has defined two sets of application programming interfaces (APIs): TEE Internal APIs (1.0) and TEE Client APIs (1.0). The TEE Internal APIs are utilized by a TA, and the TEE Client APIs support the communication interfaces that Rich OS software can use to interact with its TAs. Ericson and Nokia are two of several vendors that have developed the TEE. GlobalPlatform also promulgated a compliance testing process and issued a protection profile to allow certifying that a TEE meets the target security level. The advantage of working within a community, rather than "going it alone," is that multiple vendors can contribute to the security assurance efforts, and products are interoperable and interchangeable. Open-source TEE supporting standard interfaces have become available recently; these are sometimes known as OP(en)-TEE (Bech, 2014). Figure 16.5 depicts the software environment of a TEE-based IoT device; it shows that the standard Rich OS can securely delegate some key functions to the trusted side of the software kernel.

Chipmakers and devices using these chipsets utilize TEEs to deliver platforms that have trust (i.e., the assurance that the device is only running legitimate, uncorrupted firmware or software) built in from the get-go; in turn, service and content providers rely on integral trust to build critical applications (e.g., financial, insurance, and e-health). Typical TEE-oriented applications to date have included digital rights management (e.g., digital content, films, and music), mCommerce and mPayment credentials and transactions, and enterprise data (which can include insurance data). While these concepts are described here as supporting the end-user device (e.g., the medical monitoring gear), they can also apply to the processing end of the path (e.g., the analytics engines, portals, or software as a service [SaaS] cloud-based applications). Looking specifically at security, the TEE environment can increase the level of assurance of the medical devices as relates to the following: user authentication, trusted processing and isolation, transaction validation, usage of secure resources, and certification. For example, a TEE-based environment ensures that a rouge device that presents into the environment is not accepted as a legitimate user—the rouge device may have the nefarious intent to inject some malware at some point in time. Also, the TEE makes sure that a device (or user) does not escalate its privileges beyond what is the intent of the system administrator

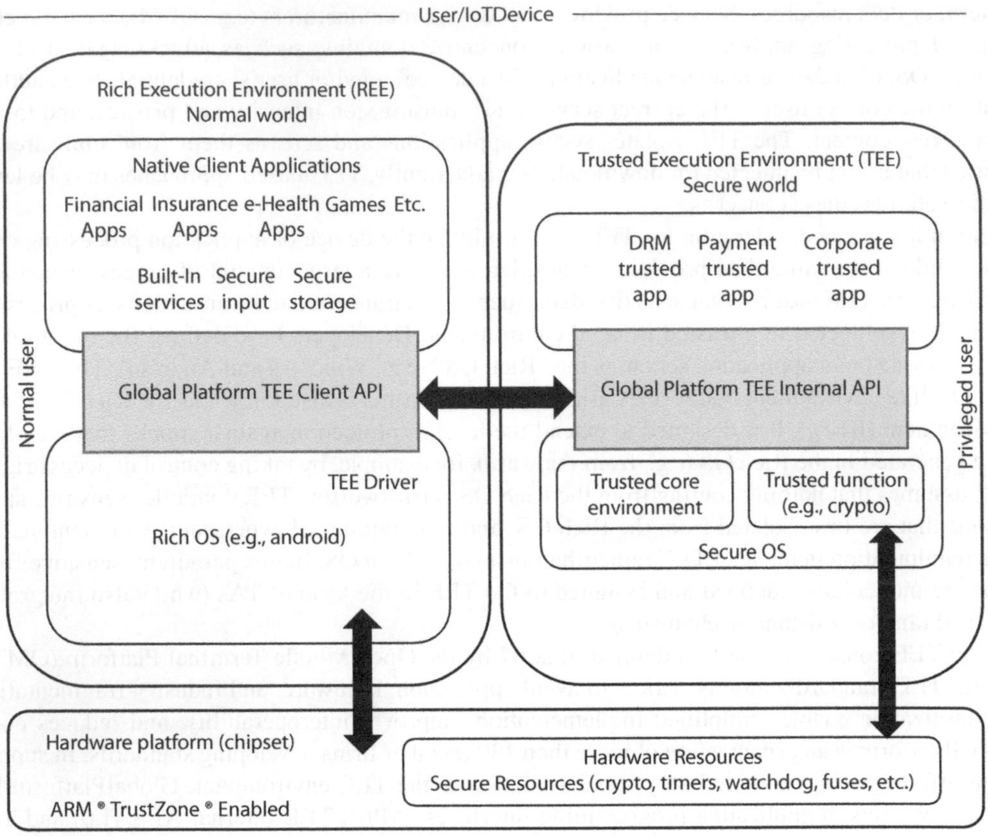

FIGURE 16.5 OP-TEE architecture. Note: The ARM TrustZone® provides isolation from the Rich OS. OP-TEE targets ARM cores and thus includes a secure monitor code for TrustZone (the code executed when the core switches between TrustZone and non-TrustZone modes). OP-TEE can also support architectures other than ARM TrustZone (e.g., the Cortex-M and Cortex-R range of ARM cores).

(by providing isolation that limits functionality or access to data or some function—e.g., a function controlling automatic medication delivery from a body-worn pump).

As a side note, TEEs can increase the capabilities of secure elements (SEs). SEs are secure components that consist of autonomous, tamper-resistant hardware within which secure applications and related confidential cryptographic data (e.g., key management) are stored and executed. While enjoying a high level of security, these devices have limited functionality; an example includes an NFC device. SEs can work in conjunction with the TEE to enhance their capabilities.

16.5.3.2 Intel TXT

Intel® TXT is a scalable architecture that specifies hardware-based security protection (Intel Corporation, 2012). These methodologies are built into Intel's chipsets to address threats across physical and virtual infrastructures; the technology is designed to harden platforms to better deal with threats of hypervisor attacks, BIOS or other firmware attacks, malicious rootkit installations, or other software-based attacks. TXT aims to increase protection by allowing greater control of the launch stack through a measured launch environment (MLE) and enabling isolation in the boot process. It extends the virtual machine extensions (VMX) environment of Intel Virtualization Technology to provide a verifiably secure installation, launch, and use of a hypervisor or OS. MLE enables an accurate comparison of all the critical elements of the launch environment against a "known good" source. TXT creates a cryptographically unique identifier for each approved launch-enabled component and then provides hardware-based enforcement mechanisms to block the launch of code that does not match approved code. This hardware-based solution provides the foundation on which trusted platform solutions can be built to protect against the software-based attacks; the technology is broadly applicable to servers and IoT devices. More specifically, TXT provides (Intel Corporation, 2012)

- *Verified launch*: A hardware-based chain of trust that enables launch of the MLE into a known good state. Changes to the MLE can be detected through cryptographic (hash-based or signed) measurements.
- *Launch control policy (LCP)*: A policy engine for the creation and implementation of enforceable lists of known good or approved, executable code.
- *Secret protection*: Hardware-assisted methods that remove residual data at an improper MLE shutdown, protecting data from memory-snooping software and reset attacks.
- *Attestation*: The ability to provide platform measurement credentials to local or remote users or systems to complete the trust verification process and support compliance and audit activities.

Figure 16.6 depicts the TXT functional flow in a Trusted Platform Module (TPM). It shows that after the system goes through a boot-up stage, various state variables are compared with prestored data; only if a match in state parameters is established is the device allowed to move to the next stage, which would entail loading the application code, at which point other state checks are made. If the checks fail, reporting of the condition occurs (to some administrator); there would be an indication that "trust" of the platform cannot be established. TXT can be used in IoT environments,

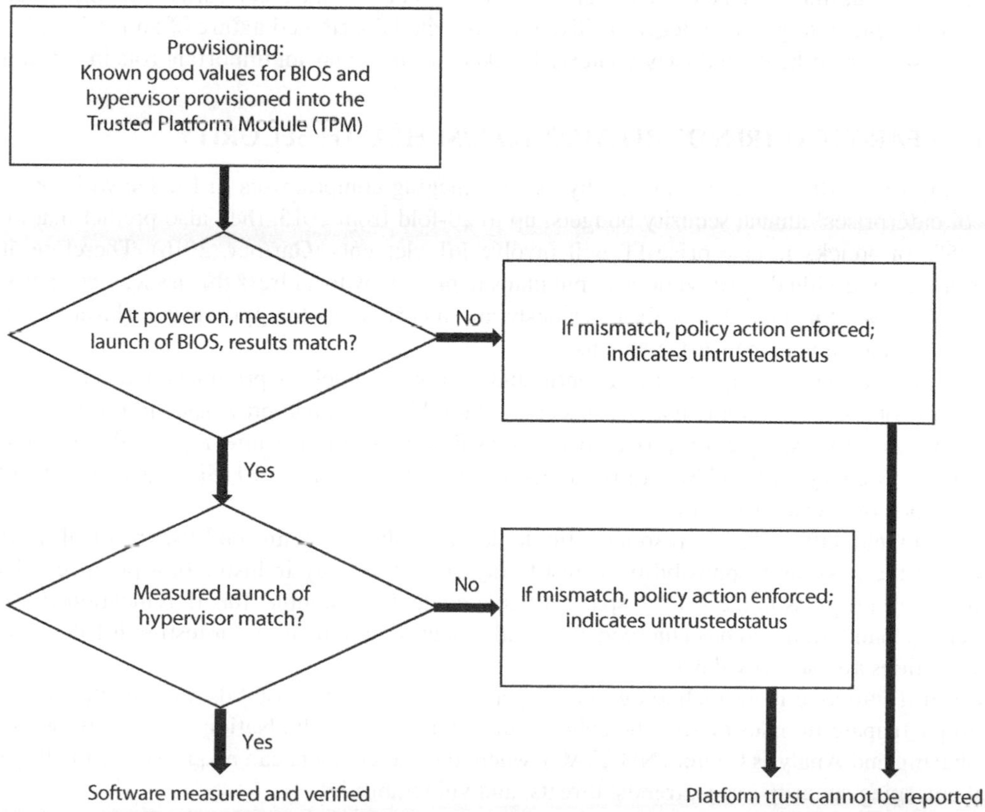

FIGURE 16.6 Basic operation of TXT (modeled after Intel materials).

particularly at the server end, to ascertain that the medical analytics engines and the storage systems (e.g., storing patient health data) are secure. These processes can also be embedded in the medical IoT device to ascertain, among other validations, that the device is legitimate—in the sense that it runs only official code, performing only special functions, and not exceeding its level of functional authority. The TPM is an International Organization for Standardization (ISO) standard for a secure cryptoprocessor, which is a dedicated microcontroller designed to secure hardware by integrating cryptographic keys into devices (ISO/IEC 11889 was published in 2009—the technical specification for TPM was developed by the Trusted Computing Group [TCG]).

16.5.3.3 Other Approaches

Some have sought to implement security mechanisms at a much higher level than the physical underlying facilities, both in an IoT context and in the broader IT context. One such mechanism is the "blockchain." A blockchain is a form of distributed ledger (a distributed database) that retains an expanding list of records, while precluding revision or tampering (an untamperable ledger). The blockchain encompasses a data structure of "child" (aka successor) blocks; each block includes sets of transactions, time stamps, and links to a "parent" (aka predecessor) block. The linked blocks constitute a chain. It intrinsically provides universal accessibility, incorruptibility, openness, and the ability to store and transfer data in a secure manner. The original application was as a ledger for bitcoins. Users are able to add transactions, verify transactions, and add new blocks. Proponents see opportunities for the use of blockchains for e-health companies (and banks and many other industries), as they allow a replacement of the common centralized data paradigm, thus fostering additional process disintermediation (Jones et al., 2016). Possible applications of interest to the e-health industry include claims filing and processing; claim fraud detection, for example, spotting multiple claims from a claimant (medical office) for the same procedure; data decentralization; and cybersecurity management (e.g., data integrity). Given the expected distributed nature of an IoT-based ecosystem (also in the e-health industry context), blockchains may play an important role in the future.

16.6 NEAR-TERM TRENDS RELATED TO E/M-HEALTH SECURITY

According to a 2016 report by Gartner, by 2020 managing compromises in IoTSec will consume 20% of enterprises' annual security budgets, up to 20-fold from 2015; they also predict that more than 25% of attacks in enterprise IT will involve IoT elements (Gartner, 2016). Therefore, it is important for individual organizations to put plans in place now to address this issue. One approach is to select a vendor that takes IoTSec seriously and incorporates the appropriate (layer-by-layer) security mechanisms in their IoT elements.

Naturally, security management is a continuum for both technology providers and end users. As an example of positive movement along this continuum, Microsoft had previously introduced an OS for IoT devices, for example, for processors such as Raspberry Pi; it is now reportedly planning to add BitLocker encryption and Secure Boot technology to the Windows 10 IoT (to ascertain that the device boots use trusted software).

While manufacturers have a responsibility to develop robust systems, end-use (medical) institutions have the ongoing responsibility of instituting and abiding by industry best practices. Some (commercial) progress in the IoTSec space is being made. For example, the aforementioned IoTSF (iotsecurityfoundation.org) has emerged as a vendor-neutral organization for testing IoT devices for vulnerabilities and security flaws.

A useful initiative is for technology developers and users (e.g., hospitals and healthcare facilities) to participate in industry-specific cybersecurity fora, such as the National Health Information and Sharing and Analysis Center (NH-ISAC), where the stakeholders can acquire early intelligence about emerging security issues, trends, threats, and vulnerabilities.

Continued research in scalable and efficient IoTSec tools and mechanisms, and continued incorporation and deployment of these techniques, particularly in the e/m-health environment, are needed.

16.7 CONCLUSION

This chapter assessed the critical need of IoTSec in the specific context of e/m-health. IoT/MBAN-based devices can greatly improve the delivery of healthcare services in the near future, but security is a fundamental requirement for this vertical application, as, in fact, it is for other critical infrastructure monitoring IoT applications.

The plethora of IoTSec challenges faced by e/m-health applications was discussed. This was followed by a review of key regulatory regimens for e/m-health (in the United States), all of which have security guidelines and compliance statements. This was followed by a review of the need for IoT architectures and related security architectures, as drivers for interoperability, reduced costs, and accelerated deployment. A number of available security mechanisms that can be used to support e/m-health IoTSec at various architecture layers were discussed, followed by a quick assessment of some near-term trends related to e/m-health security.

REFERENCES

Al-Fuqaha, M. G., et al. 2015. Internet of Things: A survey on enabling technologies, protocols, and applications. *IEEE Communication Surveys and Tutorials* 17 (4): 2347.

Bech, J. 2014. OP-TEE, Open-Source Security for the Mass-Market. *Core Dump* online magazine, September 3, 2014. http://www.linaro.org/blog/core-dump/op-tee-open-source-security-mass-market/.

Buckiewicz, B. 2016. Overview of Medical Body Area Networks, White Paper, LSR/Laird Business, Cedarburg, WI. www.lsr.com/white-papers/overview-of-medical-body-area-networks.

CMS Program Statistics. 2015. U.S. Center for Medicare & Medical Services. https://www.cms.gov

Computer Security Resource Center. n.d. Federal Information Security Management Act (FISMA). http://csrc.nist.gov/drivers/documents/FISMA-final.pdf.

Department of Health and Human Services. n.d. Health Insurance Portability and Accountability Act of 1996 (HIPAA). https://www.hhs.gov/hipaa/for-professionals/privacy/laws-regulations/.

Department of Health and Human Services. n.d. Proposed HIPAA Regulations Implementing HITECH. http://www.hhs.gov/hipaa/for-professionals/security/guidance/proposed-rulemaking-to-implement-HITECH-act-modifications/index.html

Department of Health and Human Services. The Common Rule. http://www.hhs.gov/ohrp/regulations-and-policy/regulations/common-rule/.

Department of Health and Human Services. n.d. The Enforcement Rule Table (Part 160, Subparts C–E). http://www.hhs.gov/hipaa/for-professionals/special-topics/enforcement-rule/index.html.

Department of Health and Human Services. n.d. The Health Information Technology for Economic and Clinical Health (HITECH) Act. http://www.hhs.gov/hipaa/for-professionals/special-topics/HITECH-act-enforcement-interim-final-rule/index.html.

Department of Health and Human Services. n.d. The Privacy Rule Table (Part 164, Subpart E). http://www.hhs.gov/hipaa/for-professionals/privacy/.

Department of Health and Human Services. n.d. The Security Rule Table (Part 164, Subpart C). http://www.hhs.gov/hipaa/for-professionals/security/.

Department of Health and Human Services. n.d. Patient Protection and Affordable Care Act. http://www.hhs.gov/healthcare/about-the-law/read-the-law/.

Electronic Code of Federal Regulations. n.d. 42 CFR (Code of Federal Regulations) Part 2. http://www.ecfr.gov/cgi-bin/text-idx?rgn=div5;node=42%3A1.0.1.1.2.

Equal Employment Opportunity Commission. n.d. Genetic Information Nondiscrimination Act (GINA) of 2008. https://www.eeoc.gov/laws/statutes/gina.cfm.

Ericsson Mobility Report, Stockholm, Sweden, 2016.

Ericsson. 2016. Cellular Networks for Massive IoT, Ericsson White Paper, UEN 284 23-3278, January 2016.

Family Educational Rights and Privacy Act (FERPA). n.d. http://www2.ed.gov/policy/gen/guid/fpco/ferpa/index.html.

Federal Register. n.d. The Breach Notification Rule Table (Part 164, Subpart D). https://www.federalregister.gov/articles/2013/01/25/2013-01073/modifications-to-the-hipaa-privacy-security-enforcement-and-breach-notification-rules-under-the.

Gartner. 2016. Gartner Says Worldwide IoT Security Spending to Reach $348 Million in 2016 and by 2020, More Than 25 Percent of Identified Attacks in Enterprises Will Involve IoT, Press Release, Stamford, CT, April 25, 2016.

Gartner. 2016. Gartner Says Worldwide Wearable Devices Sales to Grow 18.4 Percent in 2016, Press Release, February 2016.

George, T. 2016. Is the FDA's Cybersecurity Guidance Improving Cyber Resilience? *SecurityInfoWatch* Online Magazine, June 24, 2016.

GlobalPlatform. 2015. Trusted Execution Environment (TEE) Guide, White Paper, GlobalPlatform, Redwood City, CA.

Government Publishing Office. n.d. General Provisions Applicable to Part 164—The Security, Breach Notification, and Privacy Rules (Part 164, Subpart A). https://www.gpo.gov/fdsys/pkg/CFR-2005-title45-vol1/pdf/CFR-2005-title45-vol1-part164.pdf.

Granjal, J., E. Monteiro, and J. S. Silva. 2015. Security for the Internet of Things: A survey of existing protocols and open research issues. *IEEE Communication Surveys and Tutorials* 17 (3): 1294.

Intel Corporation. 2012. Intel® Trusted Execution Technology, White Paper, Intel, Santa Clara.

IoT Security Foundation. n.d. iotsecurityfoundation.org.

Jones, M. C., et al. 2016. Insurers, New Kids on the Blockchain? White Paper, FC Business Intelligence Ltd., London.

Lai, C., R. Lu, D. Zheng, H. Li, and X. Shen. 2015. Toward Secure Large-Scale Machine-to-Machine Communications in 3GPP Networks: Challenges and Solutions. *IEEE Communications Magazine Communications Standards Supplement*, December 2015: 12.

Materials collected/archived at www.riskbasedsecurity.com.

Minoli, D. and K. Sohraby. 2017. IoT Security (IoTSec) Considerations, requirements, and architectures. *14th Annual IEEE Consumer Communications and Networking Conference (CCNC 2017)*, Las Vegas, January 8–11, 2017.

Mordor Intelligence. 2016. Global Wearable Medical Devices Market—Growth, Trends and Forecasts (2015–2020), August 2016.

Moyer, B. 2015. Low power, wide area: A survey of longer-range IoT wireless protocols. *Electronic Engineering Journal*, September 7, 2015. http://www.eejournal.com/arti.

NIST (National Institute of Standards and Technology). 2014. *Framework for Improving Critical Infrastructure Cybersecurity*, Version 1.0. NIST, Gaithersburg, MD, February 12, 2014.

Social Security Administration. n.d. Medicaid (Title XIX of the Social Security Act), https://www.ssa.gov/OP_Home/ssact/title19/1900.htm.

Staff. 2015. Building a Trusted Foundation for the Internet of Things, Gemalto/SafeNet Whitepaper, Meudon, France.

Staff. 2016. IoT in Healthcare Market to be Worth $409.9 Billion by 2022: Grand View Research, Inc., PR Newswire Europe via COMTEX, May 24, 2016.

Sternberg, S. 2016. Medical Errors Are Third Leading Cause of Death in the U.S., *U.S. News and World Report* (online), May 3, 2016.

U.S. Department of Health and Human Services, Food and Drug Administration, Center for Devices and Radiological Health, Office of the Center Director, Center for Biologics Evaluation and Research. 2016. Postmarket Management of Cybersecurity in Medical Devices Draft Guidance for Industry and Food and Drug Administration Staff, Draft Guidance, January 22, 2016.

Wallace, J. 2016. Securing the Embedded IoT World, Guest Blog, June 2016, IoT Security Foundation Website. https://www.iotsecurityfoundation.org/.

17 IoT Considerations, Requirements, and Architectures for Insurance Applications

Daniel Minoli, Benedict Occhiogrosso, Kazem Sohraby, James Gleason, and Jake Kouns

CONTENTS

17.1 Introduction 347
17.2 IoT Applications in the Insurance Industry 349
17.3 IoT Challenges in Insurance Applications 351
17.4 IoT Architectures and Layer-Oriented IoTSec Mechanisms 352
17.4.1 IoT Architectures 352
17.4.2 Layered Security 354
17.4.2.1 Overall (All Layers) 357
17.4.2.2 Lower Layers (Fog Networking Layer) 357
17.4.2.3 Hardware Level 357
17.4.2.4 Other Approaches 358
17.5 Traffic Characteristics 358
17.6 Conclusion 360
References 361

17.1 INTRODUCTION

IoT is entering the daily operation of many industries and, in many instances, offering disruptive redesign of the underlying business processes. Applications include not only "obvious use cases," such as smart grids, smart cities, smart homes, physical security, e-health, asset management, and logistics, but also use cases that are somewhat more esoteric, such as banking, insurance, business process management, and government process optimization. Interestingly, in reality, the obvious use cases all have a bearing on the insurance business itself, since, for example, a smart city (say, with an intelligent transportation system and improved traffic flows) may lead to fewer accidents. The same is true for smart homes, in addition to reducing risks of fire, theft, and vandalism. IoT-enabled physical security (at the commercial building level, city level, or residential home level) can lower overall losses due to various contingencies, while also reducing crime. Loss mitigation improvements may also be expected in the context of asset management and logistics. Finally, improvements in health obtained via IoT-enabled outpatient e-health monitoring can not only improve the quality of life for patients, but also, importantly, reduce treatment and insurance costs. Such automation can be ideally supported by the IoT.

According to a 2016 Gartner report, by 2020 more than 50% of newly instituted business processes will incorporate IoT principles and concepts (Gartner, 2016). This trend will certainly be applicable to the insurance industry. Insurance is a form of risk management aimed at transferring or hedging the risk of a class of losses from one entity to another in exchange for financial considerations. The insurance firm sells the insurance; the insured, or policyholder, purchases the insurance policy. The premium is the amount of financial value charged for a certain amount of insurance coverage. As a point of reference, the annual revenue of the U.S. insurance industry (insurance premiums) now exceeds \$1.2 trillion—the United States is the largest global market for insurance in terms of revenue, with this revenue being in the range of 7% of the U.S. gross domestic product (GDP). The industry is broadly composed of the life and health (L/H) sector, the property and casualty (P/C) sector, and the health (H) sector, although other sectors and segments also exist. In 2014, there were 1031 L/H insurance entities, 2718 P/C insurance entities, and 1060 health insurance entities in the United States. Net income has fluctuated somewhat in recent years, although the industry remains profitable. In 2014, L/H had a net income of \$38 billion and the P/C sector had a net income of \$65 billion (\$103 billion total, or 8%–9% of the top line) (Annual Report on the Insurance Industry, 2015) (Figure 17.1). Given the revenue pool under discussion, a technology

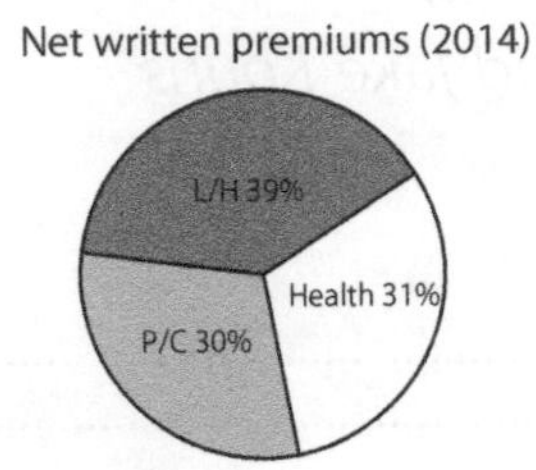

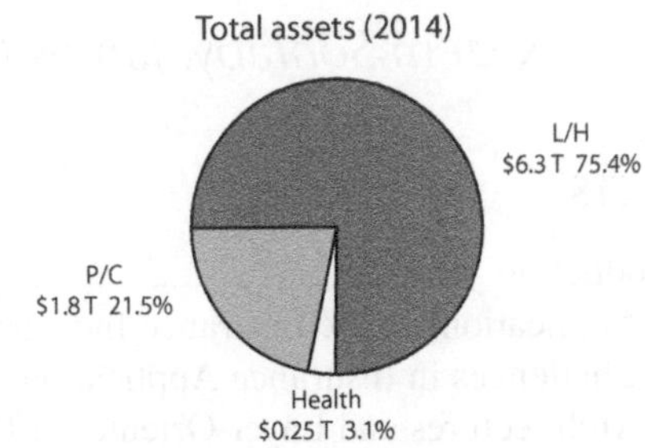

L/H insurance groups by 2014 U.S. life insurance lines direct premiums written

2014 rank	Insurance group	2014 direct premiums written ($000)
1	MetLife Inc.	$ 95,331,132
2	Prudential Financial Inc.	44,720,129
3	New York Life Lnsurance Group	28,393,849
4	Jackson National Life Group	26,708,218
5	AEGON	25,339,180
6	Lincoln National Corp.	24,329,107
7	American International Group	23,279,901
8	Principal Financial Group Inc.	18,891,511
9	Manulife Financial Corp.	18,513,758
10	Massachussets Mutual Life Insurance Co.	16,818,431
	Combined Top 10	$ 322,325,216
	Combined Top 100	$ 579,129,131
	Total U.S. Life Insurance Lines	$ 590,581,992

L/H insurance groups by 2014 U.S. A and H lines direct premiums written

2014 rank	Insurance group	2014 direct premiums written ($000)
1	United Health Group Inc.	$ 43,507,881
2	Aetna Inc.	23,151,559
3	Aflac Inc.	14,601,368
4	Cigna Corp.	13,410,940
5	MetLife Inc.	6,657,580
6	Unum Group	5,259,763
7	Mutual of Omaha Insurance Co.	3,262,797
8	Guardian Life Ins. Co. of America	3,214,961
9	Assurant Inc.	2,843,114
10	Genworth Financial Inc.	2,605,503
	Combined Top 10	$ 118,521,467
	Combined Top 100	$ 164,277,711
	Total U.S. A and H lines	$ 166,128,398

P/C insurance groups by 2014 U.S. combined lines direct premiums written

2014 rank	Insurance group	2014 direct premiums written ($000)
1	State Farm Mutual Automobile Insurance	$ 58,508,587
2	Liberty Mutual Insurance	29,364,559
3	Allsate Corp.	28,892,088
4	Berkshire Hathaway Inc.	26,395,906
5	Travelers Companies Inc.	22,790,776
6	Nationwide Mutual Group	18,935,862
7	Progressive Corp.	18,914,866
8	American International Group	18,653,981
9	Farmers Insurance Group of Companies	18,611,695
10	USAA Insurance Group	15,678,176
	Combined top 10	$ 256,746,495
	Combined top 100	$ 482,038,028
	Total U.S. P/C sector	$ 565,933,448

Health insurance groups by 2014 U.S. health lines direct premiums written

2014 rank	Insurance group	2014 direct premiums written ($000)
1	United Health Group Inc.	$ 54,968,422
2	Anthem Inc.	52,217,860
3	Humana Inc.	45,598,914
4	Healthcare Service Corp.	28,933,352
5	Aetna Inc.	23,099,513
6	Centene Corp.	13,499,981
7	Independence Health Group Inc.	12,249,432
8	Highmark Insurance Group	11,649,152
9	Wellcare Health Plans Inc.	11,161,715
10	Guidewell Mutual Holding Corp.	10,673,671
	Combined top 10	$ 264,052,013
	Combined top 100	$ 482,216,927
	Total U.S. health lines	$ 510,136,609

FIGURE 17.1 U.S. insurance industry statistics, to position IoT value added.

that can help the industry save an outlay of funds by providing better risk management (better visibility and better predictive or preventive mechanisms) will certainly impact the bottom line in a demonstrable way. For example, a 1% improvement in profitability equates to about $10 billion per year, which is quite significant.

Insurance firms in the L/H sector offer life insurance as well as annuities and accident and health (A&H) products. Policies in the former case intend to protect against the financial risk and loss of the insured person that result from that person's death and provide income streams for retirement; policies in the latter case cover expenses for health and long-term care and/or provide income in the case of full or partial disability. Insurance firms in the P/C sector offer products that mitigate the risk of financial loss caused by damage to property or exposure by individuals and businesses to various liabilities. Health insurers offer health insurance products only.

As discussed elsewhere in this text, the basic concept of IoT is the instrumentation of all sorts of things (wearables, health monitors, home appliances, smartphones, and infrastructure control elements, e.g., for smart grid and smart cities). Applications such as smart grids, smart cities, smart homes, physical security, e-health, asset management, and logistics all have relevance to the insurance business itself. Therefore, advances in the IoT field are of keen interest to the insurance industry.

Unfortunately, a number of deployment-limiting issues currently impact the scope of IoT utilization. These retarding issues include a lack of comprehensive end-to-end standards, fragmented cybersecurity solutions, and a dearth of fully developed vertical applications. These issues also impact the adoption of the technology in the insurance industry. This chapter covers some of the issues and possible solutions that can be leveraged by the insurance industry as it contemplates the deployment of IoT-based profitability-enhancing technologies.

The rest of this chapter is organized as follows: Section 17.2 describes a number of specific applications of IoT technology in the insurance industry. Section 17.3 provides a quick view to some of the IoT-related challenges that have to be addressed in order to facilitate broader deployment of this technology in this vertical sector. Building on Section 17.3, Section 17.4 discusses the important topics of IoT architectures and IoT security (IoTSec). Finally, Section 17.5 makes some important observations about IoT traffic flows and related implications.

17.2 IoT APPLICATIONS IN THE INSURANCE INDUSTRY

We have already alluded to possible "piggyback" applications of the IoT for the insurance industry in the previous section. Some of the general IoT applications are a clear case of automation, facilitation, and broad data collection functions (in various vertical domains); while these applications tend to have "coattail" implications for the insurance industry, they are not per se so-called *primary* applications, but *secondary* or "dependent" applications, although still opportunity-rich. Because one does not immediately think of the insurance application as being "ostensibly ripe" for IoT automation *at the subscriber end of the business process*, a number of such possible "direct" applications are discussed below. The normal information and communication technology (ICT) data management, data crunching and data mining, of the insurance customers' data has obviously gone on for decades—the discussion here is to extend automation out to the insured party with the goal of further understanding risk management on the part of the insurer (e.g., IoT-based connectivity empowers insurers—especially P/C insurers but also L/H and H insurers—to predict risk and respond to it before contingencies occur and claims are filed, thus mitigating the severity of an adverse event or its frequency) (Haller, 2011; Internet of Things, 2016; Meyer et al., 2013).

The IoT automation is happening in the context of some other trends impacting the insurance industry. These include but are not limited to the following (Reifel et al., 2014): the P/C sector is facing near-term pressure regarding profitability and premium growth; the financial underpinning, whereby net income cannot be easily generated via reasonably-safe market investments, having to be generated by underwriting savvy and riskier investments; increased competitive pressures capping price increases; some generational shifts regarding the ownership of housing, second homes, cars, and so forth; advances in automotive safety systems (lowering accident rates, losses, and consequently premiums); the emergence of autonomous vehicles; and the "smart consumer," including agent disintermediation and the available body of online data allowing the consumer to do "comparison shopping" when purchasing insurance. The combined weight of these factors, along with a crescendo of forces, will become a serious business challenge for the industry in the second part of this decade. The other sectors of the industry face similar challenges. Some observers believe that while the IoT will likely disrupt the existing insurance business model, it will at the same time "unlock growth opportunities" (U.S. Government Center for Medicare & Medical Services, 2016).

The intrinsic addition of end-point sensors, data gathering, and data analytics already opens up major new horizons for insurers and underwriters. P/C-relevant applications in this context include the smart or connected home (especially when using universal or integrated hubs), the smart or connected car, and the individual's pervasive and ubiquitous computing and connectivity (also known as the "connected self"); all these effectively enhance the "lines of communication" between the user and the insurer, well beyond the usual billing and customer service interaction. Vertical IoT solutions, as might be beneficial to an insurance-oriented environment, would typically comprise an integrated end-to-end (sensors-to-analytics) suite that, by definition, includes the insurance-focused end-system devices, the fog–concentration–core connectivity, the bridging middleware, and the analytics application processing systems, especially those focused on insurance (risk management–related) functions.

A corollary to this automation is that insurers can use the technology to influence the behavior (thus risk carriage) of the customers (e.g., car speed monitors, blood pressure, or exercise activity monitors). This applies not only to lifestyle and possible behavior modification, but also to managing emergency situations to affect some kind of intervention and reduce the contingency or loss associated with the event. One example might be a remote set of cameras to provide some oversight of an older parent living alone.

In addition to the management actions and activities as they relate to a specific individual, the aggregate sensor-generated data can provide, with analytics, macrolevel statistics of group behavior to refine risk models (allowing insurers to predict risk and thus adjust premiums) for entire sets of events. Also, the process of reviewing claim data may be simplified or expedited with the use of data collected by sensors, including multimedia-based IoT devices (e.g., video streams, images, and even social media and/or crowd sensing), to reduce operational costs.

In addition to risk management for individuals (in various phases and aspects of life), IoT mechanisms (e.g., real-time data collection and analysis) as applied to the "Industrial Internet" allow insurers to get granular details about risk associated with the manufacturing and production process, distribution process, logistics and transportation processes, and even product liability (by monitoring how the product behaves or responds once placed in its target environment).

A key IoT-based insurance company application will clearly be in the e-health/mobile-health (e/m-health) space. The increased prevalence of chronic diseases, such as diabetes, heart conditions, and high blood pressure, driven by an aging worldwide population, is animating the demand for medical devices that facilitate advanced mobile monitoring. IoT-powered devices allow real-time patient monitoring (e.g., vital signs, blood pressure, and medication delivery), and also enable healthcare personnel to respond to the patient's transient medical situation. In 2014, healthcare expenditures represented 17.5% of the U.S. GDP (about $3.0 trillion, or $9523 per capita), and that share of GDP is expected to grow to 19.6% of GDP by 2024 (Reifel et al., 2014). Reportedly, 80% of all large insurers are coupling their plans to wellness programs to manage costs (Kearny, 2014).

The pervasive penetration of smartphones among consumers—in addition to other IoT end-point mechanisms—is fostering the rollout of IoT-based automation, and insurers are now earnestly exploring the opportunities offered by the "connected health and life" paradigm. Insurers such as Zurich, USAA, Swiss Re, State Farm, Progressive, Liberty Mutual, John Hancock, Esurance, Chubb, Berkshire Hathaway, American Family, Allstate, and AIG have already made forays into the IoT-enhanced field. A short list of some documented applications and/or initiatives includes the following:

- Introduction of usage-based insurance (UBI) for automotive applications. Many insurers now offer this option or are close to operationalizing the concept. Eventually, advanced driver-assisted systems (ADASs), in conjunction with autonomous driving (with vehicle-to-infrastructure and vehicle-to-vehicle connectivity), will take this to a full denouement.
- Progressive Corporation offers a monitoring program (called Snapshot) that rewards customers with lower insurance rates if they avoid aggressive driving and braking, minimize late-night travel, and drive relatively few miles on a yearly basis. A device (dongle) connects to the On-Board Diagnostics II (OBD-II) port in the automobile and tracks the driving episodes for a number of measured parameters. In addition, the driver can see, on a daily basis, their driving "habits" on a website. (OBD-II—introduced in the mid-1990s—is an improvement over OBD-I, which was introduced in the early 1990s, in terms of capabilities and de facto standardization; onboard diagnostic ports were first introduced commercially in the late 1970s (Goodwin, 2010).)
- State Farm recently launched a smart home policy plan that includes premium discounts for various monitoring and alarm systems. Typical systems include but are not limited to water sensors, smoke detectors, and connected doorbells.
- American Modern Insurance recently prototyped a smart home that incorporates smart devices (such as but not limited to a smart thermostat, a keyless entry doorway, water sensors, and carbon monoxide and smoke detectors), with the goal of studying and assessing how the IoT technology modulates risk and claim response.
- John Hancock Life Insurance has a program to offer premium discounts (up to 15%) for policyholders who track and do well on metrics related to daily exercise. Customers are given a Fitbit Inc. fitness tracker to monitor their daily activities.
- Firms such as AXA, American Family, Aviva, and Liberty Mutual are providing venture capital funds for several IoT start-up companies.

Fundamentally, however, the insurance industry needs to reassess the business model to be able to capitalize on the emerging customer opportunities afforded by the automation brought about by the IoT. Insurers will be forced to "find new ways of doing business," especially given the expected rapid pace of technological and lifestyle changes driven by the IoT.

17.3 IoT CHALLENGES IN INSURANCE APPLICATIONS

Industry stakeholders have recognized that there are challenges to the broader injection of IoT technologies in the insurance field, including the engagement model (how to get the customer to participate), connectivity (how to achieve cost-effective, always-on, typically mobile, region-wide, real-time networking), standardization (vendor-independent, industry-wide architectures, frameworks, and protocols), and critically, cybersecurity challenges. Market research indicates that financial, insurance, and e-health providers seek increased transactional cybersecurity and increased standardization to expedite service development, while at the same time reducing costs associated with the maintenance of proprietary environments. IoT devices, including those used in the insurance industry, are subject to intrusion. For example, the IoT Security Foundation (IoTSF) asserts that less than 10% of all IoT products on the market are designed with adequate IoTSec (iotsecurityfoundation.org), and the issue is well publicized (Al-Fuqaha et al., 2015; Granjal et al., 2015; Lai et al., 2015) (even the UBI dongle mentioned earlier lacks reliable security: it has no reach-back network authentication mechanisms or encryption). Hence, the intrinsic IoTSec challenge associated with the IoT needs to be addressed; in particular, IoTSec is critically relevant to e/m-health applications, as discussed in Chapter 16. Figure 17.2 enumerates some of IoT-related challenges that have to be addressed by the stakeholder, as well by the insurance industry; challenges include

- *Intrinsic IoTSec issues*: Many attack points exist in the overall ecosystem.
- *Use of (vulnerable) gateways and concentration points*: The gateway is a "high-value" target point.
- *Low-complexity devices*: Devices may not be able to run a firewall script or function.
- *Limited onboard power*: Devices may not be able to do a lot on "number crunching."
- *Open environment, allows tampering*: Devices in the field may not be physically secure.
- *Device mobility*: Devices may roam into nonsecure subnetworks—for example, an open Wi-Fi system.
- *Always-connected/always-on mode of operation*: In this mode of operation, devices are subject to repeated and/or persistent attacks).
- *Lack of agreed-upon end-to-end standards*: This predicament makes it more difficult to support interworking.
- *Lack of agreed-upon end-to-end architecture*: This predicament makes it more difficult to utilize multiple technology suppliers).
- *Devices universe by type and cardinality*: The large population of devices may impact the scalability of traditional networking and security mechanisms.

17.4 IoT ARCHITECTURES AND LAYER-ORIENTED IoTSec MECHANISMS

The business and technical challenges implied by the discussion above are best mitigated by the introduction of a system architecture. Security in general and security architectures in particular depend on the availability of an underlying overall ICT architecture, to build required capabilities on a common, well-defined baseline. An architecture is a formal description of the structure of the ecosystem (in this case, the IoT ecosystem), including system components, the externally visible properties of these components, and the relationships (e.g., the behavior) between them (Roebuck, 2011; Minoli, 2006a,b). Architectures can lead to standardization, which in turn leads to commoditization, which almost invariably results in lower overall system costs and expedited rollout of a capability. These conditions are important to the rollout of IoT automation in the insurance industry (and elsewhere).

The American National Standards Institute/Institute of Electrical and Electronics Engineers (ANSI/IEEE) Standard 1471-2000 states that an architecture is "the fundamental organization of a system, embodied in its components, their relationships to each other and the environment, and the principles governing its design and evolution." Thus, an architecture can be seen as a blueprint for the optimal placement of resources in the ICT environment for the effective support of the system functions. A metaphor can be drawn by thinking of a corporate or ICT blueprint for the planning of a city or a large development where the blueprint provides the macroview of how elements (roads, lots, and utilities) fit, particularly in relation with one another. There are many approaches and architecture frameworks that can be utilized. An IoT architecture is an accepted overview description of the elements, functions, interactions, and interfaces of and between IoT components in the ecosystems, namely, devices, gateways, network elements, analytics engines, and cloud entities.

17.4.1 IoT Architectures

In the context of the discussion at hand, a number of IoT architectures have emerged in the recent past, but there is not yet an industry-wide acceptance of any particular framework. The list includes the Arrowhead Framework, the European Telecommunications Standards Institute

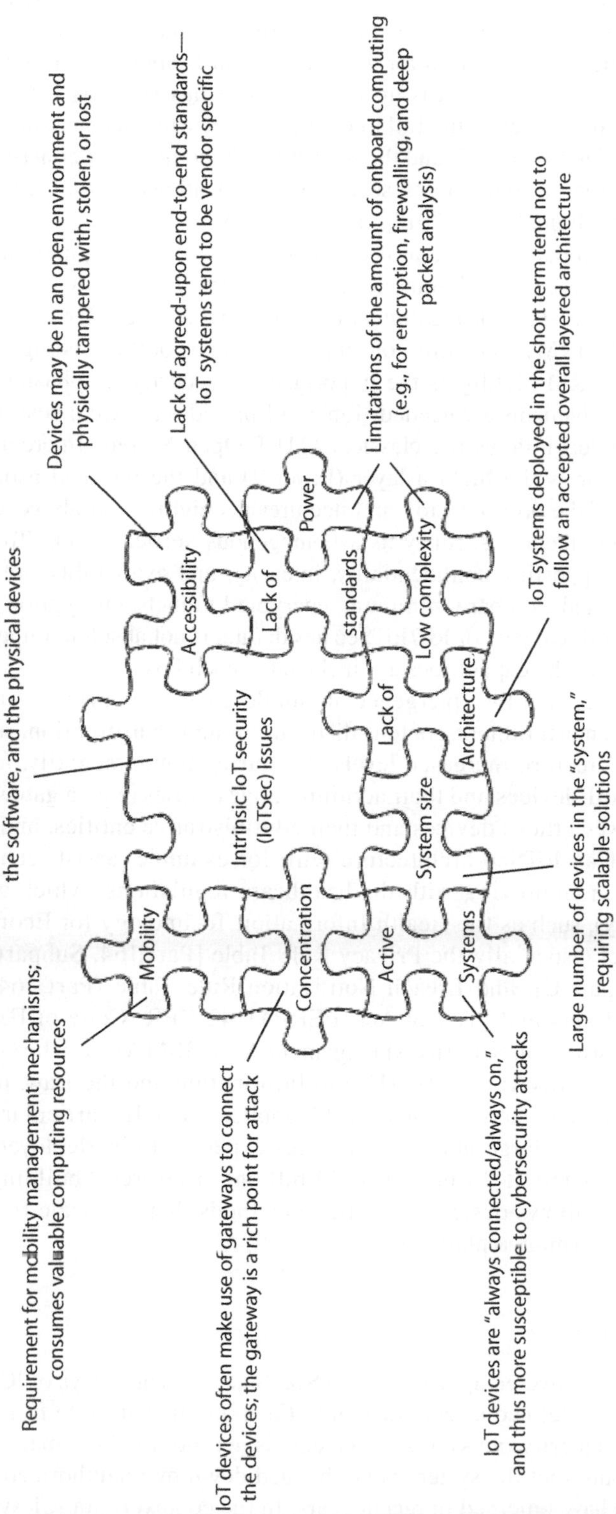

FIGURE 17.2 Challenges impacting deployment of IoT in general and in the insurance industry in particular.

(ETSI) high-level architecture for machine-to-machine (M2M), the Industrial Internet Reference Architecture (IIRA), the Internet of Things Architecture (IoT-A), the evolving ISO/IEC WD 30141 Internet of Things Reference Architecture (IoT RA), the Reference Architecture Model Industrie 4.0 (RAMI 4.0), and the IEEE standard for an architectural framework for the IoT. The dearth of agreed-upon architectures and standards up to the present has not only impacted the broad IoT deployment per se, but also impeded the full integration of security mechanisms in IoT applications. IoTSec is a major consideration in IoT, and the security architecture must consistently support a system state comprised of secure components, secure communications, and secure asset access control (Al-Fuqaha et al., 2015; Granjal et al., 2015; Lai et al., 2015).

Architectures, frameworks, and standards enable seamless, even plug-and-play connectivity and operation. To assist the process of IoT standardization and advance the security agenda, the authors recently introduced a seven-layer IoT architecture model, which we refer to as the Open Systems IoT Reference Model (OSiRM), that builds on and extends some of the existing frameworks (Minoli and Sohraby, 2017). OSiRM highlights the importance of security, emphasizing that in practical terms, layer-specific mechanisms are needed. Table 17.1 provides a textual description of the model. The table (and the model) follows the classical ITU-T Open System Interconnection Reference Model, where application is the highest layer (Layer 7) and the physical aspects are the lowest layer (Layer 1). Some of the proposed IoT architectures (as highlighted above) do include security considerations, but most include security as a homogenous vertical stack. To be truly effective, security mechanisms supporting confidentiality, integrity, and availability are needed at each of the architecture layers, and as a bare minimum, encrypted tunnels, encryption of data at rest, and key management are a critical part of IoT/IoTSec desiderata, if not absolute imperatives (Kouns and Minoli, 2010). In OSiRM, the capabilities are included at each layer.

We have already noted that the emergence of mobile financial services—including insurance applications—online entertainment content distribution and consumption, and cloud software as a service (SaaS) all require increased levels of security. Fundamentally, IoTSec requires the ability to (1) identify IoT devices and their administrative entities (e.g., a gateway), (2) protect the information flow between those devices and their administrative entities, and (3) prevent device hijacking. The use of an IoTSec architecture will, for example, assist technology developers and user institutions in complying with the healthcare regulations (which also clearly impact the insurance industry), such as the Health Information Technology for Economic and Clinical Health (HITECH) Act (especially the Privacy Rule Table [Part 164, Subpart E], Security Rule Table [Part 164, Subpart C], and Breach Notification Rule Table [Part 164, Subpart D]), the Family Educational Rights and Privacy Act (FERPA), 42 CFR (Code of Federal Regulations) Part 2, the Federal Information Security Management Act (FISMA), and the Genetic Information Nondiscrimination Act (GINA) of 2008 (Health Information and the Law, n.d.) (for additional details on regulations, see Chapter 16 on e-health applications). In particular, scalable solutions are needed considering the large number of devices expected to be deployed by the end of the decade (variously estimated at 20 billion to 30 billion): a layered "building block" approach intrinsic in the OSiRM allows designers to utilize methods that can scale from low-cost microcontrollers to high-performance platforms.

17.4.2 Layered Security

This section discusses the layered approach to IoTSec. Since the early days of ICT, security requirements have been broadly defined as confidentiality (keeping the data safe from being divulged by unauthorized agents), integrity (making sure the data is not modified by unauthorized agents), and availability (making sure that the system is not bogged down by unauthorized agents). Additional security requirements have emerged in recent years. In the context of an IoT system, the following layer-oriented approach and armamentarium may be useful.

TABLE 17.1
Tabular View of the Open Systems IoT Reference Model

Layer	Description	Elements/Examples	In-Layer Security Mechanism
Layer 7 (L7)	This is the applications layer. It encompasses a vast array of horizontal and/or vertical applications or "application domains" (also described in terms of use cases). As is the case for Layer 1, effectively the list of applications is "unlimited" in scope. Applications include healthcare, building management, energy, production, transportation, smart cities, security, and retail, to name just a few.	e-Health/assisted living Smart building (building function automation) Smart meters/grid; green systems Production line management Intelligent transportation systems, logistics Smart cities, waste/water/lighting management Surveillance, sensing Supply chain automation	L7 security L7 authentication and authorization L7 encryption and key management L7 trust and identity management
Layer 6 (L6)	This layer encompasses the data analytics and storage functions.	Institutional engine Cloud SaaS/Big Data Internet and App Store apps Storage	L6 security L6 authentication and authorization L6 encryption and key management L6 trust and identity management
Layer 5 (L5)	This layer supports the data centralization function. The purpose of this function is to transport the data that requires centralized analysis or storage to a point in the network where such processing can take place. Such transport (transfer) should be achieved in an efficient and economical manner. This corresponds to the traditional core networking functions of modern networks. It includes institutionally owned (core) networks, industry-specific extranets, public/private/hybrid cloud-oriented connectivity, and Internet tunnels. These networks are generally composed of carrier-provided connectivity services and infrastructure and entail wireline and/or wireless links.	Institutional (core) network Extranet Public cloud Private cloud Hybrid cloud Internet	L5 security L5 authentication and authorization L5 encryption and key management L5 trust and identity management
Layer 4 (L4)	This layer supports the data aggregation function. The purpose of this function is to collect the data that exists at the edge, that requires further processing somewhere else in the ecosystem. Such collection should be achieved in an efficient and economical manner. This function may entail some kind of data summarization or protocol conversion (e.g., mapping from a thin, low-complexity protocol used by the IoT clients in consideration of low-power predicaments, to a more standard networking protocol), as well as the edge networking capabilities. The data aggregation function is typically handled in a "gateway" device. Edge networking represents the outer tier of a traditional network infrastructure, the access tier, employing well-known networking protocols.	Edge networking Gateway	L4 security L4 authentication and authorization L4 encryption and key management L4 trust and identity management

(Continued)

TABLE 17.1 (CONTINUED)
Tabular View of the Open Systems IoT Reference Model

Layer	Description	Elements/Examples	In-Layer Security Mechanism
Layer 3 (L3)	This layer supports fog networking, that is, the localized (site- or neighborhood-specific) network that is the first hop of the IoT client ("device cloud") connectivity. Typically, fog networking is optimized to the IoT clients' operating environment and may use specialized protocols. It could be a wired link (e.g., on a factory LAN, say in a robotics application) or wireless (on a wireless LAN, also optionally including infrared links, e.g., Li-Fi).	Wired network, for example, LAN Wireless network, for example, PAN, LAN, MAN, and WAN	L3 security L3 authentication and authorization L3 encryption and key management L3 trust and identity management
Layer 2 (L2)	This layer encompasses the data acquisition capabilities. It is composed of sensors (appropriate to the "thing" and the higher-layer "application"), embedded devices, embedded electronics, sensor hubs, and so on. Layers 1 and 2 could be seen as being in symbiosis in the IoT world in the sense that things "married" with sensors become the IoT clients or end points. The collected information might be data parameters, voice, video, multimedia, localization data, and so on.	Sensors Embedded devices	Layer 2 security L2 authentication and authorization L2 encryption and key management L2 Trust and identity management
Layer 1 (L1)	This layer is comprised of the universe of things that are subject to the automation offered by the IoT. Clearly this is a large domain, including Body Area Networks (BANs) (e.g., people with wearables and e/m-health medical monitoring devices), smartphones, appliances (e.g., refrigerators, washing machines, and air conditioners), homes and buildings (including HVAC and lighting systems), surveillance cameras, vehicles (cars, trucks, planes, and construction machinery), and utility grid elements. Effectively, this list is unlimited in scope.	People Smartphones Appliances Shopping carts Industrial equipment Cars, trucks, buses, trains Homes, buildings, thermostats, water meters Video cameras	L1 security L1 authentication and authorization L1 encryption and key management L1 trust and identity management L1 antitampering L1 Trusted Execution Environment

Note: LAN, local area network; PAN, personal area network; MAN, metropolitan area network; WAN, wide area network.

17.4.2.1 Overall (All Layers)

OSiRM includes three security-related mechanisms that effectively exist independently at each layer, as needed:

- Encryption and key management: This mechanism supports the confidentiality requirement discussed earlier (keeping data from being read by unauthorized agents). As part of that process, one needs to protect cryptography keys from being misappropriated.
- Trust and identity management: This mechanism supports the integrity requirement (e.g., can the data or user be trusted). As part of that process, one needs to control how software is modified (e.g., during a system upgrade) and also how data is modified by a legitimate entity (e.g., at a concentration or summarization gateway).
- Authorization and authentication: This mechanism supports part of the integrity requirement (who is the "user" and what kind of data can this user read, write, and modify). It also supports part of the availability requirement (avoiding denial of service [DoS] incidents) (e.g., can this user multicast? can this user send data to point x in the network? can this user send more than *y* packets per second to point *z* in the network?).

Other realms and mechanisms can be added to the OSiRM IoTSec model if deemed appropriate (e.g., antitampering, trusted software, and secure ledgers). Thus, the following mechanisms are intrinsically supported in a layer-oriented manner by the model: tamper resistance by specifying the physical protection of devices, user identification by confirmation of the entities involved in a transaction, ensured services with protection against DoS, system-wide secure communication via strong encryption, system-wide management of secure content via data integrity mechanisms, and system-wide secure network access, avoiding man-in-the-middle attacks. Additionally, in this OSiRM model there will be optimized differences for a given security function at different layers, as well as specializations that may occur with the type of thing and/or type of application.

17.4.2.2 Lower Layers (Fog Networking Layer)

A number of connectivity solutions exist for the fog area, along with some security mechanisms (particularly for transmission confidentiality). For example, ZigBee, Bluetooth Low Energy (BLE), and Wi-Fi HaLow offer MAC layer encryption in support of first-hop confidentiality, while others do not, and thus the developer or technology provider must provide encryption tools. Even when providing first-hop confidentiality, end-to end confidentiality must be ensured. As implied in the OSiRM model discussed above (and illustrated in Table 17.1), strong security measures for authorization and authentication, encryption and key management, and trust and identity management must be implemented at each layer of the model and end to end.

17.4.2.3 Hardware Level

In some instances, security capabilities are ideally handled at the hardware level. The Trusted Execution Environment (TEE) (GlobalPlatform, n.d.) is of particular importance to (IoT) applications that deal with sensitive user data, including insurance-related data (such as user real estate locations, user real estate contents [e.g., jewelry], and medical claims). TEE provides end-to-end security by enforcing protected execution of authenticated code: it ensures that sensitive data, such as insurance-related transactions, are processed, stored, and protected in a trusted isolated environment. The TEE concept was first defined in the 2007 time frame by the Open Mobile Terminal Platform (OMTP) forum. The work has continued with GlobalPlatform, which is an organization of more than 130 member firms developing standards. So far, GlobalPlatform has defined two sets of application program interfaces (APIs) between elements of the TEE ecosystem: TEE Internal APIs and TEE Client APIs (Bech, 2014). Intel Trusted Execution Technology (TXT) is a system hardware technology that deals with verifying and/or maintaining a trusted operating system (OS); the technology is designed to harden platforms to better deal with threats of hypervisor attacks, BIOS or other firmware attacks, malicious root kit installations, or other software-based attacks. Among other capabilities, TXT provides what is called "Verified Launch," that is, the use of a "known goods" boot-up code verified through cryptographic (hash-based or signed) measurements (Intel Corporation, 2012).

17.4.2.4 Other Approaches

Security mechanisms can also be implemented at a much higher level than the physical underlying facilities. One such mechanism is the "blockchain," which is a form of distributed ledger (a distributed database) that retains an expanding list of records, while precluding revision or tampering (tamper-resistant ledger) (Jones et al., 2016). More specifically, a blockchain is a time-stamped database that retains the entire logged history of transactions on the system; each transaction processor on the network/system maintains its own local copy of this database and consensus formation algorithms enable every copy, no matter where it is, to remain synchronized. An early application of blockchains is bitcoins (cryptocurrency), but proponents advocate for a much a larger set of applications, including cybersecurity.

As described, a blockchain operates as a large, decentralized ledger that records each and every transaction related to the object in question, and stores this transaction information on a distributed (public) network to prevent content alteration. Thus, blockchains allow a distributed population of users to keep track of a specified group of transactions, independently of their individual (personal) stake in the underlying activity represented by the ledger, although in many instances they do, in fact, have a personal stake in the transactions.

It is a known fact that many goods manufacturers and shippers have concerns about physical security, authenticity, and/or privacy of goods on the move; such goods on the move typically carry insurance provided by an insurance company. For example, there often are concerns related to the difficulty of verifying the authenticity of items and systems through multistage, multinational supply, distribution, and service chains (which, say, might raise concerns about counterfeit items and/or the requirement of tracking legally controlled items, such as medicines, medical devices, controlled pharmaceutical substances, arms, and negotiable bonds). An IoT-based blockchain platform can be employed to optimize the logistics process to track goods movement over multiple distribution points, to achieve regulatory compliance (for both control products and a controlled state—e.g., content temperature, vibration, and acceleration), to maintain a chain of control, and to support other contractual compliance. Intrinsic in the use of the blockchain for this application is the integrity (security) of the records associated with this transaction.

It should be clear (e.g., in the context of the previous example) that insurance companies can benefit from mechanisms that reduce risk, improve efficiency, and increase the assurance of integrity. IoT-based methods, in conjunction with blockchain techniques, can enhance the effectiveness of the control, improve physical and cybersecurity, and lower overall risk.

17.5 TRAFFIC CHARACTERISTICS

A question that has emerged is whether the carrier networks are actually ready to take on the onslaught of traffic that is expected to be generated by the IoT, at the tune of 20 billion to 30 billion end-point devices by the early part of the next decade, and handle that traffic with reliability, prioritization, and security. The ability to manage this traffic in the network is important for all IoT applications in general and insurance industry applications in particular. Just consider the importance of being able to reliably deliver e/m-health data in real time (and, of course, securely). The ability to reliably deliver traffic is important to insurance-related IoT applications, and insurance companies, not only in the context of medical offices reliably receiving real-time information from an e-health monitor worn by a patient, thus being able to possibly alert the medical staff of an impending medical emergency, but also in the context of control centers monitoring a power grid, dam, home alarm, or set of security cameras, all of which have implications on the risk borne by the insurance carrier. If relevant events are not registered by the IoT-based analytics system because the traffic did not get delivered due to network congestion, this could have financial and liability implications for the insurance carrier (as well as for the IoT system designer).

The IoT has distinct intrinsic traffic characteristics in terms of both the sensor node traffic origination and the upstream transmission of that traffic. The sensor node traffic is typically composed of short bursts (although some multimedia applications are emerging, particularly in the context of surveillance); there may be mobility involved, as well as dealing with low transmission power and range (which is relevant to wireless applications). Additionally, the node may run a stripped-down protocol with very basic encapsulation of data within the protocol data unit (PDU). Three factors need to be taken into consideration regarding the traffic stream generated by the IoT device, including devices supporting insurance applications:

- Some devices and applications collect and process a fair amount of data, but only a small portion is actually passed back to the central point and the cloud. For example, a sensor may collect the real-time calories consumed by a jogging person (e.g., step-by-step) and display that data on a smart watch or smartphone, but may only send the total calories consumed when the jog is completed. Another example might be a UBI sensor collecting the real-time speed of a moving car, but only transmitting up into the cloud daily (or at the end of the journey) the maximum speed (or the cases where the speed limit was exceeded when a digital or smart road is available) and the total distance traveled. That case is depicted in the rightmost "state machine" of Figure 17.3 (only upstream traffic shown). This application does not necessarily impose a significant burden on a typical provider's network. The leftmost state machine of Figure 17.3, on the other hand, depicts a situation where a lot more of the traffic is sent into the core or cloud portion of the network.
- The second consideration is how much data is actually collected by the sensor. One application might have sensors collecting a lot of data, passing up only a small portion (e.g., some summary or extract), while another application may collect less data but pass it all along to the analytics engine.
- Lastly, as depicted in Figure 17.4, some applications may entail tens of thousands of nodes in one administrative domain (such as a city) with hundreds of gateway nodes and core

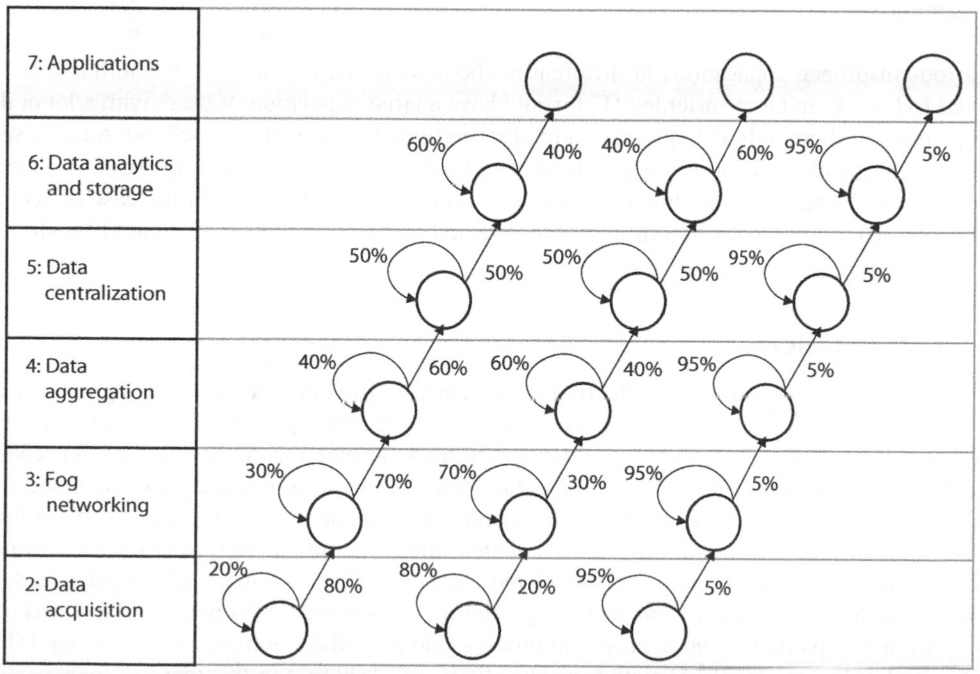

FIGURE 17.3 Traffic handling examples for different vertical applications, impacting the provider's network traffic (only upstream traffic flow shown).

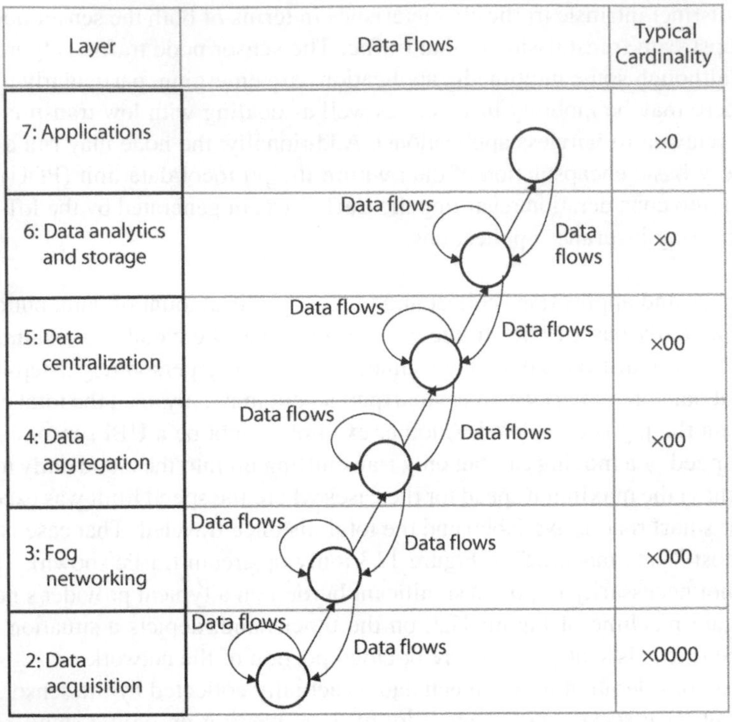

FIGURE 17.4 Population of end nodes in a vertical application (e.g., insurance).

network nodes. In this case, even if the individual end-point traffic is relatively low or a lot of the data is processed locally, the aggregate volume from the large population of end nodes can be significant. Other applications could have fewer end nodes (such as weather stations).

Various insurance applications fit different traffic models and profiles. An e-health application targeted to fitness, and a car-oriented UBI would have a large population of users with a lot of data sent upstream (although the extent of the individual end-point data itself may be low). A home security system may process a lot of the data locally and only send a small report upstream (e.g., is the temperature in Zone 1 fine, or should the air conditioner be turned on—or in the case of security monitoring, where only the video frames with detected motion are sent into the cloud for storage). And the list goes on.

17.6 CONCLUSION

The insurance industry can make effective use of emerging IoT-empowered paradigms. The IoT will impact smart cities, smart grids, smart homes, e/m-health, transportation, and many aspects of human life. It behooves the industry stakeholders to track development in this space. This chapter described the applicability of evolving IoT technologies to the insurance industry and some early applications. IoT systems can enhance personal and environment monitoring, thus enabling insurance companies to develop better-tailored risk models, therefore mitigating risk and improving profitability. It also discussed some of the IoT-related challenges that have to be addressed in order to facilitate broader deployment of this technology in this vertical sector. As noted, IoT architectures are important in this rollout and deployment process, and the reliable introduction of strong IoTSec is critically fundamental to the initiative, as also discussed elsewhere in this text.

REFERENCES

Al-Fuqaha, A., M. Guizani, M. Mohammadi, M. Aledhari, and M. Ayyash. 2015. Internet of Things: A survey on enabling technologies, protocols, and applications. *IEEE Communication Surveys & Tutorials* 17 (4): 2347.

Annual Report on the Insurance Industry, Federal Insurance Office, U.S. Department of the Treasury, Washington, DC, September 2015.

Bech, J. 2014. OP-TEE, Open-Source Security for the Mass-Market. *Core Dump* online magazine, September 3, 2014. http://www.linaro.org/blog/core-dump/op-tee-open-source-security-mass-market/.

Gartner. 2016. Gartner Says by 2020, More Than Half of Major New Business Processes and Systems Will Incorporate Some Element of the Internet of Things, Press Release, Gartner, Stamford, CT, January 14, 2016.

GlobalPlatform. n.d. Trusted Execution Environment (TEE) Guide, White Paper, GlobalPlatform Inc., Redwood City, CA.

Goodwin, A. 2010. A Brief Intro to OBD-II Technology. *CNET* online magazine, April 14, 2010. https://www.cnet.com/roadshow/news/a-brief-intro-to-obd-ii-technology/.

Granjal, J., E. Monteiro, and J. S. Silva. 2015. Security for the Internet of Things: A survey of existing protocols and open research issues. *IEEE Communication Surveys & Tutorials* 17 (3): 1294.

Haller, S. 2011. The Real-Time Enterprise: IoT-Enabled Business Processes. *Presented at the IETF IAB Workshop on Interconnecting Smart Objects with the Internet*, Zrich, March 2011.

Health Information and the Law, a Project of the George Washington University's Hirsh Health Law and Policy Program and the Robert Wood Johnson Foundation. n.d. http://www.healthinfolaw.org/ (specifically http://www.healthinfolaw.org/topics/67)

Intel Corporation. 2012. Intel Trusted Execution Technology, White Paper, Intel, Santa Clara, CA.

Internet of Things: From Sensing to Doing, in *Tech Trends 2016: Innovating in the Digital Era*, Deloitte University Press, Westlake, TX, 2016.

IoT Security Foundation. iotsecurityfoundation.org.

Jones, M. C., et al. 2016. Insurers, New Kids on the Blockchain? White Paper, FC Business Intelligence Ltd., London.

Kearny, A. T. 2014. The Internet of Things: Opportunity for Insurers, White Paper. https://www.atkearney.co.uk/documents/10192/5320720/Internet+of+Things+-+Opportunity+for+Insurers.pdf/4654e400-958a-40d5-bb65-1cc7ae64bc72.

Kouns, J. and D. Minoli. 2010. *Information Technology Risk Management in Enterprise Environments: A Review of Industry Practices and a Practical Guide to Risk Management Teams*. Hoboken, NJ: Wiley.

Lai, C., R. Lu, D. Zheng, H. Li, and X. Shen. 2015. Toward Secure Large-Scale Machine-to-Machine Communications in 3GPP Networks: Challenges and Solutions. *IEEE Communications Magazine Communications Standards Supplement*, December 2015, 12.

Meyer, S., A. Ruppen, and C. Magerkurth. 2013. Internet of Things-Aware process modeling: Integrating IoT devices as business process resources. *Presented at the Conference on Advanced Information Systems Engineering*, June 17–21, 2013, Valencia, Spain.

Minoli, D. 2006a. Enterprise Architecture Is Not Rocket Science, *Network World*, July 24, 2006.

Minoli, D. 2006b. Needed: An Enterprise Security Architecture, *Network World*, May 8, 2006.

Minoli, D. and K. Sohraby. 2017. IoT Security (IoTSec) considerations, requirements, and architectures. *14th Annual IEEE Consumer Communications and Networking Conference*, Las Vegas, Nevada, January 2017.

Reifel, J., A. Pei, N. Bhardwaj, M. Hales, S. Lala. 2014. "The Internet of Things: Opportunity for Insurers", ATKearny, Whitepaper. www.atkearney.com

Roebuck, K. 2011. *Enterprise Architecture*, Emereo Pty. Ltd, Brisbane, Australia.

U.S. Government Center for Medicare & Medical Services. 2016. Top 8 Insurance Internet of Things Projects. *Insurance Networking News* online magazine.

18 The Internet of Things and the Automotive Industry

A Shift from a Vehicle-Centric to Data-Centric Paradigm

Zahra Saleh and Steve Cayzer

CONTENTS

18.1 Introduction 363
18.2 Sources of Data 364
18.3 Digital Transformation as a Disruptive Force 365
18.4 Digital Transformation in Product-Based Industries 366
18.5 Digital Transformation in the Automotive Industry 368
18.5.1 Connected Car 368
18.5.2 Ecosystem of the Connected Car 369
18.6 Iot-Related Trends in the Automotive Industry 370
18.6.1 Trend 1: The Connected Car Moves into the Mainstream 371
18.6.2 Trend 2: From Product to Service 373
18.6.3 Trend 3: New Business Models 376
18.6.4 Trend 4: Emergence of Nontraditional Industry Entrants 378
18.7 Challenges 383
18.7.1 Innovation Cycle Time 383
18.7.2 Network Effect 383
18.7.3 Privacy and Security 383
18.7.4 Business Models and Control Points 384
18.7.5 Customer Relationship 385
18.8 Conclusions 385
References 385

18.1 INTRODUCTION

According to Mary Barra, CEO of General Motors, "The auto industry is poised for more change in the next 5–10 years than it's seen in the past 50" (Wollschlaeger et al., 2015). The automotive industry represents a major industrial and economic force globally, which started about 100 years ago in Germany and France and came of age in North America in the era of mass production (Papatheodorou and Harris, 2007). For years, the automotive industry remained relatively closed and mainly controlled by car manufacturers with relatively high barriers to entry (Wollschlaeger et al., 2015; Stanley, 2016). However, a majority of industry players and experts have argued that the automotive industry is currently on the brink of a transformation driven by the IoT (Gao et al., 2014; Hansen, 2015b; Hientz et al., 2015; Ninan et al., 2015).

IoT has many definitions, but one important theme involves physical objects interacting with each other to achieve common goals (Giusto et al., 2010). Thus, objects can become connected and smart (Ninan et al., 2015), uniting elements from both the physical and digital worlds (Fleisch et al., 2014). When such smart objects are combined with equally smart physical environments, it becomes possible to deliver new cyber-physical services (Fortino and Trunfio, 2014). As a consequence, products have turned ever more into complex systems that merge hardware, sensors, data storage, microprocessors, software, and connectivity in numerous ways (Porter and Heppelmann, 2014).

The importance of the IoT to the automotive industry is widely acknowledged (Andersson and Mattsson, 2015, 89). Cars could be considered as potentially the ultimate connected device (Schuhmacher, 2015) and, as such, could form a key element of the IoT (Gartner, 2015). Vehicles of the future are expected to be able to communicate, socialize, and interact with other things, such as other cars, roadside infrastructure, and retailers, thus coming to be a part in a broader "system of systems" (Wollschlaeger et al., 2015). Analysts vary in their estimations, but all agree that connected cars are on the rise and increasingly gaining importance (Mohr et al., 2013; Wee et al., 2015). Gartner estimates that by 2020, more than 250 million cars will be connected worldwide, with the amount of installed connectivity units in cars growing globally by 67%, while the number of consumers spending on in-vehicle connectivity will double (Ninan et al., 2015). Moreover, future generations of drivers will require the car to perform as "smartphones on wheels" and stay connected and productive on the go. Customers are even ready to pay a sizable amount for a car that fulfills all their technology needs and requirements. The number of connected cars worldwide has shown a steady growth of 30% a year, leading to one in five cars projected to be connected to the Internet by 2020 (Mohr et al., 2013).

The connected car, which has about 1 million lines of software codes and generates up to 25 GB of data per hour (Wollschlaeger et al., 2015), has the potential to become one of the most valuable computational and data platforms (Hansen, 2015a), allowing for new economic value. In the IoT, customer value is derived from the combination of connected products and digital services (Hientz et al., 2015). Future value creation in the automotive industry is expected to increase considerably and diversify with new services that could create up to $1.5 trillion, or 30% more, in additional revenue potential by 2030 (Gao et al., 2014).

Companies are currently under pressure to become part of a strong network where vast data exchange and related real-time analyses occur to secure their share of the automotive value pool. The creation of ecosystems and company-internal capability improvements is expected to become critical for companies to monetize this trend (Wee et al., 2015). These changes are likely to represent a considerable paradigm shift in the car industry, which actually derives part of its success from its traditional hierarchical structure and its risk-averse culture (Hientz et al., 2015).

It seems worthwhile to explore how new changes caused by the IoT, connectivity, and related information technology (IT)–enabled services will shape this well-established product-based value stream. While research on the IoT and car connectivity is increasing, this research needs to be grounded in a discussion about digital transformation and service innovation in product-centric firms. This chapter provides such a discussion.

The chapter begins by describing the trend toward digitization that has transformed other sectors, such as the mobile phone industry and digital music. The implications for product-based sectors are explored, noting the role of the IoT as a key bridge between the physical and digital domain. The concept of the connected car is introduced, and its surrounding ecosystem described. Four trends are identified: the adoption life cycle of the connected car, the shift from product to service, new business models, and the emergence of nontraditional industry entrants. The chapter concludes by summarizing the main findings and suggesting an outline for future work on the topic.

18.2 SOURCES OF DATA

This chapter is informed by contemporary literature, market data, and expert interviews. Table 18.1 shows details of experts interviewed for this study. The interviewees are specialists in their field and have extensive knowledge about topics related to digitization and connectivity in the automotive industry. They were chosen based on their experience in different industries, such as automotive, IT, and telecommunications. Thus, the analysis presented here integrates insights from people coming from organizations that represent significant actors within the connected car market.

TABLE 18.1
Details of Interviewees Consulted for This Work

Interviewee Code	Type	Position and Company	Interesting Because the Expert
Tech1	Technical	Senior managing Consultant at IBM Global Automotive Centre of Competence	• Is responsible for connected vehicle, after sales, and analytics
Tech2	Technical	Key account manager at Google	• Is responsible for advertising for the biggest clients from the media and entertainment industry
Cons1	Consultant	Business development technical consulting at P3 Communications	• Is responsible for the Digital Services Consulting Department • Deals with the meaning and significance of connectivity for all industries from transportation, energy, and automotive to healthcare
Cons2	Consultant	Competence center lead of "connected mobility" at P3 Automotive	• Deals with connected mobility • Was technical consultant in concurrent engineering, data management, and processes
Cons3	Consultant	Manager at Deloitte	• Is responsible for the operational implementation of projects, specifically connected cars and mobility services • Has a clear automotive focus • Dealt with business model innovation, design changes, innovation management, and software architectures
Data1	CEO	CEO of the subsidiary P3 Insight GmbH	• Built the subsidiary P3 Insight, where they collect data to build market intelligence
Auto1	OEM	Product manager at BMW	• Is responsible for all technology relations to the product, such as vehicle responsibilities for new drivetrain rollouts, cross-platforms, and connectivity, for autonomous cars in the future

18.3 DIGITAL TRANSFORMATION AS A DISRUPTIVE FORCE

In order to understand the potential of digital transformation in the automotive industry, it is instructive to review the historical pattern of digital disruption in other industries. These industries provide evidence that digitization can lead to profound shifts in customer behavior and in all stages of the related value chain. Significant industries in which digital transformation caused vast disruption are the music and the mobile phone industry. Consider first the music industry. In the past, musicians largely generated revenue from the sale of records (Mueller, 2013) and, later, digital downloads. Yet today ownership models are giving way to streaming services (Hientz, 2015; Wedeniwski, 2015). From a retailer's perspective, these on-demand services, accessible on Internet-connected devices, have the benefit of learning about a user's taste and listening habits to deliver a desired song at a desired time (Olivarez-Giles, 2015; Popper, 2015; Pullen, 2015). It is also predicted that future digital music services will even use the context of a situation to automatically deliver an appropriate song for a client (Luckerson, 2014).

The mobile phone industry is another good example of digital transformation (Pai, 2015; Frommer, 2011). By the time Google's open handset alliance started the Android mobile software platform and Apple launched its iPhone with touchscreen and application-based operating system, Nokia was one of the world's leading manufacturers of mobile telephone handsets. Nevertheless, at the time Nokia hardly realized upcoming platform threats, such as the shifting of consumer preferences and technical advances in mobile processors that were reshaping the industry (Pai, 2015). "We don't see this as a threat. We are the ones with real phones, real phone platforms and a wealth of volume built up over years" was Nokia's response in 2007 (Pai, 2015, 276). Smartphones always incorporated more features than basic cell phones and feature phones; however, after the introduction of Android and the iPhone, the very definition of smartphones changed, as handsets are now required to also function as mobile computers to sustain market demand and expectations. Other platforms were incapable of matching Apple's and Google's software competence, which increased the complexity in selling handsets and attracting third-party developers to create apps. Thus, organizations have had to either align themselves with Google and Android or develop their own mobile computing platforms (Woyke, 2014). Nokia's failure could be ascribed to institutional inertia (Surowiecki, 2013). In other words, Nokia was a hardware company that misjudged the importance of software, including the apps that run on smartphones (Surowiecki, 2013). The company was unable to challenge itself and failed to recognize the changing trend and consumer perceptions, remaining stuck with the idea that handsets were primarily about making phone calls. It ignored the potential of adding value via access to digital services, such as detecting restaurants or updating Twitter pages (Pai, 2015). While Nokia put emphasis on technical integration and excellence, the market and ecosystems moved toward the Internet, information providers, and software services (Bouwman, 2014).

18.4 DIGITAL TRANSFORMATION IN PRODUCT-BASED INDUSTRIES

It is worth considering to what extent the patterns of digital transformation apply to product-based industries. Here, the role of physical objects is more prominent, and hence the IoT is highly relevant. Indeed, customer value in a digital business world involves a combination of both the digital and physical world, resulting in a hybrid concept. Thus, the digital transformation of a physical item happens through a series of certain layers. The concept can be explained by using or showing an example of a LED light bulb and its transformation into a smart version. Increasing value is unlocked as the physical object is enriched with successive layers of digital functionality (Fleisch et al., 2014):

1. *Level 1—Physical item*: The object itself creates value for the user (e.g., provision of light). As the light bulb is a physical entity, it is tethered to a location and is able to supply benefits at this stage only in immediate surroundings.
2. *Level 2—Sensors/actuators*: The physical thing includes a minicomputer with sensor technology and actuating elements. Sensor technology is responsible for measuring local data, whereas actuating elements provide local services, and therefore create local benefits. In terms of the LED light bulb example, the microwave sensor continually, reliably, and inexpensively measures whether individuals are present in space. When human presence is detected, the actuator turns the light on automatically and off again when not, thus delivering local benefits.
3. *Level 3—Connectivity*: The sensor technology and actuator elements are connected to the Internet and are consequently globally accessible. The light bulb can now be addressed via an embedded radio module and communicate its status to authorized subscribers worldwide at low marginal costs. However, connectivity by itself does not provide added value.
4. *Level 4—Analytics*: Sensor data is gathered, stored, tested for plausibility, and classified. Subsequently, findings of other web services are integrated with it in order to arrive at consequences for the actuator elements. In the LED example, the on-and-off times in a household are collected, motion patterns are detected, and the operating hours of individual light bulbs are recorded.
5. *Level 5—Digital service*: The options afforded by the previous levels are structured as digital services, packed in an appropriate form, for instance, as a web service or mobile application, and made accessible worldwide. For example, the LED light with a presence sensor can now

be transformed to a safety lamp that, at the user's initiative and/or at the click of an app, signals an alarm to the owner or security guards in the event of an unsolicited person. In "fight-back mode," it attempts to drive the person out with a flashing red light.

At an abstract level, the value on the manufacturing side is identified in the physical product, previously not connected to the Internet, which is now equipped with digital and connected functionality. From the customer's perspective (at Level 5), value is perceived as benefits from both the physical object and the digital services associated with it. The model combines physical products at relatively low cost with proprietary as well as external digital services. However, these five levels cannot be built independently of each other. The concept is considered a bidirectional flow, a process of integration extending into the physical level rather than the mere addition of layers. The construction of the hardware, for instance, is increasingly affected by subsequent digital levels, making it essential to develop hardware in close interconnection with Internet solutions (Fleisch et al., 2014).

Thus, connectivity can stimulate service innovation in product-based companies (Fleisch et al., 2014). Digital business model patterns merge with those from the nondigital world and thus create a hybrid construct. As a result, customer value comprises certain levels from both worlds, digital and physical. In other words, IT-enabled services are expanding the traditional product-based value stream with a complementary mode of value creation (Fleisch et al., 2014). Figure 18.1 illustrates some examples of such services.

Smart, connected products are characterized by three core elements (Porter and Heppelmann, 2014): physical components, like mechanical and electrical parts; smart components, like sensors, microprocessors, data storage, controls, software, an embedded operating system, and a digital user interface; and connectivity components, such as ports, antennae, protocols, and networks, that facilitate communication between the product and the product cloud, consisting of software running on a remote server (Figure 18.2). Products are becoming capable of monitoring and reporting on their condition and environment, allowing for insights into product performance and use. In addition, product operation control can be realized by users via various remote access options, providing customization of the product's function, performance, and interface. Moreover, the combination of monitoring data and remote control capability generates novel opportunities for optimization. Algorithms can enhance product performance, utilization, and uptime, and control how products interact with related products in broader systems. Finally, the combination of monitoring data, remote control, and optimization algorithms facilitates autonomy. Thus, products become able to learn, adjust to environment and user

Connected physical item	Function of physical item	Function when equipped with layers 2 – 5
	Time	e-Call
	Drive	Insurance, traffic, charging, behavior
Every-thing	Whatever the thing can do?	Warranty service, installation guide, repair history ...

FIGURE 18.1 Physical products and digital services merge in hybrid solutions. (Simplified from Fleisch, E., et al., Business models and the Internet of things, Bosch IoT Lab, St. Gallen, Switzerland, 2014.)

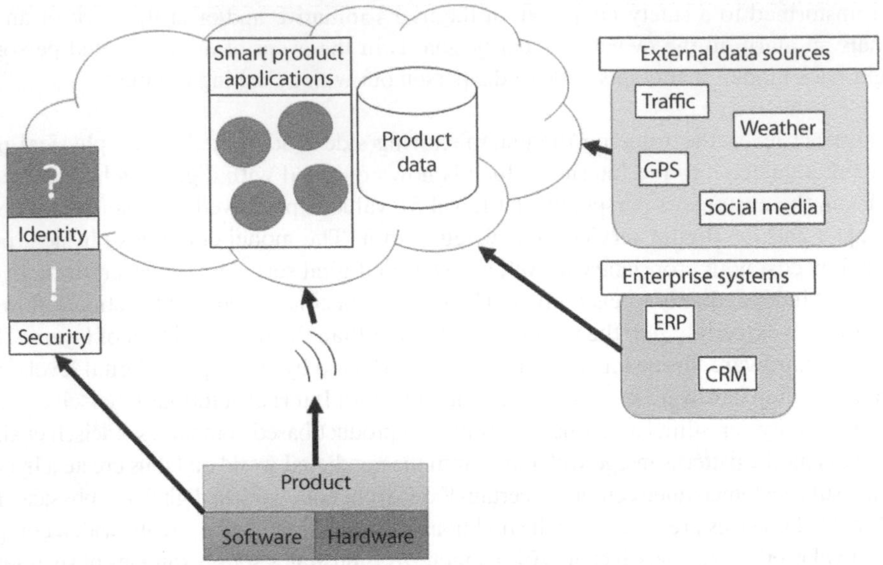

FIGURE 18.2 New technology infrastructure. The product includes sensors and connectivity (hardware) augmented with user interface, control, and operating system (software). The product communicates with the cloud using recognized protocols. The cloud provides a variety of applications running on a common platform, using analysis tools that generate insights from aggregated product data. The applications have access to additional sources of data and may interface with enterprise systems. (ERP stands for enterprise resource planning and it refers to systems and software packages employed by companies to manage day-to-day business activities, such as manufacturing, procurement, marketing, etc.; CRM is an abbreviation of customer relationship management. This system manages and analyzes customer interactions and data throughout the customer lifecycle.) Identity and security services will needed to manage authentication and access. (Adapted from Porter, M. E., Heppelmann, J. E., *Harvard Business Review*, November 2014, pp. 97–114.)

preferences, service themselves, and operate individually (Porter and Heppelmann, 2014). In the context of the automotive industry, an example is the car manufacturer Tesla with its Model S, which receives software updates through the cloud, allowing the connected car to be continually enhanced and optimized without visits to a dealer (Porter and Heppelmann, 2014).

18.5 DIGITAL TRANSFORMATION IN THE AUTOMOTIVE INDUSTRY

As shown in the previous section, the effects of digitization are increasingly influencing every sphere of life (Hientz et al., 2015; Wedeniwski, 2015). This section focuses on the automotive industry specifically (Gissler et al., 2016). Digitization here implies a connected car that is capable of monitoring, in real time, its own working parts and the safety of conditions around it. It is expected to communicate with other vehicles and with an increasingly intelligent roadway infrastructure (Gao et al., 2014). In this sense, the connected car is classified as a major element of the growing IoT (Davidson, 2015).

18.5.1 Connected Car

Table 18.2 shows three main connectivity solutions in the vehicle (Ropert, 2014; GSMA, 2013). In the embedded system, both connectivity and intelligence are directly built into the vehicle, which focus on car-centric, high-reliability, and high-availability apps, like security and safety-related services. Embedded solutions have typically been applied in premium cars, but with some exceptions. Volume brand producers like BMW, General Motors, Peugeot, Renault, and Roewe provide embedded solutions in entry models and up. Tethered solutions utilize the user's subscriber identity

TABLE 18.2
Three Main Models for Connectivity and Intelligence in the Connected Car

Model	Explanation	Typical Use
Embedded	Connectivity and HMI as part of vehicle	High-end models; brand experience Vehicle-centric, high-reliability, and high-availability apps, e.g., e-Call and breakdown call services or b-Call
Tethered	Connectivity via mobile device; HMI resides in vehicle	Navigation and Internet-based infotainment features
Smartphone integration	Connectivity and HMI in mobile device	Mostly used for higher bandwidth and personalized apps, e.g., on-demand music and social networking

module (SIM), phone, or USB key to realize connectivity, while intelligence is embedded in the vehicle. This option can be used for Internet-based infotainment features and connected navigation. In relation to the latter, connectivity is based on integrating the car and the phone, in which the communication module (e.g., SIM), on the one hand, and intelligence, on the other, are delivered by the phone. However, the human–machine interface (HMI) mostly stays in the car. The smartphone integration is also suitable for infotainment and navigation, but here the provider is external. This solution implies car manufacturers ceding control of at least some of the applications and services used in the vehicle. In addition, it is also unreliable in terms of security and safety solutions, given the requirement for the user to activate their phone (GSMA, 2013). In practice, most automobile manufacturers are building strategies that rely on several connectivity options for different segments (e.g., embedded for high-end models and tethered for entry-level vehicles) or for distinctive applications (e.g., embedded for safety and smartphone integration for infotainment) (GSMA, 2012).

The customer value unlocked by the connected car goes beyond streaming videos or music, and encompasses car safety, including diagnostics of mechanical problems and modifying current driving patterns (Craig, 2015).

Vehicle connectivity involves the set of functions and capabilities that digitally link cars to drivers, services, and other cars. The multiple features aim to optimize the operation and maintenance of the car, as well as the comfort and convenience of the driver (Habeck et al., 2014). Furthermore, the connected car is described to be equipped with communication technology that enables the direct flow of data to and from the vehicle (Wee et al., 2015).

To gain a clearer vision of connected vehicle capabilities, PwC, a global professional services company, provides a first overview of the technologies involved in the connected car and classifies them into the categories displayed in Figure 18.3 (Viereckl et al., 2015). These technologies are expected to be modified and extended over the coming years.

The number of connected car features is increasing by the day (Habeck et al., 2014), as both premium and volume automobile manufacturers begin to consider car connectivity technologies as fundamental to their futures (PwC, 2015). Cars are converting into integrators of multiple technologies, productive data centers, and components of a larger mobility network (Gao et al., 2014). However, it is estimated that these developments still require about a decade to be fully deployed (Knight, 2015).

18.5.2 Ecosystem of the Connected Car

From a data point of view, the vehicle is merely a node on a network and a further "thing" in the grand scheme of IoT (Wollschlaeger et al., 2015). The connected car could be described as the "poster child" of the IoT paradigm (Wollschlaeger et al., 2015) since it represents a part of a wider system of systems, consisting of not only vehicles but also cities, physical infrastructure, retail,

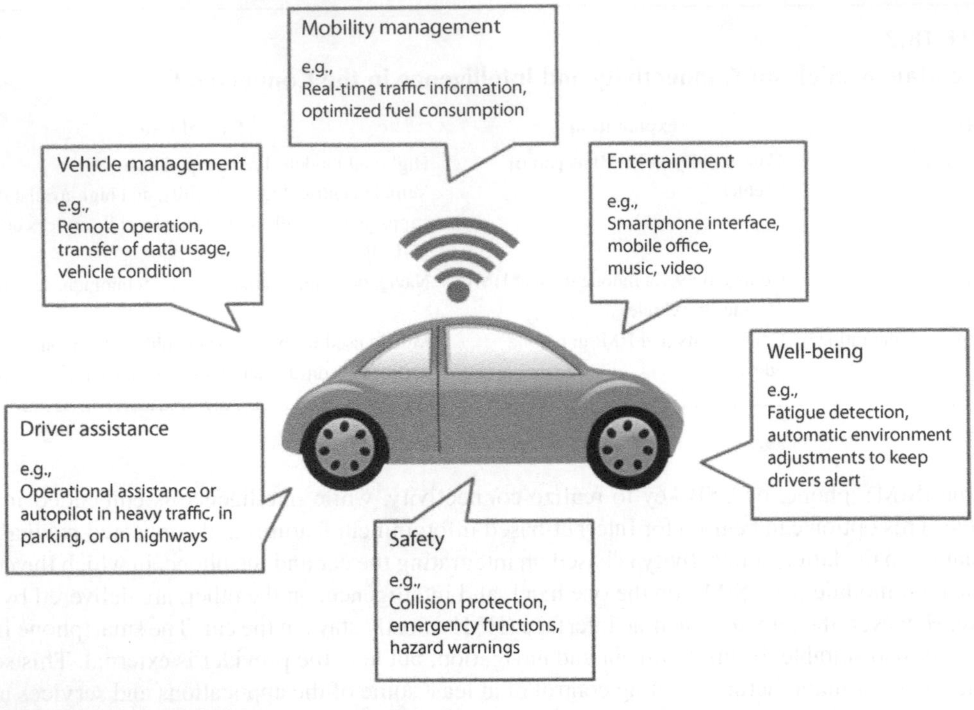

FIGURE 18.3 Some examples of functional areas of connected car development. (Adapted from PwC, My transport, connected, 2015. Available at https://www.strategyand.pwc.com/media/file/Strategyand_In-the-Fast-Lane.pdf.)

insurance, and so forth, and leverages essential IoT-enabling technologies like sensors, analytics, big data, and natural language processing. Connected cars enable a wide range of different connectivity modes, such as vehicle-to-infrastructure (V2I), vehicle-to-vehicle (V2V), and vehicle-to-services (V2S) (Hientz et al., 2015). By means of this technological capability, cars can share information with each other, as well as their surroundings, in order to enhance safer and efficient driving (Qualcomm, 2015). V2I represents a mode in which the vehicle is able to interact with traffic lights and other infrastructure elements, for instance. V2V involves an information exchange between vehicles about actual road and weather conditions (user, car, Internet, etc.). In V2S, various vehicles are transmitting data to web servers to conduct real-time big data analysis; accordingly, the user can obtain data-based service offers (Hientz et al., 2015). The introduction of such information exchange compels an agreement among automakers and suppliers in terms of communication technology, protocols, and the like, but some efforts have already been made in this direction (e.g., the Car2Car Consortium*) (IEE Control Systems Society, 2011).

18.6 IOT-RELATED TRENDS IN THE AUTOMOTIVE INDUSTRY

Analysis of this research and the findings from expert industries revealed the following basic narrative for the automotive industry. The broad adoption of the connected car signals a shift from product to service offerings, which in turn enables new business models and the emergence of

* The Car2Car consortium (C2C 14) is a nonprofit, industry-driven organization initiated by European vehicle manufacturers and supported by equipment suppliers, research organizations, and other partners. Car2Car is specifically focused on cooperative intelligent transport systems (C-ITSs), where the word *cooperative* is related to exchange of information through wireless communication technology among telematics devices, which are installed in both vehicles and roadside infrastructure (Camara and Nikaein, 2015).

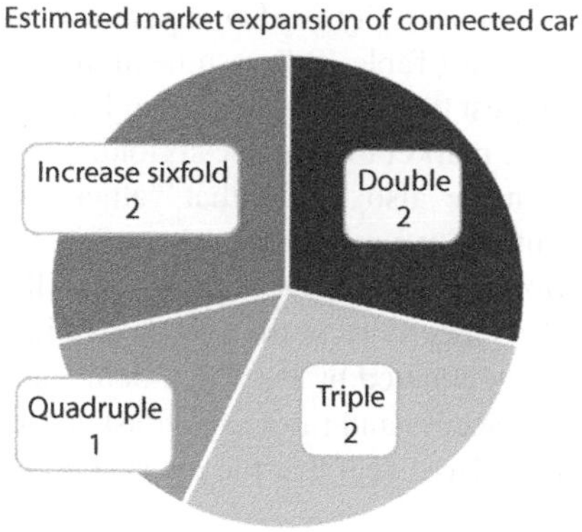

FIGURE 18.4 Global connected car market expected to increase by 2020. Survey of seven experts. The number of experts choosing each option is shown in the chart.

nontraditional industry entrants. This narrative is deconstructed into four key trends in the following sections.

18.6.1 Trend 1: The Connected Car Moves into the Mainstream

With the help of the S-curve, it is possible to detect the situation in which a specific technology is found at a specific point in time, as well as predict its future evolution and development limits (Nieto et al., 1998). The consultancy company PwC has used this approach to model the future growth of the connected car (PwC, 2015). The connected car is projected to enter a rapid period of growth as the technology moves toward full market acceptance. PwC predicts that the market will grow 30% from 2014 to 2020, to reach US$125 billion (PwC, 2015).

According to the experts interviewed, companies are not ignoring the growing importance of connectivity to the automotive industry. In particular, car manufacturers are aware of this shift and have started to react. They also underscore that connectivity is becoming essential to maintain future value creation, and that there is actually no car manufacturer ignoring this shift. However, Cons2* points out that connectivity is not yet changing the automotive industry substantially, as connectivity functionalities in vehicles still lack quality and user-friendliness. Although some customers are focusing on connectivity features and may choose one original equipment manufacturer (OEM) over the other based on these features, conventional vehicle sales remain prevalent. Cons2 suggests that connectivity is still not critical for selling cars. However, he predicts that this existing condition is likely to shift dramatically in the coming 5–10 years. Furthermore, all experts suggest that companies invest in connectivity in order to remain in the market since companies that do not incorporate these new features will lack competitiveness. As Data1 contends, "If you don't adjust to these new upcoming situations, your existing business model can break away. When you have such a significant evolution like we have now with digitalization, it's so easy to be outperformed by others."

Looking to the future, one might expect that connectivity will lead to a more fundamental change in the traditional business of car manufacturers. Cons3 predicts that in terms of connectivity, "2020 is the year, the game changer year." In other words, car manufacturers will have to disrupt their current business model in order to secure future survival. This is also confirmed by PwC, who find that premium and volume car manufacturers consider the connected car to be essential for future existence (Viereckl et al., 2015).

* The interviewees are identified by abbreviated code: Tech = technology (IT) provider; Cons = consultancy; Data = data analytics company; Auto = automotive manufacturer. See Table 18.1 for full details of the interviewees.

The increasing significance of connectivity is also apparent from the questionnaire conducted for this study (Figure 18.4). All experts (Table 18.1) assume an increase in the global connected car market by 2020. The responses suggest that the market will at least double in the coming years. Two out of seven experts even expect the market to increase sixfold.

The results from the questionnaire also reveal that vehicles with connectivity features are expected to penetrate the global automotive market by at least 25% before 2020. The majority of the experts even presume a penetration rate of 50% or more (Figure 18.5). In other words, all experts expect an appreciable increase of connectivity in the automotive industry in the next 3 years. This growth is also proposed in a study conducted by PwC that demonstrates that by 2020, 90% of new cars will be equipped with Internet access and interactive dashboards (PwC, 2015).

The results of the questionnaires also reveal that the interviewees expect connectivity to become indispensable in vehicles (Figure 18.6).

This trend can also be observed in McKinsey's consumer survey results, which demonstrate that the number of people who would switch from their current OEM to another manufacturer if it were the only one that offered a car with connectivity features increased globally from 20% in 2014 to 37% in 2015. According to McKinsey, connectivity is shifting from a "should-have" to a "must-have" feature for every OEM (Wee et al., 2015).

Regulation may also drive this trend and push the auto industry beyond traditional competencies. Historically, regulation focused on safety. The process began with seat belts and padded dashboards and moved on to airbags, automotive "black boxes," and rigorous structural standards for crashworthiness. Today, regulations include increasingly stringent requirements for emissions and

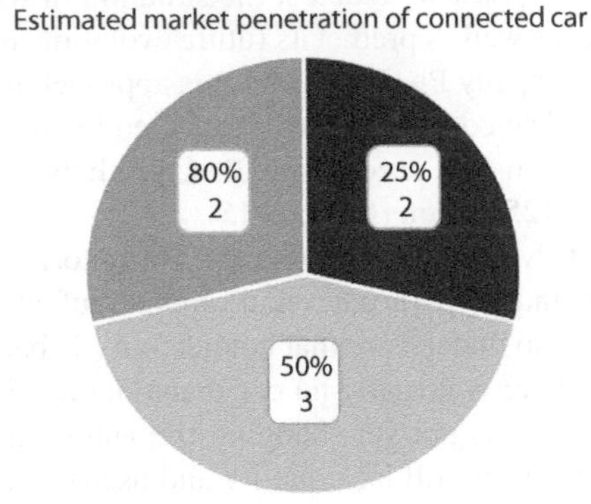

FIGURE 18.5 Penetration rate of cars with connectivity features in the global automotive market by 2020. Survey of seven experts. The number of experts choosing each option is shown in the chart.

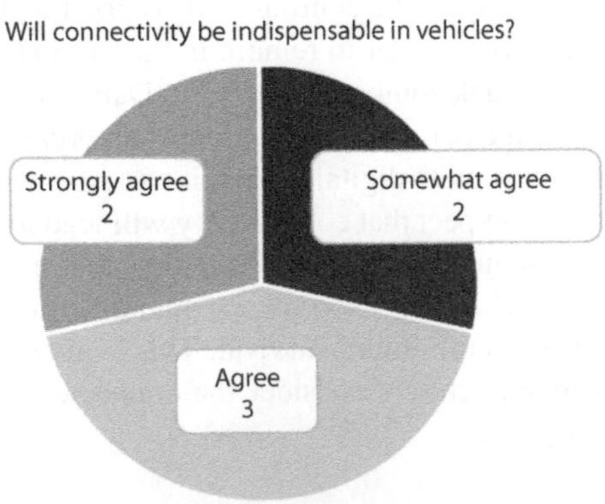

FIGURE 18.6 Will connectivity be indispensable in vehicles? Survey of seven experts. The number of experts choosing each option is shown in the chart.

fuel economy. Regulatory scrutiny and future emissions standards may require car manufacturers to adopt some form of IoT (Gao et al., 2014). For instance, connected cars present an opportunity to avoid wasting gas idling in traffic or circling for parking spots. Instead, they would be rerouted to traffic-less streets and communicate with parking spots ahead of time to find an open spot. Furthermore, connected cars are expected to improve the abilities of road operators and emergency services (IMS, 2016). Recently, the European parliament has voted in favor of e-Call regulation, which obliges all new cars to be equipped with e-Call systems* from April 2018 (Kollaikal, 2015).

18.6.2 Trend 2: From Product to Service

The connected car will become increasingly important in the automotive industry and will push the customer value toward service over product. As Cons1 contends, "The way of using a car changed completely." This is a pattern seen in other product-based industries, which are increasingly shifting toward services (Hientz et al., 2015) under the influence of digital transformation. There is a need to adjust business processes, organizational structures, and IT, to realize this new paradigm. The growing significance of service innovation in the automotive industry can be observed in the questionnaire results, as shown in Figures 18.7 and 18.8.

* e-Call is an emergency alert system that will notify rescue services automatically (Bethany, 2015).

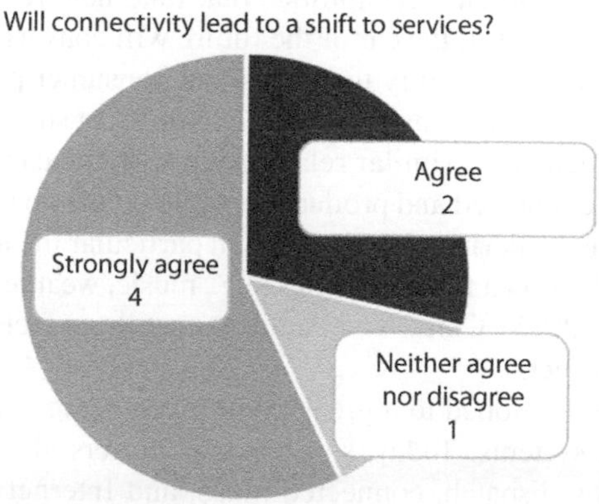

FIGURE 18.7 Connectivity will lead to a shift from one-time vehicle sales to more service-oriented models. Survey of seven experts. The number of experts choosing each option is shown in the chart.

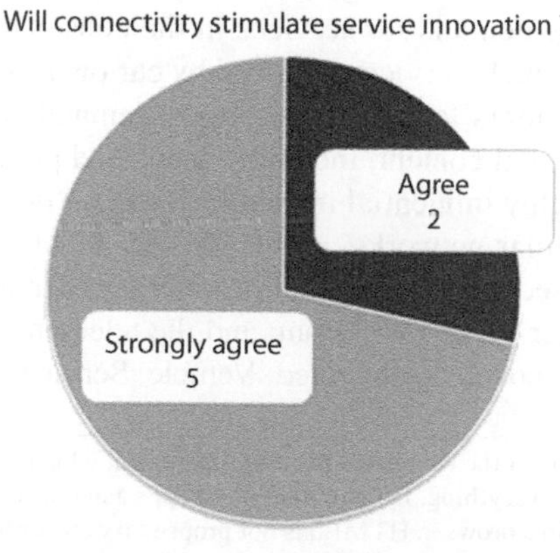

FIGURE 18.8 Connectivity will stimulate service innovation. Survey of seven experts. The number of experts choosing each option is shown in the chart.

The experts observe a fundamental change in the relationship between the car and the consumer. They agree that the car of the past was largely considered a status symbol, which was mostly owned. The interviewees suggest that vehicles in current Western urban societies have progressively lost their status symbol, and that consumers increasingly show a reduction in brand loyalty. Connectivity represents a key enabler of mobility services that could accelerate these shifts by intensifying attractive alternatives to car ownership. As a result, cars may become less interlinked with a consumer's personality and lifestyle and be perceived more as a means to an end. Auto1 comments that cars are increasingly considered as "just a solution to get from A to B." This agrees with a study recently conducted by IBM, which shows that consumers tend to highly value the convenience of cars without the need to be involved in a traditional ownership model (Stanley, 2016).

In addition, experts stress that due to connectivity, new opportunities are emerging that allow customers to spend time in a car in a completely different way than before. Tech2 explains that the meaning of driving is now going much beyond "steering a car." As Tech2 states, "When you look at the car in the past, it was mainly about driving. … The car of the future will be much more focused on entertainment, of how you spend your time best in the car." Both experts, Tech2 and Cons3, refer to the potential of connectivity supporting a shift in mind-set in terms of "use of time" in the car resulting from new product capabilities (e.g., semi- or fully autonomous driving and navigation systems).

Tech1 and Auto1 draw attention to the growing technological complexity and intelligence of the car, which allows for novel product capabilities that take new relationships between car and consumer further. According to Auto1, the car of the future will convert to an integral part of a consumer's digital life and will have the ability to learn about consumer preferences, and thus change the very concept of transport. As demonstrated by Deloitte's consumer research of 2015, drivers of the upcoming generation demand a similar relationship with the car as they currently have with their smartphones to remain connected and productive while on the go (Ninan et al., 2015). A report by IBM maintains that the vast majority of consumers, in particular the new upcoming generation of drivers, require integrating their own functionalities (e.g., music, weather, e-mail, or navigation) into the car (Wollschlaeger et al., 2015). Cons2 and Auto1 expect these factors to transform the vehicle into a highly customizable object.

In other words, cars have developed to a great extent from being entirely mechanical machines to more digitally integrated systems. Today, some manufacturers already offer customers convenient services like emergency dispatch, connected maps, and Internet radio. Recently, Tesla even launched semiautonomous driving capabilities in their cars (Kollaikal et al., 2015).

According to GSMA, various factors affect the development of in-vehicle services, with regard to both the type of services and the speed of evolution (GSMA, 2012). These factors embrace the evolution of adjacent industries and the technological developments enabling novel services. Advances in adjacent industries, including demand for services in the consumer electronics industry, tend to highly inspire the forms of in-vehicle services required by car owners. The report of GSMA reveals that, for example, car manufacturers' clients increasingly demand access to a growing volume of travel-related user-generated digital content, including maps and points of interest. Emerging technologies that are considered highly influential in enabling service development include cloud-based platforms, high-bandwidth cellular networks, and HTML5,* which allows for the development of new services using browser-based apps rather than proprietary carmaker platforms (GSMA, 2012).

In 2012, the car manufacturer Volvo Car Group and the telecommunications company Ericsson partnered in terms of taking novel "Connected Vehicle Services" to market (Andersson and

* HTML5 represents the newest version of the Hypertext Markup Language, which is the standard language used to create web pages. HTML5 includes almost everything, from animation to apps and music to movies, and can also be utilized to build complex applications that run in a browser. HTML5 is not proprietary and cross-platform, so it can be used on multiple devices (Marshall, 2011). It can help car manufacturers deliver the content and capabilities required by customers; keep pace with the release of novel consumer devices, applications, and services; and provide a quality app experience in cars while maintaining lower costs (O'Shea, 2013).

Mattsson, 2015; Ericsson, 2012). This concept represents a step toward the commercialization of technical innovation and related, new services. Ericsson's "Multiservice Delivery Platform" and "Connected Vehicle Cloud" solution created the technical base for offering novel services in Volvo's new vehicles. They enable drivers and passengers to access applications for, among other things, information, navigation, and entertainment from a screen in the car. Simultaneously, Volvo Car Group is able to open parts of the platform to additional actors of the car industry's ecosystem. Content providers are able to obtain agreements with Volvo and other members of the ecosystem, like Internet radio providers, road authorities, city governments, and toll road operators. Thus, car vehicles are turning into a new hub for numerous connected services, by enabling them to share Internet access to further devices, inside as well as outside the car (Andersson and Mattsson, 2015; Ericsson, 2013).

IoT-based services offer further benefits, such as automatic notification of accidents (Andersson and Mattsson, 2015). Moreover, interactions between the car and the driver's smartphone and apps may be available from a distance. Drivers may unlock their cars, monitor the status of batteries in electric cars, detect the location of the car, and so forth. In addition, continuous, remote diagnostics and interactions with numerous car functions may be enabled by various sensors (Andersson and Mattsson, 2015). Drivers can download applications and interact with several novel partners via the Connected Vehicle Cloud* built on Ericsson's Service Enablement Platform.† Accordingly, the new technical platforms foster the emergence of novel ecosystems of players in distinct sectors coming together as elements of the new interconnected infrastructure.

The example of Volvo and Ericsson's platform represents the first step toward entirely connected and integrated vehicle and infrastructure services, positioning the company as a key player in the connected car market (Just-auto global news, 2014; Volvo Car Group, 2012). However, connected car services are still in their early development stages (e.g., plans, visions, technical feasibility, and pilot projects) and are still not in widespread use (Andersson and Mattsson, 2015).

Figure 18.9 shows how a physical-digital model can be applied to the automotive industry. Note that customer value is derived from the service, but the service is made possible by the physical asset (the vehicle and associated electronics).

The experts agree that connectivity by itself does not deliver any added value. What makes it valuable is the sensor data of the object, which can now be integrated with findings of other web services in order to arrive at consequences for the actuator elements (normally in a cloud-based back-end system). With the help of analytics, patterns can be filtered out to identify valuable digital services for the customer at later stages. As stated by Data1, "If you can offer intelligence … it's power." Cons3 asserts that companies are already heavily investing in analytics (e.g., pattern recognition and predictive analytics) that will enable them to analyze large amounts of data and identify patterns that enable aligning products with customer preferences and needs. Furthermore, it is strongly recommended to develop hardware in close interconnection with Internet solutions: "Big data by itself is nothing. Only if you can cleverly analyze it, then you really have a lever of improvement" (Tech2).

According to Cons3, the value of service offers depends on consumer generations, as different generations may have different preferences and requirements. It is necessary to understand customer needs and to reflect them from different angles. Cons3 also refers to customer profiling as a possibility to identify customer value. Customer profiling involves an evaluation of customers with respect to the lengths of time they have been customers, the stage of development of their business and product lines, and the life cycle stage of their products. It includes companies focusing on the specific and changing needs of their customers (Dunk, 2004). Also, McKinsey stresses the importance of customer needs and expectations in detail (McKinsey & Company, 2014).

* http://fortune.com/2015/06/23/auto-tech-race/.

† https://www.daimler.com/produkte/services/mercedes-me/.

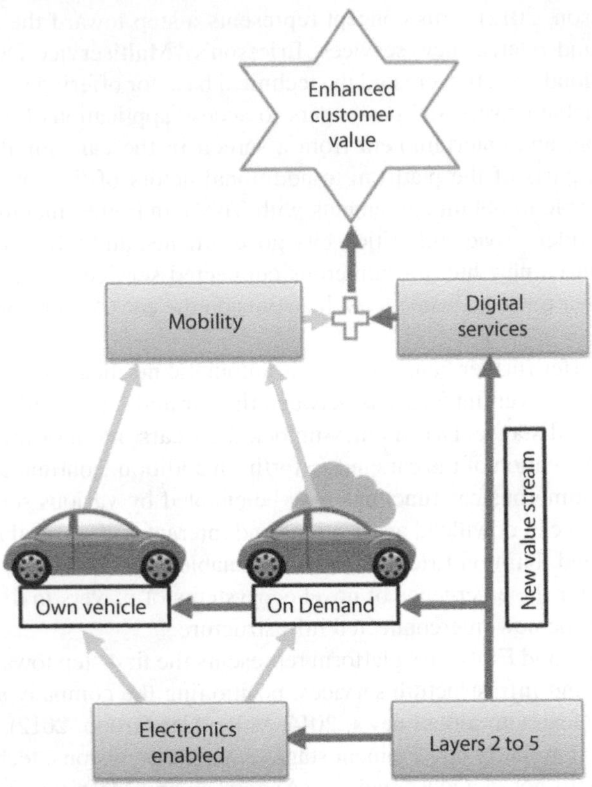

FIGURE 18.9 Complementary model of value creation. (Adapted from Hientz, H., et al., Automotive E/E development 2030: Software drives, Kugler-Maag CIE, Troy, MI, 2015, pp. 2–93.)

According to several experts, a further aspect to consider is the brand image of a company. Some experts highlight that connected car services heavily depend on brand image and related customers. As stated by Cons3, services necessarily have to adapt to the brand image, as customers have certain connections with, as well as expectations from, a brand. Thus, companies are required to evaluate what kind of service should be offered in conformance with the brand image. "In terms of old-timers, no one would think that 'connected services' would play a big role and match the brand image. Nevertheless, it remains a relevant market at which relevant sales are achieved" (Cons3).

18.6.3 Trend 3: New Business Models

The shift to services leads naturally to new business models. The hardware behind the cars is decreasing in importance (Hientz et al., 2015), and in order to extract value from the connected car, Cons1 and Tech2 stress the necessity to increasingly act and operate like data companies. Tech1 refers to the fact that many car manufacturers are still at the very beginning of this transformation and are making first experiments, such as BMW's "Connected Drive"* and Daimler's "me connect."†

"In the last 5 days, we have launched a new product … a cloud-based app. So, this learns where you normally go, the time you normally leave for work, … and will give you a notification when to leave based on traffic flow along the route, for instance" (Auto1).

* BMW Connected Drive is a multitiered infotainment system (BMW, 2013).

† Daimler's me connect comprises multiple connectivity services allowing one to connect everywhere and all the time with the car (Daimler, 2017).

As demonstrated by Auto1, connectivity enables car manufacturers to increasingly learn about consumer habits and to realize personalized service offerings.

The majority of experts emphasize that the degree of change and reaction may vary from company to company. As stated by Tech1, premium car manufacturers have an advanced approach to address the current changes caused by connectivity. Auto1 draws attention to the fact that BMW, for instance, is aiming at market leadership in the connected car market and has taken a different approach to other manufacturers, being the first in the United Kingdom to integrate an embedded SIM card and telematics control unit into every car in 2014. Furthermore, it has been found that car manufacturers and suppliers increasingly employ external consultants and companies to form appropriate strategies supporting to acquire competitiveness in the connected car market. As believed by the majority of experts, companies have begun to change their entire corporate structure, set up task forces, create new departments, or even enter new partnerships. As stated by Tech2, despite the awareness that the majority of traditional automotive organizations have, several companies, like Two-Tier manufacturers, lack the knowledge as well as the resources to shift the organization into the new paradigm.

According to Data1, car manufacturers are entering novel dimensions by accessing new data streams enabled through connectivity. He claims that these data streams will uncover a user's behavior and identify patterns that help to build and optimize service offers, as well as better adapt to consumer preferences. Tech1 describes this incident as the biggest business opportunity in the connected car market. Nevertheless, as reported by Tech1, only a small number of car manufacturers have understood the actual value behind connectivity and have started to explore new opportunities. Auto1 sees the connected car for BMW as a possibility to offer even more personalized service to customers. The potential of learning about customer preferences allows car manufacturers to offer a customer the right service at the right time. Auto1 emphasizes that BMW is already taking steps to better position itself as a service provider: "We act as a mobility provider in numerous forms and not just as a car manufacturer" (Auto1).

According to Auto1, connectivity already allows BMW to move from "just selling a car" to the provision of new customer value as an additional layer to the physical product. As stated by Auto1, BMW, for instance, already represents one of the largest worldwide providers of parking spaces, which allow for digital service offerings such as ParkNow.* The experts believe that the service component in the automotive industry will continue to grow in importance in the coming years and ascertain that it will lead to a diversification of carmakers' offerings and increase in profit opportunities on top of car sales. It is estimated that BMW's parking services realize annual global revenues of $24 billion (Parkmobile, 2014).

Furthermore, it is argued that connected cars tend to be associated with shared cars (Andersson and Mattsson, 2015). As reported by the majority of interviewees, connectivity is the essential premise for mobility service providers to be able to develop service-based business models. Car sharing is expected to show further growth and expansion into novel markets, with an estimated global revenue of $34 billion from 2015 to 2024 (BusinessWire, 2016).

Cons2 concludes that creating value from mere car sales will decline considerably in the future; instead, a significant proportion of the revenue will come from digital services: "Tesla is already able to push updates out that enable autonomous driving functionalities and charge $3000 for it."

Industries with advanced connectivity can serve as an example to understand the potential and importance to move toward digital services. In the music industry, global revenues from streaming services have now surpassed the sales generated by traditional music formats, accounting for 45% compared with 39% of the total for physical formats in 2015 (Ellis-Petersen, 2016). Also, the mobile phone industry shows how quickly revenue streams can shift toward service and content providers,

* ParkNow uses data to provide consumer services for on- and off-street parking through its app Parkmobile (Parkmobile, 2017), as well as business intelligence to councils and other parking providers (BMW Group, 2017).

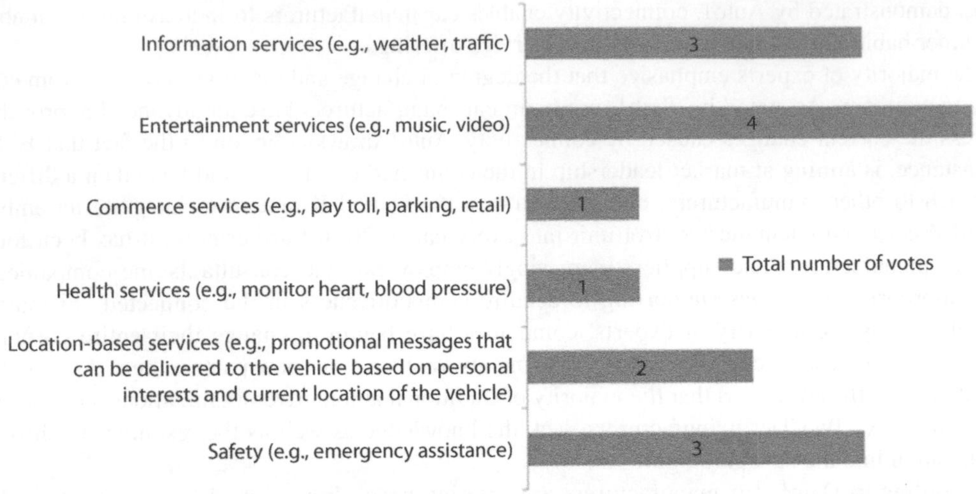

FIGURE 18.10 What connectivity service will be most valuable for the customer in the future? Survey results from experts using a multiple-choice format. The bars show total number of votes cast for each service.

with all but a few mobile phone manufacturers making losses from selling handsets. In 2013, Apple and Samsung were the only companies able to profit from handset sales (Habeck et al., 2014).

Among connected car services, most of the interviewees perceive safety, entertainment, and information services as the most valuable in the future (Figure 18.10). However, Cons2 and Tech2 feel that the degree of the shift from transactional selling toward services is still unclear.

18.6.4 Trend 4: Emergence of Nontraditional Industry Entrants

Improved connectivity and the shift to services enable new business models based around information. This naturally allows an entry point for those companies whose core competencies lie in this space; such companies are sometimes referred to as digital players (Habeck et al., 2014). Cons3 points out that the growing emergence of digital players in the connected car market forces car manufacturers and suppliers to react and to take connectivity into stronger consideration. Tech2 feels that the growth of the share economy is also stimulating the rise of connectivity in the automotive sector, as cars, which are not owned but utilized on a time basis, are highly dependent on connectivity features. Since many car manufacturers also rely on car-sharing models, connectivity becomes fundamental to exploit the potential of the sharing economy. Cons2, however, refers to electrical mobility as further driving a rise in connectivity. The majority of the experts predict that electric or hybrid vehicles will come into widespread use by 2025. Cons2 explains that these vehicles represent mobility concepts that can only be implemented appropriately and become commercially viable by means of connectivity.

Connecting the car to the Internet is thus introducing digital players to a traditional industry segment (Hansen, 2015; c. Weakly connected or totally disconnected companies and industries are now drawn into closer cooperation, as confirmed by several scholars (Hientz et al., 2015; Andersson and Mattsson, 2015; Habeck et al., 2014). Car manufacturers are no longer the only ones expected to benefit from the development of the connected car, as digital players and others have the opportunity to widen their business boundaries by entering the automobile sector. This refers in particular to the telecommunications industry and the IT sector (Hientz et al., 2015). J. N. Habeck, et al. 2014, provides an overview of traditional and nontraditional actors extending their activities from different starting points (Table 18.3).

Car manufacturers increasingly partner with smartphone manufacturers and mobile operators to build collaborative platforms (Andersson and Mattsson, 2015; Habeck et al., 2014; Korosec, 2015;

TABLE 18.3
Traditional and Nontraditional Industry Players Are Extending Their Activities from Different Starting Points

Player	CRM/Payment	Services, Content, and Apps	Software (OS/Platform, App Store)	Systems (Car–Software–User Interface)	Hardware (Components, Car Design)	Infrastructure (Mobile Network, Mobility Infrastructure)
OEM		○	○	●	●	
Supplier		○	○	○	●	
Digital	●	●	●	○	○●	○●
Telecom	●	○●				○●
Automotive insurers	●	○●				

Source: J. N. Habeck, M. Bertoncello, M. Kässer, F. Weig, M. Hehensteiger, J. Hölz, R. Plattfaut, C. Wegner, M. Guminski, and Z. Yan, Connected car, automotive value chain unbound, Advanced Industries, Chelsea, MI, September 2014, pp. 7–50

Note: Closed circles indicate core activities; open circles represent new areas of focus.

Newcomb, 2014). About 10 manufacturers have recently cooperated with Apple to give it access to in-car screens that allow for apps to be utilized by drivers inside the car (Korosec, 2015). The Car Connectivity Consortium,* established by some of the largest auto and smartphone manufacturers, has created Mirrorlink,† a device interoperability standard connecting smartphone content to car dashboards. Furthermore, Google's Open Automotive Alliance‡ agreed with Audi, GM, Honda, and Hyundai on a common platform for Android integration with connected vehicles (Newman, 2014; TNS, 2016). Also, digital players like music-streaming services (e.g., Pandora, Spotify, and Deezer) have already built partnerships with specific OEMs (Habeck et al., 2014). Despite those collaborative strategies, each single major manufacturer is building its own connected car system, such as Renault's R-Link, GM's Onstar, BMW's Connected Drive, Audi's Connect, and Toyota's Entune (Andersson and Mattsson, 2015; TNS, 2016). This parallel, proprietary approach is aimed at reducing, on the one hand, high dependency on a single smartphone platform, and on the other, the growing loss of control over the in-car environment to technology companies (Schuhmacher, 2015; TNS, 2016).

There is a consensus among all the experts interviewed that the technological abilities of the connected car have opened up a gap in the market for novel business models based on the provision of Internet skills, which will also attract organizations outside of the automotive industry. According to Data1, car components are increasingly digitized, from speedometer pointers to displays. The car's core now utilizes more electronics and data than ever before. Data1 highlights that technology players, such as Google and Apple, in particular, possess the capabilities to manage and control these complex processes and position themselves to deliver the electronic heart of the vehicle.

All experts agree that connected cars will form a further area in the consumer's digital life in which companies acquire the possibility to obtain vast amounts of data. As emphasized by Cons3, "The oil of the future is data; you can earn money with it." Cons2, Cons3, Tech1, and Auto1 point out that one reason for these new players to enter the market may be to push their operating systems (Android Auto§ and Apple Carplay¶) into vehicles in exchange for access to user data. Experts assume that the data generated in a vehicle would support technology companies like Google to acquire more specific information from users that help them, for instance, to improve their advertising campaigns and sell the data profiles to third parties. According to Tech1 and Cons2, Google and Apple tend to force OEMs into partnerships by creating threatening scenarios in terms of "If you don't let me into your vehicles, I will build my own ones." One could argue that they are in fact already pursuing this strategy through initiatives such as the Google Car, also known as Waymo,** although such initiatives are currently more akin to suppliers than to automobile manufacturers (Bhuiyan, 2017).

Cons2 supposes that technology companies consider the automotive market as an attractive opportunity to benefit from the time consumers spend in vehicles. As claimed by Auto1, in 2014, the amount of time an average person spent in the car significantly exceeded the time they would spend on watching TV. Cons2 also refers to the large potential of interacting with customers via autonomous cars and presents the following scenario: "For example, you could be sitting in the vehicle and McDonald's has a promotion which pushes that promotion through to Google and Google through their autonomous vehicles. So, you are sitting in a Google vehicle and in the windscreen you can see via augmented reality, for example, promotional offers, such as pay for one burger and get another one for free."

Tech2 refers to Google's ability of adding substantial value to the data stemming from the connected car: "At Google, our key asset is information, and I think that our biggest leaver is to put that information to use and to add different information sources and better driving intelligence or

* http://carconnectivity.org/.
† http://www.mirrorlink.com/.
‡ http://www.openautoalliance.net/.
§ https://www.android.com/intl/en_uk/auto/.
¶ http://www.apple.com/uk/ios/carplay/.
** https://www.google.com/selfdrivingcar/.

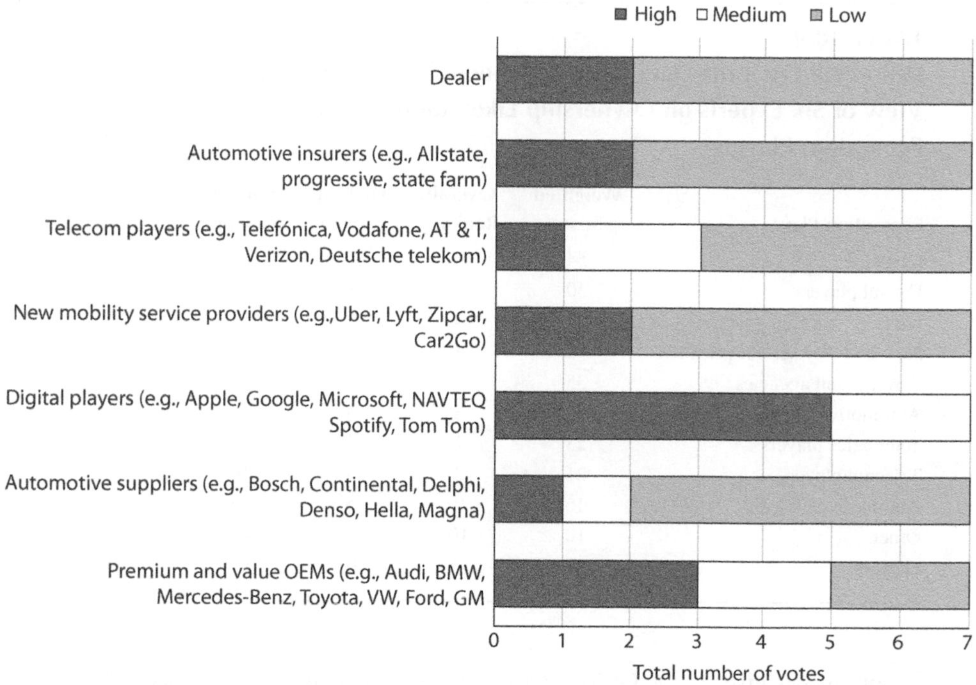

FIGURE 18.11 Estimated influence in the connected car market. Experts were asked to what extent the player would "influence revenue streams." The answers were categorized as high, medium, or low influence.

mobility intelligence." Tech1 and Tech2 stress the example of Google's investment in Waze,* which is a GPS-based geographical navigation application program that aggregates driving information and collects data mainly from moving smartphones. As a result, user information can be employed to create additional benefits for the navigation function (e.g., display of emerging traffic jams). Tech1 affirms that technology companies increasingly aim to transform the connected car into a service unit to extend their current portfolio.

However, Tech1 also mentions the appearance of other players aiming for the data in the car. Among these are telecommunication companies such as Vodafone and Telefónica, whose objective is to sell their mobile data packages; insurance companies; financial service providers; automobile clubs; and new mobility service providers. McKinsey relates the online transportation network company Uber, for instance, to new types of software-enhanced mobility functions, which has disrupted the taxi business in several cities by means of a simplified business model based on superior software algorithms that enable, among other things, short waiting times and price-matching supply and demand (Wee et al., 2015).

When asked which players will have the highest influence over the main revenue streams, the interviewees predicted that the digital players (e.g., Google and Apple) will have the upper hand over OEMs (Figure 18.11). New mobility service providers such as Uber and Lyft also have a big opportunity in this space. One of the key battlegrounds may be data ownership (Table 18.4). In Table 18.4, experts were asked to rank the players in terms of likelihood of owning the data. Each player's ranking was scored, with a ranking of 1 allocated 10 points, down to a rank of 11, allocated 0 points. The rankings were summed across the experts. Thus, the maximum score attainable was 60 (every expert ranks this player 1st place) and the minimum 0 (every expert ranks this player 11th place).

* https://www.waze.com/.

TABLE 18.4
Who Will Own the Data in the Connected Car? Combined View of Six Experts on Ownership Likelihood for Various Ecosystem Players

Ecosystem Player	Weighted Score	Overall Ranking	Highest Ranking	Lowest Ranking
OEMs	54	1	1	4
Digital players	50	2	1	6
Users	44	3	1	9
New mobility service providers	40	4	3	7
Government agencies	35	5	3	8
Automotive suppliers	32	6	4	7
Telematics players	23	7	6	8
Telecom players	21	8	4	9
Automotive insurers	16	9	5	10
Other	10	10	1	11
Dealers	5	11	10	11

The overall top rank is given to the OEMs, with digital players coming a close second. The experts were in agreement that dealers had the least likelihood of owning the data.

Both Table 18.4 and Figure 18.11 serve to illustrate the growing importance and power of non-traditional market participants. The responses show a tendency of technology companies to acquire significant revenue streams and high influence over vehicle data.

On the one hand, most of the interviewed experts consider digital players as potential suppliers and business partners for OEMs in the connected car market. On the other hand, experts also mention the possibility of digital companies to transform into direct competitors, when they aim to create their own hardware. As claimed by Tech2, "There are rumors about Apple developing their own car." Furthermore, they could also represent potential customers of OEMs, if they decide to operate as mobility service providers and thus acquire the necessary vehicle fleets directly from OEMs. According to the experts, Google and Apple could additionally utilize mobility services and related vehicle fleets as a further means for data collection.

Cons1 considers both Google and Apple as the ones with the greatest influence to shift the entire automotive industry toward services: "They will turn the market upside down if the automotive industry is not watching out. They understand that service is actually a business, and I think there is only a time frame of 3 years where automobile companies need to transform into data companies in order to remain successful in the market."

Cons2 also refers to the powerful role of the multinational conglomerate corporation LG, which is building a connectivity platform in collaboration with the car manufacturer Volkswagen. Besides, it is said that the company already delivers the entire electric power train for Volkswagen. As a result, LG owns the most complex technology elements in the vehicle, namely, the telematics platform and the electric vehicle drivetrain. Cons2 asserts that in the coming years, most of the innovation will happen around digital technologies, and observes that LG's portfolio also involves smartphones, smart homes, lithium-ion batteries, and the construction of charging stations. According to Cons2, organizations such as LG represent a big threat for car manufacturers, as they have already mastered many components of the digital customer ecosystem of the future.

Auto1 stresses the need to recognize threats like this and shift organizations toward the new digital ecosystem. Several experts compare the current situation in the automotive industry to the events that occurred in the mobile phone industry during the last decade. Cons2 points out that in contrast to Nokia, Apple and Google have encouraged numerous companies to utilize their platforms to offer services on hardware devices. Although Nokia was one of the largest mobile phone companies, it failed due to its weak position in the ecosystem (Bouwman et al., 2014).

Auto1 underlines the necessity to add new value to products in order to encourage customers to enter the ecosystem. However, Tech1 is concerned that digital players will overtake car manufacturers as they are "too slow and too confident right now." Furthermore, Cons3 portends that car manufacturers could become highly dependent on technology providers, as they have the skills and capabilities to manage and administer the entire IT infrastructure and ensure the technical realization of services. In order to be able to sustain the market, McKinsey suggests the creation of ecosystems and improvement of a company's internal capability, such as improvement of internal software capabilities (Wee et al., 2015).

18.7 CHALLENGES

The four trends outlined create a plausible future narrative that is supported by a range of evidence. However, this path is not inevitable, and there are some challenges for the IoT transformation of the automotive industry. Broad adoption of IoT may be held back until institutional ethical considerations (regulations and standards, trust, or security) and current network structures and processes (barriers to reconfigure existing business networks and develop new business models) are relieved (Andersson and Mattsson, 2015). The main challenges are discussed in more detail below.

18.7.1 Innovation Cycle Time

The experts note, among other things, that the combination of physical products and digital services involves a clash between different innovation cycles. New features, like operating system upgrades and novel applications, are offered almost constantly, while car manufacturers work on 5-year cycles (Andersson and Mattsson, 2015; Hansen, 2015). Car manufacturers need to develop an innovation model with small updates being rolled out rapidly and as needed in between larger and less frequent updates; as previously noted, this approach may realize new revenue channels by providing clients with tailored additional services for a fee (Wee et al., 2015). Tesla, for instance, develops its software in-house and is already able to realize frequent upgrades for its customer's cars. According to McKinsey, the last upgrade was released in 2015 and delivers multiple services to its customers, such as advanced driver-assistance systems (ADASs) features like automatic emergency braking, improved maps, and navigation.

18.7.2 Network Effect

Tech2 highlights that certain services also depend on a network effect in which the value of a service is highly determined by the number of users. As stated by Tech2, a real customer value or unique selling point (USP) can only be achieved if the company gains the majority of the market share (e.g., online social networks like Facebook and Twitter).

18.7.3 Privacy and Security

IoT technologies are expected to progress fast, resulting in high levels of data volume and variety available to the companies involved, bringing opportunity but also challenges. Experts consider data privacy and protection, as well as new business model development, to be one of the biggest challenges in terms of enabling service innovation within the automotive industry. Auto1 and Tech2 refer to the need for legal changes and the support of legislation about data. Consumers are concerned about digital safety and data privacy. According to McKinsey's consumer survey, about

37% of respondents would not even consider a connected vehicle. In terms of vehicles being hacked, consumers also show significant concerns (McKinsey & Company, 2014).

18.7.4 Business Models and Control Points

A further consideration is that value from a digital service can only be realized with an appropriate business model. According to the majority of the experts, creation of a new business model represents one of the key challenges in enabling service innovation (Figure 18.12). In the telecommunications industry, Google and Amazon, for example, were able to introduce new business models and successful monetizing services by securing critical control points. Google, for instance, generates on average a $3.40 profit per Android device per year, which is attributable to its control over the operating system, the app store, and the ad platform. Although Amazon and Apple generate far more profit margin per device, the Google revenue stream comes from a far broader ecosystem of devices, most made by other companies (Habeck et al., 2014).

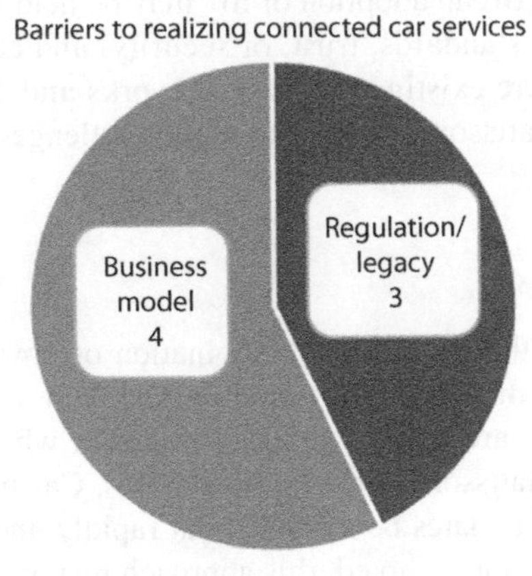

FIGURE 18.12 Main barriers of realizing or monetizing connected car services. Survey of seven experts. The number of experts choosing each option is shown in the chart.

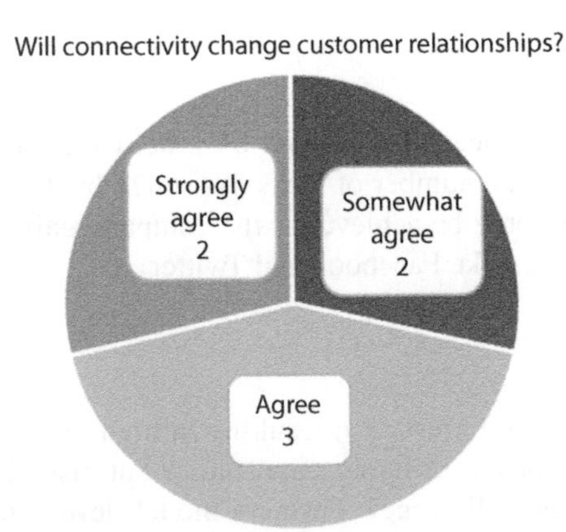

FIGURE 18.13 Connectivity will change customer relationships. Survey of seven experts. The number of experts choosing each option is shown in the chart.

18.7.5 Customer Relationship

Moreover, connectivity will allow moving from bundled packages to services on demand and will require novel pricing models, such as micropayments on a per-feature, per-use, per-mile, or per-minute basis (Andersson and Mattsson, 2015). A survey by McKinsey reveals that consumers show little willingness to pay for connectivity features. Solely 35% of new car buyers are willing to spend an additional $100 for smartphone integration, while only 21% are willing to pay for subscription-based services (McKinsey & Company, 2014). Most of the experts interviewed note that services innovation leads to a change in customer relationship structures (Figure 18.13). Cons1 and Cons2 highlight that various car manufacturers still lack the knowledge, resources, and skills to deal with new forms of customer relations.

18.8 CONCLUSIONS

The automotive industry, like other product-based industries, is liable to disruption by digitization. The IoT is likely to shape this disruption, and lead to a hybrid model where physical and digital elements combine to provide new service-based business models. As has been discussed, connectivity is not just a future vision: connectivity is already on streets and in cars. Today, connectivity is helping drivers of electric vehicles to identify available charging stations and provide battery charge status updates (Navigant Research, 2016). The IoT could radically transform the automobile industry by allowing cars to share vast amounts of real-time data. In principle, this new paradigm will enable cars to learn, heal, drive, and socialize with other cars and their environment.

This chapter provided a narrative showing how connected cars can be viewed as a node in the IoT. This connectivity drives a product to service transformation that is already starting to occur in the sector. The implication of this is that new business models are enabled, meaning that the automotive sector has become open to nontraditional entrants, such as digital players and mobility service providers. The control points and loci of value generation are also likely to shift.

There are many challenges facing an IoT transformation, quite apart from the myriad technical issues to be addressed. For example, the connected car requires a more agile, on-demand innovation cycle than a traditional vehicle. The changing industry landscape and power structure will prove difficult for all players to navigate. Privacy and security concerns will shape customer relationships. Brand loyalty cannot be guaranteed, and it may not be to the carmaker.

So, the automotive industry is likely to change from heavily product based to more service based, with new opportunities to provide significant added value. Value will no longer be solely derived from the physical car and primarily one company, but mainly by digital services delivered by a network of participating companies. The focus on services allows both established and novel industry entrants to enter new customer segments with IT-enabled services, leading to a shift in balance of power in the industry, one based on control of data. The central position of the automobile is no longer ensured in the IoT-enabled automotive industry.

REFERENCES

Andersson, P. and Mattsson, L.-G. 2015. Service innovations enabled by the "Internet of things". *IMP Journal* 9 (1): 85–106.

Bethany, B. 2015. EU to introduce eCall alert device for car crashes. Accessed January 27, 2017. http://www.bbc.com/news/technology-32495058.

Betz, F. 1987. Managing Technology: *Competing through New Ventures, Innovation, and Corporate Research*. Upper Saddle River, NJ: Prentice-Hall.

Bhuiyan, J. 2017. The Google Car was supposed to disrupt the car industry. Now Waymo is taking on suppliers, January 27, 2017. Accessed January 30, 2017. http://www.recode.net/2017/1/27/14399770/waymo-google-supplier-self-driving-car.

BMW Group. 2013 BMW Connected Drive. Accessed January 27, 2017. http://www.bmw.com/com/en/insights/technology/connecteddrive/2013/..

BMW. 2017. Find, park, pay—One app for everything. Accessed January 27, 2017. https://www.bmwgroup.com/en/brands-and-services/park-now.html

Bouwman, H., C. Carlsson, J. Carlsson, S. Nikou, A. Sell, and P. Walden. 2014. How Nokia failed to nail the smartphone market, presented at 25th European Regional Conference of the International Telecommunications Society, Brussels.

BusinessWire. 2016. Global revenue from carsharing services is expected to total $34.6 billion from 2015 to 2024, according to Navigant Research, May 4, 2016. Accessed August 9, 2016. http://www.businesswire.com/news/home/20160504005014/en/Global-Revenue-Carsharing-Services-Expected-Total-34.6.

Camara, D. and N. Nikaein. 2015. *Wireless Public Safety Network 1: Overview and Challenges*. Amsterdam: Elsevier.

CarConnectivity Consortium. 2016a. Home page. Accessed January 29, 2017. http://carconnectivity.org/.

CarConnectivity Consortium. 2016b. Mirror Link home page. Accessed January 29, 2017. http://www.mirrorlink.com/.

Choi, S., F. Thalmayr, D. Wee, and F. Weig. 2016. Advanced driver-assistance systems: Challenges and opportunities ahead, McKinsey & Company, Washington, DC. Accessed January 29, 2017. http://www.mckinsey.com/industries/semiconductors/our-insights/advanced-driver-assistance-systems-challenges-and-opportunities-ahead.

Craig, W. 2015. Investing in the future of the auto industry: The connected car, *The Street*, December 22, 2015. Accessed August 5, 2016. https://www.thestreet.com/story/13405393/1/investing-in-the-future-of-the-auto-industry-the-connected-car.html.

Crotti, M., F. Gringoli, and L. Salgarelli. 2010. Classification of emerging protocols in the presence of asymmetric routing. In *The Internet of Things: 20th Tyrrhenian Workshop on Digital Communications*, edited by D. Giusto, A. Iera, G. Morabito, and L. Atzori, 13–25. New York: Springer.

Daimler. 2017. Mercedes me. Accessed January 27, 2017. https://www.daimler.com/produkte/services/mercedes-me/.

Datson, J. 2016. Mobility as a service: Exploring the opportunity for mobility as a service in the UK, Transport Systems Catapult, Milton Keynes, UK.

Davidson, L. 2015. How connected cars are driving the Internet of things, January 27, 2015. Accessed November 8, 2016. http://www.telegraph.co.uk/finance/newsbysector/industry/engineering/11372205/How-connected-cars-are-driving-the-Internet-of-Things.html.

Denton, T. 2004. *Automobile Electrical and Electronic Systems*, 3rd ed. Oxford: Elsevier.

Dunk, S. 2004. Product life cycle cost analysis: the impact of customer profiling, competitive advantage, and quality of IS information. *Management Accounting Research* 15: 401–414.

Ellis-Petersen, H. 2016. Streaming growth helps digital music revenues surpass physical sales, *Guardian*, April 12, 2016. Accessed August 16, 2016. https://www.theguardian.com/music/2016/apr/12/streaming-revenues-bring-big-boost-to-global-music-industry.

Ericsson. 2012. Connected car services come to market with Volvo Car Group and Ericsson, December 17, 2012. Accessed July 28, 2016. http://hugin.info/1061/R/1665573/540444.pdf.

Ericsson. 2013. Connected Vehicle Cloud: Under the hood. http://archive.ericsson.net/service/internet/picov/get?DocNo=28701-FGD101192&Lang=EN&HighestFree=Y.

Fleisch, E., M. Weinberger, and F. Wortmann. 2014. Business models and the Internet of things. St. Gallen, Switzerland: Bosch IoT Lab.

Fortino, G. and P. Trunfio, eds. 2014. *Internet of Things Based Smart Objects: Technology, Middleware and Applications*. Heidelberg: Springer.

Frommer, D. 2011. History lesson: How the iPhone changed smartphones forever, *Business Insider*, June 6, 2011. Accessed August 8, 2016. http://www.businessinsider.com/iphone-android-smartphones-2011-6?IR=T.

Gao, P., R. Hensley, and A. Zielke. 2014. A road map to the future for the auto industry, McKinsey & Company, Washington, DC, October 2014. Accessed November 11, 2016. http://www.mckinsey.com/industries/automotive-and-assembly/our-insights/a-road-map-to-the-future-for-the-auto-industry.

Gartner. 2015. Gartner says by 2020, a quarter billion connected vehicles will enable new in-vehicle services and automated driving capabilities, January 26, 2015. Accessed July 22, 2016. http://www.gartner.com/newsroom/id/2970017.

Gissler, C. O., C. Knackfuß, and F. Kupferschmidt. Are you ready for pole position? Driving digitization in the auto industry, Accenture Strategy, Arlington, VA, 2016, 2–6.

Giusto, D., A. Iera, G. Morabito, and L. Atzori, eds. 2010. *The Internet of Things: 20th Tyrrhenian Workshop on Digital Communications*. New York: Springer.

GSMA. 2012. 2025 every car connected: Forecasting the growth and opportunity, GSMA Connected Living Programme: MAutomotive, London, 1–28.

GSMA. 2012. Connected cars: Business model innovation, GSMA Connected Living Programme: MAutomotive, London.

GSMA. 2013. Connecting cars: The technology roadmap, GSMA Connected Living Programme: MAutomotive, London, 1–61.

Habeck, J. N., M. Bertoncello, M. Kässer, F. Weig, M. Hehensteiger, J. Hölz, R. Plattfaut, C. Wegner, M. Guminski, and Z. Yan. 2014. Connected car, automotive value chain unbound, Advanced Industries, Chelsea, MI, September 2014, 7–50.

Hansen, P. 2015a. 5G networks coming to autos, possibly by 2025. *Hansen Report on Automotive Electronics* 28 (6): 2.

Hansen, P. 2015b. IoT will change the auto industry, says Kugler Maag Study. *Hansen Report on Automotive Electronics* 28 (6).

Hientz, H., H.-J. Kugler, B. Maag, and D. Strube. 2015. Automotive E/E development 2030: Software drives, Kugler-Maag CIE, Troy, MI, 2–93.

IEE Control Systems Society. 2011. Vehicle-to-vehicle/vehicle-to-infrastructure control. Accessed July 26, 2016. http://www.ieeecss.org/sites/ieeecss.org/files/documents/IoCT-Part4-13VehicleToVehicle-HR.pdf.

IMS. 2016. Intelligent transportation systems—Next wave of the connected car. Accessed November 17, 2016. http://www.intellimec.com/intelligent-transportation-systems-next-wave-connected-car/.

Just-auto global news. 2014. Sweden: Volvo Car, Ericsson partnership delivers global connected car services, *just-auto*, January 8, 2014.

Knight, W. 2015. Car-to-car communication: A simple wireless technology promises to make driving much safer. *MIT Technology Review*. Accessed August 6, 2016. https://www.technologyreview.com/s/534981/car-to-car-communication/.

Kollaikal, P., S. Ravuri, and E. Ruvinsky. 2015. Connected cars, Technical Report, Sutardja Center for Entrepreneurship and Technology, Berkeley, CA, 1–21.

Korosec, K. 2015. Who will win the great auto tech race? *Fortune*, June 23, 2015. Accessed July 22, 2016. http://fortune.com/2015/06/23/auto-tech-race/.

Luckerson, V. 2014. Spotify just bought access to your online music listening habits, *Time*, August 3, 2014. Accessed July 22, 2016. http://time.com/14237/spotify-just-bought-access-to-your-online-music-listening-habits/.

Marshall, G. 2011. HTML5: What is it? Accessed October 25, 2016 http://www.techradar.com/news/internet/web/html5-what-is-it-1047393.

McKinsey & Company. 2014. What's driving the connected car, September 2014. Accessed August 6, 2016. http://www.mckinsey.com/industries/automotive-and-assembly/our-insights/whats-driving-the-connected-car.

Mohr, D., N. Müller, A. Krieg, P. Gao, H.-W. Kaas, A. Krieger, and R. Hensley. 2013. *The road to 2020 and beyond: What's driving the global automotive industry*. Washington, DC: McKinsey & Company.

Mueller, M. 2013. *Challenges of Digitalization for the Music Industry*. Münster: Grin Verlag.

Navigant Research. 2016. The future is now: Smart cars and IoT in cities, *Forbes*. Accessed November 17, 2016. http://www.forbes.com/sites/pikeresearch/2016/06/13/the-future-is-now-smart-cars/#1b8a6fda48c9.

Newcomb, D. 2014. Who should control connected car data? *PC Mag*, November 7, 2014. Accessed August 16, 2016. http://uk.pcmag.com/cars-products/37284/opinion/who-should-control-connected-car-data.

Newman, J. 2014. How Google and Apple plan to invade your next car, *Time*, June 30, 2014. Accessed July 16, 2016. http://time.com/2941556/apple-carplay-android-auto/.

Nieto, M., F. Lopez, and F. Cruz. 1998. Performance analysis of technology using the S curve model: The case of digital signal processing (DSP) technologies. *Technovation* 18 (6/7): 439–457.

Ninan, S., B. Gangula, M. von Alten, and B. Snidermann. 2015. Who owns the road? The IoT-connected car of today—and tomorrow, Deloitte University Press, August 18, 2015. Accessed July 17, 2016. http://dupress.com/articles/internet-of-things-iot-in-automotive-industry/.

O'Shea, M. 2013. Native vs. HTML5: How the auto industry is forcing change for the mobile web. Accessed October 25, 2016. http://venturebeat.com/2013/10/05/native-vs-html5-how-the-auto-industry-is-forcing-change-for-the-mobile-web/.

Olivarez-Giles, N. 2015. Spotify aims to automate music discovery with personalized weekly playlists, *The Wall Street Journal*, July 20, 2015.

Open Automotive Alliance. 2014. Introducing the Open Automotive Alliance. Accessed January 29, 2017. http://www.openautoalliance.net.

Pai, V. S. 2015. Nokia Ltd: Travails of a market leader. *Vision* 19 (3): 276–285.

Papatheodorou, Y. and M. Harris. 2007. The automotive industry: Economic impact and location issues, January 5, 2007. Accessed August 8 http://www.industryweek.com/global-economy/automotive-industry-economic-impact-and-location-issues.

Parkmobile. 2014. Parkmobile announces a substantial investment agreement with the BMW Group, September 8, 2014. Accessed August 8, 2016. http://us.parkmobile.com/members/parkmobile-the-leading-provider-of-mobile-payment-solutions-to-municipalities-and-parking-providers-announced-today-that-it-has-signed-a-substantial-investment-agreement-with-the-bmw-group.

Parkmobile. 2017. Home page. Accessed January 27, 2017. http://us.parkmobile.com/.

Popper, B. 2015. Tastemaker: How Spotify's Discover Weekly cracked human curation at Internet scale, *The Verge*, September 30. Accessed July 22, 2016. http://www.theverge.com/2015/9/30/9416579/spotify-discover-weekly-online-music-curation-interview.

Porter, M. E. and J. E. Heppelmann. 2014. How smart connected products are transforming companies. *Harvard Business Review*, November 2014, 97–114.

Pullen, P. The big streaming showdown—Beats vs. Spotify, *Fortune*, March 3, 2015. Accessed July 22, 2016. http://fortune.com/2015/03/03/the-big-streaming-showdown-beats-vs-spotify/.

PwC. 2013. Top suppliers, *Automotive News*, June 17, 3–15.

PwC. 2015. My transport, connected. Accessed August 11, 2016. http://www.pwc.com/gx/en/ceo-agenda/podcast/episode-3.html.

Qualcomm. 2015. Snapdragon automotive solutions: Connected car platforms for all types of vehicle, June 2015. Accessed August 5, 2016. https://www.qualcomm.com/news/snapdragon/2015/06/04/snapdragon-automotive-solutions-connected-car-platforms-all-types-vehicle.

Rellermeyer, J. S., M. Duller, K. Gilmer, D. Maragkos, D. Papageorgiou, and G. Alonso. 2008. The software fabric for the Internet of things. In *The Internet of Things*, edited by C. Floerkemeier, M. Langheinrich, E. Fleisch, F. Mattern, and S. E. Sarma, 87–104. Heidelberg: Springer.

Ropert, S. 2014. Technical innovations: Connected cars: Overview and trends. *Digiworld Economic Journal* 4 (96): 153.

Schuhmacher, S. 2015. How the Internet of things is transforming the automotive industry, August 21, 2015. Accessed July 26, 2016. http://www.ibmbigdatahub.com/blog/how-internet-things-transforming-automotive-industry.

Stanley, B. 2016. How consumers want their cars to fit into their lives, *Forbes*, January 12, 2016. Accessed August 16. http://www.forbes.com/sites/ibm/2016/01/12/how-consumers-want-their-cars-to-fit-into-their-lives/#7f489d5e3a6f.

Surowiecki, J. 2013. Where Nokia went wrong, *The New Yorker*, September 3, 2013. Accessed August 5, 2016. http://www.newyorker.com/business/currency/where-nokia-went-wrong.

TNS. 2016. Time to set the connected car free. Accessed August 6, 2016. http://www.tnsglobal.com/intelligence-applied/whos-best-driver-connected-car.

Viereckl, R., A. Koster, E. Hirsh, and D. Ahlemann. 2016. Connected car report 2016: Opportunities, risk, and turmoil on the road to autonomous vehicles, PwC, London.

Viereckl, R., D. Ahlemann, A. Koster, and S. Jursch. 2015. Connected car study 2015: Racing ahead with autonomous cars and digital innovation, September 16, 2015. Accessed July 26, 2016. http://www.strategyand.pwc.com/reports/connected-car-2015-study.

Volvo Car Group. 2012. Volvo Car Group and Ericsson join forces to deliver global connected car services, December 17, 2012. Accessed July 18, 2016. https://www.media.volvocars.com/global/en-gb/media/pressreleases/47168.

Wedeniwski, S. 2015. *The Mobility Revolution in the Automotive Industry: How Not to Miss the Digital Turnpike*. Heidelberg: Springer-Verlag.

Wee, D., M. Kässer, M. Bertoncello, H. Kersten, G. Eckhard, J. Hölz, F. Saupe, and T. Müller. 2015. Competing for the connected customer—Perspectives on the opportunities created by car connectivity and automation. Advanced Industries, Chelsea, MI, September 2015, 6–42.

Wollschlaeger, D., M. Foden, R. Cave, and M. Stent. 2015. Digital disruption and the future of the automotive industry: Mapping new routes for customer-centric connected journeys, IBM Center for Applied Insights, Armonk, NY, September 2015, 2–15.

Woyke, E. 2014. *The Smartphone: The Anatomy of an Industry*. New York: New Press.

Glossary

4G: State-of-the-art mobile wireless broadband communication system with peak data rates of 100 Mb/s in high-speed mobility environments (up to 350 km/h) and 1 Gb/s in stationary and pedestrian environments (up to 10 km/h).

5G: Next generation of mobile wireless broadband communication system, which will be able to magnify the capacity of wireless access networks for up to 10 Gbps (and hopefully beyond), increase the area spectral efficiency and energy efficiency, and provide a uniform quality of experience regardless of the position and features of the device being used.

Active Distribution Network: A distribution network with systems in place to control a combination of distributed energy resources comprising generators and storage.

Advanced Driver-Assistance System (ADAS): System developed for safety and improved driving that supports vehicles with monitoring, warning, braking, and steering tasks.

Advanced Message Queueing Protocol (AMQP): An open standard application layer protocol for message-oriented middleware.

Ambient Energy: Type of energy that is acquired through external portable sources like solar panels, wind, backup batteries, and generators.

Application Engineering: Within the scope of the Software Product Line Engineering paradigm, application engineering encompasses deriving particular applications built on top of a common platform.

Application Programming Interface (API): A set of protocols, tools, and definitions used for building software applications.

Artifacts: In software engineering, an artifact is one of many kinds of products produced during the development of a system, such as requirements specification, design models, code, and test cases.

Authentication: This process is part of a security triad known as authentication, authorization, and accounting (AAA). Authentication provides mechanisms to identify an entity (a device, a user, or a system process). The goal is to only admit legitimate entities into the ecosystem, to whom access to resources is allowed.

Authorization: Authenticated users are permitted to operate in the ecosystem according to established authorization to access certain resources (and not other resources) and what can (and cannot) be done with, or to, those resources.

Automatic Leakage Detection: The problem of automatically detecting the occurrence of leakages in the grid under study.

Back-End Data Sharing: A pattern that is used when there is a need to analyze combined data from several sources.

Blockchain: A distributed tamperproof ledger of transactions, as they progress through an ecosystem. The concept has been suggested for use in cybersecurity.

Bluetooth Low Energy (BLE): A power-conserving variant of Bluetooth; also known as Bluetooth Smart.

Building Automation and Control Network (BACnet): A communication protocol for building automation and control applications.

Capillary network: A short-range communication network able to collect data from low-power and usually tiny devices, and to interface them to long-range networks by means of gateways (i.e., intermediate devices).

Car Dealer: A person selling new or used cars at the retail level.

Chip: A chip (also known as a microchip) is a tiny electronic circuit on a semiconductor material.

Cloud Computing: A computing model that enables convenient, ubiquitous, on-demand access to a shared pool of configurable computing resources, usually accessible over the public Internet.

Code Division Multiple Access (CDMA): A channel access method used for radio communication technologies. It allows multiplexing of multiple signals to occupy one single channel.

Computational Intelligence: A set of nature-inspired computational methodologies and approaches to address complex real-world problems for which mathematical or traditional modeling can be useless.

Confidentiality: Assurance that no sensitive information is leaked.

Connected Car: Cars having access to the Internet and a range of sensors that are capable of sending and receiving data, recognizing the physical environment around them, and interacting with other cars or entities.

Constrained Application Protocol (CoAP): An application layer protocol in the Open Systems Interconnection model designed for resource-constrained devices that supports multicast and low overhead.

Cooperative Networking: Multifolded term embracing many different technologies, protocols, and algorithms, all sharing the common ambition to improve the efficiency of communication systems, thanks to some form of interaction among network nodes.

Credentials: A username and password pair used for authentication.

Data Classification: A supervised knowledge discovery process that uses the labeled data streams as input and classifies the new but similar data streams in single-class or multiclass datasets.

Data Clustering: An unsupervised knowledge discovery process that uses unlabeled data streams as input and produces new clusters.

Data Distribution Service (DDS): An application layer protocol used for real-time systems that addresses applications like financial trading, air traffic control, smart grid, and business applications.

Design Viewpoint: In software engineering, a design viewpoint combines some patterns, templates, and conventions for representing a view of a system from a perspective. Some examples include functional viewpoint, development viewpoint, and deployment viewpoint.

Destination-Oriented Directed Acyclic Graph (DODAG): A DODAG describes a directed acyclic graph with exactly one root, where a root is a node that has no outgoing edges.

Device–Cloud Communication: Data communication between IoT devices and cloud data centers. In this communication model, all IoT devices directly upload and download data streams in cloud environments.

Device/Edge-Centric Processing: A new data processing approach whereby all computing, networking, and storage services in IoT systems are orchestrated by considering IoT devices as a primary platform for application execution.

Device-to-Cloud Communication: A communication pattern that is used when data captured by the device from the environment is uploaded to an application service provider. Communication is based on the Internet protocol, but when the device manufacturer and the application service provider are the same, the integration of other devices may be difficult.

Device-to-Device Communication: A communication pattern that is applied when two devices communicate directly, normally using a wireless network (in the licensed spectrum).

Device-to-Gateway Communication Pattern: A communication pattern that may be used when the system contains non-Internet protocol devices, when support for legacy devices is needed or additional security functionality must be implemented.

Digital Player: A technology and software or Internet company, such as Apple, Google, and Microsoft.

Digitalized Device: An item or thing augmented with some computing power, which may or may not have the capability to communicate with the Internet and/or other digitalized devices.

Disruption: An abrupt, unforeseen change from an orderly sequence of events leading to a displacement of an existing industry, market, or technology.

Distributed Control: The control architecture adopts decentralized elements or functionalities to control the distributed components in systems.

Distributed Generation: An approach that employs small-scale technologies to produce electricity close to the end users of power. Distributed generation technologies often consist of modular (and sometimes renewable energy) generators, and they offer a number of potential benefits. In many cases, distributed generators can provide lower-cost electricity and higher power reliability and security with fewer environmental consequences than can traditional power generators.

Distributed IoT Analytics: An analytical system that supports the execution of analytic applications across multiple IoT devices and systems in parallel.

Domain engineering: Within the scope of the Software Product Line Engineering paradigm, domain engineering encompasses establishing a platform with common features.

Eavesdropping: The ability to capture and view data being passed between two points in a network.

e-Health: The use of information and communication technologies in support of healthcare delivery.

Electronic Health Record (EHR): Contains all the medical history data for patients.

Embedded connected car: Both connectivity and intelligence are directly built into the vehicle and focus on car-centric, high-reliability, and high-availability apps like security and safety-related services.

Enabling Technology: A technology that by itself or together with other technologies allows for significant improvements in the performance of users or in certain applications.

Encryption: A technique by which a plaintext message is encoded into a ciphertext, using an advanced algorithm and encryption key, in a manner such that only authorized personnel can read it by a decrypting process, using a decryption key, that reverses the encryption effect.

Encryption: The creation of an unintelligible text (ciphertext) for transmission or storage that can be rendered legible (into clear text) only by the party the has the decryption key.

Energy-Efficient Protocol: A network communication standard that routes data packets in a way that causes minimum node visits and allows nodes to sleep to save energy or power.

European Commission (EC): An institution of the European Union, responsible for proposing legislation, implementing decisions, and upholding the treaties of European Union.

Extensible Markup Language (XML): XML schemas are created to define the data structure of the XML-based input and output messages exchanged by web services.

Extensible Messaging and Presence Protocol (XMPP): An XML-based message-oriented protocol used for instant messaging.

Filtenna: An implementation of an antenna and an associated filter as a single structural unit.

Fog Computing: A model where data is stored in the most efficient location, often on the edge of the organization's computing network, rather than in remote cloud devices, improving speed of data access and removing the need for distributed data centers.

Future Internet Of Things: A concept whereby in the future all devices will be digitalized and connected to the IoT.

Global System For Mobile Communications (GSM): A standard developed by the European Telecommunications Standards Institute to describe the protocols for second-generation (2G) digital cellular networks.

Governance: The process of governing and deciding which laws and regulations will be implemented. Also, the process of deciding how regulations will be applied.

Health Insurance Portability and Accountability Act (HIPAA): A legislation passed in 1996 by the U.S. legislators to provide protection and security of medical information.

Heterogeneity: Where a system is made up of different technologies (e.g., different types of connected devices and software platforms).

Hypertext Transfer Protocol (HTTP): It is the communication protocol used to exchange content and data throughout the World Wide Web.

Identity Management: An administrative process related to identifying entities or individuals in a system (such as an ecosystem, a network, or an enterprise) and controlling their access (using authorization techniques) to resources within that system.

IEEE 802.15.4: A standard in the group of wireless personal area networks defined in 2003 that specifies the physical layer and media access control commonly used for low-power wireless sensor networks.

Inductive Coupling: An occurrence between two conductors placed close to each other in which a time-varying current flowing in one conductor induces a voltage across the ends of the other conductor.

Information and Communication Technology (ICT) Architecture: As described in ANSI/IEEE Standard 1471-2000, an **architecture** is "the fundamental organization of a system, embodied in its components, their relationships to each other and the environment, and the principles governing its design and evolution."

Institute of Electrical and Electronics Engineers (IEEE): A global association and organization of professionals working toward the development, implementation, and maintenance of technology-centered products and services.

Integrity: Assurance that the data being sent arrives in the same form.

Internet Engineering Task Force (IETF): A body that defines standard Internet operating protocols, such as TCP/IP.

Internet of Biometric Things (IoBT): A network of biometric IoT devices that can be used to monitor individuals or exchange biomedical information.

Internet of Things (Long Definition): A computing model made of uniquely identifiable (physical and virtual) objects that is able to capture their context (sensors) and transmit and/or receive data over the Internet and, in the case of actuators, is able to change its own state or the state of its surroundings, with or without very little direct human intervention.

Internet of Things (Short Definition): The concept whereby digitalized devices, both physical and virtual, are connected to the Internet and each other, without requiring the mediation of humans.

Internet Protocol Version 4 (IPv4): A 32-bit address connectionless protocol used in data communication networks.

Internet Protocol Version 6 (IPv6): A 64-bit address connectionless protocol used in data communication networks.

Intrinsic Authentication: Use of an internal property or characteristic for the purpose of specifically identifying a given individual or electrical component.

IoT Architecture: A systematic definition of functions, interfaces, and terminology for the IoT system under consideration, its decompositions and design patterns, and a defined vocabulary of terms to describe the specification of implementations. This permits standard protocols for the interfaces to be developed and allows design and implementation options to be compared.

IoT Component: Piece of an embedded system.

IoT Device: Embedded system treated as a single device.

IoT Element: Embedded system components that make up an element of a device.

IoT Environment: A collection of IoT devices working as a system or network or toward a common task.

IoT Hardware Development Platform: A physical component of an IoT development kit used for implementing prototypes.

IoT Security (IoTSec): The field of cybersecurity—principles, threats, mechanisms, and tools—specifically applied to the IoT ecosystem.

IoT System: Embedded components working together as a single system.

IoT System Development Method (SDM): An approach to develop a system systematically based on directions and rules. Since IoT systems are generally complex, and include many software, hardware, and communication components, the development of IoT systems requires systematic approaches (i.e., IoT SDMs) to be followed.

IPv6 over Low-Power Wireless Personal Area Networks (6LoWPAN): A standard that defines encapsulation and header compression mechanisms that allow IPv6 packets to be sent over IEEE 802.15.4–based networks.

IPv6 over Networks of Resource-Constrained Nodes (6Lo): Defines specifications for multiple constrained devices over a network.

Java Messaging Service (JMS): Provides a common platform for Java-based messaging between devices connected over a network.

JavaScript Object Notation (JSON): An open standard that enables the transmission of data objects that consist of attribute–value pairs between web services.

Key Management: The safekeeping and secure distribution of encryption keys to the intended stakeholders.

Knowledge Patterns: Intermediate or final results produced by data analytic applications during execution.

Larger Network: A network containing more elements than the current network.

Layered Architecture: An architecture that segments the various factions into nonoverlapping hierarchical groupings.

Life and Health (L/H) Insurance Firms: Insurance companies that primarily provide policies related to life, disability, indemnity, or supplemental health insurance. (This category excludes managed healthcare companies, which are typically included in the healthcare sector.)

Load Forecasting: The problem of predicting the temporal evolution of a certain load related to the grid under study.

Long Term Evolution (LTE): A standard for high-speed communication for handheld and mobile devices.

Low Earth Orbit (LEO) constellation: Satellite systems used in telecommunication, which orbit between 400 and 1000 miles above the earth's surface.

Low Power Long Range (LoRa): One of the IoT technologies that allows low-power devices to communicate at a long range.

Low-Power Wireless Personal Area Network (LoWPAN): A wireless network that is designed for a long range with power efficiency in mind. LoWPAN is a perfect example of energy-efficient IoT.

Machine Learning: A subfield of computer science that gives computers the ability to learn without being explicitly programmed.

Machine-to-Environment Communication: Communication or interaction between a device and the surrounding environment.

Machine-to-Human Communication: Communication between a device and a specific individual human being.

Machine-to-Machine (M2M) Communication: Communication between two devices (i.e., machines) that requires absolutely no human interaction.

Malware: Software intentionally used to damage the contents of a computer.

Medical Body Area Network (MBAN): A body area network that makes use of specially allocated frequency spectrum reserved for medical applications (specific secondary frequencies are also allowed).

Medical Regulation (U.S.): Medically related statutes published by the government or a government agency that mandate specified behavior (compliance) by industry stakeholders, such as doctors, hospitals, insurance companies, pharmaceutical companies, and medical equipment manufacturers.

Message Queue Telemetry Transport (MQTT): An application layer protocol and messaging publish/subscribe protocol optimized for resource-constrained devices and low-bandwidth, high-latency, and unreliable networks. MQTT was standardized by OASIS in 2013.

Mobility Service Provider: A provider for transportation service needs that is often associated with navigation, journey information, cashless payment, and access to transport services, such as a taxi, rail, and shared transport journeys.

Multiresolution Database: A time-series data store that allows the storage of higher resolutions of recent data that is aggregated with a suitable aggregation function to produce lower resolutions of the data for long-term retention of historical data.

Multitier Architecture: A hierarchical or layered representation of application components in order to perform distributed operations across systems.

Near-field Communication (NFC): A contactless mode of data transfer between two electronic devices.

Network Congestion: Too much traffic in a network reduces the availability of nodes, hence putting several packets in wait, which would lead to deadlocks.

Next Generation: New technology, method, or technique in a given field.

Onboard Processing: A data processing strategy that ensures application execution using onboard computational resources in IoT devices.

Open Systems Interconnection (OSI): A reference model for applications communication.

Open Systems IoT Reference Model (OSiRM): Proposed IoT architecture framework comprised of seven specified layers and in-layer security at all (or most) layers.

Original Equipment Manufacturer (OEM): Original manufacturer of a vehicle's component, that is, the components assembled and installed during the construction of a car.

Peak-to-Average Power Ratio: The ratio of the peak power to the average power in a waveform.

Personal Area Network (PAN): A wireless network that provides connectivity in a small area, usually 10 m. ZigBee and Bluetooth are two examples.

Physical Objects: All kinds of everyday objects that are present in our environments and that can contain sensors, actuators, and communication capability.

Potential Risks: Any issues or vulnerabilities that introduce risk to the system.

Printed Spiral Coil: Planar coil antennas, often fabricated as conductor loop turns on a dielectric substrate.

Privacy Standards: Privacy policies and decision making used to ensure the integrity and protection of personal or private information.

Propagation Model: A radio propagation model, also known as a radio-wave propagation model or radio-frequency propagation model, is an empirical mathematical formulation for the characterization of radio-wave propagation as a function of frequency, distance, and other conditions.

Property and Casualty (P/C) Insurance Firms: Insurance companies that provide policies to protect losses to (or of) physical assets, such as damage or loss to a home, car, or motorcycle.

Publish/Subscribe: A type of message-oriented middleware providing a distributed, asynchronous, loosely coupled many-to-many communication pattern between message producers and message consumers.

Q-Factor: The ratio of a conductor's reactive self-impedance to its resistive self-impedance, which is analogous to the ratio of the energy stored in the conductor to the energy dissipated in it.

Radio-Frequency Identification (RFID): A short-range radio technology commonly used to automatically identify and track objects.

Rectenna: An implementation of an antenna, its matching network, and a rectifier as a single structural unit.

Remote Patient Monitoring (RPM): The ability to monitor the vital signs of a patient, such as his or her temperature and blood pressure, from a far using technology-based platforms, and to send this sensed data in real time to a healthcare professional.

Secure Communication: Communication that cannot be sniffed, eavesdropped, manipulated, or blocked.

Self-reliant Mechanism: Ability of a device or node to invoke a mechanism or function on its own without using any external sources.

Service Innovation: Introduction of a new service based on novel technological skills.

Simple Object Access Protocol (SOAP): It defines a common messaging format used for request and response messages exchanged by web services.

Smart City: An urban development vision to integrate multiple information and communication technology and IoT solutions in a secure fashion to manage a city's assets.

Smart Water And Gas Grids: A set of technologies aimed at achieving smart management of the water and gas grids.

Smartphone Integration: Connectivity can be realized through integration between the car and the handset in which the communication module (e.g., subscriber identity module), on the one hand, and intelligence, on the other, are delivered by the phone.

Software Product Line Engineering (SPLE): A paradigm that aims to develop software by identifying commonalities and variabilities of a family of software products.

Supervisory Control And Data Acquisition (SCADA): An industrial computer system that monitors and controls a process. SCADA operates with coded signals over communication channels (using typically one communication channel per remote station).

Supplier: A party that supplies car manufacturers with single parts or complete assembly.

Telecom Player: Telecommunications or telephone companies and Internet service providers that play a relevant role in the development of mobile communications and the information society.

Telematics: A novel electronic technology that combines communication and information processes by utilizing computer networks.

Telematics Control Unit (TCU): An embedded system fitted in a vehicle to manage and control tracking of the vehicle.

Tethered Connected Car: A subscriber identity module, phone, or USB key is utilized to realize connectivity, while intelligence is embedded in the vehicle. This option is particularly appropriate for Internet-based infotainment features and connected navigation.

Third Generation Partnership Project (3GPP): A collaborative project aimed at developing globally acceptable specifications for third-generation (3G) mobile systems.

Third Party: A group or individual that is not the client or server, and is involved in communication or authentication.

Time-Slotted Channel Hoping (TSCH): A medium access scheme for lower power and reliable networking solutions in low-power lossy networks (LLNs).

Topological Settings: The physical or logical distribution of devices across systems.

Transmission Control Protocol (TCP): A transport layer protocol in the Open Systems Interconnection model that defines data transmission between two nodes connected over a network.

Trusted Execution Environment (TEE): A secure ecosystem of the IoT device processor or of the application system, offering the capability of isolated execution of authorized security software.

Trusted Execution Technology (TXT): A scalable architecture that specifies hardware-based security protection; the capabilities are built into Intel's chipsets to address threats across physical and virtual infrastructures.

Ultra-High Frequency (UHF): Radio frequencies in the range between 300 MHz and 3 GHz.

Unique selling point (USP): A real or perceived benefit that differentiates a product or a service from its competitors.

Universal Mobile Telecommunications Service (UMTS): A third-generation (3G) broadband and packet-based transmission of text, digitized voice, video, and multimedia at data rates up to 2 Mbps.

Usage-Based Insurance (UBI): A mechanism utilized by automobile insurers that assesses driving behaviors (e.g., miles driven, time of day, average speed, and aggressiveness in acceleration and/or braking) and allows the insurer to tailor premium rates to the individual (and his/her driving behavior).

User Datagram Protocol (UDP): A transport layer protocol in the Open Systems Interconnection model and an unreliable connectionless protocol that is used to send short messages between connected devices.

Vehicular ad hoc Network (VANET): Network of interacting vehicles that offers advanced services to passengers and drivers, including mobile multimedia streaming, intelligent transportation system functionalities, and broadcasting.

Virtual Objects: Exist in the information world and can be stored, accessed, and processed.

Wireless Body Area Network (WBAN): A network that is formed by a collection of low-power small sensory units attached to the body of a person or sometimes implanted in him or her.

Wireless Energy Harvesting: A scheme to obtain usable electrical power from ambient electromagnetic radiation.

Wireless Local Area Network (WLAN): A network that allows devices to communicate wirelessly within a limited area.

Wireless Local Area Network (WLAN): Commonplace technology enabling wireless connectivity in local area networks. The leading standard in this field is IEEE 802.11, which targets high-speed communications in star topologies (albeit multihop communications can be set up too).

Wireless Power Transfer: The intentional transmission and reception of electromagnetic energy as a means of transferring electrical power.

Wireless Sensor Network (WSN): Commonplace technology enabling low-power and short-range wireless connectivity in monitoring infrastructure. The leading standard in this field is IEEE 802.15.4.

Worldwide Interoperability for Microwave Access (WiMax): WiMAX is a wireless technology used for fourth-generation (4G) networks and belongs to the IEEE 802.16 family of wireless standards.

Index

Actuation, and application layer, 221
Adaptation layer, and multilayer application architecture, 220
Addressed discipline, 157
Advanced Message Queuing Protocol (AMQP), 187
Ambient energy harvesting, 99–100
  routing challenges, 100–102
AMQP. *see* Advanced Message Queuing Protocol (AMQP)
Anomalous operational condition, 267–269
Application layer protocols, 184–185
Arduino microcontrollers
  battery life of
    past, 117–118
    presesnt, 126–127
  connectivity and I/O interfaces
    past, 115–116
    presesnt, 125
  OS support for
    past, 117
    presesnt, 126
  security features of
    past, 119
    presesnt, 128
  size and cost of
    past, 118
    presesnt, 127
Area-specific IoT, 96–97
Artifacts, 152–153
AuRA-NMS. *see* Autonomous regional active network management system (AuRA-NMS)
Authentication
  behavioral, 84–85
    localization and metadata, 85
  description, 72–73
  hardware tokens
    connected, 81–82
    contactless, 82
    disconnected, 82
  intrinsic
    human properties, 83–84
    silicon properties, 84
  next-generation
    Blockchain, 86–87
    CryptoPhoto, 86
    Fast IDentity Online (FIDO), 85–86
  one-time passwords (OTPs)
    challenge–response-based OTP (dynamic password), 75
    lockstep-based OTP, 76
    out-of-band transmission-based OTP, 75–76
    time-based OTP (TOTP), 74–75
  overview, 71–72
  shared secret (static password), 73–74
  software tokens, 77–80
    key exchange, 78–79
    SSL or certificate exchange, 77–78
    third-party, 79–80
Automotive industry, IoT in
  challenges
    business models and control points, 384
    customer relationship, 384
    innovation cycle time, 383
    network effect, 383
    privacy and security, 383–384
  data sources, 364–365
  digital transformation
    connected car, 368–369
    as disruptive force, 365–366
    ecosystem of connected car, 369–370
    in product-based industries, 366–368
  overview, 363–364
  trends in, 370–383
Autonomous regional active network management system (AuRA-NMS)
  conceptual architecture, 234–235
  distributed control in, 235
  distributed intelligence and function integration, 236–237
  and legacy systems, 237
  unification of system information and standards, 236

Backscatter RFID, 164
Behavioral authentication, 84–85
  localization and metadata, 85
Blockchain, 86–87
Body area networks, 337

CARDAP architecture
  distributed data processing, 223–224
  multiple data processing strategies, 224
Cellular solutions (4G/5G), 338–339
Certificate exchange/SSL, 77–78
Challenge-response-based OTP (dynamic password), 75
Cisco reference model, 14–15
Cloud-centric IoT
  application requirements
    capturing, 207–208
    distributing, 208–210
    network parameters, 210
  overview, 195–197
  publish/subscribe architecture, 197–201
  sensor advertisements with metadata, 201–204
  sensor aggregates, 204–205
  sensor search, 206–207
Cloud computing, and RPM, 310–313
Coil antenna, 167–169
Communication performance assessment
  anomalous and emergent operational condition, 267–269
  data traffic modeling, 264
  normal operational condition, 265–267
Communication standards and protocols, 237–240
Communication technologies
  cooperative networking techniques
    device-to-device (D2D) communications, 53–54

in 5G networks, 52–53
in 4G networks, 50–52
for electric distribution networks
communication security, 242
data availability, robustness, and redundancy, 241
flexibility, scalability, and interoperability, 241–242
timely data delivery and differentiation, 241
for smart grids
electromagnetic issues, 282–289
IEEE 802.11ah, 282
low-power wide area networking (LPWAN), 280–281
overview, 277–278
power consumption and management, 289–291
wireless metering bus, 278–280
Community-based MOSDEN development, 223
Connected car, 368–369
ecosystem of, 369–370
Connected hardware tokens, 81–82
Contactless hardware tokens, 82
Cooperative networking techniques
cellular systems
device-to-device (D2D) communications, 53–54
in 5G networks, 52–53
in 4G networks, 50–52
crowd-sourcing systems, 58
overview, 49–50
in VANETs, 58–62
wireless networks with energy harvesting capabilities, 62–64
in WLANs, 55–56
in WSNs, 56–58
Crowd-sourcing systems, 58
CryptoPhoto, 86

Data acquisition, and multilayer application architecture, 220
Data analytics, and multilayer application architecture, 220
Data availability, for electric distribution networks, 241
Data fusion, and data management layer, 220
Data management layer, 220
Data preprocessing, and multilayer application architecture, 220
Data reduction, 225
Data traffic modeling, 264
D2D. *see* Device-to-device (D2D) communications
Descriptive analytics, 216
Design viewpoints, of SDMs, 156
Device-centric data analytics, 227
Device-centric IoT systems
immobile, 217–218
mobile, 218
Device-to-device (D2D) communications, 53–54
Digital transformation
connected car, 368–369
ecosystem of, 369–370
as disruptive force, 365–366
in product-based industries, 366–368
Disconnected hardware tokens, 82
Distributed control, for smart distribution grid, 257–258
Distributed control, in AuRA-NMS, 235
Distributed data processing
CARDAP architecture, 223–224
Distributed data processing
UuniMiner architecture, 225
Distributed intelligence, and AuRA-NMS, 236–237
Distributed processing
MOSDEN architecture, 222–223
Documentation of method, 158
Dynamic password. *see* Challenge-response-based OTP (dynamic password)

Ecosystem, of connected car, 369–370
Edge-centric distributed IoT analytics
CARDAP architecture
distributed data processing, 223–224
multiple data processing strategies, 224
device-centric IoT systems
immobile, 217–218
mobile, 217–218
MOSDEN architecture
community-based development, 223
distributed processing, 222–223
scalability, 223
multilayer application architecture
actuation and application layer, 221
data acquisition and adaptation layer, 220
data analytics and knowledge integration layer, 220
data preprocessing, fusion, and data management layer, 220
data stream layer, 219
security and privacy-preserving data sharing layer, 220–221
system management, 221
overview, 213–215
role of
descriptive analytics, 216
overview, 215–216
predictive analytics, 216
prescriptive analytics, 216
preventive analytics, 216–217
UniMiner architecture
data reduction, 225
device-centric data analytics, 227
distributed data processing, 225
load balancing, 225–227
E-health system monitoring architecture, 304–305
ELDAMeth. *see* Event-driven lightweight distilled state charts-based agents methodology (ELDAMeth)
Electric distribution networks, IoT in
AuRA-NMS management
conceptual architecture, 234–235
distributed control in, 235
distributed intelligence and function integration, 236–237
and legacy systems, 237
unification of system information and standards, 236
case studies
35 kV meshed networks, 243–245
13 kV radial networks, 245–348
communication infrastructure requirements
communication security, 242
data availability, robustness, and redundancy, 241
flexibility, scalability, and interoperability, 241–242
timely data delivery and differentiation, 241
communication standards and protocols, 237–240
current control and communication provision in DNOS, 231–234
overview, 229–231

Electromagnetic issues, 282–289
13 kV radial networks, 245–348
Emergent operational condition, 267–269
E/m-health security
challenges in, 324–326
hardware level
Intel® TXT, 342–344
trusted execution environment (TEE), 339–342
layer-oriented mechanisms
body area networks, 337
cellular solutions (4G/5G), 338–339
hybrid (home–public) hotspot networks, 338
low-power wide area, 338
overall (all layers), 337
personal area networks (PANs), 337–338
security considerations, 339
near-term trends, 344
OSiRM architecture, 329–336
overview, 321–324
regulatory requirements, 326–329
Energy-efficient IoT
ambient energy harvesting, 99–100
routing challenges, 100–102
existing solutions for
area-specific IoT, 96–97
energy-efficient scheduling, 94–96
routing protocols, 93–94
wireless sensor networks (WSNs), 96
WLAN-based IoT, 98–99
WPAN-based IoT, 99
WWAN-based IoT, 97–98
open research issues for, 102–104
overview, 91–92
Energy-efficient scheduling, 94–96
Evaluation, of SDMs
addressed discipline, 157
artifacts, 152–153
coverage of elements, 156
design viewpoints, 156
documentation of method, 158
life cycle activities, 153–156
maturity of method, 157–158
metrics, 156–157
process flow, 153
process paradigm, 157
rigidity of method, 157
scope, 157
stakeholder concern coverage, 156
tool support, 158
Event-driven lightweight distilled state charts-based agents methodology (ELDAMeth), 147–148

Fast IDentity Online (FIDO), 85–86
FIDO. *see* Fast IDentity Online (FIDO)
5G networks, 52–53
Flexibility, and electric distribution networks, 241–242
4G networks, 50–52
Function integration, and AuRA-NMS, 236–237

General software engineering methodology (GSEM), 150–151
GSEM. *see* General software engineering methodology (GSEM)

Hardware development platforms
battery life of
future, 133
past, 117–118
present, 126–127
connectivity and communication interfaces of
future, 132
past, 115–116
present, 125–126
description, 109–110
evolution timeline, 134
memory and storage capacity
future, 131
past, 114–115
present, 123–125
OS support for
future, 132–133
past, 117
present, 126
overview, 108–109, 111–112, 119–122, 128–131
processing power of
future, 131
past, 112–114
present, 122–123
related work, 110–111
security features of
future, 134
past, 118–119
present, 128
size and cost of
future, 133–134
past, 118
present, 127–128
Hardware tokens
connected, 81–82
contactless, 82
disconnected, 82
High-latency and bandwidth-constrained device landscape, 186–187
Human authentication properties, 83–84
Hybrid (home-public) hotspot networks, 338

IEEE 802.11ah communication, for smart grids, 282
IEEE 802.24 (LORA), 188
IEEE P2413, 12
Ignite | IoT Methodology, 145–146
Immobile device-centric IoT systems, 217–218
Inductive coupling RFID, 163–164
wireless power transfer through, 164–172
coil antenna, 167–169
general system architecture, 165–167
link transfer efficiency, 169–170
multiobjective link considerations, 170–172
spatial freedom, 170
Insurance industry, IoT in
architectures, 352–354
challenges in, 351–352
description, 349–351
layered security
hardware level, 357–358
lower layers, 357
overall (all layers), 357
overview, 354–356
overview, 347–349
traffic characteristics, 358–360
Intel® TXT, 342–344
Internet of Things (IoT)
application areas

energy, 24
environment/smart planet, 24
healthcare, 21–23
mobility and transportation, 23
smart agriculture, 24
smart cities, 21
smart homes and smart buildings, 23
smart manufacturing, 24
architectures and reference models
Cisco reference model, 14–15
IEEE P2413, 12
industrial reference architectures, 12–14
IoT-A project, 10–11
IoT RA, 11–12
overview, 8–10
reference layered architecture, 15–16
challenges
failure handling, 27
interoperability, 25
openness, 25–26
scalability, 26–27
security, privacy, and trust, 26
communication patterns and protocols, 17–18
definition of, 5–8
devices and test beds, 18–20
identification and discovery, 16–17
overview, 3–5
Interoperability
and Advanced Message Queuing Protocol (AMQP), 187
application layer protocols, 184–185
constraints, 180–181
and electric distribution networks, 241–242
high-latency and bandwidth-constrained device landscape, 186–187
and IEEE 802.24 (LORA), 188
and Java-based platforms, 187–188
link layer protocols, 181–182
low-power device landscape, 185–186
messaging-oriented applications, 187
network layer protocols, 183
organizational implementation and management, 35–36
overview, 179–180
real-time data and multicast abilities, 186
transport layer protocols, 183–184
web-based IoT application, 186
Intrinsic authentication
human properties, 83–84
silicon properties, 84
IoT. *see* Internet of Things (IoT)
IoT-A project, 10–11
IoT Methodology, 146–147
IoT RA, 11–12
IoT system development methods (SDMs)
application development, 147
concepts, 143–144
description, 142–143
evaluation of
addressed discipline, 157
artifacts, 152–153
coverage of elements, 156
design viewpoints, 156
documentation of method, 158
life cycle activities, 153–156
maturity of method, 157–158
metrics, 156–157
process flow, 153
process paradigm, 157
rigidity of method, 157
scope, 157
stakeholder concern coverage, 156
tool support, 158
event-driven lightweight distilled state charts-based agents methodology (ELDAMeth), 147–148
general software engineering methodology (GSEM), 150–151
Ignite | IoT Methodology, 145–146
IoT Methodology, 146–147
overview, 141–142
software product line engineering (SPLE), 148–150

Java-based platforms, and interoperability, 187–188

Key exchange, 78–79
Knowledge integration layer, and multilayer application architecture, 220

Layered security
in e/m-health applications
body area networks, 337
cellular solutions (4G/5G), 338–339
hybrid (home–public) hotspot networks, 338
low-power wide area, 338
OSiRM model, 337
personal area networks (PANs), 337–338
security considerations, 339
in insurance industry
hardware level, 357–358
lower layers, 357
overall (all layers), 357
overview, 354–356
Leakage and fault detection, 294–296
Legacy systems, and AuRA-NMS, 237
LEO constellation, 258–259
LEO network model, 259–263
Life cycle activities, 153–156
Link layer protocols, 181–182
Link transfer efficiency, 169–170
Load balancing, 225–227
Load forecasting, 292–294
Localization, and metadata, 85
Lockstep-based OTP, 76
Low-power device landscape, 185–186
Low-power wide area, 338
Low-power wide area networking (LPWAN), 280–281
LPWAN. *see* Low-power wide area networking (LPWAN)

Machine learning
leakage and fault detection, 294–296
load forecasting, 292–294
Medical sensors, in RPM, 305–306
Messaging-oriented applications, 187
Metadata
and localization, 85
sensor advertisements with, 201–204
Method maturity, of SDMs, 157–158
Method rigidity, of SDMs, 157

Metrics, of SDMs, 156–157
Mobile device-centric IoT systems, 217–218
MOSDEN architecture
community-based development, 223
distributed processing, 222–223
scalability, 223
Multilayer application architecture
actuation and application layer, 221
data acquisition and adaptation layer, 220
data analytics and knowledge integration layer, 220
data preprocessing, fusion, and data management layer, 220
data stream layer, 219
security and privacy-preserving data sharing layer, 220–221
system management, 221
Multiobjective link considerations, 170–172
Multiple data processing strategies, 224

Network layer protocols, 183
Next-generation authentication
Blockchain, 86–87
CryptoPhoto, 86
Fast IDentity Online (FIDO), 85–86
Normal operational condition, 265–267

One-time passwords (OTPs)
challenge–response-based OTP (dynamic password), 75
lockstep-based OTP, 76
out-of-band transmission-based OTP, 75–76
time-based OTP (TOTP), 74–75
Open research issues, for energy-efficient IoT, 102–104
Organizational implementation and management
data management, 41–42
description, 34–35
interoperability, 35–36
legislation and governance, 42
overview, 33–34
privacy, 37–38
security, 38–39
standards, 36–37
trust, 39–41
OSiRM architecture, 329–336
OTPs. *see* One-time passwords (OTPs)
Out-of-band transmission-based OTP, 75–76

PANs. *see* Personal area networks (PANs)
Personal area networks (PANs), 337–338
Power consumption and management, 289–291
Predictive analytics, 216
Prescriptive analytics, 216
Preventive analytics, 216–217
Process flow, of SDMs, 153
Process paradigm, of SDMs, 157
Product-based industries, digital transformation in, 366–368
Publish/subscribe architecture, 197–201

Radio-frequency energy harvesting
design considerations, 173
general system architecture, 172–173
rectenna design, 173–174
Radio-frequency identification (RFID)
backscatter, 164
inductive coupling, 163–164
coil antenna, 167–169
general system architecture, 165–167
link transfer efficiency, 169–170
multiobjective link considerations, 170–172
spatial freedom, 170
wireless power transfer through, 164–172
overview, 161–162
radio-frequency energy harvesting
design considerations, 173
general system architecture, 172–173
rectenna design, 173–174
Real-time data and multicast abilities, 186
Rectenna design, 173–174
Redundancy, and electric distribution networks, 241
Reference IoT layered architecture, 15–16
Regulatory requirements, for e/m-health security, 326–329
Remote patient monitoring (RPM)
clinical applications, 307
and cloud computing, 310–313
e-health system monitoring architecture, 304–305
industrial platforms in support of, 308
medical sensors in, 305–306
overview, 303–304
penetration issues, 315–317
security and privacy in, 313–315
wireless body area network, 309–310
RFID. *see* Radio-frequency identification (RFID)
Robustness, and electric distribution networks, 241

Satellite-based IoT
communication performance assessment
anomalous and emergent operational condition, 267–269
data traffic modeling, 264
normal operational condition, 265–267
distributed control for smart distribution grid, 257–258
LEO constellation, 258–259
LEO network model, 259–263
overview, 253–257
SBCs. *see* Single-board computers (SBCs)
Scalability
MOSDEN architecture, 223
Scalability, and electric distribution networks, 241–242
SDMs. *see* IoT system development methods (SDMs)
Sensor aggregates, 204–205
Sensor metadata advertisements, 201–204
Sensor search, for cloud-centric IoT, 206–207
Shared secret (static password), 73–74
Silicon authentication properties, 84
Single-board computers (SBCs)
battery life of
past, 118
presesnt, 127
connectivity and I/O interfaces for
past, 116
presesnt, 125–126
OS support for
past, 117
presesnt, 126
security features of
past, 119
presesnt, 128

size and cost of
past, 118
presesnt, 127–128
Smart city context, 275–276
Smart distribution grid, distributed control for, 257–258
Smart gas and water grids
communication technologies for
electromagnetic issues, 282–289
IEEE 802.11ah, 282
low-power wide area networking (LPWAN), 280–281
overview, 277–278
power consumption and management, 289–291
wireless metering bus, 278–280
description, 276–277
machine learning for
leakage and fault detection, 294–296
load forecasting, 292–294
overview, 273–275
smart city context, 275–276
Software product line engineering (SPLE), 148–150
Software tokens, 77–80
key exchange, 78–79
SSL or certificate exchange, 77–78
third-party, 79–80
Spatial freedom, and RFID, 170
SPLE. *see* Software product line engineering (SPLE)
Stakeholder concern coverage, 156
Static password. *see* Shared secret (static password)

TEE. *see* Trusted execution environment (TEE)
Third-party software tokens, 79–80
35 kV meshed networks, 243–245
Time-based OTP (TOTP), 74–75
Tokens (security)
hardware
connected, 81–82
contactless, 82
disconnected, 82
software, 77–80
key exchange, 78–79
SSL or certificate exchange, 77–78
third-party, 79–80
Tool support, for SDMs, 158
TOTP. *see* Time-based OTP (TOTP)
Transport layer protocols, 183–184
Trusted execution environment (TEE), 339–342

UniMiner architecture
data reduction, 225
device-centric data analytics, 227
distributed data processing, 225
load balancing, 225–227

VANETs. *see* Vehicular ad hoc networks (VANETs)
Vehicular ad hoc networks (VANETs)
cooperative networking in, 58–62

Web-based IoT application, 186
Wireless body area network, 309–310
Wireless local area networks (WLANs)
-based IoT, 98–99
cooperative networking techniques in, 55–56
Wireless metering bus, 278–280
Wireless power transfer through RFID, 164–172
coil antenna, 167–169
general system architecture, 165–167
link transfer efficiency, 169–170
multiobjective link considerations, 170–172
spatial freedom, 170
Wireless sensor networks (WSNs), 96
cooperative networking in, 56–58
WLANs. *see* Wireless local area networks (WLANs)
WPAN-based IoT, 99
WSNs. *see* Wireless sensor networks (WSNs)
WWAN-based IoT, 97–98

About the Contributors

Raafat Aburukba earned his bachelor's degree in computer science and software engineering, and his master's degree and PhD in electrical and computer engineering from the University of Western Ontario, London, Ontario, Canada. He is currently an assistant professor of computer science and engineering at the American University of Sharjah, United Arab Emirates. His research interests include various topics in cloud computing, business intelligence for healthcare, cooperation and coordination in distributed systems, economic-based models, and approaches for decentralized scheduling and applied to cloud computing, the Internet of Things, and smart spaces.

Fadi Aloul earned his BS degree in electrical engineering (summa cum laude) from the Lawrence Technological University, Southfield, Michigan, and his MS and PhD degrees in computer science and engineering from the University of Michigan, Ann Arbor. He has published more than 110 research papers in international conferences, workshops, and journals. He is currently a professor of computer science and engineering and the director of the HP Institute at the American University of Sharjah, United Arab Emirates.

Karolina Baras has been an assistant professor at the University of Madeira, Portugal, since 2012, and a researcher at Madeira Interactive Technologies Institute, Portugal, since 2015. She earned a PhD degree in technologies and information systems in 2012, from the University of Minho, Portugal. Her research interests are in the field of ubiquitous computing in general and in the areas of the Internet of Things, smart cities, and positive technologies.

Gennaro Boggia earned the Dr.Eng. (Hons.) and PhD (Hons.) degrees from the Politecnico di Bari, Italy, in 1997 and 2001, respectively, both in electronics engineering. Since September 2002, he has been with the Department of Electrical and Information Engineering, Politecnico di Bari, where he is currently an associate professor. From May to December 1999, he was a visiting researcher at the Telecom Italia Lab (TILab), Turin, Italy, where he was involved in the study of the core network for the evolution of 3G cellular systems. In 2007, he was a visiting researcher with FTW, Vienna, Austria, where he was involved in activities on passive and active traffic monitoring in 3G networks. He has authored or coauthored more than 100 papers in international journals or conference proceedings. His research interests include wireless networking, cellular communication, information-centric networking, the Internet of Things, protocol stacks for industrial applications and smart grids, Internet measurements, and network performance evaluation. Dr. Boggia currently serves as an associate editor for the Springer *Wireless Networks* journal.

Lina M. L. P. Brito earned a degree in systems and computing engineering from the University of Madeira (1999), Portugal, and an MSc degree in electronics and telecommunication engineering from the University of Aveiro (2004), Portugal. She earned a PhD degree in distributed systems and networks, obtained in 2011, from the University of Madeira. She is an assistant professor of the Exact Sciences and Engineering Faculty of the University of Madeira and has been a researcher at Madeira Interactive Technologies Institute since 2014. Her research interests are in the areas of networking, network management, wireless sensor networks, and the Internet of Things.

Steve Cayzer, PhD, works as a senior teaching fellow at the University of Bath, United Kingdom. Steve teaches a range of courses in innovation, product development, project management, and sustainability. His main research interests include innovation (particularly sustainable innovation), knowledge management, and education (particularly MOOCs).

John A. Chandy is a professor and the associate head of the Electrical and Computer Engineering Department at the University of Connecticut. Prof. Chandy is also codirector of the Connecticut Cybersecurity Center; interim director of the University of Connecticut Center for Hardware Assurance, Security, and Engineering; and codirector of the Comcast Center for Cybersecurity Innovation. Prior to joining the University of Connecticut, he had executive and engineering positions in software companies, working particularly in the areas of clustered storage architectures, tools for the online delivery of psychotherapy and soft-skills training, distributed architectures, and unstructured data representation. His current research areas are in high-performance storage systems, reconfigurable computing, embedded systems security, distributed systems software and architectures, and multiple-valued logic. Dr. Chandy earned PhD and MS degrees in electrical engineering from the University of Illinois at Urbana–Champaign in 1996 and 1993, respectively, and an SB in 1989.

Ming Cheng earned her BS degree in automation from Zhejiang University, Hangzhou, China, in 2016. She is currently pursuing an MSc degree at Imperial College London. Her research interests include distribution energy resources, virtual power plants, and renewable energy.

Acácio Filipe Pereira Pinto Correia recently earned his master's degree in computer science and engineering, at the Universidade da Beira Interior, Covilhã, Portugal. His dissertation focused on the study of natural language processing techniques and scientific document suggestions according to the context. He previously earned a bachelor's degree in computer science and engineering from the same university. His professional interests include procedural generation, artificial intelligence, natural language processing, cryptography, and video game development.

Adelmo De Santis was born in Velletri, Rome, in 1976. He has been working as a technician (network and system administrator) at the Università Politecnica delle Marche, Italy, since 2001. In 2006, he earned the Italian laurea degree (graduate diploma) in electronic engineering, and in 2015 a PhD in electronic engineering and telecommunications from the same university. His research interests are in the fields of Extremely High Frequencies channel characterization, radio-frequency systems, measurement instruments, and Software Defined Radio. He has been a ham radio enthusiast since 1989, and he is the author of some publications in this field. Adelmo has been a Huawei Certified Network Associate (HCNA H12-211) since July 2016.

Valentina Di Mattia was born in Teramo, Italy, in 1984. She earned her MS degree (cum laude) in clinical and biomedical engineering from "Sapienza," University of Rome, in 2009 and her PhD in bioengineering and electromagnetism from the Università Politecnica delle Marche, Ancona, Italy, in 2014. Currently, she is a research fellow at the Department of Information Engineering of the same university, and her main research activities focus on the design of electromagnetic devices for contactless monitoring of human vital signs, indoor localization and tracking, and small systems to support the autonomous walking, or running, of visually impaired people and athletes. In particular, her expertise concerns the simulation, optimization, and testing of small antennas working at high frequencies. She is a member of Siem, the Italian society of electromagnetism.

Ali Ehsan earned his BS degree in electrical engineering from the University of Engineering and Technology, Lahore, Pakistan, in 2010. Afterwards, he obtained the MSc degree in renewable energy engineering from Kingston University, London, in 2012. He worked as a lecturer of the Department of Electrical Engineering, COMSATS Institute of Information Technology, Sahiwal, Pakistan, from 2013 to 2015. Currently, he is working toward his PhD degree at the College of Electrical Engineering, Zhejiang University, Hangzhou, China. His research interests include renewable energy systems, smart grids, distributed generation planning, and optimization.

Akaa Agbaeze Eteng earned a B.Eng. degree in electrical/electronic engineering from the Federal University of Technology Owerri, Nigeria, in 2002, and an M.Eng. in telecommunications and electronics from the University of Port Harcourt, Nigeria, in 2008. In 2016, he earned a PhD in electrical engineering from the Universiti Teknologi Malaysia, Skudai. Currently, he is a lecturer in the Department of Electronic and Computer Engineering, University of Port Harcourt, Nigeria. His research interests include wireless energy transfer, radio-frequency energy harvesting, and wireless-powered communications.

Marco Fagiani was born in Fermo, Italy, on June 1985. He earned his BSc and MSc degrees (with honors) in electronics engineering in 2010 and 2012, respectively, and his PhD degree in information engineering in 2016 from the Università Politecnica delle Marche, Italy. He is currently enrolled as a postdoctoral researcher at the Università Politecnica delle Marche. His current research interests are in the areas of machine learning and computational intelligence, with a focus on smart grids' load forecasting, data analysis, and energy management.

Jim Fahrny is one of the icons of security knowledge in U.S. cable, as well as the U.S. broadcast media market as a whole. Fahrny has been the driving force behind the Xfinity X-1 security architecture, Electronic Sell Through (EST), TV Everywhere (TVE), university deployments, and several other security initiatives within Comcast. He was the driving force behind the Downloadable Conditional Access System architecture within Comcast back in 2006. He also worked on military aircraft, inertial navigation and GPS, submarines, sonar, weapon systems, ballistic missiles, C3I, mission planning, computer graphics, and system architectures. Jim has an extensive background and experience in conditional access systems for video security, including hardware key management, digital rights management, home networking security, and security protocols and standards. Fahrny is the author of more than 20 patents for security related to video distribution, content protection systems, and digital rights management. Fahrny graduated with a BS in electrical engineering from California Polytechnic University, Pomona.

Mário Marques Freire earned a 5-year BS degree in electrical engineering and a 2-year MS degree in systems and automation in 1992 and 1994, respectively, from the University of Coimbra, Portugal. He earned his PhD degree in electrical engineering in 2000 and the habilitation title in computer science in 2007 from the University of Beira Interior, Portugal. He is a full professor of computer science at the University of Beira Interior, which he joined in the fall of 1994. When he was an MS student at the University of Coimbra, he was also a trainee researcher for a short period in 1993 in the Research Centre of Alcatel-SEL (now Alcatel-Lucent) in Stuttgart, Germany.

His main research interests fall within the broad area of computer systems and networks, including network forensics and Internet traffic classification, security and privacy in computer systems, peer-to-peer networks, and cloud systems. He is the coauthor of seven international patents, the coeditor of eight books published in the Springer Lecture Notes in Computer Science book series, and the author or coauthor of about 120 papers in refereed international journals and conferences. He serves as a member of the editorial board of *ACM SIGAPP Applied Computing Review*, as associate editor of the Wiley journal *Security and Communication Networks*, and as associate editor of the Wiley *International Journal of Communication Systems*. In the past, he served as an editor of *IEEE Communications Surveys and Tutorials* (2007–2011) and as a guest editor of two feature topics in *IEEE Communications Magazine* (2008) and of a special issue of the Wiley *International Journal of Communication Systems* (2009). He served as a technical program committee member for several IEEE international conferences and is cochair of the track on Networking of ACM SAC 2017. Dr. Mario Freire is a chartered engineer by the Portuguese Order of Engineers, a member of the IEEE Computer Society and the IEEE Communications Society, and a member of the Association for Computing Machinery.

Görkem Giray is a software engineer and an independent researcher based in Izmir, Turkey. Dr. Giray earned his BSc degree (1999) and PhD degree (2011) in computer engineering from Ege University, İzmir, Turkey. He obtained his MBA degree from Koç University, Istanbul, in 2001. He has been working in various software engineering positions in the private sector since 2000. He has also been delivering courses in software engineering since 2013.

James Gleason is a seasoned professional with a rich background in sales, marketing, and business development. Gleason is responsible for new business and market development. He enjoys a successful and proven track record of providing innovative technology solutions to customers across multiple mission-critical industries, such as global financial services, the public sector (including state and local government, the federal government, healthcare, and higher education), and major publishing industries. Gleason has more than 20 years of experience providing business and management consulting services encompassing leading-edge technological innovation.

Luigi Alfredo Grieco is an associate professor in telecommunications at the Politecnico di Bari, Italy. He was a visiting researcher with INRIA, Sophia Antipolis, France, in 2009, and with LAAS-CNRS, Toulouse, France, in 2013, where he was involved with Internet measurements and machine-to-machine systems, respectively. He has authored more than 100 scientific papers published in international journals and conference proceedings of great renown that gained more than 1000 citations. His research interests include Transmission Control Protocol congestion control, quality of service in wireless networks, the Internet of Things (IoT), and future Internet. Prof. Grieco is the editor in chief for *Transactions on Emerging Telecommunications Technologies* (Wiley) and serves as an editor for *IEEE Transactions on Vehicular Technology* (for which he was recognized as the top associate editor in 2012). Within the Internet Engineering Task Force and Internet Research Task Force, he actively contributes to the definition of new standard protocols for industrial IoT applications and new standard architectures for tomorrow information-centric networking–IoT systems.

Muhammad Habib ur Rehman is working on big data mining systems for the Internet of Things. His research covers a wide spectrum of application areas, including smart cities, mobile social networks, quantified self, and m-health. The key research areas of his interest are mobile computing, edge cloud computing, the Internet of Things, and mobile distributed analytics. Currently, he has 10 publications to his credit, including 09 Institute for Scientific Information (ISI) (7 Q1 and 2 Q2) journals, 4 Institute of Electrical and Electronics Engineers conference proceedings, and a book chapter.

Vlado Handziski is a senior researcher in the Telecommunication Networks Group at the Technische Universität Berlin, where he coordinates activities in the areas of sensor networks, cyber-physical systems, and the Internet of Things. He earned his doctoral degree in electrical engineering from Technische Universität Berlin (summa cum laude, 2011) and his MSc degree from Saints Cyril and Methodius University in Skopje, Macedonia (2002). Dr. Handziski's research interests are mainly focused on testing and the software architecture aspects of networked embedded systems. He has led test bed infrastructure activities in several large European research projects, like Embedded WiSeNts, CONET, EVARILOS, and EIT ICT Labs. Dr. Handziski is member of the Institute of Electrical and Electronics Engineers and the Association for Computing Machinery.

Daniel Happ earned his MSc degree in computer science in 2013 from Free University of Berlin, Germany. He is now a PhD candidate at the Telecommunication Networks Group at the Technical University of Berlin. His research interests include large-scale cloud-connected sensor networks, publish and subscribe publish-and-subscribe messaging, and fog computing.

Asad Haque has more than 25 years of progressive responsibilities in information technology specializing in enterprise technology architecture, with a focus on networks and application security, including single sign-on (Security Assertion Markup Language) and identity management, API security, Internet of Things firmware, and ecosystem. He has devised block-chain-based autonomous authentication and device association for Internet of Things devices and mobile apps. He has pioneered and managed security services for Netrex Secure Solutions, including a firewall in the cloud architecture, and managed an intrusion prevention service. He possesses extensive hands-on experience as a chief application and network architect, including experience in systems management, information security, software development, large secure network (TCP/IP) design and deployment, departmental oversight, and management.

Syed Asad Hussain is currently leading communications and networks research at the COMSATS Institute of Information Technology, Lahore. He was funded for his PhD by Nortel Networks UK at Queen's University Belfast. He was awarded a prestigious Endeavour research fellowship for his postdoctorate at the University of Sydney, Australia, in 2010, where he conducted research on vehicular ad hoc networks. He has taught at Queen's University Belfast, United Kingdom; Lahore University of Management Sciences, and the University of the Punjab, Pakistan. He is supervising PhD students at the COMSATS Institute of Information Technology and split-site PhD students at Lancaster University, United Kingdom. Professor Hussain is serving as the dean of the Faculty of Information Sciences and Technology and in the capacity of head of the Computer Science Department at COMSATS Institute of Information Technology, Lahore. He regularly reviews Institute of Electrical and Electronics Engineers, Institution of Engineering and Technology, and Association for Computing Machinery journal papers.

Pedro Ricardo Morais Inácio was born in Covilhã, Portugal, in 1982. He earned a 5-year BSc degree in mathematics and computer science and a PhD degree in computer science and engineering, from the University of Beira Interior (UBI), Portugal, in 2005 and 2009, respectively. The PhD work was performed in the enterprise environment of Nokia Siemens Networks Portugal S.A., through a PhD grant from the Portuguese Foundation for Science and Technology. He has been a professor of computer science at UBI since 2010, where he lectures on subjects related to information assurance and security, programming of mobile devices, and computer-based simulation, for graduate and undergraduate courses, namely, the BSc, MSc, and PhD courses in computer science and engineering. He is an instructor of the UBI Cisco Academy. He is an Institute of Electrical and Electronics Engineers senior member and a researcher at the Instituto de Telecomunicações.

Prem Prakash Jayaraman is working in the area of distributed systems, in particular the Internet of Things, cloud, and mobile computing. He has published more than 45 papers, including 13 journal papers (*Transactions on Cloud Computing*, Elsevier's *Computational Science*, *Transactions on Large-Scale Data- and Knowledge-Centered Systems*, *IEEE Journal on Selected Areas in Communications*, and *Scientific World Journal*) in the related areas of his research. Prior to joining RMIT, Melbourne, Australia, Dr. Jayaraman was a postdoctoral research scientist (2012–2015) in the Digital Productivity and Services Flagship of the Commonwealth Scientific and Industrial Research Organization—the Australian government's premier research agency. Prior to that, he worked as a research fellow and lecturer at the Centre for Distributed Systems and Software Engineering, Monash University, Melbourne, Australia (2010–2011). Dr. Jayaraman obtained his PhD (2011) from Monash University, Melbourne, Australia, where he worked on developing a context-aware middleware for collecting data from wireless sensors networks using mobile smartphones in smart city environments. His thesis was titled "Cost-Efficient Collection and Delivery of Sensor Data Using Mobile Devices." His PhD paper, titled "Intelligent Processing of K-Nearest Neighbors' Queries Using Mobile Data Collectors in a Location Aware 3D Wireless Sensor Network," garnered the Best Paper Award at the 23rd International Conference on Industrial, Engineering and Other Applications of Applied Intelligent Systems (2010) (http://www.rmit.edu.au/contact/staff-contacts/academic-staff/j/jayaraman-dr-prem/).

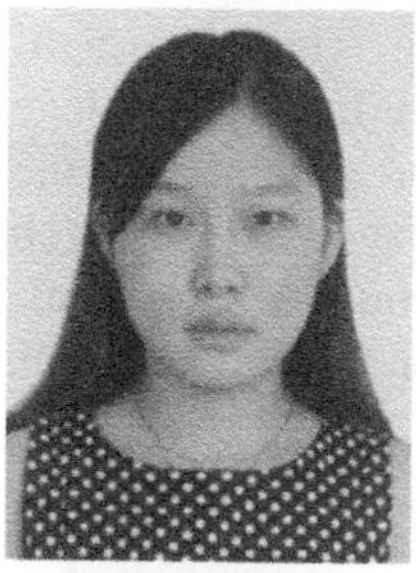

Le Jiang earned her BS degree in electrical engineering from North China Electric Power University in 2015. She is now a master student in the College of Electrical Engineering at Zhejiang University, Hangzhou. Her research interests include active distribution systems and renewable energy.

Nima Karimian earned his master's degree in electrical and computer engineering from the University of Connecticut, Storrs. He is currently pursuing a PhD degree in electrical and computer engineering with the same university. Prior to joining the University of Connecticut, he was a research assistant at the Amirkabir University of Technology (Tehran Polytechnic), Tehran, Iran. The main scope of his research lies in machine learning, deep learning, pattern recognition, biometrics authentication and identification, security of the Internet of Things, and hardware security primitives. His research in the biometrics area mainly focuses on biometrics-based key generation from noisy data. He has authored/co-authored several peer reviewed conference and journal papers. He obtained the best poster award in hardware security at the FICS annual conference in February 2016 and the best technical paper award from the VLSID conference in 2017.

Niels Karowski earned a diploma degree in computer science in 2007 from the Technical University of Berlin, Germany. He is a PhD candidate at the Telecommunication Networks Group at the Technical University of Berlin. His research interests include wireless sensor networks, neighbor discovery, and delay-tolerant networks.

Jake Kouns is the chief information security officer for risk-based security that provides vulnerability and data breach intelligence, and he also oversees the operations of OSVDB.org and DataLossDB.org. Kouns has presented at many well-known security conferences, including RSA, Black Hat, DEF CON, DerbyCon, CISO Executive Summit, EntNet Institute of Electrical and Electronics Engineers GlobeCom, FIRST, CanSecWest, InfoSecWorld, SOURCE, and SyScan, and at cyber liability forums such as AAMGA events, ACI's Cyber and Data Risk Insurance, NetDiligence's Cyber Risk and Privacy Liability Forum, and PLUS. He is the coauthor of *Information Technology Risk Management in Enterprise Environments* (Wiley, 2010) and *The Chief Information Security Officer* (IT Governance, 2011). He has briefed the Department of Homeland Security and the Pentagon on cyber liability insurance issues and is frequently interviewed as an expert in the security industry by *Information Week*, *eWeek*, *Forbes*, *PC World*, *CSO*, *CIO*, and *SC Magazine*. He earned both a bachelor of business administration and a master of business administration with a concentration in information security from James Madison University, Harrisonburg, Virginia. In addition, he holds a number of certifications, including ISC2's CISSP and ISACA's CISM, CISA, and CGEIT. He has appeared on CNN as well as on the *Brian Lehrer Show*, and was featured on the cover of *SC Magazine*.

Chee Yen Leow earned his B.Eng. degree in computer engineering from the Universiti Teknologi Malaysia, Skudai, in 2007. Since July 2007, he has been on the academic staff of the Faculty of Electrical Engineering, Universiti Teknologi Malaysia. In 2011, he obtained a PhD degree from Imperial College London. He is currently a senior lecturer in the faculty and a member of the Wireless Communication Centre, Universiti Teknologi Malaysia. His research interests include but are not limited to wireless relaying, multiple input multiple output, the physical layer security, convex optimization, communications theory, near-field wireless charging, and 5G.

Muhammad Mohsin Mehdi, a graduate in MS computer science and software engineering, is a freelance researcher by interest. His master's is from East Carolina University, Greenville, North Carolina. He also worked as a web engineer in Department of Research and Development at the same university. His bachelor's degree is from Dr. A. Q. Khan Institute of Computer Science and Information Technology, Kahuta Research Laboratories, Kahuta. He also worked for a few months at Kuwait University. His research interests include computer graphics, image processing, computer vision, and game theory. He is now a permanent faculty member at the COMSATS Institute of Information Technology, Lahore.

Dejian Meng earned his BE degree in computer science and technology from Wuhan University of Technology, Hubei, China, in 2002, and his MSc (with distinction) and PhD degrees, both in electronic engineering and computer science, from Queen Mary University of London, in 2004 and 2010, respectively. He has worked as a system engineer and consultant at the R&D Department of Mobile Technology in Nokia and Accenture London, United Kingdom, from 2008 to 2013, and has been involved in a number of high-profile industry projects and product deliveries. Currently, he is working as the director of the Department of Science and Technology, Taihu University of Wuxi, China, and has published more than 10 technical papers. His research interests over the years include mobile computing, context-aware computing, intelligent systems, and Internet of Things.

Thomas Menzel is a PhD candidate at the Telecommunication Networks Group at Technical University of Berlin, Germany, from which he also earned his diploma degree. His research is focused on energy efficiency in wireless sensor networks, especially in electrochemical batteries and their nonlinearities.

Roberto Minerva holds a Ph.D in Computer Science and Telecommunications from Telecom Sud Paris, France, and a Master Degree in Computer Science from Bari University, Italy. He was the Chairman of the IEEE IoT Initiative, an effort to nurture a technical community and to foster research in IoT. Roberto has been for several years in TIMLab, involved in activities on SDN/NFV, 5G, Big Data, architectures for IoT. Now he is a research engineer in Paris Sud Telecom and the Chief Technologist in Bitify.it, a startup aiming to drive the digitalization of businesses in several industries. He is authors of several papers published in international conferences, books and magazines.

Daniel Minoli, principal consultant, DVI Communications, has published 60 well-received technical books and 300 papers and made 85 conference presentations. He has many years of technical hands-on and managerial experience in planning, designing, deploying, and operating secure Internet Protocol (IP) and IP Version 6 (IPv6), voice over Internet Protocol, telecom, wireless, satellite, and video networks for global best-in-class carriers and financial companies. Previous roles in the past two decades have included chief technology officer at Secure Enterprise Systems Engineering, a technology assessment and enterprise cybersecurity firm he launched in the recent past; general manager and director of ground systems engineering at SES, the world's second largest satellite services provider; director of network architecture at Capital One Financial; chief technology officer at InfoPort Communication Group; and vice president of packet services at Teleport Communications Group (eventually acquired by AT&T.) Other affiliations have included Bell Labs, ITT, Prudential, and Bellcore/Telcordia/Ericsson. Over the years, Minoli has published and lectured extensively in the area of machine-to-machine/Internet of Things, network security, satellite systems, wireless networks, IP/IPv6/Metro Ethernet, video/Internet protocol television/multimedia, voice over Internet protocol, IT/enterprise architecture, and network/Internet architecture and services. Minoli has taught IT and telecommunications courses at New York University, New York; Stevens Institute of Technology, Hoboken, New Jersey; and Rutgers University, Piscataway, New Jersey. He has appeared in industry conferences as well a radio and TV technology programs. He is a graduate of the New York University Polytechnic School of Engineering (MS in computer science).

Supriya Mitra, Phd, CSCP, CPIM, is director of information technology at Schneider Electric. Supriya has a B.Tech. in mechanical engineering from Indian Institute of Technology Madras and a PhD in supply chain management from Syracuse University, New York. He was awarded the best doctoral dissertation in supply chain management by the Council of Supply Chain Management Professionals (Lombard, Illinois). He has multiple publications in leading journals such as the *Journal of Operations Management*, the *International Journal of Production Economics*, and *Advances in Business and Management Forecasting.*

Benedict Occhiogrosso is a cofounder of DVI Communications. He is a graduate of the New York University Polytechnic School of Engineering. Occhiogrosso's experience encompasses a diverse suite of technical and managerial disciplines, including sales, marketing, business development, team formation, systems development program management, procurement and contract administration budgeting, scheduling, quality assurance, and technology operational and strategic planning. As both an executive and a technologist, Occhiogrosso enjoys working with and managing multiple client engagements, as well as setting corporate objectives. Occhiogrosso is responsible for new business development, company strategy, and program management. He also, on occasion, has served as a testifying expert witness in various cases encompassing patent infringement and other legal matters.

Charith Perera is a research associate (postdoctoral research fellow) at the Open University, United Kingdom. Previously, he was a PhD student at the Australian National University, attached to the Research School of Computer Science. He also worked as a researcher at the Commonwealth Scientific and Industrial Research Organization, Canberra, Australia, during his PhD. He completed a study abroad at the Computer Lab, University of Cambridge, United Kingdom. His research interests include Internet of Things, sensing as a service, privacy, middleware platforms, sensing infrastructure, context awareness, semantic technologies, middleware, and mobile and pervasive computing (http://www.charithperera.net/).

Giuseppe Piro is an assistant professor at the Politecnico di Bari, Italy. He earned a first-level degree and a second-level degree (both cum laude) in telecommunications engineering from Politecnico di Bari in 2006 and 2008, respectively. He obtained the PhD degree in electronic engineering from Politecnico di Bari in March 2012. His main research interests include quality of service in wireless networks, network simulation tools, 4G and 5G cellular systems, information-centric networking, nanocommunications, and Internet of Things. He founded both LTE-SIM and Nano-SIM projects and is a developer of Network Simulator 3.

Sharul Kamal Abdul Rahim earned his first degree from the University of Tennessee, Knoxville, majoring in electrical engineering, in 1996; MSc in engineering (communication engineering) from the Universiti Teknologi Malaysia, Skudai, in 2001; and PhD in wireless communication systems from the University of Birmingham, United Kingdom, in 2007. Currently, Dr. Sharul is a professor at the Wireless Communication Centre, Faculty of Electrical Engineering, Universiti Teknologi Malaysia. His research interests include antenna design, radio-frequency and microwave systems, reconfigurable antennas, beam-forming networks, smart antenna systems, and antennas for wireless energy transfer. He is also a senior member of the Institute of Electrical and Electronics Engineers Malaysia Section; corporate member of the Institute of Engineer Malaysia; and member of the Institute of Electronics, Information and Communication Engineers and Eta Kappa Nu Chapter (International Electrical Engineering Honor Society, University of Tennessee). He has published a number of technical papers, including journal articles, book chapters, and conference papers.

Imran Raza has been working as an assistant professor in the Department of Computer Science, COMSATS Institute of Information Technology, Lahore, since 2003. He earned BS (CS) and MPhil degrees from Pakistan. He has been associated with Technische Universität Ilmenau, Germany as a researcher. His areas of interests include Software Defined Networking (SDN), Network Functions Virtualization (NFV), wireless sensor networks, mobile ad hoc networks, quality of service issues in networks, and routing protocols.

Paola Russo earned her PhD degree in electronic engineering from the Polytechnic of Bari, Italy, in 1999. During 1999, she worked with a research contract at the Motorola Florida Research Laboratory. From 2000 to 2004, she worked with a research contract at the University of Ancona, Italy (now the Universita Politecnica delle Marche), on the development of numerical tools applied to different electromagnetic problems. Since January 2005, she has held a tenured position as researcher at the Università Politecnica delle Marche. She teaches ElectroMagnetic Compatibility, antenna design, and the fundamentals of electromagnetics. Her current research interests include the application of numerical modeling to EMC problems, reverberation chambers, and new antenna design, such as plasma antennas. Dr. Russo is a member of the Institute of Electrical and Electronics Engineers EMC Society and the Italian Electromagnetic Society.

Assim Sagahyroon earned his BSc degree in electrical engineering from the University of Khartoum, Sudan; the MSc degree in electrical engineering from Northwestern University, Evanston, Illinois; and his PhD degree from the University of Arizona, Tucson. From 1993 to 1999, he was with the Department of Computer Science and Engineering at Northern Arizona University, Flagstaff, and then he joined the Department of Math and Computer Science at California State University. In 2003, he joined the Department of Computer Science and Engineering at the American University of Sharjah, United Arab Emirates. He served as the department head from 2009 to 2016 and currently is a professor of computer engineering. He served as a technical reviewer for the National Science Foundation and many conferences and journals. He is a member of the editorial review boards of a few journals. He is the cofounder of the Institute of Electrical and Electronics Engineers Conference on Industrial Informatics and Computer Systems and has participated in many technical program committees of international conferences. In industry, he has worked with Zhone Technologies and briefly with Lucent. He has many publications in international conferences and journals. His research interests include innovative applications of emerging technology in the medical field, power consumption and testing of digital systems, hardware design, field-programmable gate array based designs, and computer architecture.

Zahra Saleh completed her MS in innovation and technology management at the University of Bath, United Kingdom, in 2016. While studying, Zahra obtained professional experience in innovation and business development at P3 Engineering, one of the largest engineering service providers in Germany. Furthermore, she obtained her BS from the University of Applied Sciences Hamburg, Germany, in foreign trade and international management in 2014. Her main research interest includes innovation, with a focus on sustainability.

Musa Gwani Samaila earned a 5-year B.Eng. degree in electrical engineering in 1998 from the University of Maiduguri, Nigeria. He also earned a 2-year M.Eng. degree in electronic engineering from Abubakar Tafawa Balewa University, Bauchi, Nigeria, in 2012. He is currently pursuing his PhD in computer engineering at the Department of Computer Science, University of Beira Interior, Covilhã, Portugal, with research interests in Internet of Things and embedded systems security. He was a lecturer at the Department of Electrical and Electronic Engineering Technology, Federal Polytechnic Bauchi, Nigeria, from 2001 to 2009. He is an assistant chief engineer at the Centre for Geodesy and Geodynamics, National Space Research and Development Agency, Toro, Bauchi State, Nigeria. He has published two journal papers, four conference papers, and two book chapters. He is a corporate member of the Nigerian Society of Engineers and is registered with the Council for the Regulation of Engineering in Nigeria.

João Bernardo Ferreira Sequeiros is currently enrolled in a PhD program at the Universidade da Beira Interior, Covilhã, Portugal. He has a master's degree in computer science and engineering from the Universidade da Beira Interior, which he concluded in 2016. His dissertation focused on the development of a box for automated network-based security assessments. He also has a bachelor's degree in computer science and engineering from the Universidade da Beira Interior, which he concluded in 2014. His main research and interest areas are network and application security, cryptography, and game development. He enjoys developing in C, Java, Python, and C#, and likes to challenge his knowledge in network management, modeling software, and database management systems.

Marco Severini graduated in electronics engineering from the Università Politecnica delle Marche, Italy, in 2012. He is currently working as a research fellow at the Department of Information Engineering of the same university. His current research interests lie in the design and development of task and resource scheduling algorithms, energy and power management optimization, mixed integer nonlinear programming, wireless sensor networks, embedded systems programming, and smart grids.

Shalaka Shinde is a graduate engineering trainee at Schneider Electric. Shalaka earned a BE in electronics and communication engineering and an Mtech in computer science and networking from Ramaiah University of Applied Sciences, Bangalore.

Kazem Sohraby, BS, MS, PhD, MBA, is research professor of computer science at San Diego State University. He has 22 granted and pending patent applications and has published more than 300 peer-reviewed papers and 2 textbooks in the field of computer science, electrical, and computer engineering. His previous affiliations include University of Arkansas, South Dakota School of Mines and Technology, Bell Labs, and Stevens Institute of Technology. Dr. Sohraby earned an MBA from the Wharton School, University of Pennsylvania, Philadelphia, and a PhD from New York University (Polytechnic Engineering Division), New York.

Susanna Spinsante (Institute of Electrical and Electronics Engineers [IEEE] senior member, IEEE Communications Society and Consumer Electronics Society (CES) member, and IEEE SPS SiG on Internet of Things [IoT] member) was born in Ancona, Italy, in April 1976. She earned the Italian laurea with honors in electronic engineering from the University of Ancona (now Polytechnic University of Marche), Italy, in 2002. She earned her PhD in electronic engineering and telecommunications at the same university (November 2005), where she worked as a postdoctoral researcher until December 2012, when she joined the Department of Information Engineering as a non-tenure-track assistant professor in telecommunications. She has taught telecommunication courses at the University of Trento and Ente Universitario del Fermano, Italy. Her current research interests are in the areas of low-power, long-range wireless technologies for capillary networks, multiple access protocols based on spread-spectrum communications, and communication technologies integration for IoT and ambient assisted living. She coauthored more than 100 papers in peer-reviewed international journals and conferences. She is the chair and organizer of the three editions of the International Workshop on IoT for Ambient Assisted Living (2015, 2016, and 2017), member of the D4 Working Group on Age-Friendly Environments of the EiP-AHA, and Management Committee substitute member for Italy in the IC1303 COST Action. She is a regular reviewer for several (Institute of Electrical and Electronics Engineers, Springer, and Elsevier) journals, books, and conference proceedings. She is the leading investigator of the project "TWIST: Tecnologie WIreless eterogenee e Sostenibili nelle smarT cities del futuro"—Ricerca Scientifica di Ateneo 2014, promoted by the Department of Information Engineering at the Università Politecnica delle Marche.

Stefano Squartini (Institute of Electrical and Electronics Engineers [IEEE] senior member, IEEE Computational Intelligence Society member, and member of the International Speech and Communication Association and of the Audio Engineering Society member) was born in Ancona, Italy, on March 1976. He obtained the Italian laurea with honors in electronic engineering from the University of Ancona (now the Polytechnic University of Marche), Italy, in 2002. He obtained his PhD from the same university (November 2005). He also worked as a postdoctoral researcher at the Polytechnic University of Marche from June 2006 to November 2007, when he joined the Department of Information Engineering as assistant professor in circuit theory. He has been an associate professor at the Polytechnic University of Marche since November 2014. His current research interests are in the areas of computational intelligence and digital signal processing, with a special focus on speech, audio, and music processing and energy management. He is the author and coauthor of many international scientific peer-reviewed articles (more than 160) and a member of the *Cognitive Computation, Computational Intelligence and Neuroscience, Big Data Analytics*, and *Artificial Intelligence Reviews* editorial boards (starting in 2011, 2014, 2015, and 2016, respectively). He is also associate editor of *IEEE Transactions on Cybernetics* and *IEEE Transactions on Emerging Topics in Computational Intelligence* (2017–to date) and was associate editor for *IEEE Transactions on Neural Networks and Learning Systems* (2010–2016). He is a regular reviewer for several (IEEE, Springer, and Elsevier) journals, books, and conference proceedings, and in the recent past, he organized several special sessions at international conferences with peer reviewing and special issues of Web of Science indexed journals. He has joined the organizing and technical program committees of more than 60 international conferences and workshops.

Domenico Striccoli earned, with honors, his Dr.Eng. degree in electronic engineering in April 2000, and his PhD degree in April 2004, both from the Politecnico di Bari, Italy.

In 2005, he joined the DIASS Department of Politecnico di Bari in Taranto as assistant professor in telecommunications. He teaches fundamental courses in the field of telecommunications in the Department of Electrical and Information Engineering of the Politecnico di Bari. His scientific interests span different aspects of telecommunications networks, including dynamic predictive bandwidth allocation, video on demand, three-dimensional video techniques and transmission, call admission control schemes, scheduling techniques, and quality of service/quality of experience in mobile wireless networks (wireless sensor networks, Long Term Evolution, etc.).

Fatemeh Tehranipoor is an Assistant Professor of Electrical and Computer Engineering at San Francisco State University (SFSU). She received her Ph.D. degree in Electrical Engineering from University of Connecticut (UConn) in 2017. She earned her MS degree in Computer Hardware Engineering at Shahid Beheshti University, Tehran, Iran in 2013, and her bachelor's degree with the highest honors in Computer Hardware Engineering at Babol Noshirvani University of Technology, Iran, in 2011. Her research interests include hardware security primitives, embedded and Cyber Physical Systems (CPS) security, Internet-of-Things (IoT) security, and machine learning/deep learning. She is currently serving as an *Associate Editor* for the IEEE Consumer Electronics Magazine (ICEM). She also serves on the several *Technical Program Committee*; International Symposium on Hardware Oriented Security and Trust (HOST), International Conference on Consumer Electronics (ICCE), and IEEE International Symposium on Nano-electronic and Information Systems (IEEE-iNIS). Dr. Tehranipoor received "*Best Technical Paper Award*" in 30th International Conference on VLSI Design (VLSID) & 16th International Conference on Embedded Systems in January 2017.

Bedir Tekinerdogan is a full professor and chairholder of the information technology group at Wageningen University in The Netherlands. He received his MSc and PhD in computer science from the University of Twente in The Netherlands. He has graduated around 50 MSc students, supervised around 20 PhD students, and developed more than 15 academic computer science courses. He has also been very active in scientific conferences and organized more than 50 conferences/workshops on important software engineering research topics. He has around 25 years of experience in software engineering research and education. He has been active in many national and international research and consultancy projects with various large software companies. His current research at Wageningen University concerns smart system of systems engineering, with an emphasis on software engineering and information technology. More details can be found on his LinkedIn Profile: https://www.linkedin.com/in/bedir

Eray Tüzün is a productization manager at HAVELSAN A.Ş. in Ankara, Turkey, and an adjunct faculty member at Bilkent University in the computer engineering department. He has previously worked as a software design engineer at Microsoft in the Microsoft Online Services group, senior software engineer at Howard Hughes Medical Institute, and research engineer at CWRU Genomics Center. He has a Microsoft Certified Solutions Developer (MCSD) for application lifecycle management, and Professional Scrum Master (PSM) and Professional Scrum Product Owner (PSPO) certifications. His research interests include application lifecycle management, software reuse, software product line engineering, agile software development, empirical software engineering, and software engineering education.

Marta Vos completed her PhD research in 2014, studying radio-frequency identification implementation where the radio-frequency identification system crosses the boundary between public and private sectors. Since 2014, she has worked as a senior lecturer at the Whitireia Institute of Technology, where she teaches in the postgraduate program and supervises postgraduate students. Her research has continued to focus on the implementation and management of radio-frequency identification and Internet of Things (IoT) systems, and she has published a number of papers in this area. She has also expanded her interest in the IoT, publishing on the maturity of the IoT research field. Currently, Dr. Vos is investigating data governance in distributed environments with an aim to understand how governance can be understood where organizations span different jurisdictions, particularly from a data privacy perspective.

Adam Wolisz earned his degrees (diploma 1972, PhD 1976, habil. 1983) from Silesian University of Technology, Gliwice, Poland. He joined Technische Universität Berlin in 1993, where he is a chaired professor in telecommunication networks and an executive director of the Institute for Telecommunication Systems. He is also an adjunct professor at the Department of Electrical Engineering and Computer Science, University of California, Berkeley. His research interests are in architectures and protocols of communication networks. Recently, he has been focusing mainly on wireless/mobile networking and sensor networks.

Paul A. Wortman earned his BS degree in computer and electrical engineering from Trinity College, Hartford, Connecticut, in 2010. Currently, he is pursuing a PhD degree in the Department of Electric and Computer Engineering at the University of Connecticut. In tandem, he works as a technician in the CSI Lab at the University of Connecticut, performing black box penetration testing. These duties include the examination of routers, gateways, and security systems, and the organization and development of competitions for the annual CyberSEED event held by the University of Connecticut. His research interests include modeling embedded systems devices and their security design and implementation. He is currently working on methods for defining and modeling security requirements and constraints within embedded and Internet of Things devices.

Qiang Yang earned his BS degree (first-class honors) in electrical engineering in 2001, and his MSc (with distinction) and PhD degrees in both electronic engineering and computer science from Queen Mary, University of London, in 2003 and 2007, respectively. He worked as a postdoctoral research associate in the Department of Electrical and Electronic Engineering, Imperial College London, from 2007 to 2010 and was involved in a number of high-profile UK EPSRC and European IST research projects. Currently, he is an associate professor at the College of Electrical Engineering, Zhejiang University, Hangzhou, China, and has published more than 100 technical papers, coauthored 2 books, and holds more than 10 national patents. His research interests over the years include communication networks, smart energy systems, and large-scale complex network modeling, control, optimization, and simulation. He is a member of various international academic bodies, including the Institute of Electrical and Electronics Engineers, the Institution of Engineering and Technology, and Institute of Electronics, Information and Communication Engineers, as well as a senior member of China Computer Federation.

Hailin Zhao earned his BS degree in automation in 2014 from Zhejiang University, Hangzhou, China. He is currently a master candidate in the College of Electrical Engineering, Zhejiang University. His research interests are inverter control in microgrids with a large penetration of small-scale DGs and the intelligent control of smart energy networks.

Printed and bound by CPI Group (UK) Ltd, Croydon, CR0 4YY
24/10/2024
01778291-0010